I0061071

Moon-Based Synthetic Aperture Radar

Lunar explorations have received increasing attention in recent years with tremendous application values, including using the Moon as a remote sensing platform for Earth observation. As an active sensor, the Synthetic Aperture Radar (SAR) can detect changes in the atmosphere, terrain, and ocean. Moon-based SAR, complementary to the spaceborne SAR systems, expands our capabilities of watching and understanding the Earth. This book explains the Moon-Earth observation geometry, generic parameters, image focusing, and outlook using the Moon-based SAR. Written as a SAR imaging of Earth on the lunar-based platform, it makes it an essential reference to those interested in planetary and Earth sciences.

FEATURES

- Uses the Moon as a remote sensing platform for Earth observation
- Explains how to obtain a high spatial resolution with a short revisit time using the Moon-based SAR
- Covers the observation geometry, range and signal models, two-dimensional signal spectrum, and focusing algorithms for the Moon-based SAR
- Presents a detailed analysis of sources of phase errors in the Moon-based SAR signal
- Includes global case studies and introduces conceptual ideas for further research

This book is intended for senior graduate students, professional researchers, and engineers studying and working in the fields of lunar exploration and remote sensing applications, especially when dealing with high-orbit SAR studies.

Zhen Xu joined the Department of Electronic and Information Engineering at Shantou University in 2020. He earned a PhD in cartography and geographic information system at the Institute of Remote Sensing and Digital Earth, Chinese Academy of Sciences, and the University of Chinese Academy of Sciences in 2020. His research interests include the system design and signal processing of Moon-based SAR. He was awarded the Outstanding Talent of Shantou University and the High-Level Talent of Shantou City. Dr. Xu also received Young Scientist Awards from the International Conference on Space, Aeronautical, and Navigational Electronics (ICSANE) in 2018 and the General Assembly and Scientific Symposium of the International Union of Radio Science (URSI GASS) in 2023.

Kun-Shan Chen is a nationally distinguished professor at the Nanjing University in China. He earned a PhD in electrical engineering at the University of Texas at Arlington in 1990. His research interests include microwave remote sensing theory, modeling, systems, measurement, intelligent signal processing, and data analytics for radar. He has authored and co-authored over 500 refereed journals and conference papers. He is the author of several books published by CRC Press. His academic activities are numerous, spanning 30 years. Dr. Chen, an IEEE Fellow, received the 2021 IEEE GRSS Fawwaz Ulaby Distinguished Achievement Award for his contributions to microwave scattering and emission modeling of rough surface and radar image simulation and understanding.

SAR Remote Sensing

Series Editor: Jong-Sen Lee Naval Research Laboratory, Washington DC

Synthetic Aperture Radar (SAR) is an indispensable and highly capable Earth remote sensing instrument. It is a day–night and all-weather sensor for all aspects of environmental monitoring, disaster (earthquake, tsunami, forest fire, etc.) assessment, and military applications. Since the 1980s, a plethora of spaceborne SAR systems have been launched and are in operation, and new satellites are launched every day. Many SAR imaging modes have been developed and tailored to specific applications. SAR is becoming a vast and continuously evolving field of remote sensing technology and applications. This book series provides timely information on advancements, and encompasses innovative SAR scattering theory, processing techniques, and applications in all environments. Books in this series serve and benefit professionals, academics, and students in the remote sensing community.

Moon-Based Synthetic Aperture Radar: A Signal Processing Prospect
Zhen Xu and Kun-Shan Chen

Airborne Circularly Polarized SAR: Theory, System Design, Hardware Implementation, and Applications
Josaphat Tetuko Sri Sumantyo, Ming Yam Chua, Cahya Edi Santosa, and Yuta Izumi

Spatial Analysis for Radar Remote Sensing of Tropical Forests
Gianfranco D. De Grandi and Elsa Carla De Grandi

Radar Scattering and Imaging of Rough Surfaces: Modeling and Applications with MATLAB®
Kun-Shan Chen

Polarimetric SAR Imaging: Theory and Applications
Yoshio Yamaguchi

Imaging from Spaceborne and Airborne SARs, Calibration, and Applications
Masanobu Shimada

For more information about this series, please visit: www.routledge.com/SAR-Remote-Sensing/book-series/CRCSRS

Moon-Based Synthetic Aperture Radar

A Signal Processing Prospect

Zhen Xu
Kun-Shan Chen

CRC Press
Taylor & Francis Group
Boca Raton London New York

CRC Press is an imprint of the
Taylor & Francis Group, an **informa** business

Designed cover image: the original file was created by ©Visio and the background in Tiff was generated by © Celestia and © Photoshop

First edition published 2024
by CRC Press
2385 NW Executive Center Drive, Suite 320, Boca Raton FL 33431

and by CRC Press
4 Park Square, Milton Park, Abingdon, Oxon, OX14 4RN

CRC Press is an imprint of Taylor & Francis Group, LLC

© 2024 Zhen Xu and Kun-Shan Chen

Reasonable efforts have been made to publish reliable data and information, but the author and publisher cannot assume responsibility for the validity of all materials or the consequences of their use. The authors and publishers have attempted to trace the copyright holders of all material reproduced in this publication and apologize to copyright holders if permission to publish in this form has not been obtained. If any copyright material has not been acknowledged, please write to let us know so we may rectify in any future reprint.

Except as permitted under US Copyright Law, no part of this book may be reprinted, reproduced, transmitted, or utilized in any form by any electronic, mechanical, or other means, now known or hereafter invented, including photocopying, micro-filming, and recording, or in any information storage or retrieval system, without written permission from the publishers.

For permission to photocopy or use material electronically from this work, access www.copyright.com or contact the Copyright Clearance Center, Inc. (CCC), 222 Rosewood Drive, Danvers, MA 01923, 978-750-8400. For works that are not available on CCC please contact mpkbookspermissions@tandf.co.uk

Trademark notice: Product or corporate names may be trademarks or registered trademarks and are used only for identification and explanation without intent to infringe.

ISBN: 978-1-032-31168-5 (hbk)
ISBN: 978-1-032-31171-5 (pbk)
ISBN: 978-1-003-30843-0 (ebk)

DOI: 10.1201/9781003308430

Typeset in Times
by SPi Technologies India Pvt Ltd (Straive)

Contents

Preface

The Moon has recently sparked renewed global interest, leading to numerous lunar exploration missions being planned and executed with the ultimate goal of establishing a permanent lunar base. The lunar base holds tremendous scientific significance, which could serve as an enduring platform for research spanning various fields. Earth observation, in particular, would reap immense benefits from a lunar base. From this perspective, the lunar base could provide unique opportunities for onboard sensors to capture extensive and seamless imagery of our planet at consistent intervals, thereby augmenting our comprehension of Earth. As an active instrument, the Synthetic Aperture Radar (SAR) enables observing Earth irrespective of the prevailing weather or lighting conditions. This distinctive advantage renders the Moon-based SAR (MBSAR) a premier choice for conducting Earth observation from the lunar base.

The concept of MBSAR has captivated the curiosity of geoscience, lunar exploration, and related communities since the early 2000s. Compared to the current spaceborne SAR, the MBSAR bears remarkable advantages—the most notable being its capability to cover most regions of Earth daily. Additionally, the MBSAR's spatial resolution is not constrained by the high altitude; rather, the elevated orbit permits a remarkable enhancement in the azimuth resolution. Further, the MBSAR can install multiple large antennas, with great potential for forming a stable baseline.

However, the orbital characteristics and unique geometry pose a particular challenge for the Doppler shift effect and associated synthetic aperture processing, henceforth influencing the signal properties and system parameters in the MBSAR. In this regard, special care must be exercised to ensure optimal imaging capabilities. In contrast to the spaceborne SAR, the Doppler properties and associated generic parameters of the MBSAR exhibit notable dissimilarities. In addition, the stop-and-go assumption and linear path trajectory used for conventional SAR signal modeling cannot be applied to MBSAR. Moreover, the spatiotemporally varying phase errors given rise by the orbital perturbations and atmospheric effects present remarkable challenges, further complicating the image performance of the MBSAR.

To exploit the benefits and address the challenges of MBSAR, numerous scholars have dedicated themselves to developing the configuration design, signal modeling, and image formation of MBSAR. This book presents a comprehensive framework of the MBSAR by highlighting its recent developments from a signal processing viewpoint. This presentation adopts a systematic view, commencing with the SAR basics and the rationale behind MBSAR. From there, it establishes the Moon-based Earth observation geometry, which then serves as the basis for determining the genic parameters, signal properties, and coverage performances in the MBSAR. Following this, it puts forward the signal model, its two-dimensional (2-D) spectrum, along the corresponding focusing algorithm, in the context of MBSAR. By employing the established signal model, we also conduct a preliminary analysis regarding orbital perturbation effects and atmospheric effects on the imaging performance in the MBSAR.

The goal of this book is to lay out the fundamental (physics and geometry) signal and imaging properties of MBSAR. This book consists of eight chapters. Chapter 1 focuses on the background of the MBSAR. The text delves into the SAR basics and summarizes the current spaceborne SAR missions. It then identifies the inherent trade-off between spatial resolution and coverage in spaceborne SAR and explores prospective solutions to surmount such a constraint. Among the potential solutions, the MBSAR concept is highlighted, along with its rationale, benefits, and challenges being illustrated. This introductory chapter serves as a prelude to the subsequent chapters, offering readers a thorough comprehension of the MBSAR. Chapter 2 presents the time-space reference frames and corresponding reciprocal transformations. Then, the intricate details of Earth–Moon relative motion are explicated in the provided reference frames, upon which the geometric model of MBSAR is established. Chapter 3 deals with the generic parameters of the MBSAR, with a specific emphasis on azimuthal resolution through a retrospective perspective and further expanded upon by fully considering both the Earth's rotation and MBSAR's inertial motion. Chapter 4 discusses the MBSAR's spatiotemporal coverage performance by scrutinizing the swath width achieved with a single antenna and the optimal coverage attainable through multiple antennas. On this basis, we shall proceed to select the optimal site location for the MBSAR by considering its coverage performance as a focal point. Chapter 5 focuses on the signal modeling of MBSAR, explicitly emphasizing the Doppler properties and total zero-Doppler steering method. Moreover, this chapter sheds light on establishing a non-stop-and-go signal model for the MBSAR. Based on this fundamental premise, Chapter 6 puts forward the range model and 2-D signal spectrum; it subsequently places significant emphasis on the signal processing algorithms employed for imaging formation, culminating in simulations of target responses in the scenario of MBSAR. Considering that the orbital perturbations potentially affect MBSAR imaging performance, Chapter 7 develops a comprehensive approach for evaluating such effects. Following this, we scrutinize the spatiotemporal-varying phase errors resulting from lunar orbital precessions and their implications on the imaging performance and orbital determination of MBSAR. Chapter 8 takes the background ionosphere as an example to look into the spatiotemporal variation characteristics of the atmosphere and their impacts on signal propagation in the MBSAR. Leveraging the insights gleaned from this analysis, we quantitatively inspect the geometric distortion and image defocusing induced by spatiotemporally varying background ionospheric effects.

This book is specifically geared toward the scholarly community of scientists and engineers practicing in SAR and MBSAR. This volume could serve as a tool for understanding signal properties and image formation of MBSAR. As perhaps the first book of its kind, it is not error-free, all at the authors' responsibility. Many references in the field of SAR were omitted as it was not possible to include them all. We apologize for those not appropriately cited. Finally, we welcome feedback from the readership to improve our work. Then, we will be well repaid.

Acknowledgments

We are deeply grateful to pioneers in the fields of Moon-based Earth observation and the SAR community, whose contributions inspire us to explore this fascinating field. This book resulted from many researchers' pioneering works, too numerous to mention each individual. We offer our thanks. While it is infeasible to acknowledge everyone, we attempted our best to cite their works in the references of this book. Special thanks go to Professor Huadong Guo for his steadfast focus and insight into the progress of Moon-based Earth observation. This book would not have existed without his guidance.

This book was written after encouragement and inspiration from our book series editor, Dr. Jong-Sen Lee. His perception of signals is keen and reflective—our sincere thanks for his passionate support. We sincerely appreciate the wonderful CRC Press team, specifically our editor, Irma Britton, and assistant, Chelsea Reeves. They have unfailingly paved the way for the successful completion of this book.

The first author, Zhen Xu, wishes to express his utmost gratitude to Professor Kun-Shan Chen. Since the year 2015, I have consistently derived benefits from the invaluable guidance of Professor Chen in the field of radar remote sensing. It was a great honor for me to collaborate with him on this book. I am grateful to Miss Qingshan Zhang for her invaluable companionship and encouragement. My heartfelt thanks go to my parents for their unwavering support throughout my life's journey. Finally, some works featured in this book were sponsored National Natural Science Foundation of China under Grant 42101398 and Shantou University Scientific Research Foundation for Talents under Grant NTF20023.

The second author, Kun-Shan Chen, thanks Zhen Xu for his devoted and energetic efforts in compiling this book. Most material in this book is from our joint papers scattered in academic journals. Making this material into an organized and concise book demands dedication, efficiency, and patience. Together we learned about waves and signals in Moon-Earth bodies through discussion and debate. Finally, I am indebted to my wife, Jolan, for her love and understanding throughout my life and to our children, Annette, Vincent, and Lorenz, for giving me a constant source of joy.

Nomenclature

Abbreviations	Full Expansion
ACS	Antenna Coordinate System
AHRE	Advanced Hyperbolic Range Equation
AOL	Argument of Latitude
AOP	Argument of Perigee
BCRS	Barycentric Celestial Reference System
CLEP	Chinese Lunar Exploration Program
CPE	Cubic Phase Error
CRF	Celestial Reference Frame
DE	Development Ephemerides
DFMR	Doppler Frequency Modulation Rate
ECEF	Earth-Centered Earth Fixed
ECI	Earth-Centered Inertial
ECR	Earth-Centered Rotational
EHRE	Extended Hyperbolic Range Equation
EOP	Earth Orientation Parameter
ET	Ephemeris Time
FFT	Fast Fourier Transform
FM	Frequency Modulation
FMCW	Frequency Modulated Continuous Wave
FT	Fourier Transform
GAST	Greenwich Apparent Sidereal Time
GCRS	Geocentric Celestial Reference System
GEO	Geosynchronous Earth Orbit
GMST	Greenwich Mean Sidereal Time
GPS	Global Positioning System
GRS80	Geodetic Reference System 1980
HRE	Hyperbolic Range Equation
IAU	International Astronomical Union
ICRF	International Celestial Reference Frame
ICRS	International Celestial Reference System
IERS	International Earth Rotation and Reference Systems Service
IRW	Impulse Response Width
ISAR	Inverse Synthetic Aperture Radar
ISLR	Integrated Sidelobe Ratio
ITRF	International Terrestrial Reference Frame
ITRS	International Terrestrial Reference System
JPL	Jet Propulsion Laboratory
LEO	Low Earth Orbit
LFM	Linear Frequency Modulation
LLR	Lunar Laser Ranging
MBSAR	Moon-Based Synthetic Aperture Radar
MCI	Moon-Centered Inertial
MCMF	Moon-Centered Moon Fixed
MEME	Mean Equator and Mean Equinox

MER	Mean Earth Rotation
MSR	Method of Series Reversion
NASA	National Aeronautics and Space Administration
NESZ	Noise Equivalent Sigma Zero
PA	Principal Axis
PDH	Probability Distribution Histogram
POSP	Principle of Stationary Phase
PRF	Pulse Repetition Frequency
PRM	Polynomial Range Model
PRM2	Polynomial Range Model of 2nd Order
PRM3	Polynomial Range Model of 3rd Order
PRM4	Polynomial Range Model of 4th Order
PSLR	Peak to Side Lobes Ratio
QPE	Quadratic Phase Error
RAAN	Right Ascension of Ascending Node
RD	Range Doppler
RPE	Residual Phase Error
RRE	Residual Range Error
SAR	Synthetic Aperture Radar
SAT	Synthetic Aperture Time
SLAR	Side Looking Airborne Radar
SLR	Satellite Laser Ranging
SNR	Signal to Noise Ratio
STEC	Slant Total Electron Content
TAI	International Atomic Time
TCB	Barycentric Coordinate Time
TCG	Geocentric Coordinate Time
TDB	Barycentric Dynamical Time
TDT	Terrestrial Dynamical Time
TEC	Total Electron Content
TOI	Target of Interest
TT	Terrestrial Time
USNO	United States Naval Observatory
UT1	Universal Time
UTC	Coordinated Universal Time
VLBI	Very Long Baseline Interferometry
VTEC	Vertical Total Electron Content
WGS 84	World Geodetic System 1984

1 Background

1.1 SAR BASICS

Radar is an electronic device that uses modulated pulse signals to detect targets and determine their positions. Radar systems have evolved to perform a wide range of complex functions, with Earth observation being the most significant [1–3]. Unlike optical sensors, radar, as an active sensor, can penetrate clouds and rain and monitor the Earth's surface day or night, irrespective of weather conditions. Depending on the operating band, radar sensors can also penetrate terrain surfaces, vegetation, and snow. Moreover, radar scattering from a natural surface unveils electromagnetic and geometric properties, yielding distinct information in contrast to the visible and infrared sensors [4–7]. In this regard, an appropriate radar system allows us to infer Earth's geophysical parameters [8].

Early radar systems measured distance using time delays and determined the target's direction and velocity by analyzing antenna beam pointing and Doppler shift; these radar systems were quite constrained by their resolution capabilities [9]. In the 1950s, the side-looking airborne radar (SLAR) was developed to enable higher cross-track resolution (range resolution) images using a longer fixed antenna parallel to the aircraft's fuselage [10]. However, the SLAR system still encountered challenges in achieving satisfactory and consistent along-track resolution (also referred to as azimuthal resolution) at higher altitudes and wider antenna beamwidths [11]. In 1952, Wiley made a landmark breakthrough by improving the azimuth resolution of an SLAR by utilizing Doppler shifts [12]. This achievement, which he called the "Doppler beam-sharpening" system, is regarded as the first synthetic aperture radar (SAR), leading to the widespread adoption of SAR technology [13].

In principle, the SAR and SLAR differ in their approaches to processing the Doppler spectra of signals. The SAR employs the Doppler shift resulting from sensor motion with respect to the target area to synthesize an equivalent long antenna aperture [14]. As a result, the SAR has an advantage over SLAR in achieving finer and more uniform resolution in the along-track direction that surpasses the diffraction limit of the antenna [15]. To better understand synthetic aperture for achieving high resolution, Figure 1.1 depicts the geometry of the airborne Stripmap SAR system, whose beam-crossing velocity is identical to that of the airborne platform.

In the case where the radar is broadside looking, the range history from radar to target is a function of time and is expressed as:

$$R(\eta) = \sqrt{R_c^2 + V_{SAR}^2 (\eta - \eta_c)^2} \tag{1.1}$$

The induced Doppler shift takes the form of:

$$f_d(\eta) = -\frac{2}{\lambda} \frac{dR(\eta)}{d\eta} \approx -\frac{2}{\lambda} \frac{V_{SAR}^2}{R_c} \eta \tag{1.2}$$

where:
η is the azimuth time (slow time),
η_c is beam-center crossing time, which is set to zero
R_c is the slant range of the beam centerline,
λ is the wavelength of the probing signal, and
V_{SAR} is the SAR velocity.

DOI: 10.1201/9781003308430-1

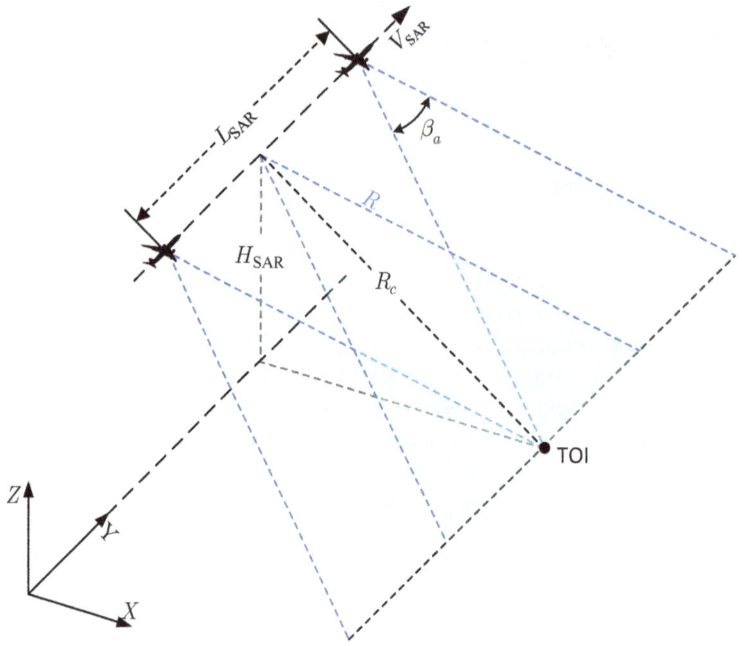

FIGURE 1.1   The SAR observation geometry (in the Stripmap mode).

As the radar moves, a linear virtual array can be created by coherently combining the complex received signals. The resulted virtual array is commonly referred to as the synthetic aperture length [16]. As demonstrated in Figure 1.1, the synthetic aperture length is associated with the central slant range R_c and real aperture beamwidth β_a along the azimuth, as:

$$L_{SAR} = R_c \beta_a \tag{1.3}$$

with the 3-dB beamwidth defined by:

$$\beta_a = 0.886 \cdot \lambda \ell_a^{-1} \tag{1.4}$$

where ℓ_a is the real aperture length along the azimuth direction, and the factor 0.886 is related to the 3-dB beamwidth.

The synthetic aperture length is also correlated to the epoch when the radar beam is directed toward the target of interest (TOI). This epoch, referred to as the synthetic aperture time (SAT), begins when the TOI enters the radar beam and ends when it moves out of the radar beam [17]. Mathematically, the SAT is obtained by:

$$T_{SAR} = 2R_c \cdot \tan\left(\beta_a/2\right) \cdot V_g^{-1} \tag{1.5}$$

With a small beamwidth, the SAT can be approximated to:

$$T_{SAR} \approx \frac{2R_c}{V_g} \cdot \frac{\beta_a}{2} = 0.886 \frac{\lambda R_c}{V_g \ell_a} \tag{1.6}$$

The Doppler bandwidth, representing the maximum change in Doppler frequency across the SAT, is given by:

$$B_D = \left| f_d \big|_{\eta=0.5T_{SAR}} - f_d \big|_{\eta=-0.5T_{SAR}} \right| \approx \frac{2}{\lambda} \frac{V_{SAR}^2}{R_c} \cdot T_{SAR} = 0.886 \cdot \frac{2}{\ell_a} \cdot \frac{V_{SAR}^2}{V_g} \tag{1.7}$$

Assuming a constant radar velocity, we then determine the time and spatial resolutions along the SAR track as follows, respectively:

$$t_a = \frac{1}{B_D} = \frac{1}{0.886} \cdot \frac{\ell_a}{2} \cdot \frac{V_g}{V_{SAR}^2} \tag{1.8}$$

$$\rho_a = 0.886 \cdot V_g t_a = \frac{\ell_a}{2} \frac{V_g^2}{V_{SAR}^2} \tag{1.9}$$

For the airborne SAR in the Stripmap mode, the beam-crossing velocity is identical to SAR velocity; hence, the synthetic aperture processing enables airborne Stripmap SAR systems to attain an azimuthal resolution that is precisely one-half the real aperture length. This characteristic renders the azimuthal resolution independent of the range and probing wavelength, making it available for realizing a high-resolution radar imaging with a small antenna.

It is crucial to acknowledge that attaining an infinitely fine azimuthal resolution is not feasible due to various practical constraints, including radiation power, swath width, signal fidelity, and pulse repetition frequency (PRF), among others. Indeed, the PRF holds practical significance for the SAR system, which presents conflicting requirements for swath coverage and Doppler bandwidth [18]. Such a requirement is expressed as follows:

$$B_D < f_{PRF} < \frac{c}{2W_{sr}} \tag{1.10}$$

where c and W_{sr} are the signal propagation velocity and slant range swath, respectively. The enforcement of this restriction can be elucidated as a wider Doppler bandwidth necessitates a high PRF to prevent azimuth ambiguity. Nevertheless, adopting a high PRF imposes certain restrictions on the unambiguous swath width because, in part, the signal propagated across the slant range swath must fit between two successive signal pulses to avoid range ambiguity.

The aforementioned requirements place further constraints on the swath width with respect to the azimuthal resolution. In light of Eqs. (1.9) and (1.10), we impose the slant range swath to the azimuth resolution as follows:

$$\frac{W_{sr}}{\rho_a} < T_{covres} \tag{1.11}$$

with

$$T_{covres} = \frac{1}{0.886} \cdot \frac{c}{2V_g} \tag{1.12}$$

Subsequently, Eq. (1.12) can be transformed into a function dependent on the platform's altitude in the case of spaceborne SAR. To maintain simplicity without sacrificing generality, we assume the spaceborne SAR follows a circular orbit, the SAR's inertial velocity can be determined by [19]:

$$V_{\text{SAR}} = \mu^{0.5} R_{\text{SAR}}^{-0.5} \tag{1.13}$$

In a Low Earth Orbit (LEO) scenario where the SAR's inertial motion is the primary contributor to the beam-crossing velocity, one can set aside the Earth's rotation and solely put on the SAR's inertial motion. Then, the beam-crossing velocity of the spaceborne SAR with the beam centerline directed toward the zero-Doppler plane can be approximated to [20]:

$$V_g \approx \frac{R_{\text{TOI}}}{R_{\text{SAR}}} V_{\text{SAR}} = \frac{\mu^{0.5} R_{\text{TOI}}}{R_{\text{SAR}}^{1.5}} \tag{1.14}$$

where

R_{TOI} represents the Earth's radius at the TOI's location,

μ is the Earth's gravitational coefficient, and

R_{SAR} is the distance from the Earth's center to the SAR system.

As a result, the threshold pertaining to the restriction of the slant swath width versus azimuthal resolution, as presented in Eq. (1.12), can be revised in the following manner:

$$T_{\text{covres}} \approx \frac{1}{0.886} \cdot \frac{c}{2} \cdot \frac{R_{\text{SAR}}^{1.5}}{\mu^{0.5} R_{\text{TOI}}} \tag{1.15}$$

Eqs. (1.11)–(1.15) indicate that the SAR system inherently imposes limitations on the trade-off between the unambiguous swath width and azimuthal resolution. From this perspective, it is apparent that enhancing the azimuth resolution in a spaceborne SAR at a particular orbital altitude will inevitably result in a reduced swath width. Contrarily, increasing the swath width degrades the azimuth resolution.

It is feasible to attain high slant range resolution in both SAR and SLAR systems by implementing broadband signals, such as the chirp signal. Also, appropriate matched filtering techniques for pulse compression may be applied to achieve a fine range resolution [21]. The resulting slant range resolution is inversely proportional to the signal bandwidth, namely:

$$\rho_r = 0.886 \cdot \frac{1}{2} \cdot \frac{c}{B_r} \tag{1.16}$$

where:

B_r is signal bandwidth,

the ratio of 1/2 is due to a round-trip delay of the radio signal, and

the 3-dB bandwidth along the range direction is accounted for by the factor of 0.886.

In terms of spatial resolution, SAR represents a significant improvement over SLAR, which enables producing fine-resolution radar images from either aircraft or spacecraft without compromising the azimuthal resolution. In this regard, SAR is capable of producing two-dimensional scatter images using received echoes and revealing hitherto unknown features on the Earth's surface. As such, SAR has become an indispensable tool in remote sensing and has promising yet underused applications in Earth observation [3, 22].

1.2   CURRENT SAR MISSIONS FOR EARTH OBSERVATIONS

As early as the 1960s, proposals were put forward for spaceborne SAR systems for Earth observations [23]. It was only in June 1978 that the first successful launch of spaceborne SAR mounted on SEASAT was achieved [24, 25]. The SEASAT acquired millions of square kilometers of high-resolution radar images, opening a milestone for Earth-observing SAR missions in the ensuing decades [26–28]. As a result, the spaceborne SAR gradually garnered the interest of scientists from across the globe and exhibited remarkable progress.

Both governmental and commercial sectors have either launched or are preparing to launch a series of spaceborne Earth-observing SAR missions. These missions have evolved from single imaging modes with single polarizations to multiple imaging modes with multiple polarizations, encompassing carrier frequency from the Ultra High Frequency (UHF) to Super High Frequency (SHF) [29]. Table 1.1 highlights current operating spaceborne SAR missions to offer an overview of their main system parameters.

Thus far, the deployment of spaceborne SAR missions has proven to be of immeasurable worth, catering to a plethora of Earth observation applications and becoming an indispensable facet of remote sensing [30]. The following sections briefly introduce some of the current operating spaceborne SAR missions.

1.2.1 TerraSAR-X/TanDEM-X

The TerraSAR-X/TanDEM-X mission is an endeavor in Earth observation via radar, jointly conducted by the German Aerospace Center (DLR) and Airbus Defence and Space (ADS) [31]. This mission employs two X-band radar satellites that work in tandem to capture high-resolution radar images of the Earth's surface. TerraSAR-X was launched in June 2007, whereas TanDEM-X was launched three years later, in June 2010. These twin satellites share the same Sun-synchronous orbit at an altitude of 514 km, with an inclination angle of 97.44° and a repeat cycle of 11 days/167 orbits; moreover, the twin satellites maintain proximity, ranging between 250 and 500 m along the orbit [32]. A visual representation illustrating the twin satellites in orbit around Earth can be accessed at: https://www.dlr.de/en/research-and-transfer/projects-and-missions/tandem-x.

TerraSAR-X/TanDEM-X provides a variety of imaging modes, such as the spotlight, Stripmap, scanSAR, and extended scanSAR modes, to select. Such features facilitate acquiring high-resolution images over a small area and low-resolution images over a large region with a swath width of up to 260 km. Besides, each spaceborne SAR possesses a suite of polarimetric combinations, ranging from single or dual polarization to full polarizations, making it adaptable for many applications. Additionally, the satellites establish a high-speed laser link, enabling them to communicate and synchronize their measurements [33]. Through their collective acquisitions, the twin flying in parallel satellites have archived a globally consistent digital elevation model (DEM) of high accuracy, released in 2014 [34].

The TerraSAR-X/TanDEM-X mission has demonstrated a multitude of valuable applications, encompassing but not limited to [35]:

1) Topography and Cartography: The primary goal of the TerraSAR-X/TanDEM-X mission is to generate high-resolution, global DEMs. These data are useful for creating accurate maps, conducting terrain analysis, and supporting applications in geology, hydrology, and climate research.
2) Land surface monitoring: TerraSAR-X/TanDEM-X provides high-resolution images that can be used to detect changes in the landscape over time and to monitor the effects of human activities such as mining, urbanization, and deforestation.
3) Disaster response: TerraSAR-X/TanDEM-X provides rapid and accurate information about the extent of damage caused by natural disasters such as earthquakes, floods, and landslides, helping to guide relief efforts.
4) Oceanography monitoring: TerraSAR-X/TanDEM-X can penetrate through clouds and provide data about sea surface roughness, currents, and waves. Such information is useful for monitoring glaciers, oil spills, and ocean circulation patterns.

TABLE 1.1

Overview of Spaceborne SAR Missions and Corresponding Technical Specifications

Sensor	Mission Duration	Freq. Band (Pol.)	Orbit Altitude	Operation Mode	Swath Width	Spatial Resolution	Organization, Country
SEASAT	1978.06–1978.10	L (HH)	769–799 km	Stripmap	100 km	25 m	NASA, USA
ERS-1/2	1991–2000/1995–2001	C (VV)	782–785 km	Stripmap	102.5 km	Az: 30 m; Rg: 26.3 m	ESA, Europe
J-ERS-1	1992–1998	L (HH)	568 km	Stripmap	75 km	18 m	JAXA, Japan
Radarsat-1	1995–2013	C (HH)	793–821 km	Fine	45 km	8 m	CSA, Canada
				Standard	100 km	30 m	
				Wide	150 km	30 m	
				ScanSAR Narrow	300 km	50 m	
				ScanSAR Wide	500 km	100 m	
				Extended High	75 km	18–27 m	
				Extended Low	170 km	30 m	
ENVISAT/ASAR	2002–2012	C (dual)	772–774 km	Image mode	100 km	28 m	ESA, Europe
				Wide Swath	400 km	150 m	
				Alternating	100 km	Az: 29 m; Rg: 30 m	
				Wave mode	5 km	Az: 28 m; Rg: 30 m	
				Global Monitoring	≥400 km	Az: 950 m; Rg: 980 m	
ALOS/PALSAR	2006–2011	L (quad)	692 km	Fine beam (sing-pol)	40–70 km	7–44 m	JAXA, Japan
				Fine beam (dual-pol)		14–88 m	
				Direct downlink		14–88 m	
				ScanSAR	250–350 km	100 m (multi-look)	
				Polarimetry	30 km	30 m	
TerraSAR-X/TanDEM-X	2007–today/2010–today	X (quad)	514 km	High-resolution spotlight mode	10 km	Az: 1.1 m; Grg: 0.74–1.77 m	DLR&ADS, Germany
				Spotlight mode	10 km	Az: 1.7 m; Grg: 1.48–3.49 m	
				Staring Spotlight Mode	4.6–7.5 km	Az: 0.24 m; Grg: 0.85–1.77 m	
				Stripmap (SM)	30 km	Az: 3.3 m; Grg: 1.7–3.49 m	
				ScanSAR (4 beam)	100 km	Az: 18.5 m; Grg: 1.7–3.49 m	
				ScanSAR (6 beam)	194–266 km	Az: 40 m; Grg: <7 m	

Satellite	Period	Band	Altitude	Mode	Swath	Resolution	Agency
Radarsat-2	2007–today	C (quad)	798 km	Fine	50 km	8 m	CSA, Canada
				Wide Fine	150 km	8 m	
				Standard	100 km	25 m	
				Wide	150 km	25 m	
				ScanSAR Narrow	300 km	50 m	
				ScanSAR Wide	500 km	100 m	
				Ocean Surveillance	530 km	Variable	
				Fine quad-pol	25 km	12 m	
				Wide Fine quad-pol	50 km	12 m	
				Standard quad-pol	25 km	25 m	
				Wide standard quad-pol	50 km	25 m	
				Extended High	75 km	25 m	
				Extended Low	170 km	60 m	
				Spotlight	18 km	1 m	
				Ultrafine	20 km	3 m	
				Wide Ultrafine	50 km	3 m	
				Extrafine	125 km	5 m	
				Multi-Look Fine	50	8 m	
				Wide Multi-Look Fine	90	8 m	
				Ship Detection	450	Variable	
COSMO-SkyMed (CKS)-1/4	2007...2010–today (CSK-1, 2 and 4 remain operational)	X (dual)	619 km	Spotlight	10 km	<= 1 m	ASI, Italy
				HIMAGE	40 km	3–15 m	
				PingPong	30 km	15 m	
				WideRegion	100 km	30 m	
				HugeRegion	200 km	100 m	

(Continued)

TABLE 1.1 (CONTINUED)

Sensor	Mission Duration	Freq. Band (Pol.)	Orbit Altitude	Operation Mode	Swath Width	Spatial Resolution	Organization, Country
RISAT-1	2012–2017	C (quad)	536 km	Coarse Resolution ScanSAR Mode	223 km	50 m	ISRO, India
				Medium Resolution ScanSAR Mode	115 km	25 m	
				Fine-Resolution Stripmap Mode-1	25 km	3 m	
				Fine-Resolution Stripmap Mode-2	25 km	9 m	
				High-Resolution Spotlight Mode (HRS)	10 km	1 m	
HJ-1C	2012–2023	S (VV)	500 km	Stripmap	40 km	5 m (single look); 20 m (four looks)	NDRCC/SEPA, China
				ScanSAR	100 km		
Kompsat-5	2013–2022	X (dual)	550 km	High resolution	5 km	1 m	KARI, Korea
				Standard mode	30 km	3 m	
				Wide swath	100 km	20 m	
ALOS-2	2014–today	L (quad)	628 km	Spotlight	25 km	Az: 1 m; Rg: 3 m	JAXA, Japan
				Ultrafine Stripmap	50 km	3 m	
				High sensitive Stripmap (CP/DP/SP)	50 km	6 m	
				High sensitive Stripmap (FP)	30 km	6 m	
				Fine Stripmap (CP/DP/SP)	70 km	10 m	
				Fine Stripmap (FP)	30 km	10 m	
				ScanSAR	350 km (5 scans)	100 m	
Sentinel-1a/1b	2014/2016–today	C (dual)	693 km	Stripmap (SM)	80 km	Az: 4.3–4.9 m; Rg: 1.7–3.6 m	ESA, Europe
				Interferometric Wide swath (IW)	250 km	Az: 22 m; Rg: 2.7–3.5 m	
				Extra Wide swath (EW)	400 km	Az: 43 m; Rg: 7.9–15 m	
				Wave (WV)	20 km	Az: 4.8 m; Rg: 2.0–3.1 m	
PAZ	2018–today	X (quad)	514 km	StripMap-S	30 km	3 m	CDTI, Spain
				StripMap-D	15 km	6 m	
				ScanSAR	100 km	16 m	
				Spotlight-S	10 km	1 m	
				Spotlight-D	10 km	2 m	
				High-Resolution Spotlight-S	5 km	< 1 m	
				High-Resolution Spotlight-D	5 km	< 2 m	

Mission	Years	Band (pol.)	Altitude	Mode	Resolution	Swath	Agency
SAOCOM	2018–today	L (quad)	619.6 km	Stripmap	< 10 m	> 40 km	CONAE, Argentine
				TopSAR narrow (SP/DP)	< 30 m	> 150 km	
				TopSAR narrow (QP)	< 50 m	> 100 km	
				TopSAR wide (SP/DP/CL-POL)	< 50 m	> 350 km	
				TopSAR wide (QP)	< 100 m	> 220 km	
ICEYE X1	2018–today	X (VV)	570 km	Dwell	1 m	5 km	ICEYE, Finland
				Spot	1 m	5 km	
				Spot extended area	1 m	15 km	
				Stripmap	3 m	30 km	
				ScanSAR	15 m	100 km	
RADARSAT Constellation	2019–today	C (quad)	586–615 km	Low Res. 100 m	100 m	500 km	CSA, Canada
				Medium Res. 50 m	50 m	350 km	
				Medium Res. 30 m	30 m	125 km	
				Medium Res. 16 m	16 m	30 km	
				High Res. 5 m	5 m	30 km	
				Very High Res. 3 m	3 m	20 km	
				Low Noise	100 m	350 km	
				Ship Detection	Variable	350 km	
				Spotlight	Az: 1 m/Rg: 3 m	20 km	
				Quad-Polarization	9 m	20 km	
BIOMASS	Launch scheduled in 2024	P (quad)	660 km	Stripmap	200 m	50 km	ESA, Europe
TanDEM-L	Launch scheduled in 2024	L (quad)	745 km	Stripmap, sing/dual-pol	Az: 3 m–7 m; Grg: 2.4 m–4.0 m	350 km	DLR, Germany
				Stripmap, quad-pol	Az: 3 m–7 m; Grg: 2.8 m–3.8 m	175 km	

Note: 'Az' means azimuthal resolution, 'Rg' stands for range resolution, and 'Grg' indicates ground range resolution.

The TerraSAR-X/TanDEM-X spaceborne SAR mission represents a momentous scientific under-taking that bestows high-resolution imaging capabilities and precise digital elevation models. This possesses exceptional potential for enriching our understanding of the planet. The TerraSAR-X/TanDEM-X has been impacting cartography, geology, and environmental studies, to name some.

1.2.2   TanDEM-L

Tandem-L is a forthcoming spaceborne SAR mission developed by the DLR in collaboration with various partners. This mission is unique in its deployment of two cooperative L-band spaceborne radars that fly in close formation with variable adjustable spacing, with one satellite trailing the other [36]. A schematic diagram depicting Tandem-L dual satellites flying in formation is available at the following link: https://www.dlr.de/en/images/2019/4/tandem-l-proposal. The arrangement of this fly formation configuration empowers the spaceborne bistatic SAR to gather up-to-date Earth's information on a global scale, resulting in exceeding accuracy when observing dynamic Earth phenomena [37].

The Tandem-L satellites are planned to operate in a Sun-synchronous orbit at 745 km and a repeat cycle of 16 days/231 cycles, proffering a swath width of hundreds of kilometers. Besides, the Tandem-L mission would operate at a relatively longer wavelength (23.6 cm) than the Tandem-X mission for tomographic measurements of the three-dimensional structure of vegetation and ice regions [38]. Moreover, the Tandem-L mission features polarimetric SAR interferometry that enables the measurement of forest height and uses multi-pass coherence tomography that facilitates the determination of the vertical structure of vegetation and ice. The mission also employs the latest digital beam-forming techniques, which not only increase the swath width, but also enhance the imaging resolution [39].

The Tandem-L mission is designed to operate for at least 11 years, with primary mission goals comprising [37, 40]:

1) Monitoring and Measuring Earth's Surface: The Tandem-L mission shall aid in the surveillance and quantification of various attributes of the Earth's surface, such as forests, agriculture, urban landscapes, and water bodies.
2) Deforestation and Reforestation: The Tandem-L can quantify forest biomass, thereby offering information regarding the dynamics of deforestation and reforestation and facilitating the development of forest conservation and management strategies.
3) Glaciology and Cryosphere Studies: The Tandem-L mission can measure glacier movements and melting processes, offering crucial information for comprehending climate change and its effects on the glaciers, ice caps, and permafrost.
4) Natural Disaster Management: Tandem-L can capture surface deformations of the Earth with an accuracy of millimeters. The resulting data can be utilized to evaluate the extent of destruction caused by natural disasters and to support disaster relief operations.
5) Ecological Studies: Tandem-L data can be used to study various ecosystems, including wetlands, coastal zones, and coral reefs, providing vital information for environmental management and conservation efforts.

Once operational, the Tandem-L mission is expected to acquire images of the Earth's landmass on a weekly basis, providing crucial information currently unavailable in Earth observation, thereby improving scientific forecasting, and associated socio-political recommendations. Such information on dynamic processes is essential for climate change studies.

1.2.3   Sentinel-1 A/B

Sentinel-1 is a spaceborne SAR mission operated by the European Space Agency (ESA) as part of the European Union's Copernicus program. Sentinel-1 mission consists of a pair of satellites,

Sentinel-1A and Sentinel-1B. The Sentinel-1A was launched in 2014, and its counterpart, Sentinel-1B, was launched two years later, in 2016 [41]. The two satellites are strategically situated in the Sun-synchronous orbit at a height of 693 km above the Earth's surface, such orbital configurations allows the radar to image the globe every 6 days [42, 43]. The observation scenario for the Sentinel-1 mission can be accessed on the ESA's website at: https://sentinel.esa.int/web/sentinel/missions/sentinel-1/observation-scenario.

Using a C-band SAR, Sentinel-1 can image the Earth by various modes, facilitating swath widths spanning from dozens to hundreds of kilometers. The Sentinel-1 data set is freely accessible to global users through the Copernicus Open Access Hub (https://scihub.copernicus.eu/). The Sentinel-1 A/B mission presents a plethora of applications, spanning diverse fields encompassing [44]:

1) Environmental monitoring: Sentinel-1 furnishes data for monitoring environmental metamorphoses, such as deforestation, land use changes, and soil moisture.
2) Disaster response: Sentinel-1 can map floods, landslides, and other natural disasters, providing information for emergency responders and aid organizations. Additionally, it can be used to monitor the areas stricken by earthquakes and other geohazards.
3) Agriculture monitoring: Data from Sentinel-1 can be leveraged to monitor crop growth and health, estimate crop yields, and identify drought or other environmental stress areas.
4) Maritime surveillance: Sentinel-1 can be utilized to monitor ship traffic, track oil spills, and detect illegal fishing activities. It can also be utilized to trace the fluctuations in ice sheets and glaciers that potentially affect marine navigation.

The Sentinel-1 mission's contributions to Earth observation are profound, it is a crucial data source for various applications. The mission's freely available data ensures accessibility to a diverse pool of users, who can utilize it for environmental monitoring, disaster response, and scientific research. As such, with many applications ahead, the Sentinel-1 mission's contributions continue to be far-reaching, holding immense potential for future research and development.

1.2.4   RADARSAT CONSTELLATION MISSION (RCM)

The RADARSAT Constellation Mission (RCM) is Canada's commitment to Earth observation satellites managed by the Canadian Space Agency (CSA). Launched in June 2019, the RCM consists of three identical satellites, each weighing around 1400 kg, and equipped with a state-of-the-art C-band SAR instrument [45]. One can find a schematic illustration of RCM at: https://www.asc-csa.gc.ca/eng/satellites/radarsat.

Each satellite of the RCM orbits the Earth in a Sun-synchronous orbit, with an altitude of roughly 600 km and an inclination of 97.74°. This orbital configuration allows for an orbital cycle of approximately 96 minutes, thereby enabling the RCM to repeat data collection on the Earth's surface every 4 days [46]. Besides, the RCM operates in multiple imaging modes, including high-resolution imaging, wide-area surveillance, and ship detection modes, allowing the RCM to gather data with a resolution of up to 1 m for various Earth observation applications [47]. Some of the sample applications are outlined as follows [48]:

1) Maritime Surveillance: The RCM has the ability to detect and track ships in maritime surveillance and security.
2) Disaster Management: The RCM is capable of providing fast-repeat and high-resolution images of disaster-affected areas so as to enable emergency response.
3) Resource Management: The RCM can be used to monitor changes in land use, vegetation, and water resources.
4) Arctic Monitoring: The RCM is instrumental in monitoring changes in sea ice and providing essential data on the Arctic environment, which is crucial for climate research and resource development.

The RCM is a vital tool for Earth observation, offering comprehensive coverage of landmass and coastal areas, with a particular emphasis on the Arctic region that is vulnerable to climate change. The provided data are valuable for improving our understanding of the Earth's systems and processes. Moreover, the RCM contributes to global efforts to monitor and manage the Earth's resources and environment while supporting disaster management and response efforts.

1.2.5 BIOMASS

The BIOMASS mission is an Earth observation initiative crafted by the ESA. With a scheduled launch in 2024, its primary aim is to provide precise and frequent information on forest properties on a global scale [49, 50]. The upcoming BIOMASS Mission will be equipped with a pioneering P-band SAR (P-SAR), facilitating space-based Earth observation at the P-band for the first time [51]. The BIOMASS mission is designed to function for a period of five years, during which it will monitor a minimum of eight growth cycles of the global forests. This endeavor will yield invaluable insights into the planet's forest ecosystems and unlock a wide range of unforeseen applications [52].

The Biomass P-SAR is scheduled to work in the Stripmap mode, which employs a solitary antenna beam to illuminate a swath width of roughly 50 km at a spatial resolution of 200 m. The P-SAR mechanism shall employ tripartite, complementary swaths that will interlace, featuring an incidence range spanning from 23° to 35° [53]. Further, the satellite would be positioned in a near-circular dawn–dusk orbit that is Sun-synchronous, with an altitude of roughly 660 km and an inclination of 97.97°. The orbital configurations are designed to minimize the ionospheric effects and to facilitate baseline observation based on the dual-baseline interferometry at a repeated cycle of 17 days [54]. For a comprehensive overview of the Biomass Mission, and the associated schematic diagrams, please visit the following link: https://earth.esa.int/eogateway/missions/biomass.

The P-SAR instrument has a high sensitivity to diverse components of forest environments, allowing three distinct measurement techniques in the BIOMASS mission: horizontal mapping, height mapping, and three-dimensional (3D) mapping. This innovative capability will address various critical issues that have far-reaching scientific implications, which will serve the following functions [51–54]:

1) Forest monitoring: The BIOMASS endeavors to create accurate biomass maps of forests in tropical, temperate, and boreal areas. Such maps are crucial in monitoring changes in forest ecosystems and comprehending their impacts on the global carbon cycle.
2) Forest management: The BIOMASS would provide refined forest biomass data. Such information would enable us to evaluate the status and variations within the Earth's forests, and aid us in directing forest management practices.
3) Climate change monitoring: The BIOMASS mission will supply data for modeling and forecasting the impacts of climate change on forest ecosystems and the global carbon cycle. It will also kickstart and authenticate the terrestrial segment of Earth system models.
4) Advance remote sensing technology: The BIOMASS shall employ cutting-edge radar technology to quantify forest biomass and delineate surface morphology of forest regions, thereby enhancing remote sensing techniques to effectively monitor global forests.

The BIOMASS mission harbors the potential to augment our understanding of forests and their role in Earth's carbon cycle, thereby illuminating forest management strategies and improving climate change modeling. Further, the Biomass mission is poised to embark on the pioneering endeavor of exploring the Earth's surface at the P-band, unearthing observations that hold the potential to have a multitude of unknown and diverse applications.

1.3 WHY MOON-BASED SAR?

Presently, the spaceborne SAR has been offering constant mapping of Earth's surface. Still, the SAR technologies necessitate further advancements to meet the needs of diverse applications, a core on two extremes – the 'small' and the 'large'. The small pole concerns the augmentation of SAR spatial resolution, which is pivotal for detecting ground objects and quantitatively inverting surface parameters by remote sensing data [55–58]. On the other hand, the large pole pertains to the substantial spatiotemporal coverage performance required to monitor the Earth's surface proficiently [59]. This aspect is gaining importance as scientific inquiry increasingly recognizes Earth as an integrated entity [60, 61]. Hence, augmenting the SAR system's swath width and spatial-temporal resolution in tandem can effectively tackle the geoscience challenges posed by the Earth's interconnected systems [62]. However, the existing spaceborne SAR poses some constraints in line with the spatiotemporal coverage, which, in effect, impedes the data applications from multiple perspectives [63], as briefly outlined below.

1.3.1 Constraints of Spaceborne SAR

Research tasks necessitate acquiring geo-parameters from SAR sensors spanning the Earth's disk while maintaining temporal and spatial continuity at a consistent level [64, 65]. The current spaceborne SAR missions are confronted with the daunting challenge of fulfilling this exacting requirement as they encounter the inherent trade-off between azimuthal resolution and swath width [14, 20]. Various SAR imaging modes have been suggested to compromise the coverage and resolution, as demonstrated in Table 1.1. Nevertheless, none of those imaging modes can attain extensive coverage while maintaining a fine spatial resolution simultaneously. In addition, due to the LEO altitude, the revisit period of spaceborne SAR is restricted to several to tens of days [66]. The coarse temporal resolution, coupled with the coverage versus resolution restriction, limits some applications that require uncompromised spatiotemporal resolution, such as, but not limited to:

1) Rapid disaster response: Fast detection and swift reaction are paramount in the face of calamitous events, such as deluges, seismic activities, or oil spills [67]. Such use cases necessitate more frequent monitoring with extensive coverage at the temporal resolution of hours to days [68]. The intervals between successive flyovers of spaceborne SAR systems, and their narrow swath width, restrict the capacities of spaceborne SAR systems to detect and respond to events across vast territories promptly. This limitation might be why most SAR systems available today still find it challenging to offer a quick and timely response for efficient emergency response applications [69], This evaluating the damage in the aftermath of earthquakes, hurricanes, or volcanic eruptions. In this regard, spaceborne SAR systems may not be the optimal solution for fulfilling the requirement of swift disaster response.

2) Ecosystem disturbance monitoring: Numerous ecological disturbances, such as deforestation, necessitate high-resolution, wide-area SAR imaging with frequent revisits to identify and track changes [70]. To illustrate, detecting deforestation and forest degradation demands frequent monitoring to capture alterations between observations [71, 72]. Nonetheless, existing SAR systems encounter difficulties in achieving broad coverage and high temporal resolution. Consequently, many cases involving illegal logging or forest clearing may evade detection by the spaceborne SAR. Moreover, high-spatial resolution could be vitally helpful in detecting small-scale forest loss.

3) Glacier and ice sheet monitoring: Regular and repetitive observations are necessary to consistently monitor glaciers' movement and melting rates throughout the ablation season [73]. To accurately monitor the rapid changes occurring in Glacier regions, it is essential to utilize SAR imaging with both high temporal resolution and wide coverage. Indeed, it demands tracking fast-moving glacier flow rates by SAR interferometry [74]. In this

regard, the temporal resolution of current spaceborne SAR is usually insufficient to capture the dynamics of glaciers. Further, the swath width limits an extensive coverage of ice sheets and ice fields. Due to the long revisit times, the ice dynamics might be overlooked without operational ice monitoring and forecasting. The long temporal baselines and sparse acquisitions usually deteriorate the measurement accuracy.

4) Permafrost monitoring: Monitoring permafrost is vital in studying and comprehending the ramifications of climate change in the Arctic and similar regions [75–77]. In doing so, employing high-spatial resolution wide-area SAR with frequent revisits is necessary, as the time window for permafrost monitoring is quite short [78]. To attain the objective of capturing variation in permafrost coverage during this time window necessitates SAR systems that provide high-spatial resolution imaging and possess a wide-swath width to cover vast areas [79, 80]. Supposing that a SAR system can capture such observations, it would enhance our comprehension regarding the influence of climate change on permafrost and its corresponding ecological systems, particularly during the fleeting summer season. Unfortunately, most present-day spaceborne SAR systems encounter difficulty in supporting such observation.

5) Oceanic environment monitoring: Monitoring the oceanic environment involves the observation of physical, chemical, and biological parameters, as well as marine pollutants, disasters, and climate change [81]. High-spatial resolution and continuous monitoring with frequent coverage are necessary to depict the state and changes of the oceanic environment accurately [82]. Further, the estimation of wind fields in terms of wind speeds and directions over the ocean requires frequent SAR imaging [83]. However, the long revisit time and narrow swath of most spaceborne SAR systems hinder their use for operational monitoring. Specifically, they restrict the ability to obtain high spatiotemporal resolution data and detect transient processes or high-frequency environmental change signals, which may result in missing critical information, such as environmental changes or emergencies in local sea areas [84]. Additionally, the discontinuous observation and finite operational lifespan of spaceborne SAR pose constraints on monitoring long-term slow changes in the marine environment within specific geographic locations. Furthermore, the low spatial resolution makes capturing small-scale changes or details in the marine environment difficult, if not impossible. To enhance monitoring of the oceanic environment, it is imperative to secure SAR time series with high precision and long life-cycle, broad coverage, and satisfactory spatial resolution.

1.3.2 POTENTIAL SOLUTIONS TO THE COVERAGE-RESOLUTION CONSTRAINT

In the context of a single spaceborne SAR, it is highly challenging to achieve both high azimuthal resolution and wide-swath width simultaneously during SAR data acquisition. The two factors are subject to a trade-off, and optimizing one comes at the cost of the other. Meanwhile, the temporal resolution is also subjected to the LEO altitude, resulting in less frequent revisit observation of the target region, making it difficult to track dynamic geo-phenomena. To surmount the constraints of spaceborne SAR for Earth observation, we may deploy the distributed spaceborne SAR system or employ a high-orbit SAR system.

Implementing a distributed SAR system is reliant on a constellation comprising multiple satellites [65]. This approach is characterized by its high reliability, ease of implementation, and cost-effectiveness [85]. Moreover, the distributed SAR system has the potential to collaborate with pre-existing spaceborne SAR systems [86]. To enhance the performance of Earth observation, specifically launched or upcoming spaceborne SAR missions integrate satellite constellations and advanced imaging modes [87].

On the other hand, deploying the distributed SAR amounted on the constellation of satellites can potentially bring about unexpected technical intricacies, such as system synchronization, inter-satellite and satellite-ground station communications, formation design and relative navigation, data acquisition, and signal processing [65, 88]. The other challenge in the distributed SAR pertains to its spatiotemporal coverage performance. Specifically, it remains a difficult task to ensure the temporal consistency and spatial continuity in the coverage of distributed SAR [89]. In this regard, such a system is unfavorable for observing dynamic, large-scale geo-phenomena, a distributed SAR technology breakthrough shall be required.

An alternative is to employ high-orbit SAR for Earth observation to ensure uninterrupted monitoring of an extensive region for a sustainable period [90]. According to Eq. (1.15), when the orbital altitude increases, the trade-off between the swath width and azimuthal resolution becomes more flexible as the threshold for restricting both increases. By deploying a SAR on a platform orbiting at a high altitude, it is feasible to attain both the high spatial resolution and wide-swath imaging simultaneously, thereby ensuring the spatiotemporal congruity of the observed region [91]. Moreover, the high orbit also confers the advantage of generating data with satisfactory temporal resolution [88].

An exemplary high-orbit SAR is the Geosynchronous Earth Orbit (GEO) SAR, which operates in an inclined geosynchronous orbit at an altitude of around 36,000 km [92]. The GEO SAR possesses the potential to cover an extensive area with high temporal resolution, thus facilitating Earth observation with a shorter revisit period [93]. Despite the advantages of GEO SAR, the required substantial antenna puts an enormous weight burden on the artificial satellite, given that the satellite's payload capacity is typically constrained [94]. Furthermore, it is worth noting that the GEO SAR can only attain a partial view of Earth, leaving a large portion of Earth's surface area unobservable [95]. Therefore, although the GEO SAR represents a promising technology on the horizon, pursuing a more efficient high-orbit SAR system remains to be ongoing.

As the global fascination with lunar exploration gains renewed momentum, establishing a sustainable lunar base has emerged as an inevitable trend [96, 97]. Establishing a lunar base has further stimulated the curiosity to deploy radar, referred to as Moon-Based SAR (MBSAR), for Earth observation [98]. In contrast to the artificial satellite platform, the Moon provides a distinctive opportunity to conduct SAR Earth observation without many limitations of installing large antennas [99, 100]. In this regard, the MBSAR is endowed with the formidable ability to monitor Earth with an extensive swath width, as illustrated in Figure 1.2. Further, owing to the relative motion between the Moon and Earth, the sensors (including radar) installed on a lunar base can provide a holistic view of our planet comprehensively while also revisiting it in a shorter period (approximately 24.8 hours) [101].

In the following sections, there is a brief overview of lunar exploration, followed by an analysis of how the lunar exploration has motivated observing Earth from the Moon.

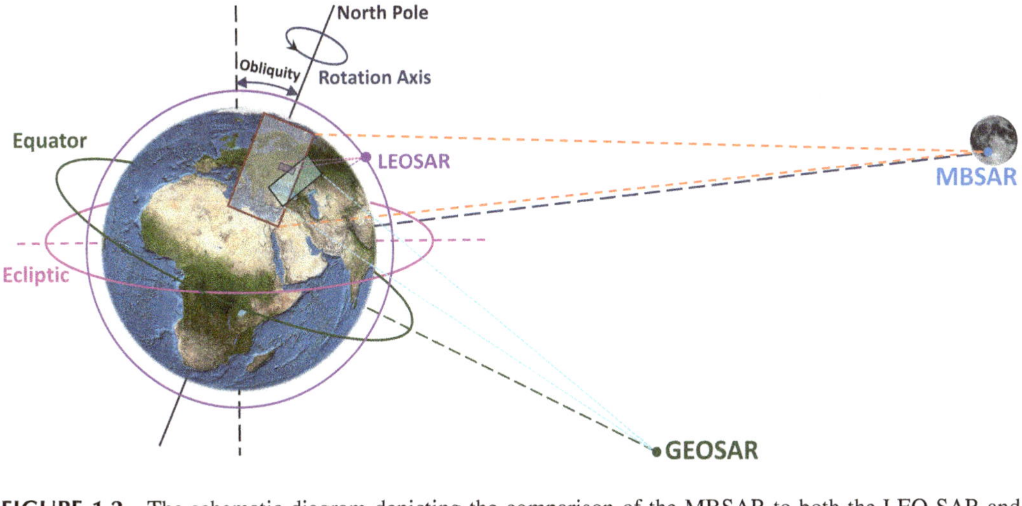

FIGURE 1.2 The schematic diagram depicting the comparison of the MBSAR to both the LEO SAR and GEO SAR systems.

Figure credit: Drs. Hanlin Ye and Zhen Xu.

1.3.3 The Lunar Exploration and Moon-Based Earth Observation

Exploring the Moon has been an ambition of humankind for centuries [102]. Since its inception in 1959, contemporary and physical lunar exploration has been underway using spacecraft. Luna 1–3 series hold distinctions of being pioneering spacecraft to embark on lunar exploration, wherein Luna 1 was the first spacecraft to approach the Moon, Luna 2 accomplished a hard landing on the lunar surface, and Luna 3 captured the photographs that provided the first glimpse of the lunar surface on its far side [103].

In the 1960s, the space race intensified as efforts were accelerated and competed to achieve a lunar landing to extend human presence to the Moon. Launched in 1966, Luna 9 accomplished the first-ever soft landing on the Moon [104]. During the period spanning from 1961 to 1972, the Apollo program, spearheaded by the National Aeronautics and Space Administration (NASA) of the USA, accomplished a series of momentous accomplishments in the field of lunar exploration [105]. The Apollo program comprised 17 missions, during which a series of photographs of our planet taken from the Moon's vantage point were obtained [106]. For instance, the Apollo 8 mission pioneered to capture a color image of Earth from the lunar view, famously dubbed the "Earthrise" image [107]. Further, during Apollo 17, two noteworthy photographs were taken: the first fully illuminated color image of our planet, and the widely recognized "Blue Marble" picture [108]. The feats mentioned above marked the first instance of conducting Earth observation from the lunar perspective.

Following the groundbreaking Apollo and Luna missions, lunar exploration advanced at a steadier pace thereafter. Although lunar missions are not as commonplace as they were in the 1960s and 1970s, this period was significant for advancing lunar discovery and technology [109]. Notable achievements included the evidence for the existence of lunar polar ice and the inaugural deep space missions reached in lunar exploration by several space agencies. For example, Japan's successful launch of the Hiten spacecraft in 1990 stands as a testament to the achievement of becoming the third nation to place an object in the Moon's orbit [110]. Additionally, the Clementine probe conducted a comprehensive mapping of the lunar surface in 1994 and detected the potential presence of water ice in a permanently shadowed crater at the south pole of the Moon [111]. In 1998, the privately-funded Lunar Prospector mapped the lunar surface composition and the magnetic and gravity fields. Further, it detected hydrogen, assumed to be in the form of water, at both lunar poles, with higher concentrations near the south pole [112]. During this epoch, we witnessed emerging space powers undertaking and achieving lunar exploration independently, a notable development in lunar exploration.

During the first decade of the twenty-first century, several prominent spacefaring nations were actively engaged in lunar exploration missions, with the objectives of mapping the lunar surface, monitoring its environment, and prospecting for resources [113]. The European Space Agency's SMART-1 orbiter, launched in 2003, spent nearly two years orbiting the Moon, carrying out technological demonstrations on the lunar terrain [114]. Inaugurated in 2007, Japan's Selenological and Engineering Explorer (SELENE), also recognized as Kaguya in Japan, generated top-quality maps of lunar terrain for two years [115]. India's Chandrayaan-1 orbiter, launched in 2008, performed remote sensing observations of the Moon for one year during its orbit [116]. In 2009, NASA's Lunar Crater Observation and Sensing Satellite (LCROSS) mission employed a rocket stage to impact the Moon and detected significant amounts of water ice, and other volatiles in the lunar soil [117]. The Lunar Reconnaissance Orbiter also launched in 2009, is still in operation, producing high-resolution maps of the lunar terrain [118].

Numerous countries and organizations have recently declared their intentions to embark on lunar explorations. India has set its sights on launching the Chandrayaan-3 lander mission in 2023 [119]. India is also seeking collaboration with Japan to launch the Lunar Polar Exploration Mission, an endeavor aimed at sending a lander and rover to explore the lunar south pole and serving as a steppingstone for future human expeditions to the Moon [120]. In 2018, NASA launched the Artemis program, which seeks to return to the Moon with the support of commercial and international partners [121]. The Moonlight initiative of ESA aims to launch three or four lunar artificial satellites into orbits circling the Moon and deploy them to form a network of communication and navigation satellites to facilitate the following lunar missions [122].

China embarked on the Chinese Lunar Exploration Program (CLEP) with the aim of conducting a comprehensive study of the Moon and investigating the potential for lunar mining, encompassing a range of sophisticated spacecraft [123]. Notably, the Chang'E 1 and Chang'E 2 lunar orbiters were launched by China in 2007 and 2010 respectively, paving the way for subsequent missions. In 2013, China made history by successfully landing the Yutu rover and Chang'E 3 lander on the Moon, becoming the third country to achieve this feat. The Chang'E 3 spacecraft was particularly noteworthy as it was the first to make a soft landing on the lunar surface since Luna 24 in 1976 [124]. Building on this success, China launched the Chang'E 4 mission in 2019, which made a soft landing on the far side of the Moon, and the Yutu-2 Moon rover deployed by Chang'E 4 subsequently broke the record for lunar surface travel [125]. The Chang'E 5 mission, launched in 2020, was a sample return mission that brought back precious lunar samples to Earth [126].

China has recently announced its forthcoming lunar exploration plan, consisting of a series of missions investigating the Moon's surface and resources [123, 127]. Specifically, Chang'E 6 is scheduled to launch in 2025. Its primary objective is to analyze the topography, composition, and subsurface structure of the South Pole-Aitken basin while also collecting samples for further analysis. Chang'E 7 is set to launch in 2026 and will focus on exploring the Moon's South Pole region for resources, assessing the potential for future human settlements, and testing new technologies required for lunar exploration. Its objectives include studying the lunar environment and conducting experiments on in situ resource utilization. Chang'E 8 mission is slated for launch in 2028, primarily focusing on verifying the in situ lunar resource development and utilization technologies. Moreover, it will conduct experiments on critical technologies required to build a lunar base.

Propelled by competition and humankind's perpetual fascination with the unknown, the lunar exploration era that peaked with the Apollo and Luna missions showcased the extraordinary accomplishments that can be attained with grand visions and global ambitions. After lying dormant for decades following the final Apollo landing, both governmental and private entities have revitalized ambitious plans to dispatch robots and astronauts back to the lunar surface [109]. Revisiting the Moon bears immense significance, and the following points highlight this notion [96, 98, 113]:

1) Scientific significance: Studying the Moon's surface and rocks can offer insights into the formation and evolution of our solar system, as well as the early history of our planet. Analysis of the lunar surface can also provide valuable information about the geological processes on the Moon. By examining aspects such as the lunar soil, we can uncover secrets about the origin of Earth and the solar system. Additionally, the Moon's unique characteristics make it an ideal platform for deploying sensors, which could open up new opportunities for Earth and astronomical observations.

2) Technological Advancements: The lunar exploration drives innovation by advancing technological capabilities. Developing the necessary infrastructure and spacecraft for lunar missions pushes the boundaries of engineering and innovation. A human presence on the Moon would enable us to push forward in life support systems, fuel production, 3D printing, and other necessary technologies for extended space missions. Further, the Moon could function as a launchpad for assembling spacecraft for future crewed missions to Mars and beyond. The advancements made during these missions have the potential to benefit various industries.

3) International Collaboration: Revisiting the Moon presents a promising international collaboration and cooperation avenue. The participating parties can effectively pool resources, tap into collective expertise, and share costs by engaging in collaborative missions and forging partnerships. This fosters a sense of global unity and facilitates peaceful exploration. Furthermore, lunar exploration not only allows us to reinforce current partnerships,

but also to forge new ones, thus advancing diplomacy and fostering an atmosphere of cooperation in the realm of space exploration.

4) Inspiring humanity: The Apollo missions were a source of wonder and curiosity for humanity. Their achievements had the power to inspire and captivate us. Future lunar exploration,
including human missions, has the potential to do the same and ignite the passion for science, technology, engineering, and mathematics in younger generations. Embarking on this
exploration journey could inspire a new generation of scientists and deep space explorers.

5) Strategic benefits: The Moon is rich in resources that hold immense value for both scientific research and space missions. Returning to the Moon is a steppingstone toward future
deep space exploration. Establishing a durable human settlement on the Moon could offer
benefits for space observation, communication, and beyond. A lunar base would facilitate
monitoring Earth's environment, space weather occurrences, and any unsanctioned space
operations.

6) Safeguard humanity preservation: Establishing a permanent human presence on the Moon
could potentially ensure the long-term survival of humanity. In catastrophic events on
Earth, lunar colonies could be a backup for civilization and help preserve and sustain
human society. Moreover, lunar bases could yield strategic advantages, serving as space-
ports for journeys to Mars and beyond, and enabling surveillance of both Earth and space.
Thus, creating an off-world contingency plan for human existence is imperative in ensuring
the longevity of our species for future generations.

In essence, the pursuit of lunar exploration holds enormous significance and substantial benefits for
our society. Recent technological advancements have rendered it feasible to explore the Moon in
more detail than ever before [128]. The new generation of explorers is reigniting the spirit of discovery for prolonged missions to establish a sustainable human presence on and around the Moon
in the coming decades [129].

A notable announcement regarding the lunar exploration prospects was made by NASA, alongside 13 other international space agencies, during the Global Exploration Strategy workshop in
2006 [130]. These prospects, referred to as the lunar exploration objectives, encompass six distinct
themes and almost 200 objectives, categorized into 23 groups. Earth observation is one of those
groups and comprises a range of objectives [131]. It is indeed a challenging endeavor to expound
upon the intricate details of each lunar exploration objective, which is also beyond the purview of
this book. Henceforth, we shall direct the attention toward the aspect of Earth observation involved
in the lunar exploration objectives with the primary features detailed in Table 1.2.

Lunar exploration with a long-term objective of establishing a sustainable lunar base would open
up novel prospects for Earth observation [132]. Further, implementing a remote sensing platform
on the lunar base (also known as the Moon-based platform) requires cutting-edge advancements
in lunar exploration technologies [98]. In turn, Moon-based remote sensing can be an invaluable
complement to lunar exploration efforts: conducting Moon-based remote sensing can provide
highly accurate Moon orbit parameters and related data, thus accelerating the development of lunar
and deep space exploration technologies [133]. Establishing the Moon-based platform presents a
groundbreaking avenue for remote sensing complementing and supplementing current Earth observations [134–137].

Compared to spaceborne sensors, Earth observation conducted by sensors placed on a Moon-
based platform poses distinct challenges [138]. For passive sensors, an increase in distance between
the device and the ground target could lead to a significant reduction in spatial resolution. Besides, the
coverage effectiveness for certain passive sensors, such as optical sensors, depends on the extent of
overlap between the sensor's observation area and the sunlit region [139, 140]. This overlap is spatiotemporally variable, which poses a challenge to the consistency and immediacy of Earth observation.
In addition, atmospheric disturbances pose a serious hurdle: the presence of clouds and fog frequently
obscures a substantial portion of the Earth's surface, further limiting the coverage provided by optical

TABLE 1.2

Overview of Earth Observation Goals in the Lunar Exploration Objectives Proposed by NASA

Objective Name	Summary	Value
Observing magnetosphere	1) Study impacts of solar activity on Earth's electromagnetic behavior and the potential effects; 2) Investigate magnetosphere interactions with the lower atmosphere.	Understanding magnetosphere-driven event impacts can improve predictive strategies and guide mitigation priorities and responses.
Creating solid Earth topography, altimetry, tomography, and vegetation maps	1) Form SAR images of Earth using Earth-Moon relative movement, the bistatic configuration is also available; 2) Use multiple antennas to achieve long and stable baselines for the SAR interferometry.	The MBSAR provides ever-present observing of the entire Earth's disk under all-weather conditions; a Dual-band SAR signal allows global observation of the ionosphere; the MBSAR interferometry allows for creating accurate topographic maps.
Observing the atmospheric composition	Employ passive sensors ranging from UVA to TIR at 1 km horizontal resolution for global mapping of tropospheric and stratospheric compositions.	Offering a unique viewpoint for characterizing surface fluxes of gases, global-scale transport of pollution, as well as ozone and aerosol dynamics.
Monitoring the Sun-Earth system and atmospheric reactions to solar activity	Conduct contemporaneous monitoring of Earth and Sun from 60 nm to 1 micron, aiming at examining impacts of solar flares and coronal mass ejections on the Earth's atmosphere, particularly the mesosphere, stratospheric ozone, and troposphere.	Understanding the Earth's atmospheric responses to the solar activity from a lunar base view; such insights are instrumental in assessing impacts of long-term solar changes on the climate and atmospheric compositions.
Determining Earth's BRDF (bi-directional reflectivity distribution function)	Observe Earth at wavelengths between 0.34 and 1.5 microns to determine Earth's BRDF from the available lunar view, specifically when the Sun, Moon, and Earth are aligned.	The Moon-based observation might fill knowledge gaps in Earth's BRDF. Such information is needed for the radiative balance of Earth in climate studies, which is lacking from views of spaceborne sensors.
Measuring ocean color	Measure oceanic phytoplankton fluorescence by utilizing active lidar from the lunar base.	Providing insight into the physiological and health conditions of marine ecosystems, particularly those submerged.
Mapping surface composition of the Earth	Conduct moderate resolution UV-VIS-SWIR-TIR multi- to hyper-spectral mapping of Earth from a Moon perspective.	Capturing multi- to hyper-spectral data sets of Earth's disk, aiding in monitoring the compositions of the Earth's surface, oceans, and atmosphere, as well as the alterations they undergo over time.
Measuring paleo solar constant	Monitor changes in solar radiation output over a century to millennium timescales in boreholes with temperature string.	Gaining insight into past climate variability as influenced by solar radiation.
Observing global ice	Measure flow rates of major ice bodies, and track sea ice coverage and density using InSAR.	Understanding the response of major ice masses to climate change; accessing sea level rise, sea ice transport, and ice concentration as a surrogate of climate forcing.
Monitoring "hot spots" of the Earth	1) Conduct multi-spectral TIR measurements of the entire Earth's disk at a spatial resolution of 1 km; 2) Enable near instantaneous identification and tracking of volcanic eruptions and wildfires.	The whole Earth's thermal data with a high temporal resolution is critical for responding to large volcanic eruptions. The Moon-based sensors are also beneficial for atmospheric monitoring and surface composition/radiant flux measurements.

(Continued)

TABLE 1.2 (CONTINUED)

Objective Name	Summary	Value
Calibrating Earth's shine (brightness of the unlit portion of the Moon seen from Earth)	Measure Earth's albedo from the lunar base and calibrate with prior earthshine measurements; Earth shine depends on Earth's albedo, determine factors that impact Earth's albedo such as cloud amount and optical thickness.	It is possible to reconstruct Earth's albedo from the collection of measurements gathered over several decades; the reconstructed albedo data can be calibrated from a lunar base, providing insight into Earth's response to changes in the energy balance.
Observing Earth's lightning	Conduct uninterrupted surveillance of Earth's lightning activity from the Moon to construct a lightning climatology.	Lightning variation patterns can serve as indicators of climate variability and potential shifts in climate conditions.

sensors [141]. In this regard, the SAR system, an active instrument impervious to weather conditions, is a good candidate for Earth observation from the lunar base [89, 142, 143]. Nevertheless, deploying MBSAR also presents particular challenges as the distance increases, despite its advantageous coverage performance, which is delved into in the subsequent sections.

1.3.4 ADVANTAGES AND CHALLENGES

In this section, we shall outline the advantages and challenges of a MBSAR system. This section commences by drawing a comparison between spaceborne SAR and MBSAR.

1.3.4.1 The Comparison between the MBSAR and Spaceborne SAR

The lunar orbit's extremely high altitude allows for exceptional coverage performance of the MBSAR. For instance, Figure 1.3 showcases the correlation between a platform's altitude and swath width, with the near and far incident angle (corresponding to the near and far ranges) setting at 30° and 60°, respectively. Indeed, the swath width of the spaceborne SAR system is generally constrained to a few hundred kilometers or even narrower, whereas that in the MBSAR can potentially exceed several thousand kilometers.

Figure 1.4 illustrates the imaging region of the spaceborne SAR with incident angles ranging from 30° to 60° (referring to Chapter 5 for the detailed orbital configurations) during one Earth day. In comparison, the maximal coverage of the MBSAR over a day is also depicted. We can observe that the MBSAR exhibits superior spatiotemporal coverage compared to the spaceborne SAR. Furthermore, while the spaceborne SAR offers a sparse viewing for Earth observation, the MBSAR provides seamless coverage of Earth with an extensive swath width, thereby facilitating the acquisition of comprehensive images of the planet.

Next, we shall compare the round-trip time delay and SAT as a function of the argument of latitude (AOL) for both MBSAR and spaceborne SAR, as illustrated in Figures 1.5 and 1.6, respectively. Note that the MBSAR and spaceborne SAR's antenna beam centers point to the zero-Doppler plane, and the azimuthal resolution is set to 10 m for both.

As is evident from Figures 1.5 and 1.6, in the spaceborne SAR, the round-trip propagation delay has a magnitude of 10^{-3} s. Meanwhile, for an azimuthal resolution of 10 m, the SAT is less than 2 s. However, in the context of MBSAR, the propagation delay can extend to several seconds. Furthermore, the SAT can reach up to hundreds of seconds. From this perspective, it is evident that the employment of the *stop-and-go* and equivalent linear trajectory assumptions, typically utilized for signal modeling and processing of spaceborne SAR, is rendered inapplicable in the case of the MBSAR.

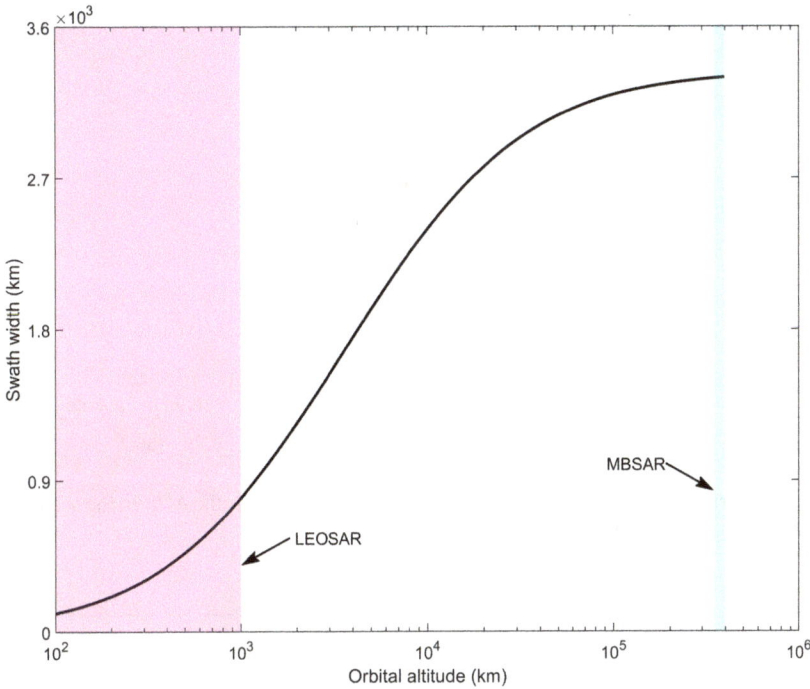

FIGURE 1.3 The swath width of the SAR system as a function of the platform's altitude.

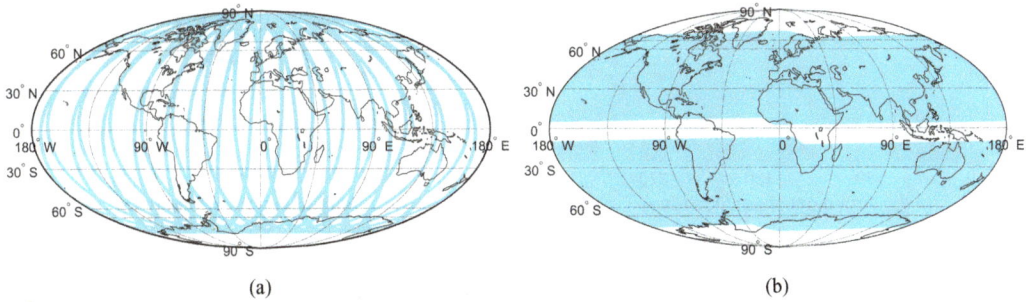

FIGURE 1.4 The daily imaging region in the scenarios of: (a) the spaceborne SAR, (b) the MBSAR. Detailed analysis regarding the imaging region of MBSAR can be found in Chapter 4.

The synthetic aperture processing of SAR depends on the Doppler effect, which in turn is influenced by the beam-crossing velocity. Thus, it is desirable to analyze the implication of beam-crossing velocity for MBSAR and spaceborne SAR in-depth. Figure 1.7 plots the beam-crossing velocity as a function of AOL in the MBSAR and spaceborne SAR scenarios.

Careful examination of Figure 1.7 reveals that the beam-crossing velocity experiences fluctuations around 7×10^3 m/s in the spaceborne SAR. Besides, the extent of these fluctuations is relatively small regardless of the AOL or incident angle. In comparison, MBSAR's beam-crossing velocity is on the order of 10^2 m/s. Moreover, when MBSAR is situated at various locations, there is a certain variation in the beam-crossing velocity. Additionally, the beam crosses velocity exhibits significant fluctuations for diverse incident angles (corresponding to distinct locations on Earth). As a result, the spatiotemporal variability of beam-crossing velocity is considerable and cannot be ignored in the MBSAR.

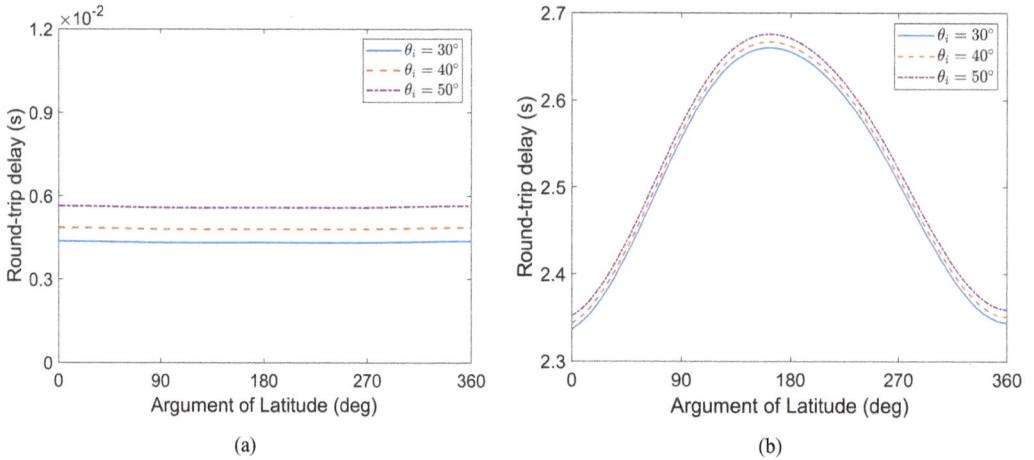

FIGURE 1.5   At various incident angles, the round-trip time delay versus the AOL in the scenarios of: (a) the spaceborne SAR, (b) the MBSAR.

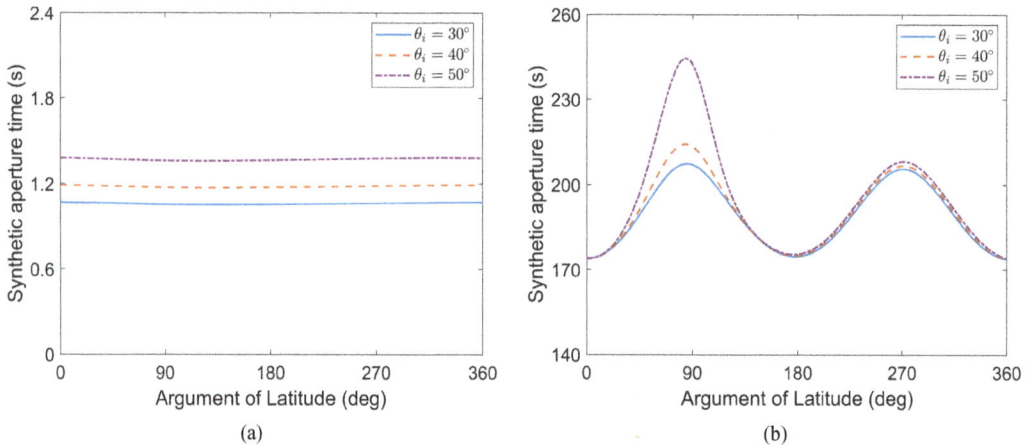

FIGURE 1.6   At various incident angles, the SAT versus AOL in the scenarios of: (a) the spaceborne SAR, (b) the MBSAR.

By employing the beam-crossing velocity illustrated in Figure 1.7, in conjunction with Eq. (1.12), one can further determine the threshold T_{covers} that pertains to limitations on the slant swath width versus azimuthal resolution. This threshold against the AOL for the spaceborne SAR and MBSAR is presented in Figure 1.8. The threshold T_{covers} of the MBSAR is far higher than that of the spaceborne SAR, suggesting the MBSAR provides greater flexibility in determining the swath width versus azimuthal resolution.

Another velocity term associated with the Doppler effect, known as the SAR's inertial velocity, is subject to orbital perturbations, exhibiting certain fluctuations across varying cycles. The fluctuation amplitudes of the SAR's inertial velocity depend on the extent of orbital perturbation. To draw a comparison regarding the degree of orbital perturbation between the MBSAR and spaceborne SAR scenarios, we shall show the difference in the SAR's inertial velocity across three consecutive cycles in MBSAR and spaceborne SAR, as depicted in Figure 1.9.

As shown in Figure 1.9, the inertial velocity of spaceborne SAR is consistent throughout various cycles with minor fluctuations in magnitude, hovering around 7580 m/s. In contrast, the inertial

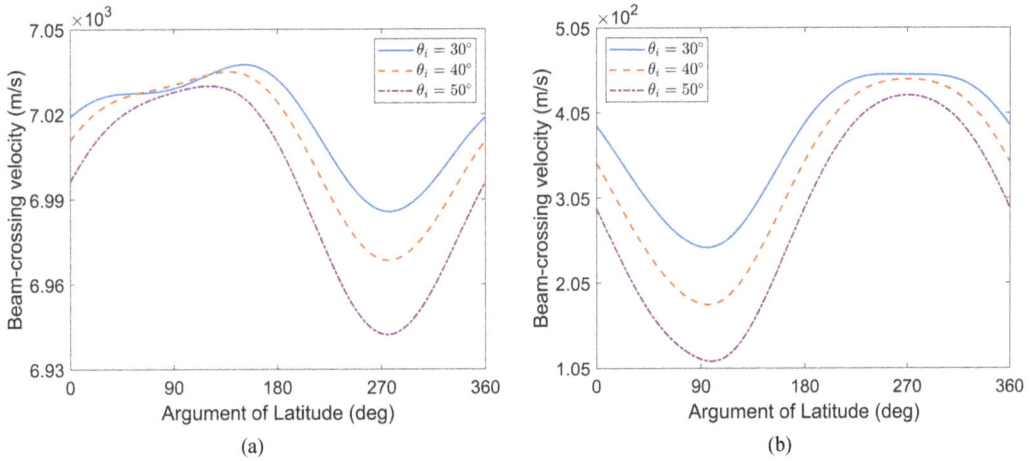

FIGURE 1.7 At various incident angles, the beam-crossing velocity versus AOL in the scenarios of: (a) the spaceborne SAR, (b) the MBSAR.

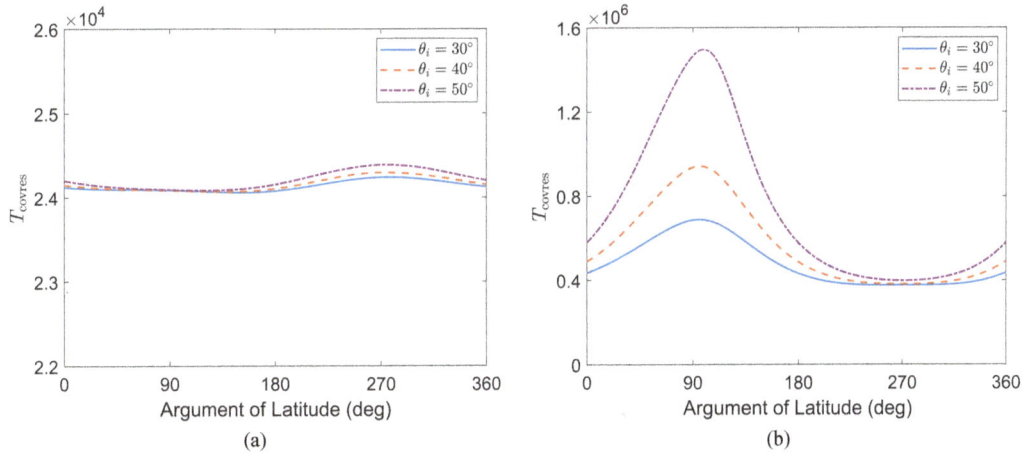

FIGURE 1.8 At various incident angles, the dimensionless threshold T_{covers} versus AOL in the scenarios of: (a) the spaceborne SAR, (b) the MBSAR.

velocity of MBSAR exhibits a relatively stronger fluctuation with respect to its location along the orbit. Further, the MBSAR inertial velocity displays diverse patterns across different cycles, suggesting that it is considerably influenced by orbital perturbations.

In the SAR domain, the synthetic aperture processing and associated azimuthal resolution depend on the Doppler frequency modulation rate (DFMR). The DFMR is the outcome generated by the Earth's rotation, SAR's inertial motion, and coupling effect. We delve into the contributions of each source to the DFMR in scenarios of the spaceborne SAR and MBSAR, Figure 1.10 depicts the pertinent results concerning those contributions.

The results in Figure 1.10 indicate that in the MBSAR, the Earth's rotation is the primary factor in producing the DFMR, whereas the component of DFMR generated by SAR's inertial motion is quite minor. Regarding the spaceborne SAR, the trend is reversed, with the DFMR predominantly derived from SAR's inertial motion. By contrast, the contribution resulting from the Earth's rotation to the DFMR is almost negligible. In both instances, the coupling effect of the Earth's rotation and SAR's inertial motion serves as a secondary source for generating the DFMR.

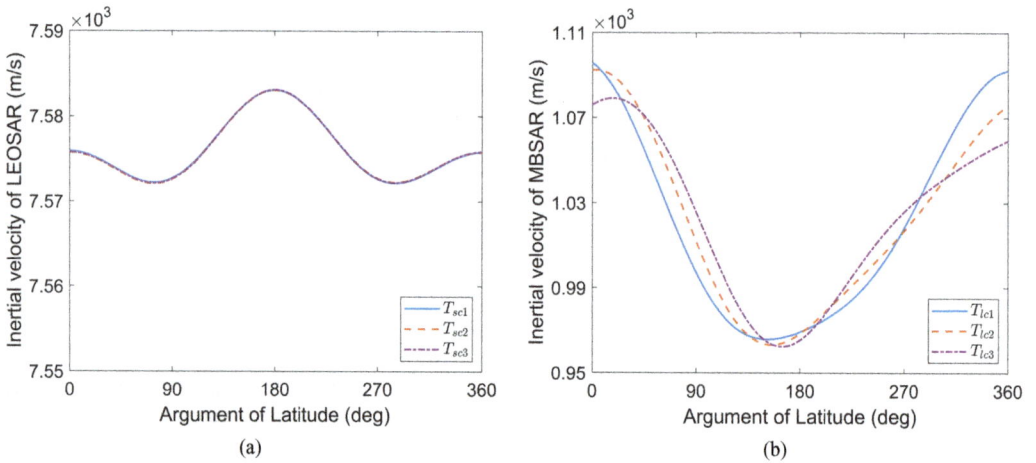

FIGURE 1.9 Within three consecutive cycles, the inertial velocity versus AOL in the scenarios of (a) the spaceborne SAR, and (b) the MBSAR, where T_{sc1}, T_{sc2}, and T_{sc3} are three consecutive cycles of spaceborne SAR, while T_{lc1}, T_{lc2}, and T_{lc3} are three consecutive cycles of MBSAR.

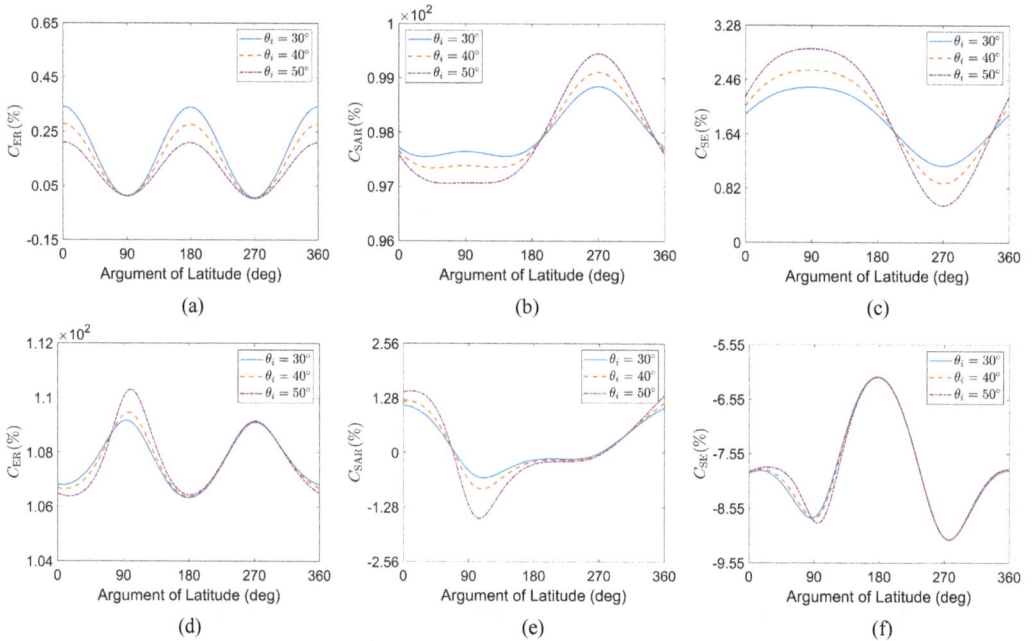

FIGURE 1.10 At various incident angles, the contributions of each factor on the DFMR versus the AOL, where the upper (a, b, c) and lower (d, e, f) rows represent the scenarios of the spaceborne SAR and MBSAR, while the 1st (a, d), 2nd (b, e), 3rd (c, f) columns indicate the contributions on the DFMR resulting from the Earth's rotation, SAR's inertial motion, and their coupling effect. The definitions of those contributions are detailed in Chapter 5.

Thus far, we have elucidated the contrasting features of the MBSAR and spaceborne SAR systems. Table 1.3 highlights these comparative results, accentuating significant differences between two distinct types of SAR systems. Also, it is evident that the MBSAR offers exceptional benefits for Earth observation compared to the spaceborne SAR. However, the inherent characteristics of the

TABLE 1.3

System Performance Comparison between the MBSAR and Spaceborne SAR

Performance Parameters	MBSAR	Spaceborne SAR
Revisit time	roughly 24–25 hours	Several days
Round-trip echo delay	~2.5 s	$<10^{-2}$ s
SAT	>180 s	~1.2 s
Beam-crossing velocity	100–450 m/s	$~7 \times 10^3$ m/s
Flexibility in designing the coverage versus resolution	High	Low
The dominant factor in determining Doppler shift	Earth's rotational motion	SAR inertial motion
The influence degree of orbital perturbation effects	Strong	Weak

Moon as a celestial object pose certain challenges to the achievement and advancement of MBSAR, as explained below.

1.3.4.2 Challenges of MBSAR

The Moon's extremely high orbit bestows remarkable capabilities upon MBSAR, yet simultaneously presents theoretical and technical challenges to its continued advancement. Foremost among these obstacles are the stringent power constraints [144]. The remote distance results in enormous transmission attenuation, necessitating the use of a transmitting power of considerable magnitude and a large antenna aperture. Notwithstanding, with the fast-paced progress of electronics and aerospace technology, the prospects of using large reflector antennas and high-power transmitters to meet future MBSAR requirements are promising [89]. One distinct advantage of MBSAR is its ability to cover a vast area of the lunar surface, offering the possibility of performing a collaborative observation using the antenna array, thereby reducing the constraints on power while monitoring Earth with MBSAR [145].

On this basis, the primary challenges in MBSAR pertain to signal acquisition, modeling, and processing [146]. The synthetic aperture time of MBSAR may span hundreds of seconds, while the round-trip propagation delay of radio signals could reach several seconds. Furthermore, the observation swath covers thousands of kilometers. In this context, the linear trajectory assumption and hyperbolic range equation model, which the spaceborne SAR imaging algorithm relies upon, may not provide accurate results. The signal and SAR models need to be substantially modified [147, 148]. For example, the *stop-and-go assumption* is no longer valid due to the long round-trip delay time for which neither the target nor the MBSAR is "frozen" in the course of signal transmission and reception [149, 150].

The other issue of concern for the MBSAR pertains to lunar orbital perturbations. It is well-known that the orbit of Moon is influenced by various perturbing forces, which give rise to significant fluctuations in the orbital elements over time [151]. In the case of MBSAR, the orbital perturbation effects on the imaging performance diverge significantly from those encountered by spaceborne SAR systems [152]. When dealing with orbital perturbations, an accurate orbital measurement might be effective in upholding good image quality in MBSAR. The orbital determination of the MBSAR, however, also exhibits certain distinctions compared to the spaceborne SAR [153]. Consequently, conducting a meticulous analysis of orbital perturbation effects on imaging performance and accuracy requirement for orbital determination in MBSAR is the first step to moving forward [154].

The atmospheric effects also pose considerable challenges to the MBSAR [155]. The propagation of radio signals is subject to the atmosphere phase delay, primarily originating from the troposphere and ionosphere [156]. In the context of MBSAR, the severity of atmospheric effect becomes more pronounced, whose impacts necessitate re-evaluations: Due to the long synthetic aperture time, the atmospheric freezing model, which is typically employed in spaceborne SAR, is unsuitable for MBSAR [157]. The temporally-varying and spatially-varying characteristics of the atmosphere must be taken into account when analyzing the atmospheric effects on the MBSAR imaging [146, 158]. Further, the atmosphere exhibits noteworthy diversities in its spatiotemporal variation within the SAR's coverage; such diversities further compound atmospheric effects, further inducing varying degrees of image distortions [159]. Therefore, the imperative lies in undertaking a quantitative analysis to assess these effects for effectively utilizing MBSAR data.

1.3.4.3 Advantages of MBSAR

Despite the challenges that may arise, manifold potential benefits can be gained from employing an MBSAR system, which merits further investigation [89, 101, 136, 144, 160]. Let us highlight some of these benefits below.

1) Unobstructed view: With a carrier frequency like that of the L-band, the MBSAR system is able to view the Earth's surface with little interruption. Such a probing signal is less susceptible to atmospheric absorption and scattering, in contrast to optical bands. which mitigates the influences of signal attenuation, scattering loss, and phase distortion.

2) Stabilizable platform: The Moon offers a steadfast surface for placing SAR antennas and other sensors, as well as an extensive baseline for observational networks. Further, the Moon's stable movement endures the ability for repeat-pass interferometry to enhance the Earth's observation in three dimensions.

3) Long life-cycle monitoring: Unlike artificial satellites, which have a limited lifespan of a few years, the Moon offers a sustainable observation platform where SAR can be perpetually maintained and maneuvered. Thus, the MBSAR harbors the great potential for Earth watch continuously, supplying invaluable data for diagnosing our Earth's health.

4) Integral observation: The Moon provides full visibility of the entire Earth disk, enabling MBSAR to capture comprehensive views of the planet's surface within a day. The MBSAR just requires a narrow beam to achieve global-scale coverage due to its exceptionally high orbit. With an antenna array, MBSAR can image the Earth's surface within one exposure time with multiple frequency bands, polarizations, and angle observations.

5) Coordinated observations: The MBSAR technology can be integrated with terrestrial, aerial, and space-based SAR equipment and other devices to enable more extensive and nuanced observations of the Earth's surface. The relatively shorter revisit time and continued observation of MBSAR render it an optimal tool for conducting long-term time series analysis in climate change research, particularly when combined with SAR systems mounted on other platforms. Further, the data collected through MBSAR could be invaluable in calibrating the other SAR data.

6) Unique perspective: The Moon exerts a significant influence on a wide range of terrestrial phenomena. One of the most notable examples is the tidal force, which is believed to act as a trigger for earthquakes and a source of internal waves. The Moon's unique position in space presents an unparalleled opportunity to monitor these phenomena continuously. MBSAR makes it possible to observe many geophysical phenomena associated with lunar motion, including tidal effects, tracing their entire life-cycle from initiation to dissipation. From this point of view, the MBSAR system can provide a unique viewpoint that is not achievable with spaceborne systems, allowing for fresh insights into phenomena such as ocean currents, weather patterns, and the long-term dynamic process of the Earth's surface.

The benefits are a direct result of the distinctive features of the Moon-based platform, which distinguishes the MBSAR from the spaceborne SAR [161]. While various technical obstacles still impede the practical implementation of MBSAR, the knowledge acquired from current SAR systems, coupled with meticulous derivation and analysis in this book, sets a precedent for laying the foundation of forthcoming MBSAR systems. This book emphasizes MBSAR from a signal processing perspective, providing guidelines for SAR configuration, methods for inspecting system properties, and references for signal processing in the MBSAR. All subjects tackled in this work are of great interest to many scientific investigations. We hope this book will inspire further research interest and exploration, in both depth and breadth, into the field of MBSAR.

REFERENCES

[1] F. M. Henderson and A. J. Lewis, *Manual of Remote Sensing: Principles and Applications of Imaging Radar*. New York: Wiley, 1998.

[2] I. H. Woodhouse, *Introduction to Microwave Remote Sensing*. Boca Raton: CRC press, 2005.

[3] K. S. Chen, *Radar Scattering and Imaging of Rough Surfaces: Modeling and Applications with MATLAB*. Boca Raton: CRC Press, 2020.

[4] A. Ishimaru, *Wave Propagation and Scattering in Random Media*. New York: Academic Press, 1978.

[5] G. Franceschetti and R. Lanari, *Synthetic Aperture Radar Processing*. Boca Raton: CRC press, 1999.

[6] A. K. Fung and K. S. Chen, *Microwave Scattering and Emission Models for Users*. Boston: Artech House, 2010.

[7] C. Elachi and J. J. Van Zyl, *Introduction to the Physics and Techniques of Remote Sensing*. New York: Wiley, 2021.

[8] J. B. Campbell and R. H. Wynne, *Introduction to Remote Sensing*, 5th ed. New York: Guilford Press, 2011.

[9] M. A. Richards, *Fundamentals of Radar Signal Processing*. New York: McGraw-Hill Education, 2014.

[10] A. P. Rowe, *One Story of Radar*. New York: Cambridge University Press, 1948.

[11] G. W. Stimson, H. D. Griffiths, C. J. Baker, and A. Dave, *Introduction to Airborne Radar*, 2nd ed. Raleigh: SciTech Publishing, 1998.

[12] C. A. Wiley, "Synthetic Aperture Radars: A Paradigm for Technology Evolution," *IEEE Transactions on Aerospace and Electronic Systems*, vol. 21, no. 3, pp. 440–443, May. 1985.

[13] F. T. Ulaby and D. G. Long, *Microwave Radar and Radiometric Remote Sensing*. Ann Arbor: Univ. Michigan Press, 2014.

[14] K. S. Chen, *Principles of Synthetic Aperture Radar: A System Simulation Approach*. Boca Raton: CRC Press, 2015.

[15] H. Maître, *Processing of Synthetic Aperture Radar Images*. Hoboken, NJ: Wiley, 2013.

[16] A. Moreira, et al., "A Tutorial on Synthetic Aperture Radar," *IEEE Geoscience and Remote Sensing Magazine*, vol. 1, no. 1, pp. 6–43, Mar. 2013.

[17] J. C. Curlander and R. N. McDonough, *Synthetic Aperture Radar: Systems and Signal Processing*. New York: Wiley, 1991.

[18] K. Tomiyasu, "Tutorial Review of Synthetic-Aperture Radar (SAR) with Applications to Imaging of the Ocean Surface," *Proceedings of the IEEE*, vol. 66, no. 5, pp. 563–583, May. 1978.

[19] O. Montenbruck, E. Gill, and F. Lutze, *Satellite Orbits: Models, Methods, and Applications*. Berlin: Springer, 2000.

[20] I. G. Cumming and F. H. Wong, *Digital Signal Processing of Synthetic Aperture Radar Data: Algorithms and Implementation*. Boston: Artech House, 2005.

[21] N. Levanon and E. Mozeson, *Radar Signals*. Hoboken: Wiley, 2004.

[22] C. Oliver and S. Quegan, *Understanding Synthetic Aperture Radar Images*. Raleigh: SciTech Publishing, 2004.

[23] R. Taylor and J. Kalish, "Background and Requirements on Radar Sensors for Spacecraft," *Proc. AIAA Guidance and Control Conference*, Cambridge, Mass, pp. 63–348, Aug. 1964.

[24] R. L. Jordan, "The SEASAT-A Synthetic Aperture Radar System," *IEEE Journal of Oceanic Engineering*, vol. 5, no. 2, pp. 154–164, Apr. 1980.

[25] D. S. Simonett, "The Utility of Radar and Other Remote Sensors in Thematic Land Use Mapping from Spacecraft," Department of Geography and Center for Research in Engineering Science, University of Kansas, Lawrence, Kansas, USGS Annual Report No. 14-08-0001-10848, Jun. 1969. [Online]. Available: https://pubs.usgs.gov/of/1969/0254/report.pdf

[26] F. Leberl, J. Raggam, C. Elachi, and W. J. Campbell, "Sea Ice Motion Measurements from SEASAT SAR Images," *Journal of Geophysical Research: Oceans*, vol. 88, no. C3, pp. 1915–1928, Feb. 1983.

[27] F. M. Henderson, "An Evaluation of SEASAT SAR Imagery for Urban Analysis," *Remote Sensing of Environment*, vol. 12, no. 6, pp. 439–461, Dec. 1982.

[28] J. C. Curlander, "Utilization of Spaceborne SAR Data for Mapping," *IEEE Transactions on Geoscience and Remote Sensing*, vol. GE-22, no. 2, pp. 106–112, Mar. 1984.

[29] M. Shimada, *Imaging from Spaceborne and Airborne SARs, Calibration, and Applications*. Boca Raton: CRC press, 2018.

[30] M. Davidson, L. Iannini, R. Torres, and D. Geudtner, "New Perspectives for Applications and Services Provided by Future Spaceborne SAR Missions at the European Space Agency," *Proc. 2022 IEEE International Geoscience and Remote Sensing Symposium*, Kuala Lumpur, Malaysia, pp. 4720–4723, Jul. 2022.

[31] P. Klenk, et al., "TerraSAR-X / TanDEM-X Mission and Calibration Status," *Proc. 14th European Conference on Synthetic Aperture Radar*, Leipzig, Germany, pp. 1–5, Jul. 2022.

[32] R. Kahle, et al., "Formation Flying for along-Track Interferometric Oceanography—First in-Flight Demonstration with Tandem-X," *Acta Astronautica*, vol. 99, pp. 130–142, Jun.–Jul. 2014.

[33] R. Werninghaus and S. Buckreuss, "The TerraSAR-X Mission and System Design," *IEEE Transactions on Geoscience and Remote Sensing*, vol. 48, no. 2, pp. 606–614, Nov. 2009.

[34] M. Zink, et al., "The Global TanDEM-X DEM – A Unique Data Set," *Proc. 2017 IEEE International Geoscience and Remote Sensing Symposium*, Fort Worth, USA, pp. 906–909, Jul. 2017.

[35] S. Buckreuss, et al., "Ten Years of TerraSAR-X Operations," *Remote Sensing*, vol. 10, no. 6, p. 873, Jun. 2018.

[36] G. Krieger, et al., "The Tandem-L Mission Proposal: Monitoring Earth's Dynamics with High Resolution SAR Interferometry," *Proc. 2009 IEEE Radar Conference*, Pasadena, USA, pp. 1–6, May. 2009.

[37] S. Huber, et al., "Tandem-L: A Technical Perspective on Future Spaceborne SAR Sensors for Earth Observation," *IEEE Transactions on Geoscience and Remote Sensing*, vol. 56, no. 8, pp. 4792–4807, Jun. 2018.

[38] S. Huber, et al., "Tandem-L: Design Concepts for a Next-Generation Spaceborne SAR System," *Proc. 11th European Conference on Synthetic Aperture Radar*, Hamburg, Germany, pp. 1–5, Jun. 2016.

[39] M. Schandri, M. Zink, and M. Bachmann, "TanDEM-X Mission Status and Outlook on the Tandem-L Mission," *Proc. 23rd International Radar Symposium*, Gdansk, Poland, pp. 443–446, Sep. 2022.

[40] A. Moreira, et al., "Tandem-L: A Highly Innovative Bistatic SAR Mission for Global Observation of Dynamic Processes on the Earth's Surface," *IEEE Geoscience and Remote Sensing Magazine*, vol. 3, no. 2, pp. 8–23, Jun. 2015.

[41] R. Torres, et al., "Sentinel-1 SAR System and Mission," *Proc. 2017 IEEE Radar Conference*, Seattle, USA, pp. 1582–1585, May. 2017.

[42] D. Geudtner, et al., "Sentinel-1 System Capabilities and Applications," *Proc. 2014 IEEE Geoscience and Remote Sensing Symposium*, Quebec City, Canada, pp. 1457–1460, Jul. 2014.

[43] P. Potin, et al., "Copernicus Sentinel-1 Constellation Mission Operations Status," *Proc. 2019 IEEE International Geoscience and Remote Sensing Symposium*, Yokohama, Japan, pp. 5385–5388, Jul.–Aug. 2019.

[44] R. Torres, et al., "GMES Sentinel-1 Mission," *Remote Sensing of Environment*, vol. 120, pp. 9–24, May. 2012.

[45] M. Dabboor, et al., "Results Update on the Performance of the Radarsat Constellation Mission," *Proc. 2022 IEEE International Geoscience and Remote Sensing Symposium*, Kuala Lumpur, Malaysia, pp. 4427–4430, Jul. 2022.

[46] A. A. Thompson, "Overview of the RADARSAT Constellation Mission," *Canadian Journal of Remote Sensing*, vol. 41, no. 5, pp. 401–407, Sep. 2015.

[47] B. Brisco, M. Mahdianpari, and F. Mohammadimanesh, "Hybrid Compact Polarimetric SAR for Environmental Monitoring with the RADARSAT Constellation Mission," *Remote Sensing*, vol. 12, no. 20, p. 3283, Oct. 2020.

[48] M. Dabboor, et al., "The RADARSAT Constellation Mission Core Applications: First Results," *Remote Sensing*, vol. 14, no. 2, p. 301, Jan. 2022.

[49] S. Quegan, et al., "The Role of the BIOMASS Mission in Carbon Cycle Science and Politics," *Proc. 2021 IEEE International Geoscience and Remote Sensing Symposium*, Brussels, Belgium, pp. 771–774, Jul. 2021.

[50] A. Leanza, et al., "Earth Explorer BIOMASS P-Band SAR Mission: Status and Calibration Concept," *Proc. 14th European Conference on Synthetic Aperture Radar*, Leipzig, Germany, pp. 1–4, Jul. 2022.

[51] T. Le Toan, et al., "The BIOMASS Mission: Mapping Global Forest Biomass to Better Understand the Terrestrial Carbon Cycle," *Remote Sensing of Environment*, vol. 115, no. 11, pp. 2850–2860, Nov. 2011.

[52] J. M. B. Carreiras, et al., "Coverage of High Biomass Forests by the ESA BIOMASS Mission under Defense Restrictions," *Remote Sensing of Environment*, vol. 196, pp. 154–162, Jul. 2017.

[53] S. Quegan, et al., "The European Space Agency BIOMASS Mission: Measuring Forest Above-Ground Biomass from Space," *Remote Sensing of Environment*, vol. 227, pp. 44–60, Jun. 2019.

[54] S. E. I. Essebtey, et al., "Long-Term Trends of P-Band Temporal Decorrelation over a Tropical Dense Forest-Experimental Results for the BIOMASS Mission," *IEEE Transactions on Geoscience and Remote Sensing*, vol. 60, Jun. 2022, doi: 10.1109/TGRS.2021.3082395

[55] C. O. Dumitru and M. Datcu, "Information Content of Very High Resolution SAR Images: Study of Feature Extraction and Imaging Parameters," *IEEE Transactions on Geoscience and Remote Sensing*, vol. 51, no. 8, pp. 4591–4610, 2013.

[56] Y. Inoue, E. Sakaiya, and C. Wang, "Capability of C-Band Backscattering Coefficients from High-Resolution Satellite SAR Sensors to Assess Biophysical Variables in Paddy Rice," *Remote Sensing of Environment*, vol. 140, pp. 257–266, Jan. 2014.

[57] J. Chen, et al., "Two-Step Accuracy Improvement of Motion Compensation for Airborne SAR with Ultrahigh Resolution and Wide Swath," *IEEE Transactions on Geoscience and Remote Sensing*, vol. 57, no. 9, pp. 7148–7160, May. 2019.

[58] J. Li, J. Chen, P. Wang, and O. Loffeld, "A Coarse-to-Fine Autofocus Approach for Very High-Resolution Airborne Stripmap SAR Imagery," *IEEE Transactions on Geoscience and Remote Sensing*, vol. 56, no. 7, pp. 3814–3829, Apr. 2018.

[59] O. Dubovik, et al., "Grand Challenges in Satellite Remote Sensing," *Frontiers in Remote Sensing*, vol. 2, p. 619818, Feb. 2021.

[60] E. Chuvieco, *Earth Observation of Global Change: The Role of Satellite Remote Sensing in Monitoring the Global Environment*. Springer, 2008.

[61] S. J. Purkis and V. V. Klemas, *Remote Sensing and Global Environmental Change*. Chichester, UK: John Wiley and Sons, 2011.

[62] J. M. Kellndorfer, L. E. Pierce, M. C. Dobson, and F. T. Ulaby, "Toward Consistent Regional-to-Global-Scale Vegetation Characterization Using Orbital SAR Systems," *IEEE Transactions on Geoscience and Remote Sensing*, vol. 36, no. 5, pp. 1396–1411, Sep. 1998.

[63] A. Moussessian, et al., "System Concepts and Technologies for High Orbit SAR," *Proc. IEEE MTT-S International Microwave Symposium Digest*, Long Beach, USA, pp. 1623–1626, Jul. 2005.

[64] M. Shimada and T. Ohtaki, "Generating Large-Scale High-Quality SAR Mosaic Datasets: Application to PALSAR Data for Global Monitoring," *IEEE Journal of Selected Topics in Applied Earth Observations and Remote Sensing*, vol. 3, no. 4, pp. 637–656, Dec. 2010.

[65] M. D'Errico, *Distributed Space Missions for Earth System Monitoring*. New York, USA: Springer, 2013.

[66] S. N. Madsen, C. Chen, and W. Edelstein, "Radar Options for Global Earthquake Monitoring," *Proc. IEEE International Geoscience and Remote Sensing Symposium*, Toronto, Canada, vol. 3, pp. 1483–1485 vol. 3, Jun. 2002.

[67] O. M. Bello and Y. A. Aina, "Satellite Remote Sensing as a Tool in Disaster Management and Sustainable Development: Towards a Synergistic Approach," *Procedia - Social and Behavioral Sciences*, vol. 120, pp. 365–373, Mar. 2014.

[68] S. Voigt, et al., "Global Trends in Satellite-Based Emergency Mapping," *Science*, vol. 353, no. 6296, pp. 247–252, Jul. 2016.

[69] S. Voigt, et al., "Satellite Image Analysis for Disaster and Crisis-Management Support," *IEEE Transactions on Geoscience and Remote Sensing*, vol. 45, no. 6, pp. 1520–1528, May. 2007.

[70] M. C. Hansen, et al., "High-Resolution Global Maps of 21st-Century Forest Cover Change," *Science*, vol. 342, no. 6160, pp. 850–853, Oct. 2013, doi: 10.1126/science.1244693

[71] J. Reiche, J. Verbesselt, D. Hoekman, and M. Herold, "Fusing Landsat and SAR Time Series to Detect Deforestation in the Tropics," *Remote Sensing of Environment*, vol. 156, pp. 276–293, Oct. 2015.

[72] M. Watanabe, et al., "Early-Stage Deforestation Detection in the Tropics With L-band SAR," *IEEE Journal of Selected Topics in Applied Earth Observations and Remote Sensing*, vol. 11, no. 6, pp. 2127–2133, Mar. 2018.

[73] V. Kumar, G. Venkataraman, K. A. Høgda, and Y. Larsen, "Estimation and Validation of Glacier Surface Motion in the Northwestern Himalayas using High-Resolution SAR Intensity Tracking," *International Journal of Remote Sensing*, vol. 34, no. 15, pp. 5518–5529, Apr. 2013.

[74] R. Caduff, F. Schlunegger, A. Kos, and A. Wiesmann, "A Review of Terrestrial Radar Interferometry for Measuring Surface Change in the Geosciences," *Earth Surface Processes and Landforms*, vol. 40, no. 2, pp. 208–228, Feb. 2015.

[75] S. Gruber, "Derivation and Analysis of a High-Resolution Estimate of Global Permafrost Zonation," *The Cryosphere*, vol. 6, no. 1, pp. 221–233, Feb. 2012.

[76] L. Liu, T. Zhang, and J. Wahr, "InSAR Measurements of Surface Deformation over Permafrost on the North Slope of Alaska," *Journal of Geophysical Research: Earth Surface*, vol. 115, no. F3, Sep. 2010.

[77] R. Zhao, et al., "Monitoring Surface Deformation over Permafrost with an Improved SBAS-InSAR Algorithm: With Emphasis on Climatic Factors Modeling," *Remote Sensing of Environment*, vol. 184, pp. 276–287, Oct. 2016.

[78] A. C. A. Rudy, et al., "Seasonal and Multi-Year Surface Displacements Measured by DInSAR in a High Arctic Permafrost Environment," *International Journal of Applied Earth Observation and Geoinformation*, vol. 64, pp. 51–61, Feb. 2018.

[79] Z. Zhang, et al., "A Review of Satellite Synthetic Aperture Radar Interferometry Applications in Permafrost Regions: Current Status, Challenges, and Trends," *IEEE Geoscience and Remote Sensing Magazine*, vol. 10, no. 3, pp. 93–114, Sep. 2022.

[80] L. Liu, K. Schaefer, T. Zhang, and J. Wahr, "Estimating 1992–2000 Average Active Layer Thickness on the Alaskan North Slope from Remotely Sensed Surface Subsidence," *Journal of Geophysical Research: Earth Surface*, vol. 117, no. F1, Mar. 2012.

[81] M. Lin and Y. Jia, "Past, Present and Future Marine Microwave Satellite Missions in China," *Remote Sensing*, vol. 14, no. 6, p. 1330, Mar. 2022.

[82] B. El Mahrad, et al., "Contribution of Remote Sensing Technologies to a Holistic Coastal and Marine Environmental Management Framework: A Review," *Remote Sensing*, vol. 12, no. 14, p. 2313, Jul. 2020.

[83] J. Horstmann and W. Koch, "Measurement of Ocean Surface Winds Using Synthetic Aperture Radars," *IEEE Journal of Oceanic Engineering*, vol. 30, no. 3, pp. 508–515, Jul. 2005.

[84] V. Kerbaol and F. Collard, "SAR-Derived Coastal and Marine Applications: From Research to Operational Products," *IEEE Journal of Oceanic Engineering*, vol. 30, no. 3, pp. 472–486, Jul. 2005.

[85] M. Cherniakov, *Bistatic Radar: Emerging Technology*. Chichester: John Wiley & Sons, 2008.

[86] A. Renga, M. D. Graziano, M. Grasso, and A. Moccia, "Evaluation of Design Parameters for Formation Flying SAR," *Proc. 13th European Conference on Synthetic Aperture Radar*, pp. 1–4, Mar.–Apr. 2021.

[87] D. Massonnet, "Capabilities and Limitations of the Interferometric Cartwheel," *IEEE Transactions on Geoscience and Remote Sensing*, vol. 39, no. 3, pp. 506–520, Mar. 2001.

[88] K. Alfriend, et al., *Spacecraft Formation Flying: Dynamics, Control and Navigation*. London: Elsevier, 2009.

[89] H. Guo, et al., "Conceptual Study of Lunar-Based SAR for Global Change Monitoring," *Science China Earth Sciences*, vol. 57, no. 8, pp. 1771–1779, Aug. 2014.

[90] D. M. Tralli, W. Foxall, and C. Schultz, "Concept for a High MEO InSAR Seismic Monitoring System," *Proc. 2007 IEEE Aerospace Conference*, Big Sky, USA, pp. 1–7, Mar. 2007.

[91] J. Matar, et al., "MEO SAR: System Concepts and Analysis," *IEEE Transactions on Geoscience and Remote Sensing*, vol. 58, no. 2, pp. 1313–1324, Feb. 2020.

[92] K. Tomiyasu and J. L. Pacelli, "Synthetic Aperture Radar Imaging from an Inclined Geosynchronous Orbit," *IEEE Transactions on Geoscience and Remote Sensing* vol. GE-21, no. 3, pp. 324–329, Jul. 1983.

[93] D. Bruno, S. E. Hobbs, and G. Ottavianelli, "Geosynchronous Synthetic Aperture Radar: Concept Design, Properties and Possible Applications," *Acta Astronautica*, vol. 59, no. 1, pp. 149–156, Jul.–Sep. 2006.

[94] S. Hobbs, et al., "System Design for Geosynchronous Synthetic Aperture Radar Missions," *IEEE Transactions on Geoscience and Remote Sensing*, vol. 52, no. 12, pp. 7750–7763, Dec. 2014.

[95] A. M. Guarnieri, et al., "Advanced Radar Geosynchronous Observation System: ARGOS," *IEEE Geoscience and Remote Sensing Letters*, vol. 12, no. 7, pp. 1406–1410, Jul. 2015.

[96] P. Eckart and B. Aldrin, *The Lunar Base Handbook: An Introduction to Lunar Base Design, Development, and Operations*. New York, USA: McGraw-Hill, 2006.

[97] R. Rugani, F. Martelli, M. Martino, and G. Salvadori, "Moon Village: Main Aspects and Open Issues in Lunar Habitat Thermoenergetics Design. A Review," *Proc. 2021 IEEE International Conference on Environment and Electrical Engineering and 2021 IEEE Industrial and Commercial Power Systems Europe*, Bari, Italy, pp. 1–6, Sep. 2021.

[98] E. Seedhouse, *Lunar Outpost: The Challenges of Establishing a Human Settlement on the Moon*. Chichester: Springer, 2009.

[99] M. Ramsey, "ESS Science Planning and Lunar Workshop Overview," presented at the *NASA Advisory Council Workshop on Science Associated with the Lunar Exploration Architecture*, Feb. 27–Mar. 2, 2007. [Online]. Available: http://www.lpi.usra.edu/meetings/LEA/presentations/OpeningPlenary/Ramsey_ESS.pdf

[100] M. Ramsey, "ESS Findings: Lunar Science Planning and Workshop Overview," presented at the *NASA Advisory Council Workshop on Science Associated with the Lunar Exploration Architecture*, Feb. 27–Mar. 2, 2007. [Online]. Available: https://www.lpi.usra.edu/meetings/LEA/presentations/closing_plenary/Ramsey_ESS_summary_20070302.pdf

[101] H. Guo, G. Liu, and Y. Ding, "Moon-Based Earth Observation: Scientific Concept and Potential Applications," *International Journal of Digital Earth* vol. 11, no. 6, pp. 546–557, Jul. 2017.

[102] M. C. Gutzwiller, "Moon-Earth-Sun: The Oldest Three-Body Problem," *Review of Modern Physics*, vol. 70, no. 2, pp. 589–639, Apr. 1998.

[103] G. H. Heiken, D. T. Vaniman, and B. M. French, *Lunar Sourcebook: A User's Guide to the Moon.* Cambridge, UK: Cambridge Univ. Press, 1991.

[104] L. D. Jaffe and R. F. Scott, "Lunar Surface Strength: Implications of Luna 9 Landing," *Science*, vol. 153, no. 3734, pp. 407–408, Jul. 1966.

[105] D. A. Beattie, *Taking Science to the Moon: Lunar Experiments and the Apollo Program.* Baltimore, USA: Johns Hopkins Univ. Press, 2003.

[106] T. E. Muir-Harmony, *Apollo to the Moon: A History in 50 Objects.* Washington, DC, USA: National Geographic Society, 2018.

[107] R. Poole, *Earthrise: How Man First Saw the Earth.* New Haven, USA: Yale Univ. Press, 2008.

[108] S. Nitzke and N. Pethes, *Imagining Earth: Concepts of Wholeness in Cultural Constructions of Our Home Planet.* Bielefeld, Germany: transcript Verlag, 2017.

[109] I. A. Crawford and K. H. Joy, "Lunar Exploration: Opening a Window into the History and Evolution of the Inner Solar System," *Philosophical Transactions of the Royal Society A: Mathematical, Physical and Engineering Sciences*, vol. 372, no. 2024, p. 20130315, Sep. 2014.

[110] K. Uesugi, "Results of the MUSES-A "HITEN" Mission," *Advances in Space Research*, vol. 18, no. 11, pp. 69–72, Nov. 1996.

[111] S. Nozette, et al., "The Clementine Mission to the Moon: Scientific Overview," *Science*, vol. 266, no. 5192, pp. 1835–1839, Dec. 1994, doi: 10.1126/science.266.5192.1835

[112] A. B. Binder, "Lunar Prospector: Overview," *Science*, vol. 281, no. 5382, pp. 1475–1476, Sep. 1998.

[113] I. A. Crawford, et al., "Back to the Moon: The Scientific Rationale for Resuming Lunar Surface Exploration," *Planetary and Space Science*, vol. 74, no. 1, pp. 3–14, Dec. 2012.

[114] P. J. Stooke, "Identification of the SMART-1 Spacecraft Impact Location on the Moon," *Icarus*, vol. 321, pp. 112–115, Mar. 2019.

[115] M. Kato, S. Sasaki, Y. Takizawa, and the Kaguya project, "The Kaguya Mission Overview," *Space Science Reviews*, vol. 154, no. 1, pp. 3–19, 2010/07/01 2010.

[116] R. Sridharan, et al., "The Sunlit Lunar Atmosphere: A Comprehensive Study by CHACE on the Moon Impact Probe of Chandrayaan-1," *Planetary and Space Science*, vol. 58, no. 12, pp. 1567–1577, Oct. 2010.

[117] A. Colaprete, R. C. Elphic, J. Heldmann, and K. Ennico, "An Overview of the Lunar Crater Observation and Sensing Satellite (LCROSS)," *Space Science Reviews*, vol. 167, no. 1, pp. 3–22, May. 2012.

[118] R. Vondrak, J. Keller, G. Chin, and J. Garvin, "Lunar Reconnaissance Orbiter (LRO): Observations for Lunar Exploration and Science," *Space Science Reviews*, vol. 150, no. 1, pp. 7–22, Feb. 2010.

[119] R. K. Sinha, A. Rani, T. Ruj, and A. Bhardwaj, "Geologic Investigation of Lobate Scarps in the Vicinity of Chandrayaan-3 Landing Site in the Southern High Latitudes of the Moon," *Icarus*, vol. 402, p. 115636, Sep. 2023.

[120] T. Hoshino, et al., "Lunar Polar Exploration Mission for Water Prospection – JAXA's Current Status of Joint Study with ISRO," *Acta Astronautica*, vol. 176, pp. 52–58, Nov 2020.

[121] S. Creech, J. Guidi, and D. Elburn, "Artemis: An Overview of NASA's Activities to Return Humans to the Moon," *Proc. 2022 IEEE Aerospace Conference*, Big Sky, MT, USA, pp. 1–7, Mar. 2022.

[122] F. T. Melman, et al., "LCNS Positioning of a Lunar Surface Rover Using a DEM-Based Altitude Constraint," *Remote Sensing*, vol. 14, no. 16, p. 3942, Aug. 2022.

[123] C. Li, C. Wang, Y. Wei, and Y. Lin, "China's Present and Future Lunar Exploration Program," *Science*, vol. 365, no. 6450, pp. 238–239, Jul. 2019.

[124] L. Xiao, et al., "A Young Multilayered Terrane of the Northern Mare Imbrium Revealed by Chang'e-3 Mission," *Science*, vol. 347, no. 6227, pp. 1226–1229, Mar. 2015.

[125] C. Li, et al., "Chang'E-4 Initial Spectroscopic Identification of Lunar Far-Side Mantle-Derived Materials," *Nature*, vol. 569, no. 7756, pp. 378–382, May. 2019.

[126] C. Li, et al., "Characteristics of the Lunar Samples Returned by the Chang'E-5 Mission," *National Science Review*, vol. 9, no. 2, Feb. 2021.

[127] L. Xu, Y. Zou, and Y. Jia, "China's Planning for Deep Space Exploration and Lunar Exploration before 2030," *Chinese Journal of Space Science*, vol. 38, no. 5, pp. 591–592, Sep. 2018.

[128] I. A. Crawford, "Back to the Moon?," *Astronomy and Geophysics*, vol. 44, no. 2, pp. 2.15–2.17, Apr. 2003.

[129] I. Crawford, "Why We Should Build a Moon Village," *Astronomy and Geophysics*, vol. 58, no. 6, pp. 6.18–6.21, Dec. 2017.

[130] National Aeronautics and Space Administration (NASA). *Why the Moon?* Accessed: May.10, 2022. [Online]. Available at: https://science.nasa.gov/resource/why-the-moon/

[131] National Aeronautics and Space Administration (NASA). *Lunar Exploration Objectives.* Accessed: Oct. 01, 2020. [Online]. Available: https://www.nasa.gov/pdf/163560main_LunarExplorationObjectives.pdf

[132] W. Parks, et al., "Lunar Based Observations of the Earth as a Planet," *Proc. Astrobiology Science Conference*, League City, Texas, p. 98195, 2010.

[133] B. H. Foing, "The Moon as a Platform for Astronomy and Space Science," *Advances in Space Research*, vol. 18, no. 11, pp. 17–23, Nov. 1996.

[134] E. Pallé and P. R. Goode, "The Lunar Terrestrial Observatory: Observing the Earth Using Photometers on the Moon's Surface," *Advances in Space Research*, vol. 43, no. 7, pp. 1083–1089, Apr. 2009.

[135] H. Guo, "Space-based Observation for Sensitive Factors of Global Change," *Bulletin of the Chinese Academy of Sciences*, vol. 23, no. 4, pp. 226–228, Jan. 2010.

[136] H. Guo, et al., "Moon-Based Earth Observation for Large Scale Geoscience Phenomena," *Proc. IEEE International Geoscience and Remote Sensing Symposium*, Beijing, China, pp. 3705–3707, Jul. 2016.

[137] T. Karalidi, et al., "Observing the Earth as an Exoplanet with LOUPE, the Lunar Observatory for Unresolved Polarimetry of Earth," *Planetary and Space Science*, vol. 74, no. 1, pp. 202–207, Dec. 2012.

[138] G. Fornaro, et al., "Potentials and Limitations of Moon-Borne SAR Imaging," *IEEE Transactions on Geoscience and Remote Sensing*, vol. 48, no. 7, pp. 3009–3019, Apr. 2010.

[139] Y. Ren, H. Guo, G. Liu, and H. Ye, "Simulation Study of Geometric Characteristics and Coverage for Moon-Based Earth Observation in the Electro-Optical Region," *IEEE Journal of Selected Topics in Applied Earth Observations and Remote Sensing*, vol. 10, no. 6, pp. 2431–2440, Jun. 2017.

[140] H. Ye, H. Guo, G. Liu, and Y. Ren, "Observation Duration Analysis for Earth Surface Features from a Moon-Based Platform," *Advances in Space Research*, vol. 62, no. 2, pp. 274–287, Jul. 2018.

[141] L. Yuan and J. Liao, "Exploring the Influence of Various Factors on Microwave Radiation Image Simulation for Moon-Based Earth Observation," *Frontiers of Earth Science*, vol. 14, no. 2, pp. 430–445, Jun. 2020.

[142] A. Renga, "Configurations and Performance of Moon-Based SAR Systems for Very High Resolution Earth Remote Sensing," *Proc. AIAA Pegasus Aerospace Conference*, Naples, Italy, pp. 12–13, Apr. 2007.

[143] A. Renga and A. Moccia, "Preliminary Analysis of a Moonbased Interferometric SAR System for Very High Resolution Earth Remote Sensing," *Proc. 9th ILEWG International Conference on Exploration and Utilisation of the Moon*, Sorrento, Italy, pp. 22–26, Oct. 2007.

[144] A. Moccia and A. Renga, "Synthetic Aperture Radar for Earth Observation from a Lunar Base: Performance and Potential Applications," *IEEE Transactions on Aerospace and Electronic Systems*, vol. 46, no. 3, pp. 1034–1051, Jul. 2010.

[145] Z. Xu, K. S. Chen, G. Liu, and H. Guo, "Spatiotemporal Coverage of a Moon-Based Synthetic Aperture Radar: Theoretical Analyses and Numerical Simulations," *IEEE Transactions on Geoscience and Remote Sensing*, vol. 58, no. 12, pp. 8735–8750, 2020.

[146] Z. Xu and K. S. Chen, "Temporal-Spatial Varying Background Ionospheric Effects on the Moon-Based Synthetic Aperture Radar Imaging: A Theoretical Analysis," *IEEE ACCESS*, vol. 6, pp. 66767–66786, Jul. 2018.

[147] Z. Xu and K. S. Chen, "Effects of the Earth's Curvature and Lunar Revolution on the Imaging Performance of the Moon-Based Synthetic Aperture Radar," *IEEE Transactions on Geoscience and Remote Sensing*, vol. 57, no. 8, pp. 5868–5882, Mar. 2019.

[148] Z. Xu, K. S. Chen, and G. Q. Zhou, "Effects of the Earth's Irregular Rotation on the Moon-Based Synthetic Aperture Radar Imaging," *IEEE ACCESS*, vol. 7, pp. 155014–155027, Oct. 2019.

[149] Z. Xu and K. S. Chen, "Effects of the Stop-and-Go Approximation on the Lunar-Based SAR Imaging," *IEEE Geoscience and Remote Sensing Letters*, Apr. 2021, doi: 10.1109/LGRS.2021.3070323

[150] Z. Xu, K. S. Chen, and H. Guo, "Doppler Estimation with "Non-Stop-and-Go" Assumption in Moon-Based SAR Imaging," *Proc. 2018 IEEE International Geoscience and Remote Sensing Symposium*, Valencia, Spain, pp. 7809–7812, Jul. 2018.

[151] J. Meeus, *Mathematical Astronomy Morsels*. Richmond, USA: Willmann-Bell, 1997.

[152] Z. Xu, K. S. Chen, Z. L. Li, and G. Y. Du, "Apsidal Precession Effects on the Lunar-Based Synthetic Aperture Radar Imaging Performance," *IEEE Geoscience and Remote Sensing Letters*, vol. 18, no. 6, pp. 1079–1083, Jun. 2021.

[153] Z. Xu, K. S. Chen, and G. Liu, "On Orbital Determination of the Lunar-Based SAR under Apsidal Precession," *IEEE Transactions on Geoscience and Remote Sensing*, vol. 60, May. 2022, doi: 10.1109/TGRS.2022.3176836

[154] Z. Xu, K. S. Chen, and G. Liu, "On Evaluating the Imaging Performance and Orbital Determination under Perturbations of Orbital Inclination and RAAN in the Lunar-Based SAR," *IEEE Transactions on Geoscience and Remote Sensing*, vol. 60, Jul. 2022, doi: 10.1109/TGRS.2022.3188294

[155] Z. Xu, K. S. Chen, P. Xu, and H. Guo, "Ionospheric Effects on the Lunar-Based Radar Imaging," *Proc. 2017 IEEE International Geoscience and Remote Sensing Symposium*, Fort Worth, USA, pp. 5390–5393, Jul. 2017.

[156] S. Quegan and J. Lamont, "Ionospheric and Tropospheric Effects on Synthetic Aperture Radar Performance," *International Journal of Remote Sensing*, vol. 7, no. 4, pp. 525–539, May. 1986.

[157] K. S. Chen and Z. Xu, "Ionospheric Effects on Satellite and Moon-Based SAR: Current Situation and Prospects," *Journal of Nanjing University of Information Science and Technology*, vol. 12, no. 2, pp. 135–149, Apr. 2020, doi: 10.13878/j.cnki.jnuist.2020.02.001

[158] Z. Xu, K. S. Chen, and H. Guo, "Effects of Temporally-Varying Tropospheric Path Delay on the Imaging Performance of Moon-based SAR," *Proc. 2019 Photonics and Electromagnetics Research Symposium - Fall*, Xiamen, China, pp. 617–623, 17–20 Dec. 2019 2019.

[159] Z. Xu and K. S. Chen, "Numerical Study of the Spatiotemporally-Varying Background Ionospheric Effects on P-Band Satellite SAR Imaging," *IEEE ACCESS*, vol. 8, pp. 123182–123199, Jul. 2020.

[160] Z. Xu and K. S. Chen, "On Signal Modeling of Moon-Based Synthetic Aperture Radar (SAR) Imaging of Earth," *Remote Sensing*, vol. 10, no. 3, p. 486, Mar. 2018.

[161] Z. Xu, K. S. Chen, and H. Guo, "On Azimuthal Resolution of the Lunar-Based SAR under the Orbital Perturbation Effects," *IEEE Transactions on Geoscience and Remote Sensing*, vol. 61, Apr. 2023, doi: 10.1109/TGRS.2023.3266548

2 Moon-Based Earth Observation Geometry

2.1 INTRODUCTION

The Moon-Based SAR (MBSAR) is an emerging concept that offers unique perspectives and insights for Earth observation. Unlike a conventional airborne or spaceborne SAR system, the MBSAR operates on the lunar surface, enabling it to observe the rotating Earth from a considerable distance. The MBSAR leverages the Earth–Moon distance to create a large beam footprint and the Earth–Moon relative motion to synthesize a substantial aperture, facilitating high-resolution imaging with an extensive swath width; this, in effect, maximizes the Earth observation capabilities by the MBSAR [1].

Before analyzing the swath width and system configuration, one must comprehensively understand the intricacies associated with MBSAR's observation geometry. That is, we must determine the MBSAR's motion relative to the Target of Interest (TOI) on Earth, along with accurate mapping of timing and position of echo signals to TOI. In this respect, the observation geometry comprises several crucial factors, including the space coordinates, shapes of Earth and Moon, Earth–Moon relative motion, and intrinsic correlation between the beam pointing direction and MBSAR-TOI relative locations [2, 3].

As the orbital motion of a satellite (whether natural or artificial) is a multifaceted process influenced by various factors, it is imperative to provide a precise description of the Moon's orbit and attitude, necessitating the employment of lunar ephemeris to model the observation geometry accurately [4]. The lunar ephemeris generally involves specific time coordinate distinct from those employed to describe the Earth's rotational motion [5, 6]. Consequently, the geometric model also includes multiple concepts and definitions for each set of time coordinates in the MBSAR, which can be employed in conjunction with one another depending on the practical applications.

In this chapter, we shall delve deeper into these components to establish the MBSAR's observational geometry. The determination of space reference frameworks depends on the time scale, which gives rise to complexity because each clock within a given time coordinate system possesses its own proper time. The interconnection between these proper times is established through a four-dimensional space-time transformation contingent upon the location and motion of gravitating masses. We start with the time coordinates, which concern different time scales.

2.2 TIME COORDINATES

Classical Newtonian physics considers time to be an absolute quantity that exists independently of the location or motion of a clock. This view is no longer valid within the framework of general relativity, where both the proper time and coordinate time are relative concepts [7]. To be more precise, the proper time refers to a local time that is measured by an ideal clock, which is significant only in the local scenario and is independent of the coordinate system. However, the proper time is subject to the varying gravitational environments and relative velocities. Thus, depending on its specific condition, each clock may possess its own proper time instead of adhering to a solitary, absolute time standard. In such cases, coordinate time could be the benchmark against which proper time is measured under a specified reference frame [8].

In the Earth observation, three fundamental time scales are referred to as: the atomic, sidereal, and dynamical time scales [9]. Atomic Time comprises a series of time scales that utilize the

DOI: 10.1201/9781003308430-2

electromagnetic oscillation frequency emanating from atoms as a reference, which is widely used due to its exceptional accuracy and stability. Sidereal time is determined by the Earth's rotational rate relative to a fixed star, and this time scale holds particular significance in astronomical observations and calculations. Dynamical time constitutes a specialized temporal framework that rests upon the law of motion and gravity and is employed to determine the orbital motion of celestial objects. The following discourse provides a comprehensive explanation of these concepts.

2.2.1 ATOMIC TIME

Atomic Time is derived from atomic clocks, which rely on electromagnetic oscillation with energy state transitions of specific atoms or molecules [10]. Cesium clocks offer the highest long-term stability among atomic clocks and serve as primary standards for implementing Atomic Time. Consequently, the International System of Units (SI) adopts the atomic second defined by the Cesium clock as its reference unit. The SI stipulates that the length of a second is the duration of 9,192,631,770 periods of radiation corresponding to the transition between two hyperfine levels of the unperturbed ground state of the Cesium-133 atom [11]. The remarkable precision and stability of the Cesium clock led to the establishment of a range of time coordinates, which include [12, 13]:

1) The International Atomic Time (TAI): TAI constitutes an exceedingly accurate temporal standard based on proper time defined by Earth's geoid. The TAI operates independently of Earth's rotational and orbital motions, relying instead on a vast network of atomic clocks worldwide. At present, TAI benefits from the contributions of a consortium comprising more than 400 atomic clocks, thus ensuring its unparalleled accuracy, with a precision of approximately one nanosecond. Moreover, its long-term stability exhibits a deviation of less than one second over the course of 300 years. As such, the TAI constitutes a continuous time system devoid of leap seconds and serves for accurate measurements and navigation, with its epoch commencing at precisely 00:00:00 on January 1, 1958.

2) Global Positioning System (GPS) time: Like TAI, GPS time is another Atomic Time standard, but it differs from TAI in terms of its reference epoch and time offset selection. The original purpose of GPS was to fulfill the demands of both navigational and geodetic measurements. To this end, the time scale holds paramount significance in the GPS. Consequently, GPS constellations can offer exceedingly precise timing signals that are instantaneously and globally accessible. In theory, GPS time can achieve precision up to tens of nanoseconds by accounting for the clock drift relative to TAI experienced by atomic clocks in GPS transmitters. Despite this, the signal interpretations by some receivers result in a loss of accuracy, leading to a precision of approximately 100 ns. In addition to the aforementioned sub-microsecond clock offsets, GPS time was initially synchronized with UTC on January 6, 1980; it has since diverged and currently maintains a constant offset of 19 s with TAI, namely:

$$GPS = TAI - 19\,s \tag{2.1}$$

2.2.2 SIDEREAL TIME

Sidereal time is a time system of measuring the progression of time based on the position of the fixed star in the sky [7]. An associated concept is solar time, defined by the duration it takes for Earth to complete a full rotation relative to the Sun. Both sidereal time and solar time can be employed to characterize the Earth's rotational motion, and their fundamental units are the sidereal and solar days, respectively. As the right ascension of the Sun undergoes a daily variation of approximately one degree when Earth revolves around the Sun, the duration of a sidereal day amounts to

$23^h56^m4^s.091\pm0^s.005$, roughly four minutes shorter than a 24^h solar day. Currently, there are widely accepted standards for sidereal and solar time systems [9, 14–16], as illustrated below.

3) Universal Time (UT1): UT1, a mean solar time standard, is currently accepted to maintain a consistent average duration of the solar day at 24 hours. As the solar day is subject to Earth's rotational motion and the Sun's apparent motion, it follows that the length of one second in the UT1 is not constant. In this regard, the UT1 second fails to correspond with the SI second precisely. Hence, UT1 is unsuitable for employment as a temporal scale in physics-related contexts. Still, the utilization of UT1 remains imperative, particularly in astronomical observation applications that pertain to Earth's rotation. Since the length of one second in UT1 is not constant, there are certain discrepancies regarding the number of seconds per day between UT1 and TAI. A novel time scale has been introduced to reconcile the imperatives of accurate timekeeping with maintaining the correlation of daylight to civil time. It is:

4) Coordinated Universal Time (UTC): UTC, the foremost global time standard that governs clocks and time, is founded on the SI (the atomic second in TAI), and generally maintains proximity to the mean solar time at the Greenwich Meridian (0° longitude). UTC maintains a finely-tuned equilibrium between UT1 and TAI through periodic adjustments, which involve the insertion of a full second, referred to as a leap second. This measure guarantees that UTC remains within a narrow margin of 0.9 s in relation to UT1. On 30 June 1972, the initial instance of a leap second was implemented. After that occasion, quite a few leap seconds with positive values were incorporated into UTC till 2022, perpetually highlighting the lag of UTC behind the TAI. From this perspective, though UTC is a uniform time coordinate system, it is discontinuous due to the insertion of leap seconds.

Based on the aforementioned arguments, the relationship between UT1, UTC, and TAI is given by:

$$UTC = TAI - \Delta AT \tag{2.2}$$

and:

$$UT1 = TAI - \Delta AT + \Delta UT1 \tag{2.3}$$

where:
ΔAT is the leap second,
$\Delta UT1$ is the time offset of UTC with respect to UT1.

In reference to UT1, the precise determination of the Earth's rotation can be accomplished through the use of sidereal time, represented by the angular separation (in either angular or temporal unit) between the local meridian and the vernal equinox. Given that the Earth's daily orbit around the sun and fluctuations in the length of a day are caused by precession and nutation, it is impossible to obtain the sidereal time from other time scales with sufficient accuracy [7]. Instead, the sidereal time must be determined through geodetic and astronomical observations. Once the high-precision observations are not strictly required, using mean sidereal time, a sidereal time disregarding the oscillations in the orientation of Earth's rotational axis caused by nutation, may be more practical [17]. Another type of sidereal time, known as apparent sidereal time, differs slightly from mean sidereal time in that it considers both precession and nutation. With reference to the Greenwich Meridian, one can arrive at the following sidereal times:

5) Greenwich Mean Sidereal Time (GMST): GMST, the hour angle between the prime meridian and mean vernal equinox, is measured westward from the Greenwich Meridian. The GMST can be determined by the precise instant when a specific value is attained, represented by Eq. (2.4), on a particular Julian date.

$$\text{GMST} = 24110^{\text{s}}.54841 + 1.002737909350795 \cdot \text{UT1}$$
$$+ 8640184^{\text{s}}.812866 \cdot T_0 + 0^{\text{s}}.093104 \cdot T_u^2 - 0^{\text{s}}.0000062 \cdot T_u^3 \tag{2.4}$$

where T_0 and T_u pertain to the number of Julian centuries that have transpired since 2000 Jan. 1.5 UT1, which are respectively represented as:

$$T_0 = \frac{\text{JD}_{\text{UT1}}\left(0^{\text{h}}\right) - 2451545}{36525} \tag{2.5}$$

$$T_u = \frac{\text{JD}_{\text{UT1}} - 2451545}{36525} \tag{2.6}$$

6) Greenwich Apparent Sidereal Time (GAST): The GMST takes no account of astronomical nutation. To utilize the apparent equator and equinox of date accurately, one shall determine the hour angle of the true equinox (i.e., the GAST) by correcting for the deviation caused by nutation. The GAST, expressed as the sum of GMST and nutation-induced deviation, is given by the following equation:

$$\text{GAST} = \text{GMST} + \Delta\psi\cos\varepsilon \tag{2.7}$$

where the nutation-induced drift can be obtained from the International Earth Rotation and Reference Systems Service (IERS), which is accessible at http://www.iers.org. In certain numerical ephemerides, the nutation model (or the approximation for nutation) is a must to ascertain the Earth's orientation, which is an essential factor in forecasting the celestial motion over extended periods.

2.2.3 DYNAMICAL TIME

The dynamical time scale, following the fundamental principles of Newtonian mechanics, is important in the field of astrophysics. Hence, the dynamical time serves as an independent variable in equations of celestial mechanics and comprehensively characterizes the orbital motion of celestial objects within an inertial reference frame [18]. This time scale can be deduced from the observed position of an astronomical object by employing a theory that describes its motion. For historical reasons, there exist various versions of dynamical time [19–21]:

7) Ephemeris time (ET): The ET, an initial implementation of the dynamical time concept, was implicitly defined based on the observed position of celestial objects. While it is formulated within a non-relativistic framework, ET was later superseded by two dynamical time scales: Barycentric Dynamical Time (TDB) and Terrestrial Dynamical Time (TDT). Subsequent redefinition and replacement of TDB and TDT were done by the International Astronomical Union (IAU). The TDT (latterly renamed Terrestrial Time) provides a smooth continuation of ET.

8) Terrestrial Time (TT): TT inherits the conceptual framework of TDT, which serves as a seamless continuation of ET, standing as an authoritative astronomical time standard. The TT provides a precise temporal reference for celestial observations gathered from the terrestrial vantage point. The ET aligned the SI second from TAI, ensuring its freedom from irregularities in the Earth's rotation. Although TT and TAI share the same time unit, they differ by a constant value of 32.184 s, with TT being ahead of TAI [7]. Consequently, we can present the ensuing relationship as shown below:

$$\text{TT} = \text{TDT} = \text{ET} = \text{TAI} + 32.184\,\text{s} \tag{2.8}$$

Besides, Eq. (2.9) presents a fundamental connection between UT1 and TT, wherein the UT1 characterizes the Earth's rotational behavior, while the TT is used as a benchmark for interconverting dynamic time scales.

$$\Delta T = \text{TT} - \text{UT1} \tag{2.9}$$

The time offset ΔT can be obtained from the United States Naval Observatory (USNO) website by accessing the link: https://maia.usno.navy.mil/ser7/deltat.data. Also, one can calculate ΔT using the Earth Orientation Parameters (EOPs) daily reported by IERS (http://www.iers.org). In cases where the period falls outside the range covered by IERS or USNO reports, one shall depend on a prediction of ΔT. One forecast is available through the annual report of IERS [22]. The USNO also produces a similar prediction (https://maia.usno.navy.mil/ser7/deltat.preds). Figure 2.1 showcases a graphical depiction of time offset ΔT as reported and predicted by USNO, alongside that derived from EOPs reported by IERS. If heightened precision is not deemed imperative, a series of polynomial equations are provided by [23] that amalgamate previous records and firsthand observations to ascertain ΔT. These polynomial equations facilitate commutating ΔT for any given point in time ranging from 2000 BCE to 3000 CE.

As the implementation of MBSAR is still to be explored, the study conducted using recorded EOP data may only partially reflect characteristics of MBSAR, as the irregular Earth–Moon relative motion may prevent a complete capture of these characteristics. Hence, it is crucial to establish a spatiotemporal transformation that factors in both the Earth's rotation and lunar revolution, encompassing past, present, and future scenarios. It highlights the importance of predicting ΔT when it comes to studying the MBSAR.

9) Geocentric Coordinate Time (TCG): TCG was defined by the IAU, serving as the time coordinates for the Geocentric Celestial Reference System (GCRS). The TCG was developed specifically to facilitate calculations related to precession, nutation, and satellites

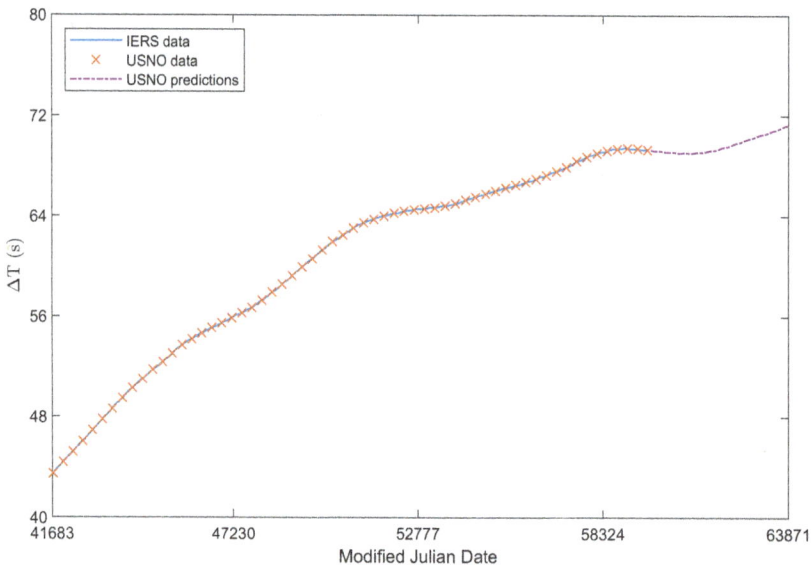

FIGURE 2.1   The time offset ΔT reported and predicted by USNO, and time offset ΔT derived from the EOPs reported by USNO.

(including the Moon) revolving around Earth. TCG is based upon a reference frame in a state of co-movement with the Earth's center, as per the framework of the general theory of relativity. In this regard, the TCG is identical to the proper time elapsed by an ideal clock situated in Earth's center. Hence, the ticking rate of TCG is faster than that of clocks on the Earth's surface. This discrepancy can be quantified by adding a constant scale factor in TT, demonstrated as:

$$\text{TCG} = \text{TT} + \left(\frac{L_G}{1-L_G}\right) \cdot \left(\text{JD}_{\text{TT}} - 2443144.5\right) \cdot 86400\,\text{s} \tag{2.10}$$

where:
JD_{TT} refers to Julian date in TT,
L_G is a scale factor that equals to $6.969290134 \times 10^{-10}$.

10) Barycentric Coordinate Time (TCB): TCB is a time standard established by the IAU, acting as the time coordinates for the Barycentric Celestial Reference System (BCRS). TCB refers to the elapsed time as measured by a clock in a reference frame that is moving together with the barycenter of the solar system. From this point of view, TCB is not influenced by gravitational time dilation resulting from the sun and other celestial bodies. The relationships between TCB and other relativistic time scales are defined using general relativistic metrics. The transformation from TCB to TCG is given by:

$$\text{TCB} = \text{TCG} + \frac{1}{c^2} \cdot \left\{ \int_{t_0}^{t} \left[\frac{v_{\text{emc}}^2}{2} + w_{\text{ext}}\left(\vec{x}_{\text{emc}}\right) \right] dt + \vec{v}_{\text{emc}} \cdot \left(\vec{x} - \vec{x}_{\text{emc}}\right) \right\} + \mathcal{O}\left(c^{-4}\right) \tag{2.11}$$

where:
c is the light velocity,
$\vec{x}$ represents the observer's barycentric position,
$\vec{x}_{\text{emc}}$ refers to the barycentric coordinate vector of Earth's center of mass,
$\vec{v}_{\text{emc}}$ stands for the velocity vectors of Earth's center of mass, and
w_{ext} is the Newtonian potential of all solar celestial bodies except for Earth evaluated at the geocenter.

11) Barycentric Dynamical Time (TDB): TDB was established in 1976 to replace the non-relativistic timescale known as ET. In 2006, TDB was redefined as an internationally recognized relativistic coordinate time scale. After this, the TDB is widely utilized in astronomy for computing planetary orbits and ephemerides. TDB can now be expressed as a linear scaling of TCB, such that:

$$\text{TDB} = \text{TCB} - L_B \cdot \left(\text{JD}_{\text{TCB}} - 2443144.5\right) \cdot 86400\,\text{s} - 6.55 \times 10^{-5}\,\text{s} \tag{2.12}$$

where JD_{TCB} is the TCB Julian date, $L_B = 1.550519768 \times 10^{-8}$.

Figure 2.2 shows differences in time offset among these time scales. For a thorough examination and additional details, one may refer to [9, 24].

In building the Moon-based Earth observation geometry, care must be exercised when applying these time scales. For example, we use the UT1 time scale to characterize Earth's rotational motion. By contrast, the TDB is commonly employed by numerical lunar

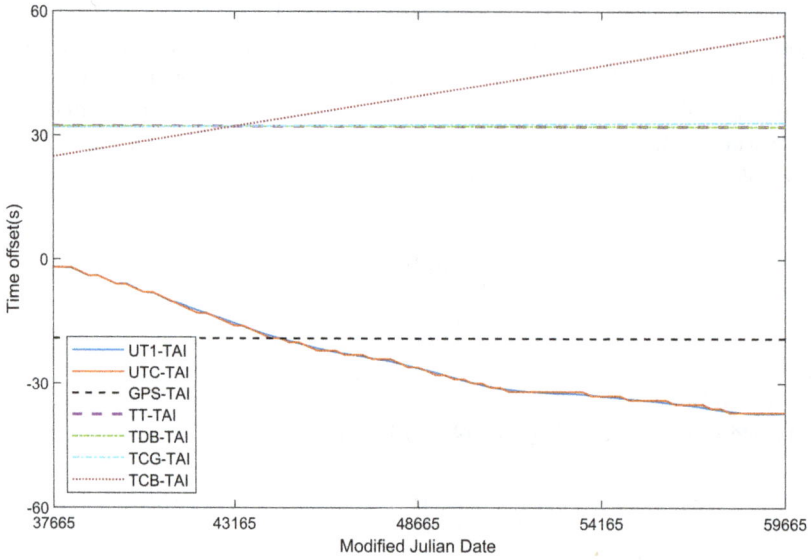

FIGURE 2.2   The disparity in time scales compared to the TAI from MJD 37665 to MJD 59665.

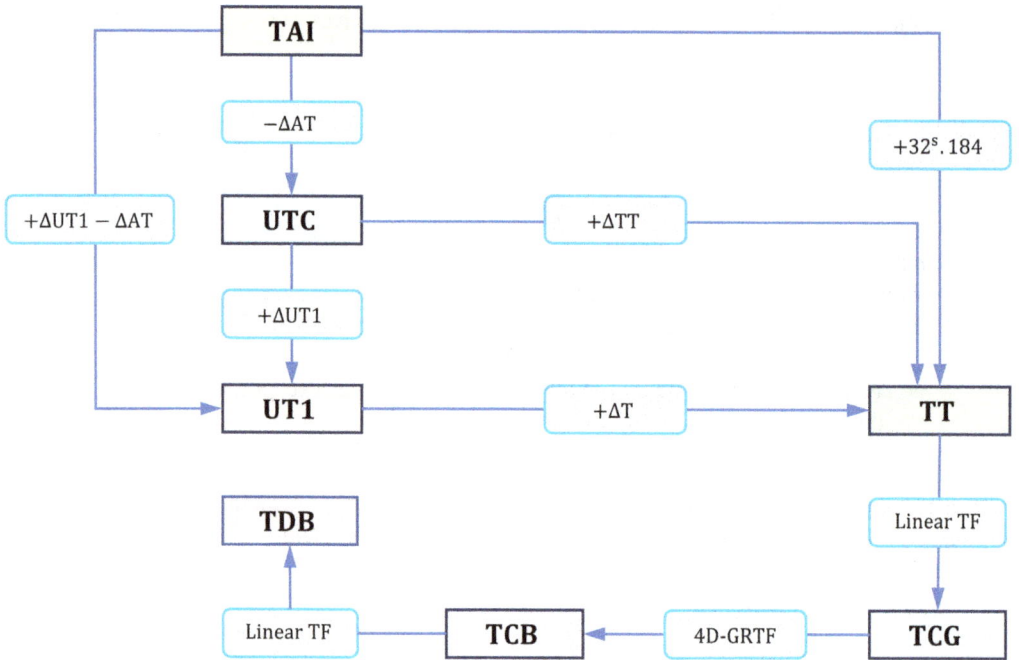

FIGURE 2.3   The relationships between various time scales and transformation routines, where the "Linear TF" indicates linear transformation, and "4D-GRTF" represents 4-dimensional general relativity transformation. The above time offsets are readily available in the IERS database.

ephemerides to determine the orbit Moon's barycenter and its attitude. Therefore, it is imperative to consider the transformation between UT1 and TDB, as depicted in Figure 2.3, into a unified time coordinate system, which may either be UT1 or TDB, contingent upon the applications in the MBSAR.

2.3 SPACE COORDINATES

In MBSAR observation geometry, we shall consider the relative motion between the SAR and its observed TOI. Hence, to establish a Moon-based observation geometry, it is imperative to establish a framework in celestial measurement theory that determines the lunar revolution and rotational motions of the Moon and Earth at a particular moment [25]. Moreover, this framework should accurately represent the geometric alignment between the MBSAR's beam pointing direction and TOI's position. However, various space coordinate systems are employed to describe the MBSAR's motion relative to the TOI, encompassing the Earth–Moon motion, MBSAR's position on the Moon, and TOI's position on Earth. To this end, it is necessary to introduce essential reference frameworks and their reciprocal transformations for establishing the MBSAR's observation geometry, as illustrated below.

2.3.1 SPACE REFERENCE FRAMEWORKS

The Earth-Centered Inertial (ECI) coordinate system, such as the GCRS (a kind of ECI created by the IAU), is usually used to describe the orbit of the spaceborne sensor, while the position of terrestrial TOI is generally specified within the Geodetic Reference System [26]. Hence, through appropriate coordinate transformations, the positions of the spaceborne SAR and TOIs can be reconciled within the Earth-Centered Earth-Fixed (ECEF) coordinate system [17], e.g., International Terrestrial Reference System (ITRS), which is a type of ECEF that was also established by the IAU in 2000.

In the MBSAR, the Earth observation geometric model differs considerably from that of the spaceborne SAR. As a natural celestial body with a mean radius of 1737.4 km, the lunar surface provides placing antenna at the location from which the Earth is observable [27]. Therefore, it is not conceivable to designate the antenna's location to the platform's barycenter, as is often done in spaceborne SAR. The MBSAR's position must be situated on the Moon's surface. Additionally, the Moon's surface topography (as illustrated in Figure 2.4), coupled with its ellipsoidal shape (though with a slight flattening), adds a layer of complexity to accurately identifying the MBSAR's position. Hence, determining the platform's position requires a reference system that precisely describes the Moon's motion and orientation with respect to the Earth. Further, it is imperative to establish a reference framework that describes the position of TOI on the Earth ellipsoid accurately.

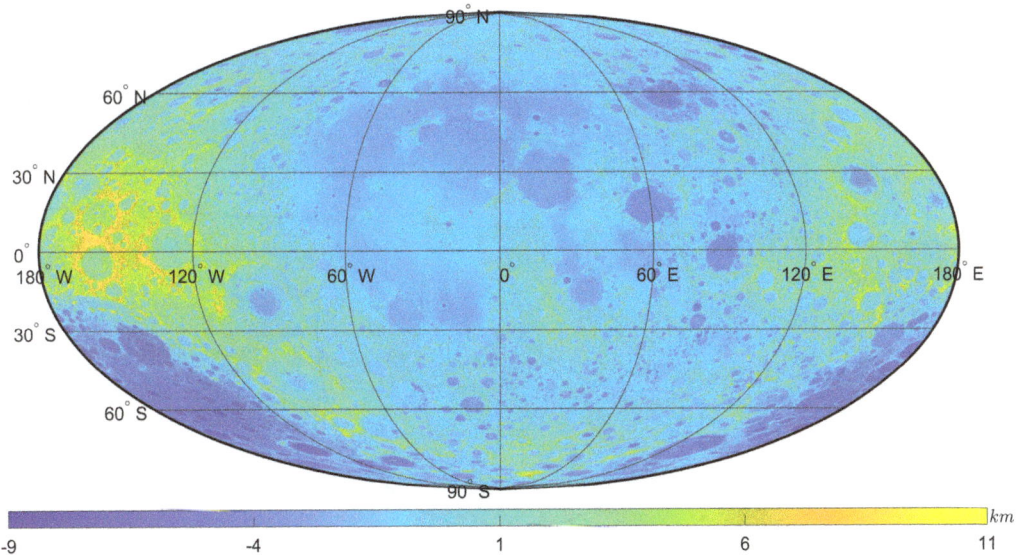

FIGURE 2.4 The lunar global topographic map obtained by the LRO (lunar reconnaissance orbiter); the lunar DEM data are taken from [28].

At this point, it becomes clear that three reference systems are required for establishing the MBSAR's geometric model [4, 29]:

1) A Moon-centered reference system determines the MBSAR's position;
2) A celestial reference system determines the Moon's orbit around Earth;
3) An Earth-centered reference system determines the TOI's location.

Each of these systems shall be explained in detail below.

2.3.1.1 Moon-Centered Reference Systems

For a MBSAR, it is necessary to determine the positions of the SAR's antenna on the lunar surface. To this end, we further need the following reference systems [3, 30–32]:

1) Selenographic Coordinate Reference (SCR coordinate system): SCR is employed to determine the geographical coordinates of MBSAR's antenna on the lunar surface. Like the geodetic coordinate system, SCR utilizes the altitude, latitude, and longitude to pinpoint a specific location on the lunar surface. The latitude is gauged in degrees north or south in relation to the Moon's equator, while longitude is measured in degrees east or west in reference to the lunar prime meridian. Note that the Moon exhibits an ellipsoidal shape, but the disparity between its polar radius (around 1737 km) and equatorial radius (around 1739 km) is insignificant [33]. Hence, it is possible to make an approximation of the Moon as a sphere, where the ellipsoidal shape could be implicit in the variations in topography [34]. In this context, altitude, the vertical distance between a particular point on the lunar surface and the reference ellipsoid, is calculated by subtracting the reference radius of the Moon from the measured value.

2) Moon-Centered Moon-Fixed (MCMF) coordinate system: MCMF is a right-hand reference framework utilized to specify an object's position and orientation in relation to the Moon's barycenter. The MCMF has its origin established at the Moon's barycenter, with the z-axis aligned with the direction of the Moon's mean rotational pole and the x-axis pointing toward the Mean Earth Rotation (MER) direction. The y-axis follows the right-hand rule. Consequently, the lunar prime meridian is aligned to coincide with the MER direction, as illustrated in Figure 2.5.

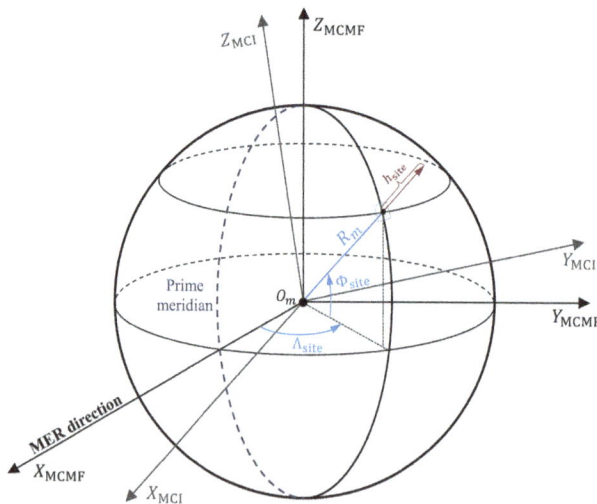

FIGURE 2.5 The schematic representation of Moon-centered reference frames, where O_m is the Moon's barycenter, R_m is the Moon's reference radius, Λ_{site} and Φ_{site} are the longitude and latitude of MBSAR's site, and h_{site} is the altitude of MBSAR's antenna above the lunar geoid.

3) Moon-Centered Inertial (MCI) coordinate system: MCI is a right-handed coordinate system utilized to specify the position and orientation of the Moon with respect to celestial objects. The International Celestial Reference System (ICRS) proposed by the IAU is now a widely accepted inertial coordinate standard, which has its origin located at the barycenter of the solar system. The MCI has the same orientation as the ICRS, but its origin point is situated at the barycenter of the Moon.

Using these reference systems makes it feasible to precisely ascertain the position and orientation of the MBSAR's antenna on the lunar surface. To provide more specific details, given that the MBSAR's antenna is situated at a site of (Λ_{site}, Φ_{site}) with an altitude of h_{site} above the geoid in the SCR, its location in the MCMF coordinate system can be described in the following manner:

$$\mathbf{R}_{SAR(MCMF)} = \left(R_m + h_{site}\right) \cdot \left[\cos \Lambda_{site} \cos \Phi_{site}, \sin \Lambda_{site} \cos \Phi_{site}, \sin \Phi_{site}\right] \tag{2.13}$$

where R_m is the Moon's reference radius.

Note that the MCMF coordinate system is defined in the MER frame, while the Principal Axis (PA) frame is generally used in certain numerical Ephemerides that describe the dynamics of the Moon's revolution and rotational motion [35]. The PA frame is distinct from the MER frame in that it aligns the axes with the principal axes of the Moon's inertia tensor. A two-step process can be employed to transform the position vector from the MCMF to MCI coordinate systems, which involves:

1) Converting the vector in the MER into PA frames;
2) Transforming the yielded vector into the MCI coordinate system by utilizing physical libration parameters expressed in terms of Euler angles.

The operation mentioned above can be mathematically represented as:

$$\mathbf{R}_{SAR(MCI)} = \mathbf{U}_{MCI}^{MCMF} \mathbf{R}_{site(MCMF)} \tag{2.14}$$

$$\mathbf{U}_{MCI}^{MCMF} = \left[\mathbf{R_z}\left(-\phi_m\right)\mathbf{R_x}\left(-\theta_m\right)\mathbf{R_z}\left(-\psi_m\right)\right]\left[\mathbf{R_z}\left(c_z\right)\mathbf{R_y}\left(c_y\right)\mathbf{R_x}\left(c_x\right)\right] \tag{2.15}$$

where:
(c_x, c_y, c_z) represent the elements on the transformation between MER and PA frames, which may vary depending on the used ephemeris [36],
$(\phi_m, \theta_m, \psi_m)$ are Euler angles that parameterize physical librations,

$\mathbf{R_x}$, $\mathbf{R_y}$, and $\mathbf{R_z}$ are the right-handed rotation matrices:

$$
\begin{aligned}
\mathbf{R_x}\left(\vartheta\right) &= \begin{bmatrix} 1 & 0 & 0 \\ 0 & \cos\vartheta & \sin\vartheta \\ 0 & -\sin\vartheta & \cos\vartheta \end{bmatrix} \\[2mm]
\mathbf{R_y}\left(\vartheta\right) &= \begin{bmatrix} \cos\vartheta & 0 & -\sin\vartheta \\ 0 & 1 & 0 \\ \sin\vartheta & 0 & \cos\vartheta \end{bmatrix} \\[2mm]
\mathbf{R_z}\left(\vartheta\right) &= \begin{bmatrix} \cos\vartheta & \sin\vartheta & 0 \\ -\sin\vartheta & \cos\vartheta & 0 \\ 0 & 0 & 1 \end{bmatrix}
\end{aligned} \tag{2.16}
$$

2.3.1.2 Celestial and Terrestrial Reference Systems

The celestial equatorial reference system is commonly used in Earth observation. This system employs the celestial equator as its fundamental plane while orienting the z-axis toward the north celestial pole and directing the x-axis toward the position of the vernal equinox [37]. The celestial equatorial reference system may be categorized based on the selected coordinate origin: the first one is centered at the geocenter (e.g., J2000.0 coordinate, GCRS), the second one has its origin at the barycenter of the solar system (e.g., ICRS), and other references (e.g., the MCI).

One typical Earth-centered celestial equatorial reference system is the J2000.0 coordinate. This coordinate, defined with the Mean Equator and Mean Equinox (MEME) of Earth, designates the x-axis as the standard epoch equinox and the z-axis as the standard epoch celestial pole. Its fundamental plane is established as the Earth's equatorial plane at the epoch J2000.0, i.e., at 12:00 on January 1, 2000 TT. Another commonly applied Earth-centered celestial system is the GCRS, which sets itself apart from the J2000.0 coordinate by incorporating a constant frame bias matrix.

The ICRS stands as the prevailing standard celestial reference system whose origin is situated at the solar system barycenter. The fundamental plane of ICRS is tightly aligned with J2000.0, with the x-axis located near the intersection point between the mean equator of J2000.0 and the mean plane of ecliptic [38]. The ICRS is ascertained through the International Celestial Reference Frame (ICRF) by observing reference celestial sources at radio frequencies. Thus, the Celestial Reference Frame (CRF) encompasses the numeric coordinates of these reference sources, which are ascertained according to the principles established by ICRS. The axes of ICRF are determined by measuring the positions of extragalactic sources, with a focus on quasars, using the Very Long Baseline Interferometry (VLBI). Currently, the IAU has endorsed three versions of ICRFs: ICRF1 [39], ICRF2 [40], and ICRF3 [41]. The newest ICRF3 was introduced in August 2018 and commenced operations on the 1st of January 2019.

The MCI coordinate system, as previously defined, shares alignment of its fundamental plane and coordinate axes with ICRS, but its origin is located at the barycenter of the Moon. While the MCI serves to locate a specified point on the lunar surface, the GCRS is employed to explicate the position of the Moon's barycenter, along with its relative motion with respect to the geocenter. Like MCI, GCRS has a fundamental plane and coordinate axes aligned with the corresponding elements of ICRS, but with an origin at the geocenter.

For depicting the TOI's position on the Earth's surface, it is also essential to introduce an ECEF coordinate system in the Moon-based Earth observation. ITRS furnishes such a reference system, wherein the origin is situated at the geocenter of the entire Earth, encompassing oceans and atmosphere [24]. ITRS adopts a length unit aligned with the SI meter, while implementing the TCG as its designated time scale. Moreover, the Bureau International del'Heure (BIH) at the epoch 1984.0 provided the initial orientation of the ITRS. The orientation's time evolution is secured through the implementation of a no-net-rotation stipulation with respect to the Earth's crust [7].

The IERS is responsible for maintaining and updating the ITRS, employing the International Terrestrial Reference Frame (ITRF) to realize ITRS. The ITRF offers estimated coordinates and velocities of specific observation stations authorized by the IERS. Their determination relies on various observational techniques such as satellite laser ranging (SLR), Lunar Laser Ranging (LLR), GPS, and VLBI measurements [42]. Updated versions of the ITRF are released every few years, showcasing global differentials with a precision reaching a centimeter level.

The ITRF is established upon the foundation of Cartesian equatorial coordinates. Once necessary, these coordinates can be converted to geodetic coordinates by incorporating a reference ellipsoid. Then, the location of TOI on the ellipsoidal Earth can be determined in terms of altitude, latitude, and longitude. Specifically, the altitude denotes the vertical distance between the TOI on the actual Earth's surface and the reference ellipsoid. Besides, the latitude is expressed in degrees north or south with respect to Earth's equator, while longitude is measured in degrees east or west relative to the Greenwich Meridian. In such a scenario, the IERS suggests making use of the Geodetic Reference System 1980 (GRS80) [24].

Bearing the aforementioned information in mind, we can define the following reference systems centered on the Earth.

2.3.1.3 Earth-Centered Reference Systems

As depicted in Figure 2.6, the position of TOI is described from various perspectives using different Earth-centered reference systems, as outlined below [3, 29, 32].

4) Geodetic reference system: Given that Earth is an ellipsoid, a reference ellipsoid provides an approximation of the ellipsoidal shape of Earth. The geodetic reference system, building upon this fundamental principle, is supplementary to the Cartesian coordinates to locate the TOI on Earth's surface. We note that several reference ellipsoids are used in geodetic applications to represent the shape of the Earth, two commonly employed ones are the GRS 80 and World Geodetic System 1984 (WGS 84) [43, 44]. The reference ellipsoids initially implemented by both GRS 80 and WGS 84 were identical. However, WGS 84 has undergone successive refinements since its inception, resulting in a subtle disparity between the reference ellipsoids of GRS 80 and WGS 84 [45]: Both accept an equatorial radius of 6378.137 km while they differ slightly in the inverse flattening values, with GRS 80 employing 298.257222101 and WGS 84 employing 298.257223563.

5) The ECEF coordinate system: ECEF, also referred to as the Earth-Centered Rotational (ECR) coordinate system, is utilized to specify the positioning of TOI with respect to the geocenter [17]. In ECEF, the x-axis is oriented toward the junction of the Greenwich Meridian and the equatorial plane, while the z-axis aligns with the North Pole of Earth, and the y-axis extends orthogonally to both the x and z axes. In the Moon-based Earth observation, ITRS is employed to provide the ECEF coordinate system.

6) The ECI coordinate system: ECI, a right-handed inertial frame of reference, is employed to specify the position of the lunar barycenter along its orbit. In the Moon-based Earth

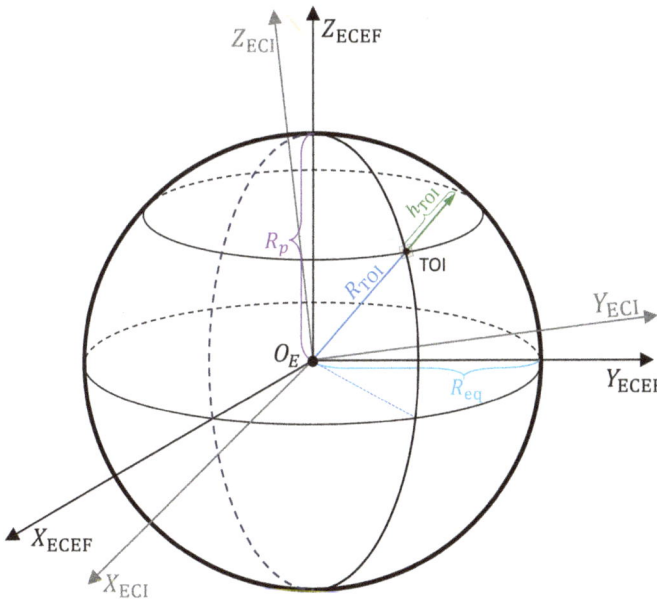

FIGURE 2.6 The schematic representation of the Earth-centered reference frames, where O_E is the Earth's center, R_{TOI} is the Earth's radius with respect to the TOI, h_{TOI} is the altitude of TOI above the geoid. R_p and R_{eq} are the polar and equatorial radii of the reference ellipsoid.

observation, the GCRS serves as the adopted standard for defining ECI, whose origin is situated at the geocenter, with its fundamental plane and coordinate axes aligned following the ICRS adopted by IAU. Using the ECI coordinate system as a foundation, it is feasible to construct a local system for determining the TOI's position as observed from the MBSAR, as elucidated in the following.

7) Antenna Coordinate System (ACS): ACS serves as an auxiliary coordinate system with respect to the Earth's center and SAR's antenna for establishing the linkage between the position of TOI and beam pointing of MBSAR [46]. In ACS, the origin refers to the antenna's position, with the z-axis coinciding with the MBSAR's vector directed toward the geocenter, the y-axis aligning with the MBSAR's angular momentum direction relative to Earth, and the x-axis is determined by the right-hand rule.

The aforementioned reference frameworks are listed in Figure 2.7. As depicted, it is possible to convert the MBSAR on the lunar surface into the ECI coordinate system:

$$\mathbf{R}_{\mathbf{SAR(ECI)}} = \mathbf{R}_{\mathbf{MC(ECI)}} + \mathbf{R}_{\mathbf{SAR(MCI)}} = \mathbf{R}_{\mathbf{MC(ECI)}} + \mathbf{U}_{\mathbf{MCI}}^{\mathbf{MCMF}} \mathbf{R}_{\mathbf{site(MCMF)}} \tag{2.17}$$

where $\mathbf{R}_{\mathbf{MC(ECI)}}$ is the position of Moon's barycenter in ECI.

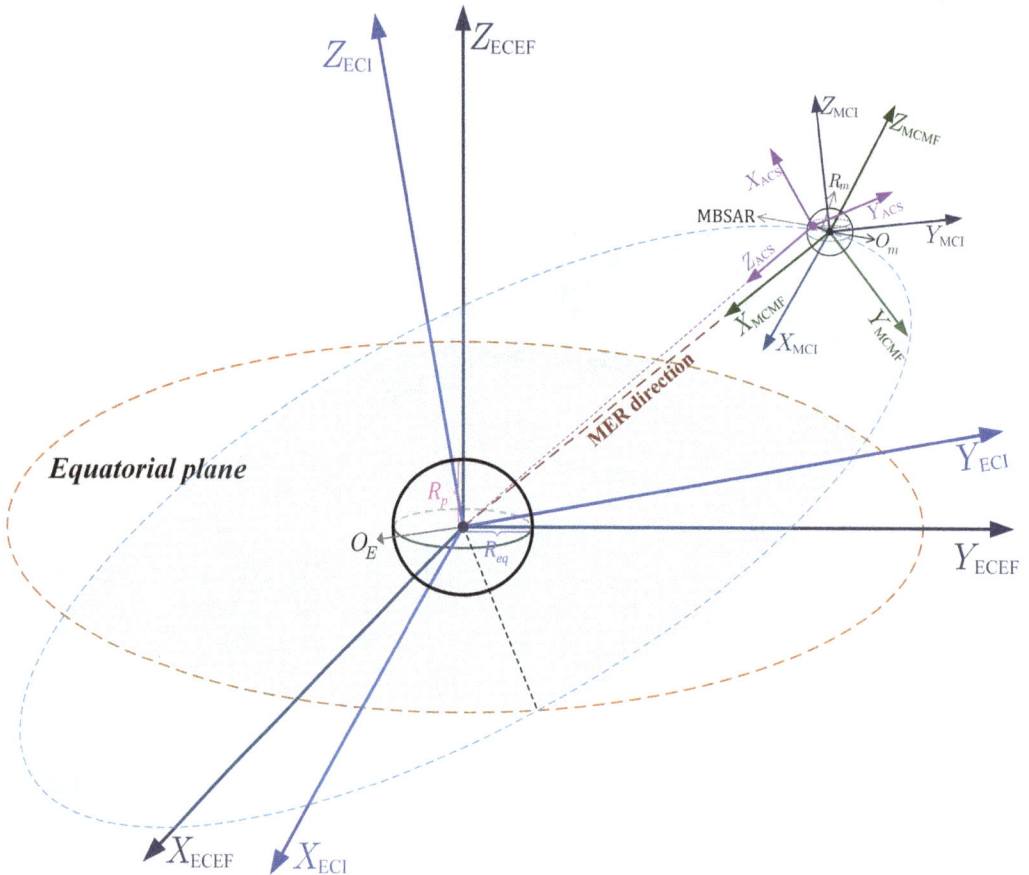

FIGURE 2.7   The schematic representation of overall reference systems employed in the MBSAR, where the SCR and geodetic systems are respectively implicit in MCMF and ECEF (Not to scale for the ake of clarity).

2.3.2 The Transformations between the ECI and ECEF Coordinates

Figure 2.8 depicts procedures for converting the MBSAR and TOI in the ECI coordinate system or the ECEF whichever is more appropriate. For instance, if the objective is to determine the coverage performance and imaging region of the MBSAR, the ECEF coordinate system is required to project the beam footprint onto the Earth's surface [1, 32]. On the other hand, it may be more advantageous to examine the Doppler properties in the ECI coordinate system, as the Doppler shift is a consequence of both the MBSAR's inertial motion and the Earth's rotational motion [46–48].

In the following, we consider the transformation between ECI and ECEF coordinate systems. Given a state vector $\mathbf{r}_{ECI}$ in the ECI coordinate system ($\mathbf{r}_{ECI}$ can be either the MBSAR's or TOI's position), its counterparts in the ECEF coordinate system can be derived by [7, 24]:

$$\mathbf{r}_{ECEF} = \mathbf{C}_{ECEF}^{ECI}\mathbf{r}_{ECI} \tag{2.18}$$

The matrix $\mathbf{C}_{ECEF}^{ECI}$, which facilitates the transformation from ECI to ECEF coordinate systems, is defined as follows:

$$\mathbf{C}_{ECEF}^{ECI} = \mathbf{\Pi\Theta NPB} \tag{2.19}$$

where $\mathbf{\Pi}$, $\mathbf{\Theta}$, $\mathbf{N}$, $\mathbf{P}$, and $\mathbf{B}$ respectively represent the polar motion, Earth's rotation, nutation, precession, and frame bias. Specific definitions of those matrices are provided below.

Due to the precession phenomenon, the rotational axis of the Earth undergoes changes, which, in turn, causes a shift in the orientation of the equatorial plane. There are several methods to address the precessional motion, one of which involves [24]:

$$\mathbf{P} = \mathbf{R}_z\left(\chi_A\right)\mathbf{R}_x\left(-\omega_A\right)\mathbf{R}_z\left(-\psi_A\right)\mathbf{R}_x\left(\varepsilon_0\right), \tag{2.20}$$

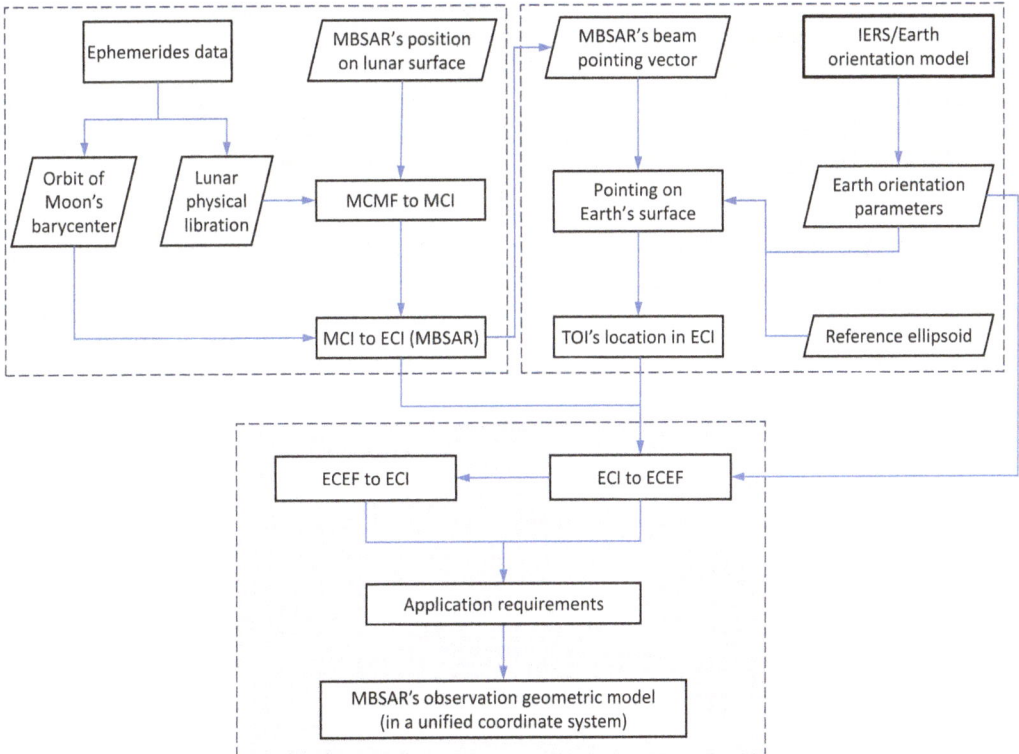

FIGURE 2.8 The general procedures for establishing the MBSAR's observation geometric model.

where:

$$
\begin{cases}
\varepsilon_0 = 84381''.406 \\
\psi_A = 5038''.481\,507 \cdot T - 1''.079\,0069 \cdot T^2 \\
\quad - 0''.001\,140\,45 \cdot T^3 + 0''.000\,132\,851 \cdot T^4 - 0''.000\,000\,0951 \cdot T^5 \\
\omega_A = \varepsilon_0 - 0''.025\,754 \cdot T + 0''.051\,2623 \cdot T^2 \\
\quad - 0''.007\,725\,03 \cdot T^3 - 0''.000\,000\,467 \cdot T^4 - 0''.000\,000\,3337 \cdot T^5 \\
\chi_A = 10''.556\,403 \cdot T - 2''.3814292 \cdot T^2 \\
\quad - 0''.001\,211\,97 \cdot T^3 + 0''.000\,170\,663 \cdot T^4 - 0''.000\,000\,056 \cdot T^5
\end{cases} \tag{2.21}
$$

$$
T = \frac{\mathrm{JD_{TT}} - 2451545}{36525} \tag{2.22}
$$

The numerical ephemeris commonly makes use of a specific precession model to depict the precessional motion of Earth. As an illustration, each version might incorporate a distinct precession model within the Development Ephemeris (DE) generated by the Jet Propulsion Laboratory (JPL). In DE 430, the matrix responsible for precession is mathematically expressed as [36]:

$$
\mathbf{P} = \mathbf{R}_z\left(-z\right) \mathbf{R}_y\left(\vartheta\right) \mathbf{R}_z\left(-\zeta\right) \tag{2.23}
$$

$$
\begin{cases}
\zeta = 2306''.2181 \cdot T + 0''.301\,88 \cdot T^2 + 0''.017\,998 \cdot T^3 \\
\vartheta = 2004''.3109 \cdot T - 0''.426\,65 \cdot T^2 - 0''.041\,833 \cdot T^3 \\
z = 2306''.2181 \cdot T + 1''.094\,68 \cdot T^2 + 0''.018\,203 \cdot T^3
\end{cases} \tag{2.24}
$$

In the subsequent version of DE, namely DE 440/441, the Vondrák precession model is adopted to characterize the precessional motion of Earth [49]; detailed information regarding this model can be found in [50].

In addition to the secular precession, the Earth's rotational axis is also subject to slight periodic perturbations commonly referred to as nutation. These perturbations result from the monthly and annual lunar and solar torque variations consolidated in the precession analysis. The corresponding transformation matrix for nutation is formulated as follows [51]:

$$
\mathbf{N} = \mathbf{R}_x\left(-\bar{\varepsilon} - \Delta\varepsilon\right) \mathbf{R}_z\left(-\Delta\psi\right) \mathbf{R}_x\left(\bar{\varepsilon}\right) \tag{2.25}
$$

where:
$\bar{\varepsilon}$ is the mean obliquity of the ecliptic,
$\bar{\varepsilon} + \Delta\varepsilon$ is the true obliquity of the ecliptic, and
$\Delta\psi$ is the periodic shift of the vernal equinox.

As mentioned previously, both parameters, $\Delta\varepsilon$ and $\Delta\psi$, are available in the IERS. The mean obliquity adopted by IAU 2006 is expressed in the following manner [52]:

$$
\begin{aligned}
\bar{\varepsilon} = \varepsilon_0 &- 46''.836\,769 \cdot T - 0''.0001831 \cdot T^2 \\
&+ 0''.002\,003\,40 \cdot T^3 - 0''.000\,000\,576 \cdot T^4 - 0''.000\,000\,0434 \cdot T^5
\end{aligned} \tag{2.26}
$$

The DE430/440 accepts the following value for the mean obliquity:

$$
\bar{\varepsilon} = 84381''.448 - 46''.815 \cdot T - 0''.000\,590 \cdot T^2 + 0''.001813 \cdot T^3 \tag{2.27}
$$

The precession and nutation phenomena furnish the current alignment of the Earth's rotational axis, while the Earth's rotation relative to its rotational axis is referred to as GAST, as specified in Eq. (2.7). This transformation can be done by the rotation of the z-axis [7]:

$$\Theta = \mathbf{R}_z \left(\text{GAST} \right) \tag{2.28}$$

Apart from precession and nutation, polar motion is also induced by variations in the Earth's mass distribution, resulting in the motion of the Earth's angular momentum axis relative to the reference frame. To account for this, we make use of the following correction [24]:

$$\Pi = \mathbf{R}_x \left(-y_p \right) \mathbf{R}_y \left(-x_p \right) \mathbf{R}_z \left(s' \right) \tag{2.29}$$

with:

$$s' = \frac{1}{2} \int_{T_0}^{T} \left(x_p \dot{y}_p - \dot{x}_p y_p \right) dt \approx 47 \cdot T \, \mu as \tag{2.30}$$

where T_0 is the epoch J2000.0, x_p and y_p are the polar motion coordinates, and both are EOPs that are provided by the IERS (accessible at http://www.iers.org). Some ephemerides also offer estimations of EOPs for predicting the long-term orbits of the Moon or planets.

Note that a slight discrepancy arises between the coordinate axes of ICRS and the J2000.0 system. The following transformation requires the use of a frame bias matrix to account for this discrepancy in the coordinate axes [53]:

$$\mathbf{B} = \mathbf{R}_x \left(-\eta_{\text{bias}} \right) \mathbf{R}_y \left(\xi_{\text{bias}} \right) \mathbf{R}_z \left(\alpha_{\text{bias}} \right) \tag{2.31}$$

with:

$$\begin{cases} \eta_{\text{bias}} = -0''.006\,8192 \pm 0''.000\,0100 \\ \xi_{\text{bias}} = -0''.016\,6170 \pm 0''.000\,0100 \\ \alpha_{\text{bias}} = -0''.014\,6000 \pm 0''.000\,5000 \end{cases} \tag{2.32}$$

The velocity vector $\mathbf{u}_{\text{ECI}}$ corresponding to the position vector $\mathbf{r}_{\text{ECI}}$ in the ECI coordinate system can be converted into the ECEF coordinate system using the following transformation [17]:

$$\mathbf{u}_{\text{ECEF}} = \mathbf{C}_{\text{ECEF}}^{\text{ECI}} \mathbf{u}_{\text{ECI}} + \frac{d\mathbf{C}_{\text{ECEF}}^{\text{ECI}}}{dt} \mathbf{r}_{\text{ECI}} \tag{2.33}$$

The transformation of position and velocity vectors from ECEF to ECI coordinate systems is expressed as follows:

$$\mathbf{r}_{\text{ECI}} = \mathbf{C}_{\text{ECI}}^{\text{ECEF}} \mathbf{r}_{\text{ECEF}} \tag{2.34}$$

$$\mathbf{u}_{\text{ECI}} = \mathbf{C}_{\text{ECI}}^{\text{ECEF}} \mathbf{u}_{\text{ECEF}} + \frac{d\mathbf{C}_{\text{ECI}}^{\text{ECEF}}}{dt} \mathbf{r}_{\text{ECEF}} \tag{2.35}$$

with:

$$\mathbf{C}_{\text{ECI}}^{\text{ECEF}} \mathbf{C}_{\text{ECEF}}^{\text{ECI}} = \mathbf{I} \tag{2.36}$$

where $\mathbf{I}$ is an identity matrix.

By coordinate transformation, the position and velocity vectors of the MBSAR and TOI can be converted into a unified coordinate system, thereby establishing the observation geometric model for the MBSAR. More details regarding this observation geometry will be presented in Section 2.6.

2.4 EARTH–MOON RELATIVE MOTION

The Doppler shift effect caused by the relative motion between the TOI and SAR is pivotal in the formation of the SAR image [54]. Understanding the Earth–Moon relative motion is of fundamental importance in ascertaining the Doppler effect and its property to assess the imaging performance of the MBSAR [55]. To start, we shall provide an overview regarding the motion characteristics of the Moon.

2.4.1 LUNAR MOTION CHARACTERISTICS

While studying the movement of the Moon relative to Earth, it is feasible to regard it as a rigid body. Analogous to other rigid bodies, the Moon orbiting Earth can be bifurcated into two components, namely: the orbital motion of the lunar barycenter around Earth and the rotational motion of the Moon [56]. In examining the motion of the lunar barycenter, we can treat the Moon and Earth as two masses concentrated at both barycenters.

The mutual gravitational attraction between two masses gives rise to a two-body system known as the Earth–Moon system. When observing this system in the BCRS, its barycenter follows an orbit around the Sun, while both the Moon and Earth orbit around this barycenter. The barycenter of the Earth–Moon system is located approximately 4641 km away from the Earth's center, along the line connecting the two celestial bodies [57]. Directing our focus toward the orbit of the Moon around Earth, we may employ orbital elements to characterize the lunar revolution around Earth, as depicted in Figure 2.9. The definitions of all those orbital elements in Figure 2.9 are listed in Table 2.1.

In the Earth–Moon system, if only the mutual attraction between Earth and Moon is considered, the orbit of the Moon would conform to Keplerian principles, and the corresponding orbital elements would remain constant. However, this ideal scenario is not representative of reality. The Moon orbiting Earth is strongly perturbed by the gravitational pull from Sun, and to a lesser extent, from planets; additionally, the non-spherical Earth shape, tidal acceleration, and other factors influence

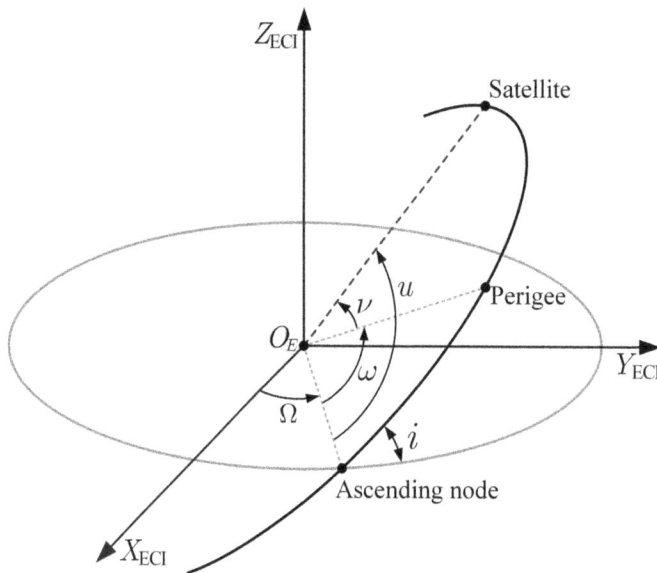

FIGURE 2.9 The schematic diagram of orbital elements.

TABLE 2.1

Definition of Orbital Elements

Symbol	Parameter	Definition
a	Semi-major axis	The longest semidiameter of an elliptical orbit extends from the center through a focus and to the perimeter.
e	Eccentricity	Eccentricity is a measure of the extent to which an elliptical orbit differs from a circular one, $0 \le e < 1$
i	Orbital Inclination	The intersection angle between the equatorial and orbital planes.
Ω	Right ascension of ascending node (RAAN)	The angle between the longitude of $0°$ and the point in the orbit where the satellite crosses the equator from south to north.
ν	True anomaly	The angle between the perigee and satellite directions.
ω	Argument of perigee (AOP)	The angle between the ascending node and perigee directions.
u	Argument of latitude (AOL)	The angle between the ascending node and satellite directions, the sum of true anomaly and AOP.

the Earth–Moon relative motion to some degree [7]. Consequently, the motion orbit of the Moon around Earth deviates from the Keplerian orbit, suggesting orbital elements of the Moon fluctuate over time rather than remain constant [56].

To illustrate the intricacy of the lunar revolution around Earth, we can provide a concise overview of various behaviors given rise by orbital perturbations that can occur in the orbital elements of the Moon, accompanied by corresponding visualizations in Figure 2.10:

1) The variation of eccentricity: The orbit of the Moon adheres to a nearly circular ellipse about Earth with an average eccentricity of 0.0549. Nevertheless, this eccentricity is not a constant; due to the orbital perturbations, there are evident oscillations in the eccentricity of the Moon's orbit. As presented in Figure 2.10(a), the eccentricity undergoes fluctuations spanning from 0.026 to 0.077.

2) The variation of semi-major axis: The semi-major axis has an average value of 38,400 km. The conjunction of this value, along with the average eccentricity, enables it to yield an average perigee of 36,300 km and an average apogee of 40,500 km. In fact, neither value represents the shortest and most extended distances between the Moon and the Earth within a certain lunar cycle. According to Figure 2.10(b), the perturbations can cause fluctuations in the semi-major axis up to thousands of kilometers, resulting in variations in the perigee and apogee.

3) The variation of RAAN: The ascending node of the Moon's orbit is not static but continuously shifts westward along the ecliptic (known as the nodal precession). Moreover, the nodal precession exhibits a cycle of approximately 18.6 years, signifying the ascending node with respect to the ecliptic experiences an average annual movement rate of $19°21'$. This phenomenon leads to periodic variations in the ascending node within the ECI frame. As a result, the corresponding RAAN exhibits a temporal variation pattern of 18.6 years, as depicted in Figure 2.10(c).

4) The variation of orbital inclination: The angle between the ecliptic and lunar equator remains at $1°32'$, while the lunar equator is inclined at an obliquity of $6°41'$ to its orbit plane on average (see Figure 2.11). Consequently, the inclination of the lunar orbital plane to the ecliptic fluctuates around $5°09'$. The interaction of this inclination with Earth's tilt ($23°27'$), alongside the nodal precession phenomenon, would induce an inclination angle ranging from $18°20'$ to $28°36'$ between the lunar orbit and Earth's equator on average. Moreover, fluctuations in the orbital inclination and RAAN exhibit a strong correlation, sharing the same cycle of 18.6 years, as portrayed in Figure 2.10(c) and (d).

5) The variation of AOP: The value of the AOP in the Moon's orbit is not fixed owing to the rotational shift of the orbital major axis. The phenomenon is commonly referred to as apsidal precession. The apsidal precession shares the same direction as the lunar revolution (eastward). Besides, it takes roughly 8.85 years, or 3232.6054 days, for the major axis to complete a full rotation. As a result, the AOP of the Moon's orbit in the ECI also manifests an 8.85-year cycle.

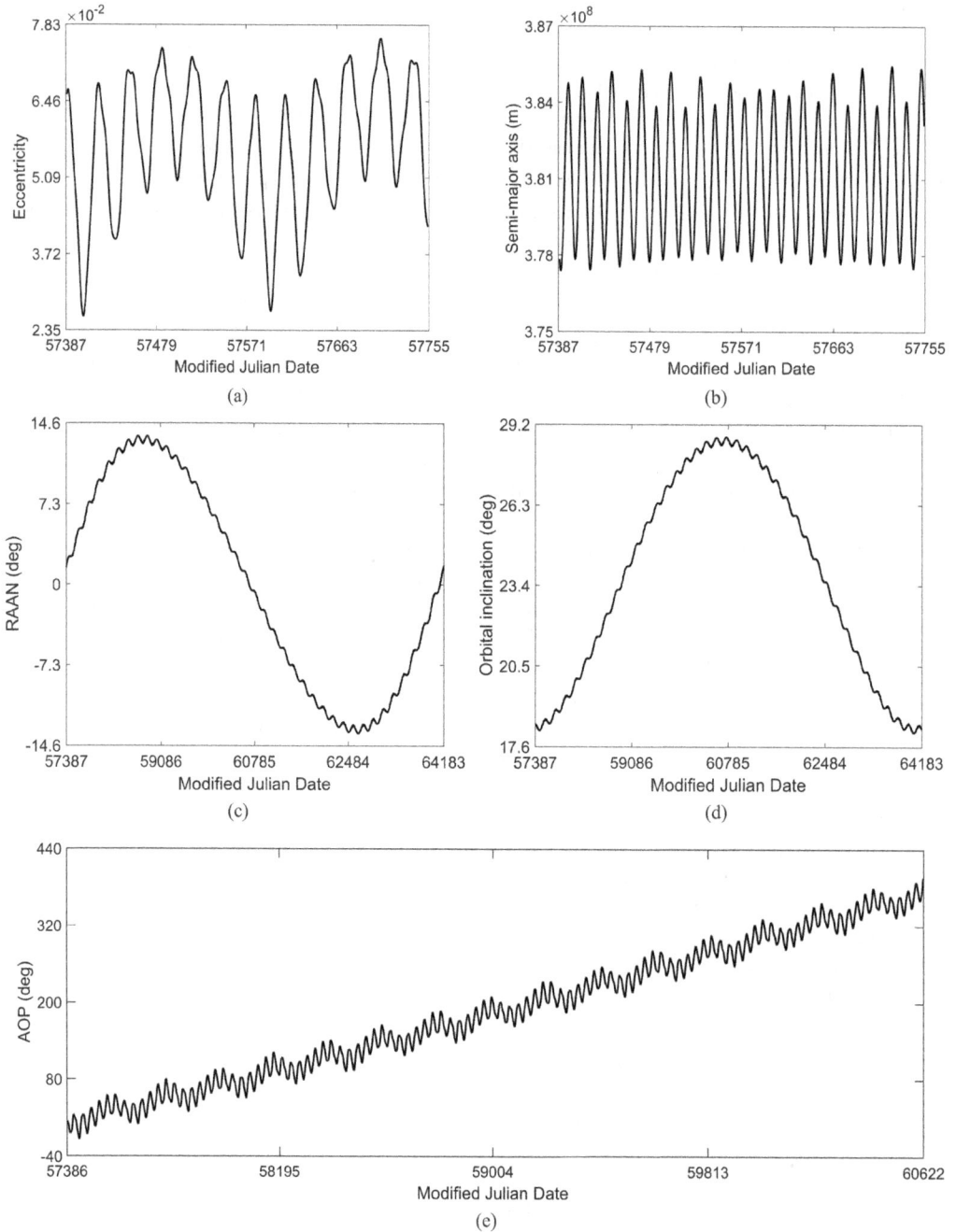

FIGURE 2.10 The temporal variations in the orbital elements of the Moon; all lunar orbital elements are given in the ECI coordinate system.

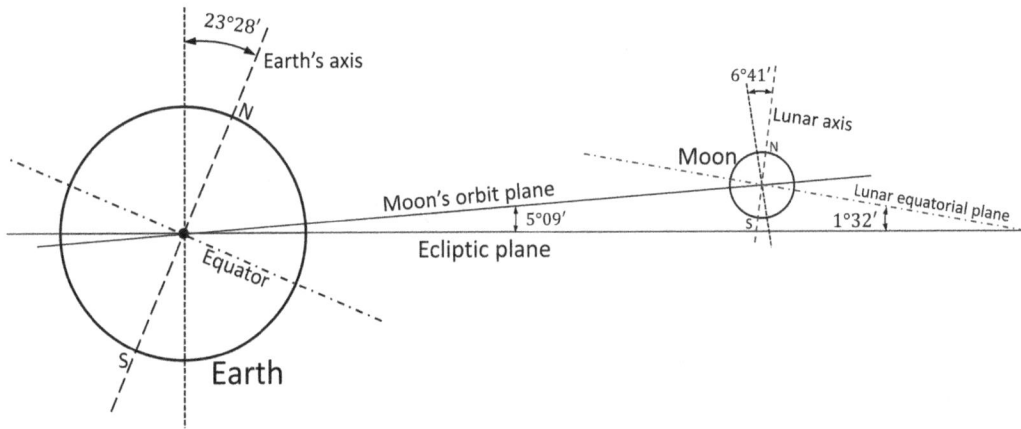

FIGURE 2.11 The schematic diagram of the Moon's orbit and orientation with respect to the Ecliptic.

The intricacies of lunar orbital elements result in multiple lunar cycles, more commonly referred to as lunar months, each characterized by unique variation regularities [58]. Table 2.2 enumerates various customary lunar months regarding the lunar revolution around Earth. All these lunar months have distinctive values and definitions for serving different purposes. The draconic month, also known as the nodical month, represents the duration between two consecutive transits of the Moon through the ascending node of the ecliptic plane [59]. This lunar cycle coincides with spaceborne SAR, which denotes the location of the SAR system through its AOL and represents one orbital cycle by AOL from 0° to 360°. Therefore, we embrace this concept to depict a lunar cycle (or an MBSAR cycle), but the reference plane (i.e., ecliptic plane) is substituted by the celestial equatorial plane, in the MBSAR. As the celestial equator is determined by vernal equinox, it follows that the averaged MBSAR cycle exhibits a proximity to the duration of a tropical month (around 27.32 days).

In terms of the rotational motion of the Moon, it exhibits unique qualities that starkly contrast with those exhibited by the Earth. A noteworthy aspect of the Moon's rotational behavior is its synchronous rotation phenomenon given rise by tidal locking, which consistently presents the same face toward the Earth [60]. The unvarying facing and stable axial inclination of the Moon render it an ideal platform for conducting extended Earth observations. Further, due to the synchronous rotation phenomenon, and a periodic oscillation referred to as libration, approximately 59% of the Moon's surface is observable from Earth.

TABLE 2.2
Definitions of Various Lunar Cycles

Cycles	Mean Value	Definition	Feature
Synodic Month	29.530589 days	The time it takes to return to the same phase.	The period of the lunar phases; determines the Moon's appearance.
Sidereal Month	27.321662 days	The time it takes to return to the same position relative to the fixed star.	Determining the Moon's position in the sky relative to stars; defining the calendrical system in some cultures.
Anomalistic Month	27.554550 days	The time it takes to return to the same point relative to the perigee.	Predicting the Moon's perigee and apogee.
Draconic Month	27.212221 days	The time it takes to return to the same point relative to the ascending node.	Also known as a nodal month or nodical month; predicting lunar and solar eclipses.
Tropical Month	27.321582 days	The time it takes to return to the same point relative to the vernal equinox.	Slightly shorter than a sidereal month due to the Earth's precession of the equinoxes.

As is well known, Cassini's laws offer a concise depiction of lunar rotational motion. Additional refinements enhancements have been devised to incorporate physical librations, positing the following statements [61, 62]:

1) The Moon exhibits a spin-orbit resonance of 1:1 with Earth. This law delineates that the Moon rotates in the same direction as Earth. Moreover, relative to the distant stars, the duration of one complete rotation of the Moon is equivalent to its average sidereal period of orbiting Earth. This period amounts to approximately 27.32 days.
2) The rotational axis remains consistently inclined at a fixed angle relative to the ecliptic plane. This principle asserts that the lunar equatorial plane maintains a steady inclination to the ecliptic plane, with a value of $1°32'$ as reported by the IAU.
3) The rotational axis of the Moon lies within a plane formed by the intersection of a normal to the ecliptic plane and a normal to the lunar orbital plane. From this perspective, the line of nodes in the orbit of the Moon is parallel to points where the lunar equatorial plane intersects with the ecliptic.

Let us now turn our attention to the lunar libration phenomenon. This phenomenon can be categorized into three types: optical libration, diurnal libration, and physical libration, which arise respectively from the non-circular and inclined nature of the Moon's orbit, the oscillation due to Earth's rotation, and the orientation of the Moon [63]. The first two are not actual oscillations of the lunar sphere; they are merely apparent librations as perceived by an observer on Earth. The third type, conversely, constitutes a true oscillation: the lunar orientation undergoes rotational motion around its celestial pole while simultaneously experiencing minor fluctuations in its direction.

The phenomenon of physical libration, also known as real libration, is correlated to the irregularity in the lunar rotational motion, resulting in a departure from its average rotational velocity. This phenomenon arises from the minor elongated shape of the lunar surface facing the Earth, which is subject to Earth's gravitational attraction and hence leads to oscillations in the lunar orientation. In this view, while tidal locking ensures that the Moon's mean rotational cycle is synchronized with its average sidereal period, an inconsistency arises between the angular velocities of the Moon's revolution and rotational motion due to physical librations and orbital perturbations.

Indeed, the physical libration of the Moon is relatively minuscule compared to rest types of librations, with its magnitude consistently remaining below $2'$ in longitude and $3'$ in latitude [56, 64]. Yet, the physical libration could modify the orientation and location of MBSAR to a certain extent, thereby affecting its signal phase history. As a result, taking no account of such an effect could potentially engender imaging deterioration in the MBSAR. Hence, the physical libration relating to lunar rotational motion necessitates particular attention in the system design and signal processing of the MBSAR.

2.4.2 Lunar Ephemerides

Configuring system parameters and processing signal of the MBSAR relies on a precise observation geometric model, thereby requiring meticulous delineation of the Moon's orbit and orientation. However, the movement and orientation of the Moon relative to Earth are exceedingly intricate, owing to a multitude of subtle perturbations [7]. Since the deployment of retroreflectors on the lunar surface as part of the Apollo 11 mission in July 1969, the LLR technique has enabled the precise measurement of lunar ranging and attitude [65], resulting in a significant precision enhancement with which lunar orbit can be measured.

The integration of modern measurements with the advent of digital computing has greatly contributed to the refinement of lunar theory, achieving an exceptional level of precision [66]. This was accomplished through numerical computations of lunar orbit using equations of motion that take account of a broader range of factors than those encompassed by classical theory [60, 67, 68]. The developed numerical lunar theory currently encompassed gravitational forces and adjustments of relativistic effects. Additionally, the theory also incorporated the effects of tidal and geophysical forces, as well as an extended exposition on the lunar libration theory. In parallel with the development of lunar theory, a series of numerical planetary ephemerides were established and proposed to demonstrate the orbit of the Moon and other celestial bodies with high accuracy.

At present, the primary numerical planetary ephemerides consist of the DE, as well as Intégrateur Numérique Planétaire de l'Observatoire de Paris (INPOP) and Ephemerides of Planets and the Moon (EPM) [69]. These ephemerides are derived through numerical integration of lunar orbits followed by fitting to observation data, resulting in a lunar orbit with relative accuracy ranging from meters to submeters. Notwithstanding, it is currently not possible to obtain a set of parameters addressing the Earth–Moon relative motion for MBSAR with absolute accuracy [70].

We may consider implementing the following approach to ameliorate the effects of uncertainty in lunar orbit on MBSAR. The primary error of source in the lunar orbit is attributed to orbital perturbations. To address this, we can establish a correlation between the errors in each orbital element and the resulting orbital errors induced by the perturbation effects. Combined with the observation geometric model, this correlation produces errors in the MBSAR signal. Through a theoretical examination of the signal errors resulting from perturbations in each orbital element, we can assess their impact on the imaging performance of MBSAR. Conversely, this analysis can also ascertain the acceptable thresholds for perturbation errors in MBSAR by taking into account tolerant thresholds for imaging performance. This process establishes the necessary criteria for achieving precision in the Earth–Moon orbit and subsequently serves as the foundation for constructing error models for the MBSAR with greater accuracy. Chapter 7 will present a preliminary study to address this issue.

The DE produced by JPL is widely used in studying Moon-based Earth observation. The DE utilizes the time system of TDB and the coordinate system of ICRS. The DE employs the Chebyshev approximation to provide the orbits and associated parameters of major celestial objects in the solar system, including the Sun, planets, and Moon [7]. Specifically, the DE provides access to various parameters regarding the Moon's orbit and orientation in the GCRS, an ECI coordinate system whose coordinate axes are aligned with the ICRS. The Moon's three-dimensional coordinates, velocity, attitudes, and other relevant data are available. Since the 1960s, JPL has developed various DE versions [71]; some versions are featured in Table 2.3.

Before the release of DE 440, E430 was renowned for hosting the most accurate lunar orbit in DE products, providing the best match to observations [36]. The achievement was made possible by incorporating a damping term between the Moon's liquid core and solid mantle. Notably, the effectiveness of this approach was limited for a few centuries. DE431 excluded the core/mantle damping term and exhibited more outstanding suitability for dozens of centuries. This advantage comes at the cost of lower accuracy in predicting the lunar orbit. After the DE430, subsequent versions, such as DE431 to DE438, were released as a minor update of DE430, but none resulted in significant improvements to the accuracy of the lunar orbit.

Significant progress was made with the release of DE440 [49]. To update the positions of celestial bodies, including the Moon, DE440 amalgamated all available data from spacecraft and LLR spanning from 2013 to 2020. Also, the revised version of DE boasts of various crucial upgrades in the dynamical model, including the incorporation of the Lense-Thirring effect, utilization of the Vondrák precession model (for Earth orientation), a new model for the Kuiper Belt, factoring in the influence of Earth's geodetic precession on lunar librations, and consequential impact of solar radiation pressure on the Earth–Moon system. Integrating the internal planetary dynamics framework with ICRF3 is an additional update. Therefore, the release of DE440 represents a significant leap forward, specifically concerning the Moon's orbit, which had not been subject to any revisions since DE430. It is preferable to employ the newest version DE (currently DE440) when conducting research related to Earth observation by MBSAR or other Moon-based sensors.

TABLE 2.3

Introductions of DE Series Products

DE Version	Release Date	Coverage Period	Reference Frame	Notable Features
DE102	1981	1410–3002	J1950	Excluded librations; First numerical long DE.
DE200	1981	1599–2169	J2000	Excluded librations; Included 5 asteroids linked to the largest perturbations.
DE 403	1993	1599–2199	ICRF	Included more data from radar & radio ranging and VLBI observation; Improved LLR accuracy; Included perturbations of 300 asteroids.
DE 404	1996	−3000–3000	ICRF	A lower precision version of DE403 Extended time coverage; Excluded nutations and librations.
DE405	1997	1600–2200	ICRF	Added data from telescopic, radar, spacecraft, and VLBI observations; Improved the perturbations model; Improved accuracies in planetary masses, positions, and velocities Oriented onto the ICRF more accurately.
DE406	1997	−3000–3000	ICRF	A lower precision version of DE405; Extended time coverage; Excluded nutations and librations.
DE409	2003	1991–2019	ICRF	Included additional data from spacecraft ranging and VLBI and telescopic observation; Improved position accuracies of Mars and Saturn.
DE410	2003	1990–2019	ICRF	Improved version of DE409; improved masses for Venus, Mars, Jupiter, Saturn, and Earth–Moon system.
DE413	2004	1899–2050	ICRF	Updated Pluto's ephemeris; Added new data from CCD telescopic observations of Pluto.
DE414	2006	1899–2050	ICRF	Updated numerical integration method; Extended ranging data of Mars Global Surveyor and Mars Odyssey spacecraft to 2005; Included CCD observations of five outer planets.
DE418	2007	1899–2051	ICRF	Included new observations of Pluto; Updated Mars spacecraft ranging and VLBI observations through 2007; Added new LLR data since DE403; Improved the orbit of Saturn.
DE421	2008	1899–2053	ICRF	Included additional ranging and VLBI observation data; Included latest estimates of planetary masses; Added data from lunar laser ranging and two more months of CCD measurement; Improved lunar theory, better Earth–Moon mass ratio. Included perturbations from 343 asteroids.
DE422	2009	−3000–3000	ICRF	Proposed to replace DE406; Extended time coverage; Improved dynamics model.
DE423	2010	1799–2200	ICRF2.0	Included position estimates of MESSENGER spacecraft; Added range and VLBI data from the Venus Express spacecraft.

(Continued)

TABLE 2.3 (CONTINUED)

DE Version	Release Date	Coverage Period	Reference Frame	Notable Features
DE430	2013	1549–2650	ICRF2.0	Improved versions of DE421; Improved planetary ephemerides, updated constants; Included more LLR, VLBI, GRAIL, and other related data for the Moon and planets. Improved lunar accuracy (to submeter level).
DE431	2013	−13,201–17,191	ICRF2.0	A lower precision version of DE430; Extended time coverage.
DE440	2020	1550–2650	ICRF3.0	Improved orbits of Jupiter, Saturn, and Pluto from recent spacecraft observations; Added new astrometric data, including new LLR data since DE 430; Improved dynamical model; Shifted barycenter relative to DE430.
DE441	2020	−1,3000–17,191	ICRF3.0	A lower precision version of DE440; Extended time coverage.

Notes:

1) Between the DE430/431 and DE440/441 series, there are also the DE432/436/438, which are minor update versions of the DE430, and the accuracies of their lunar ephemerides remain the same level or even worse in contrast to the DE 430. Hence, they are not explicitly listed.

2) Unless explicitly mentioned, otherwise, the nutation and libration are typically included in the DE products.

3) VLBI represents the very long baseline interferometry.

2.5 QUANTITATIVE ANALYSIS OF MBSAR'S MOTION

We shall investigate the motion of the MBSAR with respect to the Earth in various lunar cycles in this section. In the preceding section, we have noticed that the lunar cycle is a statistical average concept rather than a deterministic value, implying that fluctuations are present in the lunar cycle due to the orbital perturbations. For example, Figure 2.12 illustrates the fluctuation in the length of the lunar cycle over an 18.6-year interval. As seen, the lunar cycle undergoes a broad spectrum of variations, oscillating around 27.32 days. Hence, when evaluating the MBSAR's motion and associated imaging performance, analyzing each case in a certain cycle, often taken in the spaceborne SAR, is impractical and unattainable.

As illustrative examples, we shall show three scenarios throughout this book: First where the orbital inclination approaches its maximum value, second where the inclination angle is around its minimum value, and third, where the orbital inclination once again approaches its maximum that occurs at around 18.6 years after the first scenario. Each scenario encompasses three consecutive cycles, and we employ the DE440 to provide the orbit and orientation of the Moon in each cycle. Table 2.4 displays epochs in TDB corresponding to these specific cycles.

Note that the trajectory of the MBSAR differs from that of the Moon, and their relative positions are subject to temporal variations in the ECI and ECEF coordinate systems. The rationale behind this phenomenon can be ascribed to that the MBSAR's trajectory is co-affected by the lunar revolution around Earth and the rotational motion of the Moon, while the physical librations introduce a disparity between the rotational and orbital velocities of the Moon. The fact that the locations of the Moon's barycenter and MBSAR are perceived as two individual entities distinguishes MBSAR from spaceborne SAR. To demonstrate this discrepancy in detail, we can define the following position drift vector:

$$\mathbf{R}_{pd1} = \mathbf{R}_{SAR} - \mathbf{R}_{MC} \tag{2.37}$$

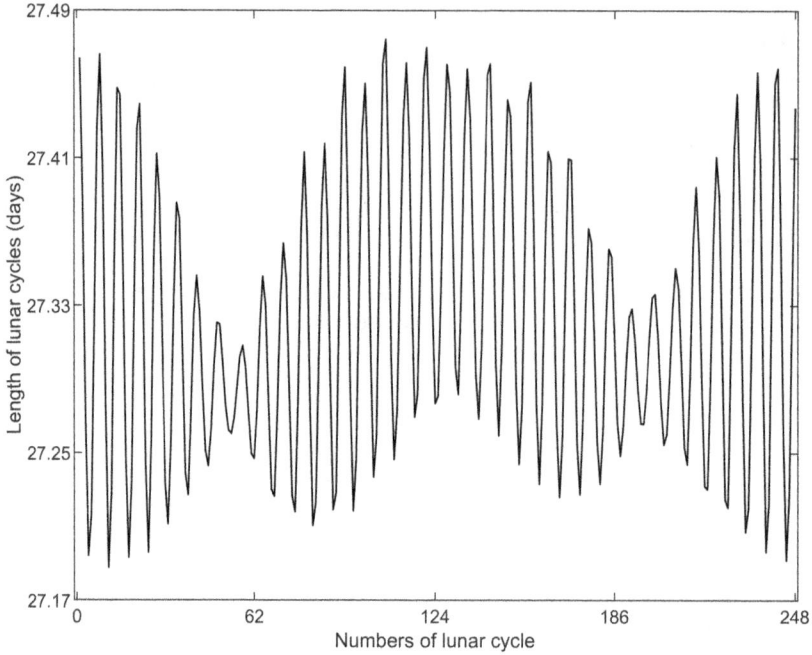

FIGURE 2.12 The length of the lunar cycle versus the number of lunar cycles within a period of 18.6 years; the initial time is the same as that of cycle t_{lc1} in Table 2.4.

TABLE 2.4

The Epochs in TDB for Selected Cycles of the Moon Orbit around Earth

Cycle	Corresponding Epoch in TDB	
T_{lc1}	From 05:05:29.061797, Mar. 11, 2024	to 16:13:49.753201, Apr. 07, 2024
T_{lc2}	From 16:13:49.753201, Apr. 07, 2024	to 01:31: 58.559204, May. 05, 2024
T_{lc3}	From 01:31:58.559204, May. 05, 2024	to 07:59:49.659816, Jun. 01, 2024
T_{lc4}	From 01:06:30.679300, Jun. 19, 2033	to 01:06:30.679300, Jul. 17, 2033
T_{lc5}	From 01:06:30.679300, Jul. 17, 2033	to 09:15:05.518484, Aug. 13, 2033
T_{lc6}	From 09:15:05.518484, Aug. 13, 2033	to 19:12:22.838122, Sep. 09, 2033
T_{lc7}	From 17:16:34.606327, Oct. 26, 2042	to 02:32:55.586200, Nov. 23, 2042
T_{lc8}	From 02:32:55.586200, Nov. 23, 2042	to 08:40:11.758487, Dec. 20, 2042
T_{lc9}	From 08:40:11.758487, Dec. 20, 2042	to 13:06:33.760675, Jan. 16, 2043

Figures 2.13 and 2.14, respectively, depict three components of vector $\mathbf{R}_{pd1}$ in the ECI and ECEF coordinate system across different cycles, with the MBSAR located at $(0°, 0°)$ in the SCR coordinate system. Given that the MBSAR's antenna has a certain height when positioned on the lunar surface, we assume the antenna is installed at a height of 1.5 m above the lunar surface for illustration without loss of generality.

Inspections of Figures 2.13 and 2.14 show that three components of vector $\mathbf{R}_{pd1}$ undergo temporal variations within a solitary lunar cycle, with each component exhibiting distinctive variation patterns. Further, the variations above depend upon lunar cycles, thereby complicating determining MBSAR's position with respect to the Moon's barycenter. When accounting for Earth's rotational motion, as depicted in Figure 2.14, a higher frequency fluctuation becomes apparent in the position difference between the MBSAR and the Moon's barycenter. Such findings highlight the necessity of considering the physical librations of the Moon and its temporal variation in MBSAR.

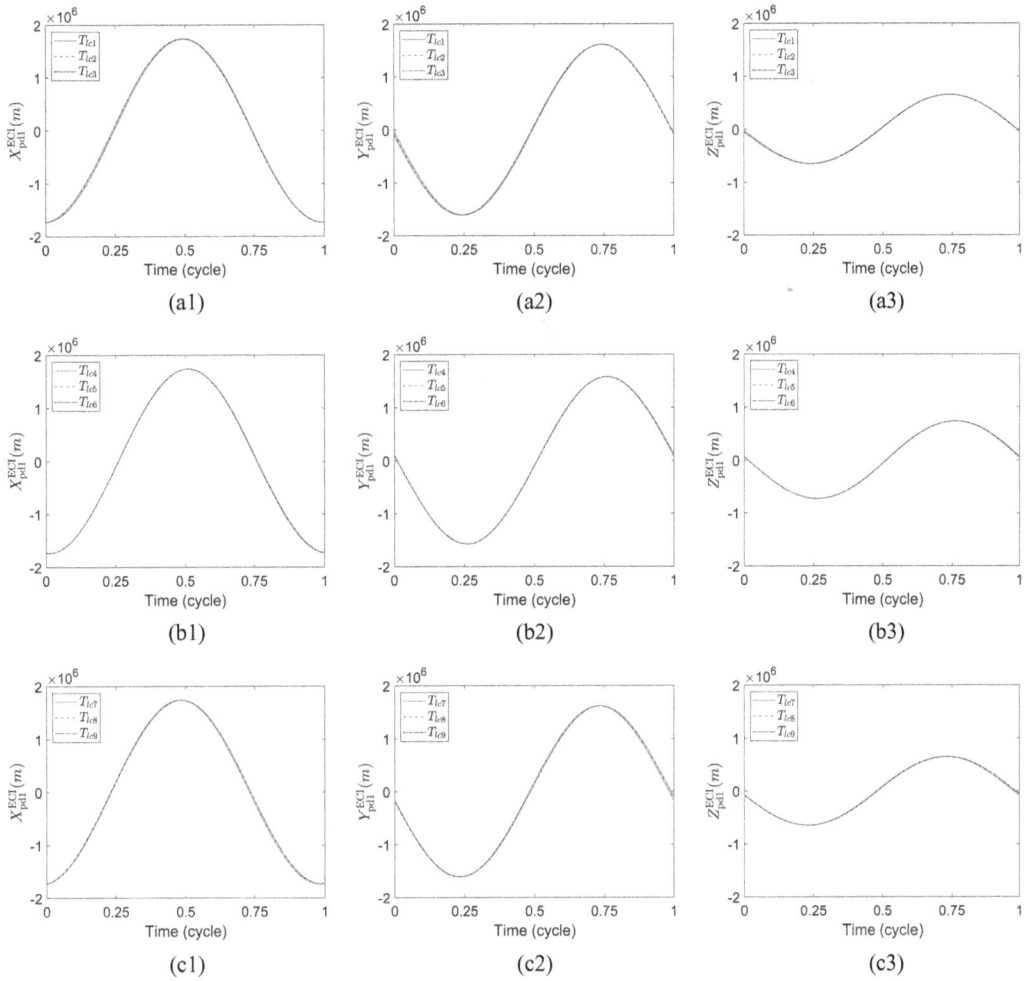

FIGURE 2.13 The temporal variations of the vector $\mathbf{R}_{pd1}$ in the ECI coordinate across the cycles from $T_{lc1} \sim T_{lc9}$, where the 1st (a1–c1), 2nd (a2–c2), and 3rd (a3–c3) columns respectively represent X, Y, and Z components of $\mathbf{R}_{pd1}$, while 1st (a1–a3), 2nd (b1–b3), and 3rd (c1–c3) rows stand for cycles of $T_{lc1} \sim T_{lc3}$, $T_{lc4} \sim T_{lc6}$, and $T_{lc7} \sim T_{lc9}$.

Section 2.4 illustrates that there are temporal variations in lunar orbital elements. Note that the orbit of the MBSAR differs from that of the Moon itself. Hence, we shall also inspect the fluctuations of the MBSAR's orbital elements over certain lunar cycles. Referring to those selected cycles, Figure 2.15 showcases the fluctuations of MBSAR's orbital elements in the ECI coordinate system. As the fluctuations in the semi-major axis, eccentricity, and AOP are implicitly contained within the variations of the Earth-MBSAR distance (the distance from the Earth's center to the MBSAR), we present the variation of the Earth-MBSAR distance across those cycles as an alternative representation for simplicity. Each set of orbital elements exhibits distinct regularities, and their fluctuations also adhere to discernible patterns over time. The complexity of the MBSAR's orbit results in significant variations in system parameters; of particular concern is the azimuthal resolution, as will be treated in Chapter 3.

The orbital perturbations give rise to time-varying orbital elements, which, in turn, lead to continuous temporal variations in the MBSAR's orbit. As examples, Figures 2.16 and 2.17 respectively depict the MBSAR's trajectory in the ECI and ECEF coordinate systems across distinct cycles. It is observed that each cycle corresponds to a distinct spatiotemporally varying trajectory of the Moon.

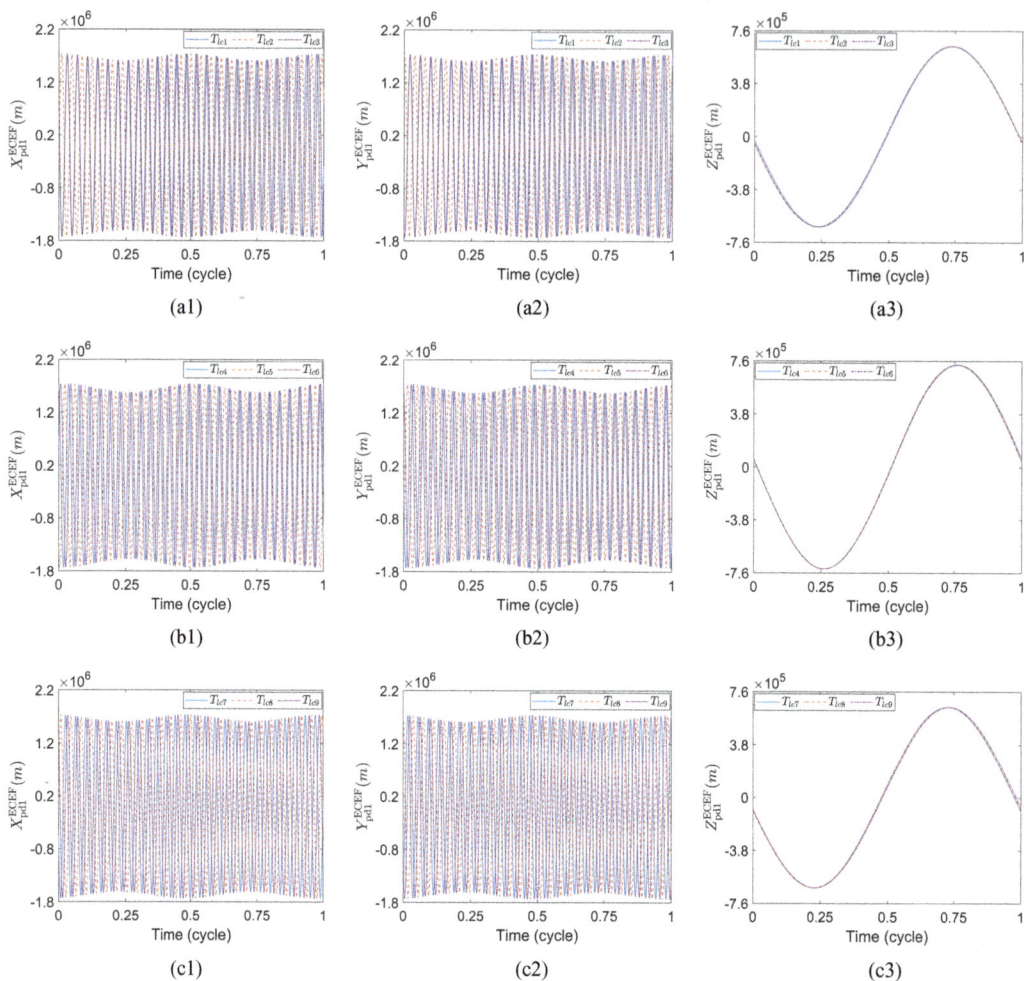

FIGURE 2.14 The temporal variations of the vector $\mathbf{R}_{pd1}$ in the ECEF coordinate across the cycles from $T_{lc1} \sim T_{lc9}$, where the 1st (a1–c1), 2nd (a2–c2), and 3rd (a3–c3) columns respectively represent X, Y, and Z components of $\mathbf{R}_{pd1}$, while 1st (a1–a3), 2nd (b1–b3), and 3rd (c1–c3) rows stand for cycles of $T_{lc1} \sim T_{lc3}$, $T_{lc4} \sim T_{lc6}$, and $T_{lc7} \sim T_{lc9}$.

Correspondingly, there would be fluctuations in MBSAR's inertial motion even over consecutive cycles, differentiating it from spaceborne SAR.

We now examine how the antenna's position on the lunar surface, in combination with lunar surface topography, impacts MBSAR's position relative to Earth. To this end, we can use the MBSAR situated at the site of $(0°, 0°)$ in the SCR coordinate system as a reference, and define another position drift vector in the ECI coordinate system, as:

$$\mathbf{R}_{pd2} = \mathbf{R}_{SAR}\left(\Lambda_{site}, \Phi_{site}\right) - \mathbf{R}_{SAR}\left(\Lambda_{site} = 0°, \Phi_{site} = 0°\right) \quad (2.38)$$

Figure 2.18 compares the vector $\mathbf{R}_{pd2}$ at various site locations on the lunar surface. In this case, three cycles, specifically T_{lc1}, T_{lc4}, and T_{lc7}, are considered.

As can be observed from Figure 2.18, the MBSAR's location in the ECI coordinate system varies when it is placed at different sites on the lunar surface; this variation is observed to be different at various locations along the MBSAR's orbit and across different cycles. Given the MBSAR's Doppler effect

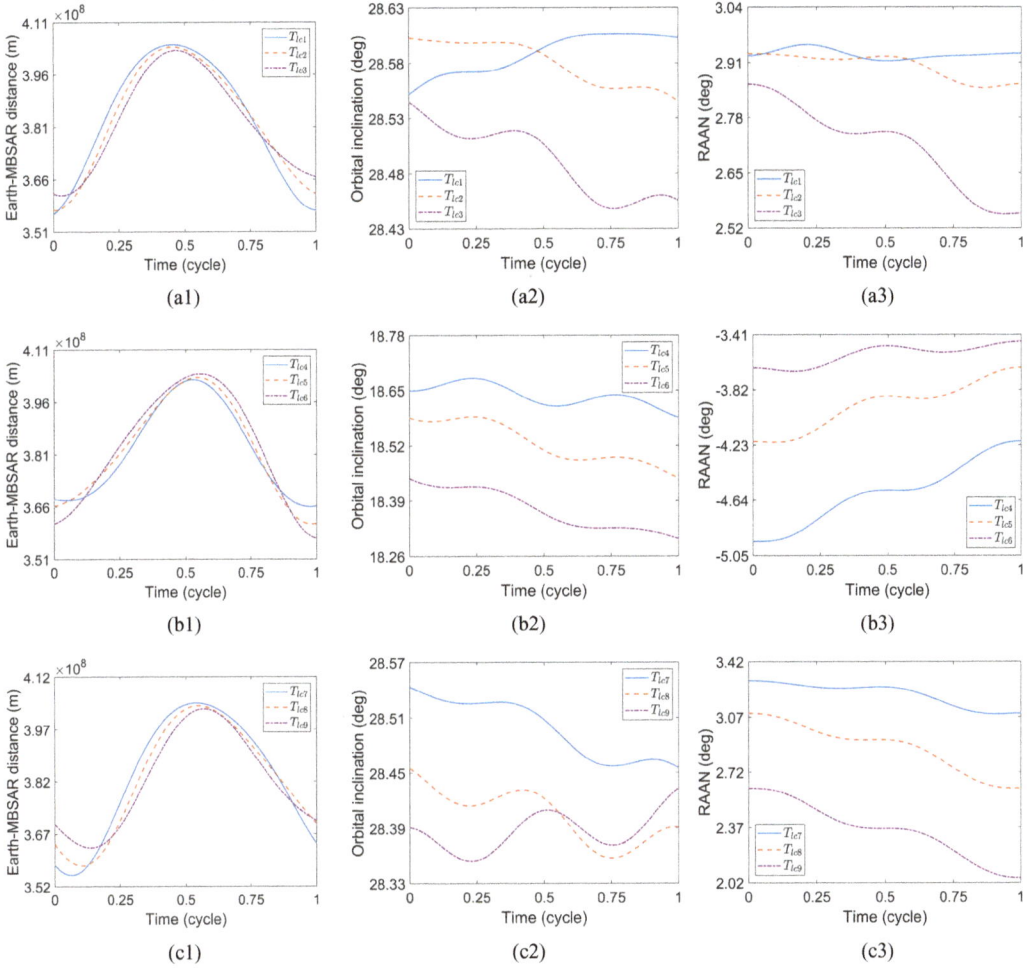

FIGURE 2.15 The variations of orbital elements across the cycles from $T_{lc1} \sim T_{lc9}$, where the 1st (a1–c1), 2nd (a2–c2), and 3rd (a3–c3) columns respectively represent Earth–Moon distance, orbital inclination, and RAAN, while 1st (a1–a3), 2nd (b1–b3), and 3rd (c1–c3) rows stand for cycles of $T_{lc1} \sim T_{lc3}$, $T_{lc4} \sim T_{lc6}$, and $T_{lc7} \sim T_{lc9}$.

and its close correlation to the relative location/motion between the TOI and MBSAR's sites, it can be inferred that the site location affects the MBSAR's imaging capability. Hence, when designing an Earth observation MBSAR system, the site location on the lunar surface and the Moon's rotational motion must be taken into account.

2.6 EARTH OBSERVATION GEOMETRIC MODEL

Now, we establish the observation geometric model that relates the beam pointing of MBSAR to the TOI's position. The SAR position vector in the perifocal frame is related to that in the ACS by:

$$\mathbf{R}_{\mathbf{SAR(PER)}} = \mathbf{R}_z\left(-\nu_{\mathrm{ECI}}\right)\mathbf{M}_{\mathbf{AE}}\left[-\mathbf{R}_{\mathbf{SAR(ACS)}}\right] = -\mathbf{R}_z\left(-\nu_{\mathrm{ECI}}\right)\mathbf{M}_{\mathbf{AE}}\mathbf{R}_{\mathbf{SAR(ACS)}}$$
$$= \left[R_{\mathrm{SAR}}\cos\nu_{\mathrm{ECI}}, R_{\mathrm{SAR}}\sin\nu_{\mathrm{ECI}}, 0\right]^T \tag{2.39}$$

where $\mathbf{R}_{\mathbf{SAR(ACS)}} = [0, 0, R_{\mathrm{SAR}}]^T$ is the distance from the geocenter to MBSAR.

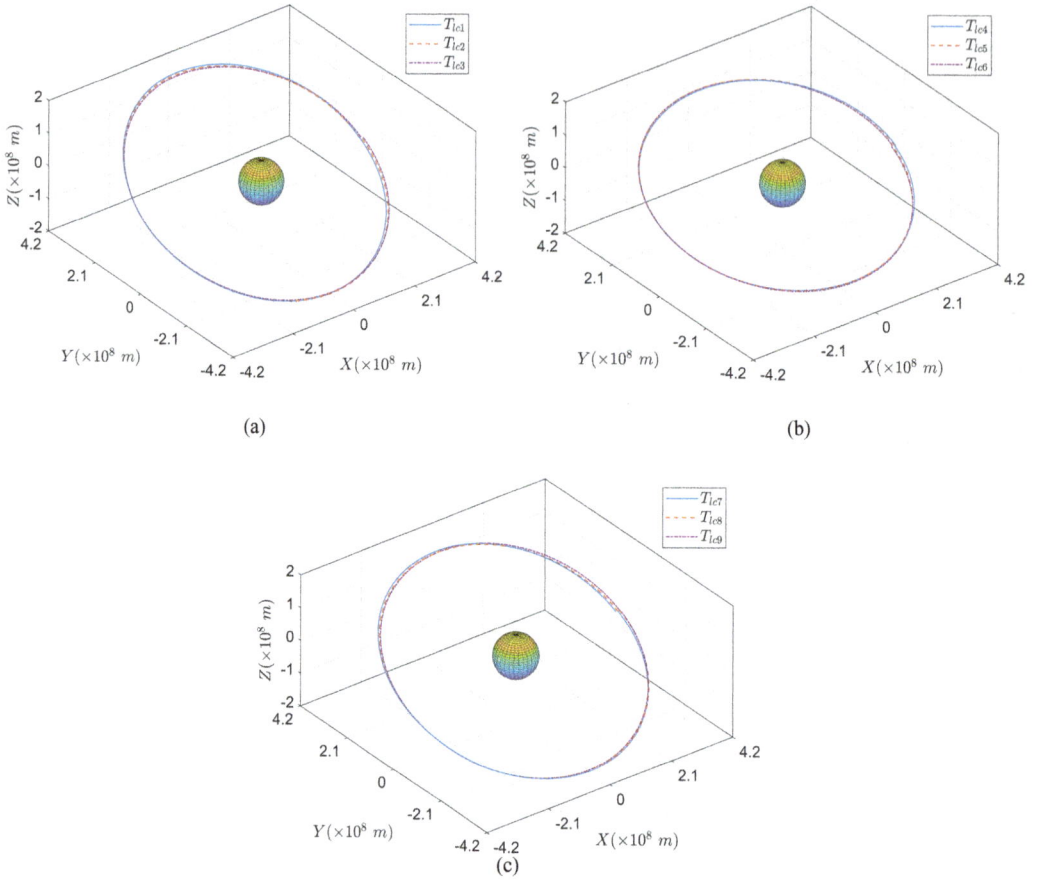

FIGURE 2.16 The trajectory of the MBSAR in the ECI coordinate across various cycles of (a) $T_{lc1} \sim T_{lc3}$, (b) $T_{lc4} \sim T_{lc6}$, (c) $T_{lc7} \sim T_{lc9}$. Note that the Earth is in rotational motion in this case, and its radius is exaggerated tenfold for clarity.

The vector $\mathbf{R}_{SAR(PER)}$ can be transformed into the ECI coordinate system using rotations $-\Omega_{ECI}$ around the z-axis, $-i_{ECI}$ around the x-axis, and $-\omega_{ECI}$ around the z-axis [72], as:

$$\mathbf{R}_{SAR(ECI)} = \mathbf{R}_z\left(-\Omega_{ECI}\right)\mathbf{R}_x\left(-i_{ECI}\right)\mathbf{R}_z\left(-\omega_{ECI}\right)\mathbf{R}_{SAR(PER)} \tag{2.40}$$

where Ω_{ECI}, i_{ECI}, ω_{ECI}, and u_{ECI} ($u_{ECI} = \omega_{ECI} + \nu_{ECI}$) represent the orbital elements of MBSAR in the ECI reference frame. Note that we can also obtain the MBSAR's position from the lunar ephemeris like DE 440 by employing Eq. (2.17).

From Eqs. (2.39) and (2.40), one can derive the following transformation:

$$\mathbf{R}_{SAR(ECI)} = \mathbf{C}_{ECI}^{ACS}\mathbf{M}_{AE}\mathbf{R}_{SAR(ACS)} \tag{2.41}$$

with

$$\begin{aligned}\mathbf{C}_{ECI}^{ACS} &= -\mathbf{R}_z\left(-\Omega_{ECI}\right)\mathbf{R}_x\left(-i_{ECI}\right)\mathbf{R}_z\left(-\omega_{ECI}\right)\mathbf{R}_z\left(-\nu_{ECI}\right)\\ &= -\mathbf{R}_z\left(-\Omega_{ECI}\right)\mathbf{R}_x\left(-i_{ECI}\right)\mathbf{R}_z\left(-u_{ECI}\right)\end{aligned} \tag{2.42}$$

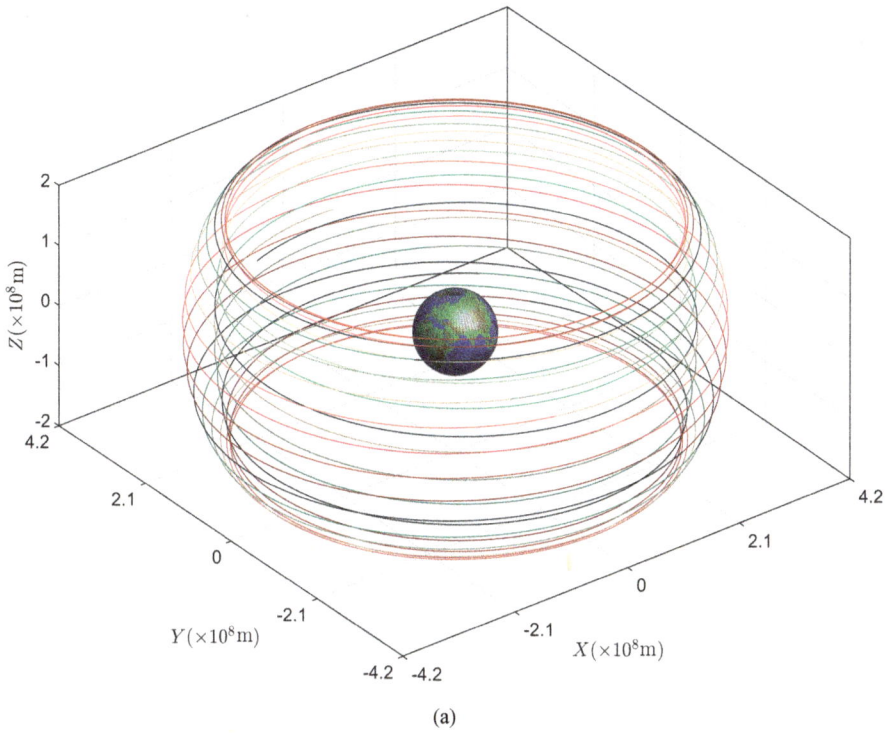

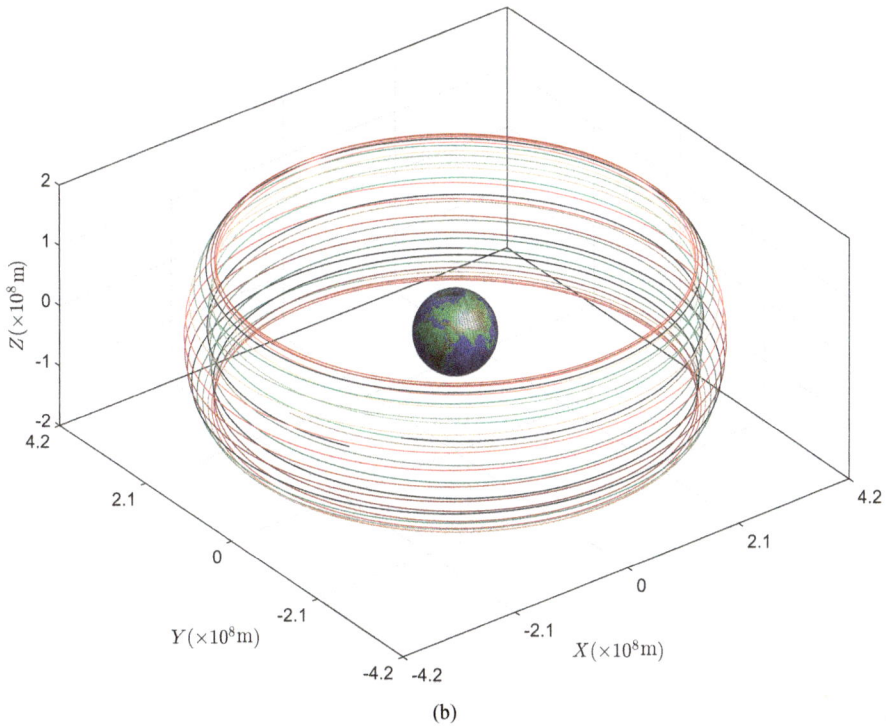

FIGURE 2.17 The trajectory of the MBSAR in the ECEF coordinate within the cycle of (a) T_{lc1}, (b) T_{lc4}. Note that the Earth is fixed in this case, and its radius is exaggerated tenfold for clarity.

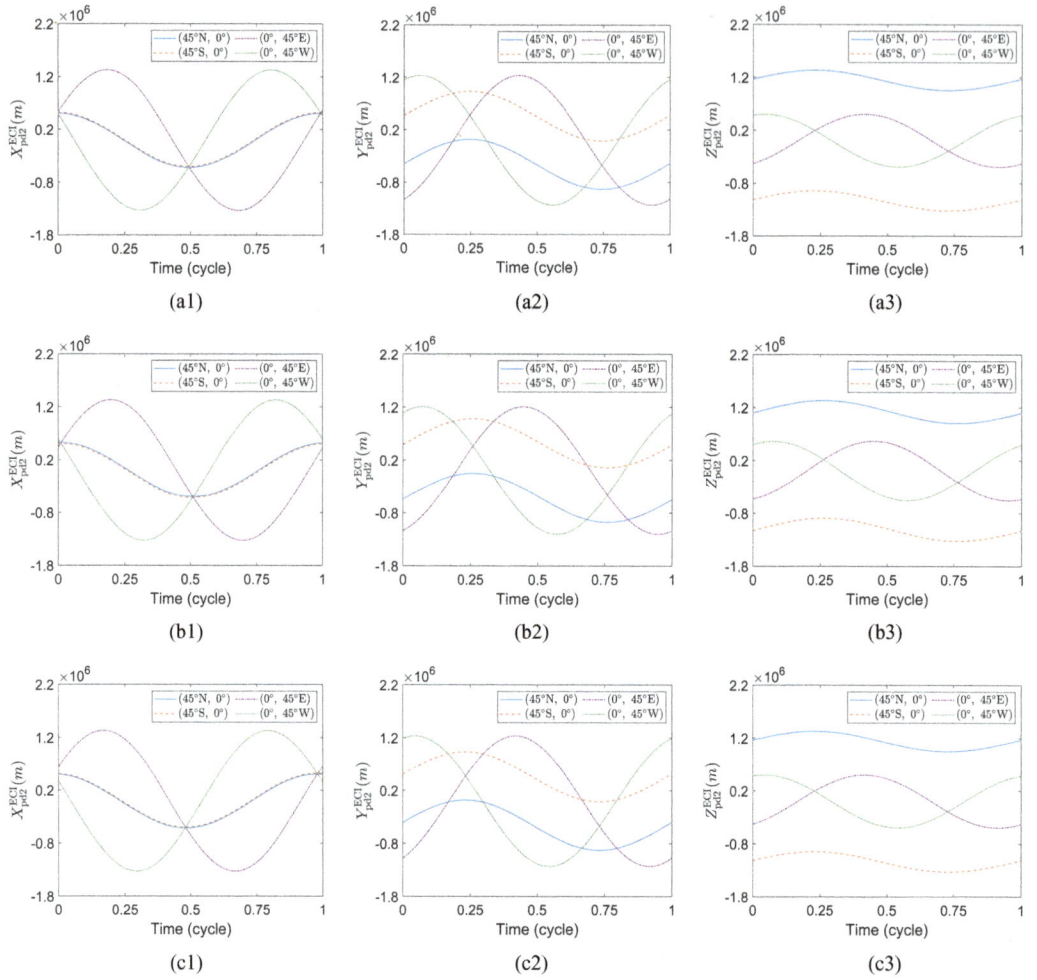

FIGURE 2.18 The temporal variations of the vector $\mathbf{R}_{pd2}$ at different site locations in the ECI coordinate across various cycles, where the 1st (a1–c1), 2nd (a2–c2), and 3rd (a3–c3) columns respectively represent X, Y, and Z components of $\mathbf{R}_{pd2}$, while 1st (a1–a3), 2nd (b1–b3), and 3rd (c1–c3) rows stand for cycles of T_{lc1}, T_{lc4}, and T_{lc7}.

In the ACS, the slant range vector can be expressed as [46, 48]:

$$\mathbf{R}_{ST(ACS)} = R_{ST} \cdot \mathbf{P}_{ACS} \tag{2.43}$$

where R_{ST} is the slant range of the beam centerline to be determined. $\mathbf{P}_{ACS}$, the unit vector aligned with the light of sight (LOS) direction, and is given by:

$$\mathbf{P}_{ACS} = \left[\xi_{sld} \cdot \sin\theta_l \cos\phi, -\xi_{sld} \cdot \sin\theta_l \sin\phi, \cos\theta_l \right]^T \tag{2.44}$$

where ϕ and θ_l are the antenna azimuth angle and look angle of MBSAR; $\xi_{sld} = \pm 1$ is related to side-looking direction, $\xi_{sld} = 1$ denotes the MBSAR is left-looking, while $\xi_{sld} = -1$ denotes the MBSAR is looking from right-hand side.

The position vector of the TOI in the ECI coordinate system is:

$$\mathbf{R}_{TOI(ECI)} = \mathbf{C}_{ECI}^{ACS} \mathbf{M}_{AE} \left(\mathbf{R}_{SAR(ACS)} - R_{ST} \mathbf{P}_{ACS} \right) \tag{2.45}$$

with:

$$\mathbf{M}_{AE} = \begin{bmatrix} 0 & 0 & -1 \\ 1 & 0 & 0 \\ 0 & -1 & 0 \end{bmatrix} \tag{2.46}$$

Next, we shall transform the position vector from ECI to ECEF coordinate systems, the counterpart of MBSAR's position vector in the ECEF coordinate system can be expressed by

$$\mathbf{R}_{SAR(ECEF)} = \mathbf{C}^{ECI}_{ECEF}\mathbf{R}_{SAR(ECI)} \tag{2.47}$$

By Eqs. (2.41) and (2.47), the following expression is given for transformation from ACS to ECEF coordinate systems:

$$\mathbf{R}_{SAR(ECEF)} = \mathbf{C}^{ECI}_{ECEF}\mathbf{C}^{ACS}_{ECI}\mathbf{M}_{AE}\mathbf{R}_{SAR(ACS)} = \mathbf{C}^{ACS}_{ECEF}\mathbf{M}_{AE}\mathbf{R}_{SAR(ACS)} \tag{2.48}$$

By comparing Eq. (2.48) with Eqs. (2.40)–(2.42), it follows that we can update MBSAR's orbital elements (Ω_{ECEF}, i_{ECEF}, and u_{ECEF}) in the $\mathbf{C}^{ACS}_{ECEF}$:

$$\mathbf{C}^{ACS}_{ECEF} = -\mathbf{R}_z\left(-\Omega_{ECEF}\right)\mathbf{R}_x\left(-i_{ECEF}\right)\mathbf{R}_z\left(-u_{ECEF}\right) \tag{2.49}$$

Similarly, the position vector of the TOI in the ECEF coordinate system can be obtained from Eq. (2.45) with revised parameters, as given below.

$$\mathbf{R}_{TOI(ECEF)} = \mathbf{C}^{ECI}_{ECEF}\mathbf{R}_{TOI(ECI)} = \mathbf{C}^{ACS}_{ECEF}\mathbf{M}_{AE}\left(\mathbf{R}_{SAR(ACS)} - R_{ST}\mathbf{P}_{ACS}\right) \tag{2.50}$$

Eq. (2.50) gives three components of TOI's position vector in the ECEF coordinate system:

$$\begin{cases} X_{TOI,ECEF} = (R_{SAR} - R_{ST}\cos\theta_l) \cdot (\cos\Omega_{ECEF}\cos u_{ECEF} - \sin\Omega_{ECEF}\cos i_{ECEF}\sin u_{ECEF}) \\ \qquad - R_{ST}\xi_{sld}\cos\phi\sin\theta_l \cdot (\cos\Omega_{ECEF}\sin u_{ECEF} + \sin\Omega_{ECEF}\cos i_{ECEF}\cos u_{ECEF}) \\ \qquad + R_{ST}\xi_{sld}\sin\phi\sin\theta_l \cdot \sin\Omega_{ECEF}\sin i_{ECEF} \\ Y_{TOI,ECEF} = (R_{SAR} - R_{ST}\cos\theta_l) \cdot (\sin\Omega_{ECEF}\cos u_{ECEF} + \cos\Omega_{ECEF}\cos i_{ECEF}\sin u_{ECEF}) \\ \qquad - R_{ST}\xi_{sld}\cos\phi\sin\theta_l \cdot (\sin\Omega_{ECEF}\sin u_{ECEF} - \cos\Omega_{ECEF}\cos i_{ECEF}\cos u_{ECEF}) \\ \qquad - R_{ST}\xi_{sld}\sin\phi\sin\theta_l \cdot \cos\Omega_{ECEF}\sin i_{ECEF} \\ Z_{TOI,ECEF} = (R_{SAR} - R_{ST}\cos\theta_l) \cdot \sin i_{ECEF}\sin u_{ECEF} + R_{ST}\xi_{sld}\sin\phi\sin\theta_l \cdot \cos i_{ECEF} \\ \qquad + R_{ST}\xi_{sld}\cos\phi\sin\theta_l \cdot \cos u_{ECEF}\sin i_{ECEF} \end{cases} \tag{2.51}$$

The Earth ellipsoid imposes the condition that (h_{TOI} is set to 0 m for simplicity) [73]:

$$\frac{X^2_{TOI,ECEF} + Y^2_{TOI,ECEF}}{R^2_{eq}} + \frac{Z^2_{TOI,ECEF}}{R^2_p} = 1 \tag{2.52}$$

Upon substituting Eq. (2.51) into Eq. (2.52), one arrives at a quadratic equation regarding the slant range R_{ST}:

$$S_1 R^2_{ST} - 2S_2 R_{ST} + S_3 = 0 \tag{2.53}$$

with:

$$\begin{cases} S_1 = R_p^{-2} s_1^2 + R_{eq}^{-2}\left(s_6^2 + s_7^2\right), \\ S_2 = R_{SAR}\left[R_{eq}^{-2}\left(s_2 s_6 + s_3 s_7\right) - R_p^{-2}\left(s_1 \sin i_{ECEF}\sin u_{ECEF}\right)\right], \\ S_3 = R_{SAR}^2\left[R_{eq}^{-2}\left(s_2^2 + f_3^2\right) + R_p^{-2}\sin^2 i_{ECEF}\sin^2 u_{ECEF}\right] - 1. \end{cases} \tag{2.54}$$

and:

$$\begin{cases} s_1 = \xi_{sld}\cos\phi\sin\theta_l \cdot \sin i_{ECEF}\cos u_{ECEF} \\ \qquad + \xi_{sld}\sin\phi\sin\theta_l \cdot \cos i_{ECEF} - \cos\theta_l \cdot \sin i_{ECEF}\sin u_{ECEF}, \\ s_2 = \cos\Omega_{ECEF}\cos u_{ECEF} - \sin\Omega_{ECEF}\cos i_{ECEF}\sin u_{ECEF}, \\ s_3 = \sin\Omega_{ECEF}\cos u_{ECEF} + \cos\Omega_{ECEF}\cos i_{ECEF}\sin u_{ECEF}, \\ s_4 = \cos\Omega_{ECEF}\sin u_{ECEF} + \sin\Omega_{ECEF}\cos i_{ECEF}\cos u_{ECEF}, \\ s_5 = \sin\Omega_{ECEF}\sin u_{ECEF} - \cos\Omega_{ECEF}\cos i_{ECEF}\cos u_{ECEF}, \\ s_6 = \cos\theta_l \cdot s_2 + \xi_{sld}\cos\phi\sin\theta_l \cdot s_4 - \xi_{sld}\sin\phi\sin\theta_l \cdot \sin\Omega_{ECEF}\sin i_{ECEF}, \\ s_7 = \cos\theta_l \cdot s_3 + \xi_{sld}\cos\phi\sin\theta_l \cdot s_5 + \xi_{sld}\sin\phi\sin\theta_l \cdot \cos\Omega_{ECEF}\sin i_{ECEF}. \end{cases} \tag{2.55}$$

At present, quite a few toolkits provide matrices for the transformation between ECI and ECEF coordinate systems. By leveraging such matrices, in conjunction with MBSAR's orbital data acquired through lunar ephemeris, a direct transition from the ACS to ECEF coordinate system can be achieved as:

$$\mathbf{C}_{ECEF}^{ACS} = \mathbf{C}_{ECEF}^{ECI}\mathbf{C}_{ECI}^{ACS} = \begin{bmatrix} C_{11} & C_{12} & C_{13} \\ C_{21} & C_{22} & C_{23} \\ C_{31} & C_{32} & C_{33} \end{bmatrix} \tag{2.56}$$

Once the transformation matrix $\mathbf{C}_{ECEF}^{ACS}$ is obtained, both the location of TOI and polynomial coefficients of Eq. (2.53) can be expressed in terms of $C_{11} \sim C_{33}$, thereby attaining the following representations:

$$\begin{cases} X_{TOI,ECEF} = C_{11}\left(R_{ST}\cos\theta_l - R_{SAR}\right) - C_{12}R_{ST}\xi_{sld}\cos\phi\sin\theta_l - C_{13}R_{ST}\xi_{sld}\sin\phi\sin\theta_l, \\ Y_{TOI,ECEF} = C_{21}\left(R_{ST}\cos\theta_l - R_{SAR}\right) - C_{22}R_{ST}\xi_{sld}\cos\phi\sin\theta_l - C_{23}R_{ST}\xi_{sld}\sin\phi\sin\theta_l, \\ Z_{TOI,ECEF} = C_{31}\left(R_{ST}\cos\theta_l - R_{SAR}\right) - C_{32}R_{ST}\xi_{sld}\cos\phi\sin\theta_l - C_{33}R_{ST}\xi_{sld}\sin\phi\sin\theta_l. \end{cases} \tag{2.57}$$

and:

$$\begin{cases} S_1 = R_p^{-2} s_{10}^2 + R_{eq}^{-2}\left(s_8^2 + s_9^2\right), \\ S_2 = -R_{SAR}\left[R_{eq}^{-2}\left(C_{11}s_8 + C_{21}s_9\right) + R_p^{-2}C_{31}s_{10}\right], \\ S_3 = R_{SAR}^2\left[R_{eq}^{-2}\left(C_{11}^2 + C_{21}^2\right) + R_p^{-2}C_{31}^2\right] - 1. \end{cases} \tag{2.58}$$

with:

$$\begin{cases} s_8 = -C_{11}\cos\theta_l + C_{12}\xi_{sld}\cos\phi\sin\theta_l + C_{13}\xi_{sld}\sin\phi\sin\theta_l, \\ s_9 = -C_{21}\cos\theta_l + C_{22}\xi_{sld}\cos\phi\sin\theta_l + C_{23}\xi_{sld}\sin\phi\sin\theta_l, \\ s_{10} = -C_{31}\cos\theta_l + C_{32}\xi_{sld}\cos\phi\sin\theta_l + C_{33}\xi_{sld}\sin\phi\sin\theta_l. \end{cases} \tag{2.59}$$

Solving Eq. (2.53), one can determine the slant range using the following equations (while discarding another solution that is deemed meaningless):

$$R_{ST} = \frac{S_2 - \sqrt{S_2^2 - S_1 S_3}}{S_1} \tag{2.60}$$

In the case where the spherical Earth assumption is made, the slant range of the beam centerline can be simplified to:

$$R_{ST} \approx R_{SAR} \cos\theta_l - \sqrt{R_E^2 - R_{SAR}^2 \sin^2\theta_l} \tag{2.61}$$

where R_E denotes the average radius of the Earth, and an authalic radius of 6371.0 km is used here for numerical illustration.

By substituting the obtained slant range into Eqs. (2.45) and (2.48), one can determine the position vectors of TOI in the ECI and ECEF coordinate systems, respectively. The associated velocity vectors can be obtained through a similar methodology in conjunction with Eqs. (2.33) and (2.35). For simplicity, the detailed procedure is not expounded in this context. It follows that the geocentric latitude and longitude of the TOI in the ECEF coordinate system can be obtained as follows:

$$\begin{cases} \Phi_{TOI,ECEF} = \tan^{-1}\left(\dfrac{Z_{TOI,ECEF}}{\sqrt{X_{TOI,ECEF}^2 + Y_{TOI,ECEF}^2}} \right) \\[3mm] \Lambda_{TOI,ECEF} = \sin^{-1}\left(\dfrac{Y_{TOI,ECEF}}{\sqrt{X_{TOI,ECEF}^2 + Y_{TOI,ECEF}^2}} \right) \end{cases} \tag{2.62}$$

Regarding the case of reference Earth ellipsoid, there is no distinction between geocentric and geodetic longitude, but a disparity does emerge in the latitude dimension. Therefore, it may be necessary to employ the geodetic latitude, which can be defined by [74]:

$$B_{TOI,ECEF} = \tan^{-1}\left[\frac{\tan\Phi_{TOI,ECEF}}{(1-\alpha_f)^2} \right] \tag{2.63}$$

where α_f is the flattening of the reference Earth ellipsoid.

In the following, Figure 2.19 presents the slant range of the beam centerline versus the look angle of the MBSAR when it is located at the AOLs of 90°, 180°, and 270°. In this case, the spherical Earth and GRS 80 Earth ellipsoidal models are considered. Additionally, Figure 2.19 depicts the discrepancies in the slant range between the two models.

Using the same set of parameters, Figure 2.20 compares the geocentric coordinates of TOI based on the spherical Earth model and that based on the GRS 80 Earth ellipsoid. In both instances, the MBSAR is positioned at (0°, 0°) in the SCR coordinate system, with the beam center pointing toward the zero-Doppler plane. Besides, the AOLs that describe the MBSAR's position are defined within the ECI coordinate system.

According to Figures 2.19 and 2.20, there exists a minor disparity in the TOI's coordinates relative to the entire Earth when comparing the ellipsoidal and spherical models, suggesting that the

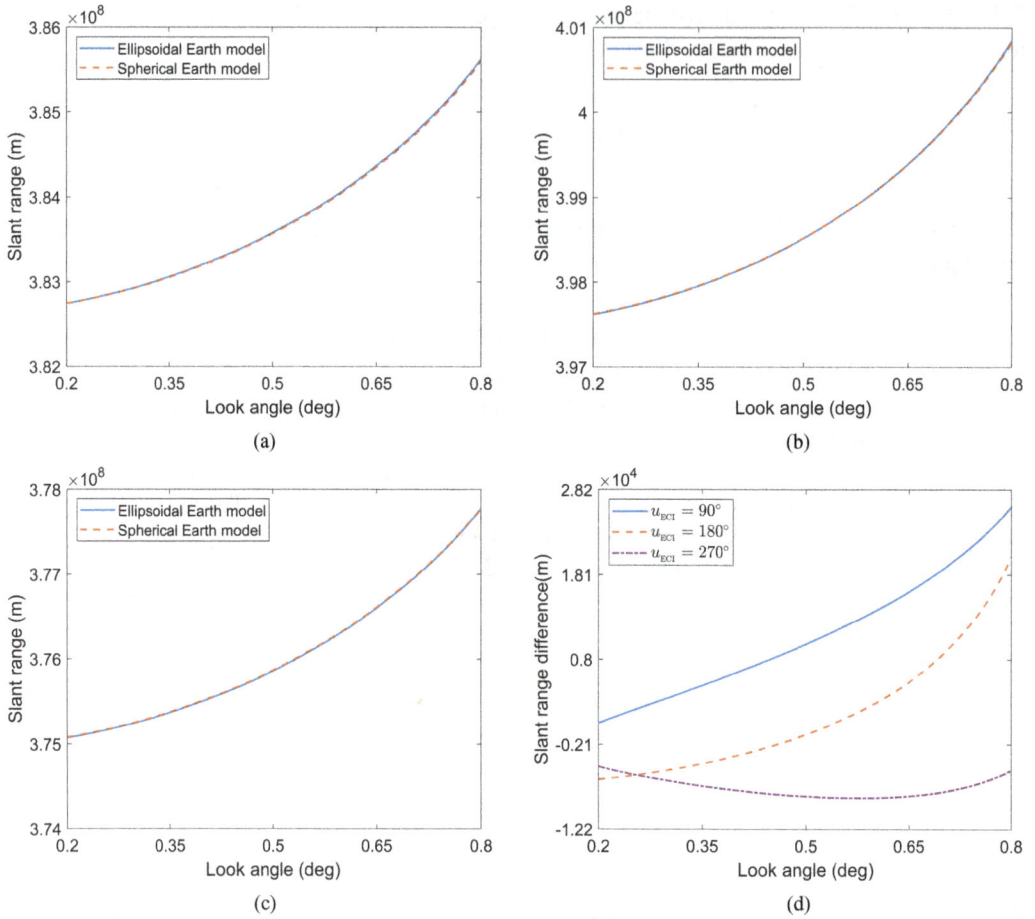

FIGURE 2.19 In the scenarios of spherical Earth and GRS 80 Earth ellipsoid, the slant range of beam centerline versus the look angle when the MBSAR is located at the AOLs of (a) 90°, (b) 180°, (c) 270°. The difference between the slant range obtained by ellipsoidal Earth and that yielded by spherical Earth.

influence of the Earth's ellipsoidal shape on the TOI's relative position is small. Consequently, in the studies regarding the relative TOI's position, such as swath width and coverage analysis, the impact of Earth's ellipsoid is minimal, permitting the adoption of a spherical Earth assumption in such investigations. In the field of signal processing, however, the variability in the slant range, resulting from Earth's ellipsoidal shape, may give rise to phase errors in the signal. Such phase errors potentially influence the imaging performance of MBSAR. Therefore, when processing signals in MBSAR, due consideration must be given to the shape of the Earth's ellipsoid.

At present, we have constructed the observation geometric model for MBSAR's Earth observations, which relates the position of the TOI/MBSAR to the antenna beam pointing vector, orbital elements, Earth orientation, and Earth's shape. The geometric model plays a pivotal role in ensuring precise TOI geolocation in interpreting the SAR data and extracting information about the Earth's surface once the MBSAR becomes operational. We can make use of the geometric model to configure the MBSAR and set criteria for its configuration, as will be discussed in the subsequent chapters.

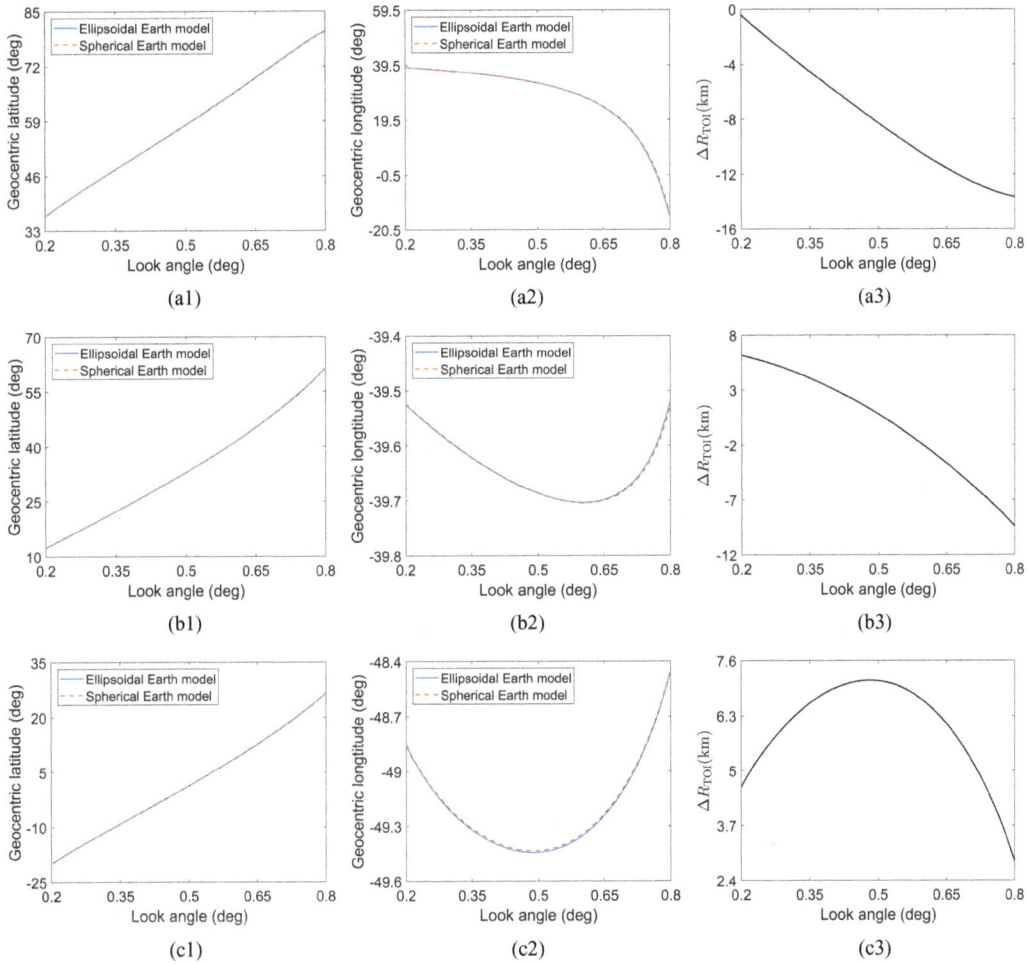

FIGURE 2.20 In the scenarios of spherical Earth and GRS 80 Earth ellipsoid, the geocentric coordinates of TOI versus the look angle of MBSAR, where the 1st (a1–a3), 2nd (b1–b3), and 3rd (c1–c3) rows stand for AOLs of 90°, 180°, and 270°, while 1st (a1–c1) and 2nd (a2–c2) columns respectively represent latitude and longitude of TOI, and the 3rd (a3–c3) column indicate the difference of Earth's radius (corresponding to TOI) in the Earth ellipsoid with respect to the authalic radius (symbolized as ΔR_{TOI}).

REFERENCES

[1] Z. Xu, K. S. Chen, G. Liu, and H. Guo, "Spatiotemporal Coverage of a Moon-Based Synthetic Aperture Radar: Theoretical Analyses and Numerical Simulations," *IEEE Transactions on Geoscience and Remote Sensing*, vol. 58, no. 12, pp. 8735–8750, 2020.

[2] H. Ye, H. Guo, G. Liu, and Y. Ren, "Observation Scope and Spatial Coverage Analysis for Earth Observation from a Moon-Based Platform," *International Journal of Remote Sensing*, vol. 39, no. 18, pp. 5809–5833, Oct. 2017.

[3] H. Ye, W. Zheng, H. Guo, and G. Liu, "Effects of Ellipsoidal Earth Model on Estimating the Sensitivity of Moon-Based Outgoing Longwave Radiation Measurements," *IEEE Geoscience and Remote Sensing Letters*, vol. 19, pp. 1–5, Jun. 2022, doi: 10.1109/LGRS.2021.3084219

[4] A. Moccia and A. Renga, "Synthetic Aperture Radar for Earth Observation from a Lunar Base: Performance and Potential Applications," *IEEE Transactions on Aerospace and Electronic Systems*, vol. 46, no. 3, pp. 1034–1051, Jul. 2010.

[5] M. Chapront-Touzé and J. Chapront, *Lunar Tables and Programs from 4000 BC to AD 8000*. Richmond: Willmann-Bell, Inc., 1991.

[6] A. W. Irwin and T. Fukushima, "A Numerical Time Ephemeris of the Earth," *Astronomy and Astrophysics*, vol. 348, pp. 642–652, Jul. 1999.

[7] O. Montenbruck, E. Gill, and F. Lutze, *Satellite Orbits: Models, Methods, and Applications*. Berlin: Springer, 2000.

[8] M. H. Soffel and M. H. Soffel, *Relativity in Astrometry, Celestial Mechanics and Geodesy*. Berlin: Springer, 1989.

[9] D. D. McCarthy and P. K. Seidelmann, *Time: From Earth Rotation to Atomic Physics*. Cambridge: Cambridge Univ. Press, 2018.

[10] D. W. Allan, J. E. Gray, and H. E. Machlan, "The National Bureau of Standards Atomic Time Scales: Generation, Dissemination, Stability, and Accuracy," *IEEE Transactions on Instrumentation and Measurement*, vol. 21, no. 4, pp. 388–391, Nov. 1972.

[11] B. N. Taylor and A. Thompson, *The International System of Units (SI)*. Washington, DC, USA: National Institute of Standards and Technology, 2008.

[12] J. Kovalevsky, I. I. Mueller, and B. Kolaczek, *Reference Frames: in Astronomy and Geophysics*. Dordrecht, Netherland: Kluwer Academic Publishers, 1989.

[13] W. Lewandowski and C. Thomas, "GPS Time Transfer," *Proceedings of the IEEE*, vol. 79, no. 7, pp. 991–1000, Jul. 1991.

[14] S. Aoki, et al., "The New Definition of Universal Time," *Astronomy and Astrophysics*, vol. 105, pp. 359–361, Jan. 1982.

[15] R. A. Nelson, et al., "The Leap Second: Its History and Possible Future," *Metrologia*, vol. 38, no. 6, pp. 509–529, Dec. 2001.

[16] R. L. Beard, "Role of the ITU-R in Time Scale Definition and Dissemination," *Metrologia*, vol. 48, no. 4, pp. S125–S131, Jul. 2011.

[17] K. S. Chen, *Principles of Synthetic Aperture Radar: A System Simulation Approach*. Boca Raton: CRC Press, 2015.

[18] S. E. Urban and P. K. Seidelmann, *Explanatory Supplement to the Astronomical Almanac*, 3rd ed. University Science Books, 2013.

[19] P. Seidelmann and T. Fukushima, "Why New Time Scales?," *Astronomy and Astrophysics*, vol. 265, pp. 833–838, Oct. 1992.

[20] B. Guinot and P. K. Seidelmann, "Time Scales-Their History, Definition and Interpretation," *Astronomy and Astrophysics*, vol. 194, pp. 304–308, Apr. 1988.

[21] M. Soffel, et al., "The IAU 2000 Resolutions for Astrometry, Celestial Mechanics, and Metrology in the Relativistic Framework: Explanatory Supplement," *The Astronomical Journal*, vol. 126, no. 6, p. 2687, Dec. 2003.

[22] W. R. Dick and D. Thaller, "IERS Annual Report 2018," Verlag des Bundesamts für Kartographie und Geodäsie, Frankfurt am Main, 207, Feb. 2020. [Online]. Available: https://www.iers.org/SharedDocs/Publikationen/EN/IERS/Publications/ar/ar2018/ar2018.pdf

[23] F. Espenak and J. Meeus, "Five Millennium Canon of Solar Eclipses: -1999 to +3000 (2000 BCE to 3000 CE)," NASA Center for AeroSpace Information, Hanover, MD, NASA Technical Publication TP-2006-214141, Oct. 2006. [Online]. Available: https://ntrs.nasa.gov/citations/20070003587

[24] G. Petit and B. Luzum, "IERS Conventions (2010)," Verlag des Bundesamts für Kartographie und Geodäsie, Frankfurt am Main, Germany, IERS Technical Note No. 36, Dec. 2010. [Online]. Available: https://iers-conventions.obspm.fr/packaged_versions/iersconventions_v1_0_0.tar.gz

[25] H. Ye, et al., "Looking Vector Direction Analysis for the Moon-Based Earth Observation Optical Sensor," *IEEE Journal of Selected Topics in Applied Earth Observations and Remote Sensing*, vol. 11, no. 11, pp. 4488–4499, Nov. 2018.

[26] H. Ye, W. Zheng, H. Guo, and G. Liu, "Effects of Temporal Sampling Interval on the Moon-Based Earth Observation Geometry," *IEEE Journal of Selected Topics in Applied Earth Observations and Remote Sensing*, vol. 13, pp. 4016–4029, 2020.

[27] G. Liu, H. Guo, and R. F. Hanssen, "Characteristics Analysis of Moon-Based Earth Observation under the Ellipsoid Model," *International Journal of Remote Sensing*, vol. 41, no. 23, pp. 9121–9139, Oct. 2020, doi: 10.1080/01431161.2020.1797220

[28] LOLA Science Team. *Moon LRO LOLA DEM 118m v1*. Accessed: Sep. 08, 2021. [Online]. Available: https://planetarymaps.usgs.gov/mosaic/Lunar_LRO_LOLA_Global_LDEM_118m_Mar2014.tif

[29] Z. Xu and K. S. Chen, "On Signal Modeling of Moon-Based Synthetic Aperture Radar (SAR) Imaging of Earth," *Remote Sensing*, vol. 10, no. 3, p. 486, Mar. 2018.

[30] J. Huang, et al., "Spatio-Temporal Characteristics for Moon-Based Earth Observations," *Remote Sensing*, vol. 12, no. 17, p. 2848, 2020.

[31] H. Ye, H. Guo, G. Liu, and Y. Ren, "Observation Duration Analysis for Earth Surface Features from a Moon-Based Platform," *Advances in Space Research*, vol. 62, no. 2, pp. 274–287, Jul. 2018.

[32] Z. Xu, K. S. Chen, and G. Q. Zhou, "Effects of the Earth's Irregular Rotation on the Moon-Based Synthetic Aperture Radar Imaging," *IEEE ACCESS*, vol. 7, pp. 155014–155027, Oct. 2019.

[33] D. E. Smith, M. T. Zuber, G. A. Neumann, and F. G. Lemoine, "Topography of the Moon from the Clementine lidar," *Journal of Geophysical Research: Planets*, vol. 102, no. E1, pp. 1591–1611, Jan. 1997.

[34] H. Araki, et al., "Lunar Global Shape and Polar Topography Derived from Kaguya-LALT Laser Altimetry," *Science*, vol. 323, no. 5916, pp. 897–900, Feb. 2009.

[35] M. Laurenti, et al., "Reference Frames Analysis for Lunar Radio Navigation System," *Proc. 2022 International Technical Meeting of the Institute of Navigation*, Long Beach, California, pp. 606–615, Jan. 2022.

[36] W. M. Folkner, et al., "The Planetary and Lunar Ephemerides DE430 and DE431," *Interplanetary Network Progress Report*, vol. 196, no. 1, pp. 1–81, Feb. 2014.

[37] M. Soffel and R. Langhans, *Space-Time Reference Systems*. Berlin: Springer, 2013.

[38] E. F. Arias, P. Charlot, M. Feissel, and J.-F. Lestrade, "The Extragalactic Reference System of the International Earth Rotation Service, ICRS," *Astronomy and Astrophysics*, vol. 303, pp. 604–608, Oct. 1995.

[39] C. Ma, et al., "The International Celestial Reference Frame as Realized by Very Long Baseline Interferometry," *The Astronomical Journal*, vol. 116, no. 1, p. 516, Jul. 1998.

[40] C. Ma, et al., "The Second Realization of the International Celestial Reference Frame by Very Long Baseline Interferometry," Verlag des Bundesamts für Kartographie und Geodäsie, Frankfurt am Main, Germany, IERS Technical Note No. 35, 2009. [Online]. Available: http://130.79.128.5/ftp/cati/i/323/tn35.pdf

[41] Gaia Collaboration, et al., "Gaia Early Data Release 3: The Celestial Reference Frame (Gaia-CRF3)," *Astronomy and Astrophysics*, vol. 667, p. A148, Nov. 2022.

[42] Z. Altamimi, P. Sillard, and C. Boucher, "ITRF2000: A New Release of the International Terrestrial Reference Frame for Earth Science Applications," *Journal of Geophysical Research: Solid Earth*, vol. 107, no. B10, pp. ETG 2-1–ETG 2-19, Oct. 2002.

[43] H. Moritz, "Geodetic Reference System 1980," *Bulletin géodésique*, vol. 54, no. 3, pp. 395–405, Sep. 1980.

[44] M. Kumar, "World Geodetic System 1984: A Modern and Accurate Global Reference Frame," *Marine Geodesy*, vol. 12, no. 2, pp. 117–126, Jan. 1988.

[45] M. Hooijberg, *Geometrical Geodesy: Using Information and Computer Technology*. Berlin, Germany: Springer, 2008.

[46] Z. Xu, K. S. Chen, and G. Q. Zhou, "Zero-Doppler Centroid Steering for the Moon-Based Synthetic Aperture Radar: A Theoretical Analysis," *IEEE Geoscience and Remote Sensing Letters*, vol. 17, no. 7, Jul. 2020.

[47] Z. Xu, K. S. Chen, and H. Guo, "Doppler Estimation with "Non-Stop-and-Go" Assumption in Moon-Based SAR Imaging," *Proc. 2018 IEEE International Geoscience and Remote Sensing Symposium*, Valencia, Spain, pp. 7809–7812, Jul. 2018.

[48] Z. Xu, K. S. Chen, and H. Guo, "On Azimuthal Resolution of the Lunar-Based SAR under the Orbital Perturbation Effects," *IEEE Transactions on Geoscience and Remote Sensing*, vol. 61, Apr. 2023, doi: 10.1109/TGRS.2023.3266548

[49] R. S. Park, W. M. Folkner, J. G. Williams, and D. H. Boggs, "The JPL Planetary and Lunar Ephemerides DE440 and DE441," *The Astronomical Journal*, vol. 161, no. 3, p. 105, Feb. 2021, doi: 10.3847/1538-3881/abd414

[50] J. Vondrák, N. Capitaine, and P. Wallace, "New Precession Expressions, Valid for Long Time Intervals," *Astronomy and Astrophysics*, vol. 534, p. A22, Oct. 2011.

[51] P. Seidelmann, "1980 IAU Theory of Nutation: The Final Report of the IAU Working Group on Nutation," *Celestial mechanics*, vol. 27, no. 1, pp. 79–106, May. 1982.

[52] P. Wallace and N. Capitaine, "Precession-Nutation Procedures Consistent with IAU 2006 Resolutions," *Astronomy and Astrophysics*, vol. 464, no. 2, pp. 793–793, Dec. 2007.

[53] N. Capitaine, J. Chapront, S. Lambert, and P. Wallace, "Expressions for the Celestial Intermediate Pole and Celestial Ephemeris Origin consistent with the IAU 2000A precession-nutation model," *Astronomy and Astrophysics*, vol. 400, no. 3, pp. 1145–1154, Jan. 2003.

[54] A. Moreira, et al., "A Tutorial on Synthetic Aperture Radar," *IEEE Geoscience and Remote Sensing Magazine*, vol. 1, no. 1, pp. 6–43, Mar. 2013.

[55] Z. Xu and K. S. Chen, "Effects of the Earth's Curvature and Lunar Revolution on the Imaging Performance of the Moon-Based Synthetic Aperture Radar," *IEEE Transactions on Geoscience and Remote Sensing*, vol. 57, no. 8, pp. 5868–5882, Mar. 2019.

[56] J. Meeus, *Mathematical Astronomy Morsels*. Richmond, USA: Willmann-Bell, 1997.

[57] M. C. Gutzwiller, "Moon-Earth-Sun: The Oldest Three-Body Problem," *Review of Modern Physics*, vol. 70, no. 2, pp. 589–639, Apr. 1998.

[58] J.-L. Simon, et al., "Numerical Expressions for Precession Formulae and Mean Elements for the Moon and the Planets," *Astronomy and Astrophysics*, vol. 282, pp. 663–683, Feb. 1994.

[59] C. D. Keeling and T. P. Whorf, "The 1,800-Year Oceanic Tidal Cycle: A Possible Cause of Rapid Climate Change," *Proceedings of the National Academy of Sciences*, vol. 97, no. 8, pp. 3814–3819, Mar. 2000.

[60] V. I. Arnol'd, V. V. Kozlov, A. I. Neishtadt, and I. Iacob, *Mathematical Aspects of Classical and Celestial Mechanics*. Berlin: Springer, 2006.

[61] S. J. Peale, "Generalized Cassini's Laws," *Astronomical Journal*, vol. 74, p. 483, 1969.

[62] V. V. Beletsky, *Essays on the Motion of Celestial Bodies*. Switzerland: Springer, 2001.

[63] D. H. Eckhardt, "Theory of the Libration of the Moon," *The Moon and the Planets*, vol. 25, no. 1, pp. 3–49, Aug. 1981.

[64] N. Rambaux and J. Williams, "The Moon's Physical Librations and Determination of Their Free Modes," *Celestial Mechanics and Dynamical Astronomy*, vol. 109, pp. 85–100, Oct. 2011.

[65] J. O. Dickey, et al., "Lunar Laser Ranging: A Continuing Legacy of the Apollo Program," *Science*, vol. 265, no. 5171, pp. 482–490, Jul. 1994.

[66] J. D. Mulholland and P. J. Shelus, "Improvement of the Numerical Lunar Ephemeris with Laser Ranging Data," *The Moon*, vol. 8, no. 4, pp. 532–538, Oct. 1973, doi: 10.1007/BF00562077

[67] J. Chapront and G. Francou, "The Lunar Theory ELP Revisited. Introduction of New Planetary Perturbations," *Astronomy and Astrophysics*, vol. 404, no. 2, pp. 735–742, Sep. 2003.

[68] J. Chapront and M. Chapront-Touzé, "Lunar Motion: Theory and Observations," *Celestial Mechanics and Dynamical Astronomy*, vol. 66, no. 1, pp. 31–38, Mar. 1996.

[69] R. G. Cionco and D. A. Pavlov, "Solar Barycentric Dynamics from a New Solar-Planetary Ephemeris," *Astronomy & Astrophysics*, vol. 615, p. A153, Jul. 2018.

[70] Z. Xu, K. S. Chen, and G. Liu, "On Orbital Determination of the Lunar-Based SAR under Apsidal Precession," *IEEE Transactions on Geoscience and Remote Sensing*, vol. 60, May. 2022, doi: 10.1109/TGRS.2022.3176836

[71] Jet Propulsion Laboratory. *JPL Planetary and Lunar Ephemerides*. Accessed: Sep. 01, 2022. [Online]. Available: https://ssd.jpl.nasa.gov/planets/eph_export.html

[72] G. B. Arfken, H. J. Weber, and F. E. Harris, *Mathematical Methods for Physicists: A Comprehensive Guide*. Oxford, UK: Academic press, 2011.

[73] J. P. Snyder, *Flattening the Earth: Two Thousand Years of Map Projections*. Chicago: Univ. of Chicago Press, 1997.

[74] J. L. Awange, E. W. Grafarend, B. Paláncz, and P. Zaletnyik, *Algebraic Geodesy and Geoinformatics*. Berlin: Springer, 2010.

3 Generic Parameters in the MBSAR System

3.1 INTRODUCTION

The Moon-Based SAR (MBSAR) has emerged as a promising advanced SAR system that produces high-resolution and extensive coverage SAR images simultaneously, surpassing the limitations of the spaceborne SAR [1, 2]. To expound upon the benefits of MBSAR for Earth observation, we assess the MBSAR's system performance quantitatively through various parameters correlated with spatial resolution, signal-to-noise ratio (SNR), and pulse repetition frequency (PRF). In a SAR image, spatial resolution refers to the ability to distinguish the minutest detectable object or minimum distance between two targets that can be resolved [3]. In the MBSAR, there is little difference in range resolution compared to the spaceborne SAR. In terms of azimuthal resolution, however, the MBSAR exhibits unique spatiotemporal variations with respect to spaceborne SAR [4]. Such a feature can be ascribed to the observation geometry of MBSAR, which deserves a close look.

The SNR is a basic metric for assessing the radar image quality, a higher SNR is desirable to improve the target detection rate and reduce the false alarm rate [5]. Besides, The SNR is related to the radiometric resolution, expressed as the Noise Equivalent Sigma Zero (NESZ) [6]. Regarding the PRF, it is a critical parameter in pulse radar: Selecting PRF that balances the spatial resolution and swath width is imperative to obtain a high-quality SAR image with extensive coverage [7]. Other generic parameters that are important to the MBSAR will also be discussed. We start with the fundamental properties of the MBSAR associated with the generic parameters based on its geometric model. In particular, the azimuthal resolution is discussed considering influences from the Earth's rotational motion and MBSAR's inertial motion.

3.2 SPATIAL RESOLUTION

Similar to the spaceborne SAR, the range resolution of an MBSAR with the chirp signal is given by [8, 9]:

$$\rho_r = 0.886 \frac{c}{2B_r} = 0.886 \frac{c}{2T_p K_r} \tag{3.1}$$

where:
  c is the propagation velocity of electromagnetic wave,
  B_r is signal bandwidth,
  K_r is the frequency modulation (FM) rate of the chirp signal, and
  T_p is the pulse duration.

From a system design point of view, selecting the relevant parameters in (3.1) for MBSAR requires deliberated efforts due to the complex observation geometry from the Moon to the Earth's surface. However, it is beyond the scope of this book; we leave it as an avenue for future study in the SAR community.

DOI: 10.1201/9781003308430-3

3.2.1 AZIMUTHAL RESOLUTION IN SPACEBORNE SAR

For a spaceborne SAR, the velocity that determines the Doppler bandwidth is not the beam-crossing velocity but an equivalent velocity [10]. Hence, in the spaceborne SAR, the Doppler bandwidth can be expressed in terms of equivalent velocity and squint angle:

$$B_D = 0.886 \frac{2}{\ell_a} \frac{V_r^2 \cos\theta_r}{V_g} \tag{3.2}$$

where:
ℓ_a is the real aperture length along the azimuth direction,
θ_r is the equivalent squint angle,
V_r is the equivalent velocity, and
V_g is the beam-crossing velocity.

The equivalent velocity in the scenario of a nearly circular satellite orbit and zero-Doppler centroid beam pointing can be approximated by [9]:

$$V_r \approx \sqrt{V_{SAR} V_g} \tag{3.3}$$

where V_{SAR} is the velocity of SAR that is determined by its platform's orbital altitude.
   Then, the azimuth resolution of a spaceborne SAR is given by [11]:

$$\rho_a \approx \frac{0.886 V_g \cos\theta_r}{B_D} = \frac{\ell_a}{2} \frac{V_g}{V_{SAR}} \tag{3.4}$$

It is apparent that the azimuth resolution of a spaceborne SAR is influenced by SAR motion, and to some extent by the orbital altitude of its platform, though its degree is comparatively less pronounced. In the MBSAR system, the Earth's rotational motion plays a vital role in contributing to the Doppler shift effect, which is fundamentally distinct from the spaceborne SAR [12]. Consequently, the MBSAR's azimuth resolution exhibits unique properties associated with the Earth's rotation.

3.2.2 BASIC EQUATIONS

Given a *stationary Moon assumption*, Moccia and Renga simplified MBSAR as an inverse SAR (ISAR) [13–15]. In light of this assumption, they depicted the MBSAR's Earth observation geometry with two imaging modes: "spotlight" and "Stripmap" modes. The derived equations governing the azimuth resolution under both imaging modes are respectively given by [15]:

$$r_a = \frac{\lambda}{2\Delta\Theta} \ \ (\text{spotlight mode}), \tag{3.5}$$

$$r_a = B_D^{-1} \Omega_E R_E \cos\phi_E \ (\text{stripmap mode}), \tag{3.6}$$

where:
λ is the radar wavelength,
$\Delta\Theta$ is the tracking angle of MBSAR,
Ω_E is the Earth's rotational angular velocity,
ϕ_E is the latitude of TOI,
R_E is Earth's radius,

B_D is the Doppler bandwidth that is given by:

$$B_D \cong 2\lambda^{-1}\Omega_E^2 R_E \cos\phi_E T \tag{3.7}$$

By Eqs. (3.6) and (3.7), the following expression is given for azimuthal resolution of the "Stripmap" MBSAR [15]:

$$r_a = \frac{\lambda}{2\Omega_E}\frac{1}{T} \tag{3.8}$$

with the synthetic aperture time:

$$T = a_r \frac{1}{\Omega_E}\frac{\lambda}{\ell_a} \tag{3.9}$$

where a_r is the aperture illumination factor.

Taking account of Eqs. (3.8) and (3.9), one obtains an expression for azimuthal resolution of "Stripmap" mode:

$$r_a = 0.5a_r^{-1}D \tag{3.10}$$

which is the same as in airborne SAR when ignoring the aperture illumination factor.

Subsequently, emphasis was given to the azimuthal resolution explicated in "Stripmap" mode. After conducting a comprehensive inquiry into the MBSAR concept, Fornaro and his team proposed a new view for the "Stripmap" MBSAR originally proposed by Moccia and Renga: they had refined and rebranded "Stripmap" mode as an "equivalent sliding spotlight" mode [16]. The team accomplished this renovation by employing the following methodology.

While the MBSAR illuminates the TOI under the *stationary Moon assumption*, the corresponding Doppler frequency resulting from Earth's rotational motion relative to MBSAR (denoted by $\mathbf{v_E}$) with the LOS vector $\mathbf{r}$ can be represented as:

$$f_d = \frac{2}{\lambda}\mathbf{v_E}\cdot\mathbf{r} \tag{3.11}$$

Consequently, the resulting Doppler bandwidth can be expressed in the following form:

$$B_D = 2\frac{2}{\lambda}\cdot\left(\mathbf{v_E}\cdot\mathbf{r}\right)_{\text{max}} \tag{3.12}$$

According to the MBSAR's observation geometry demonstrated in [16] and "equivalent sliding spotlight" mode, it is possible to rephrase the expression for the Doppler bandwidth as follows:

$$B_D = \frac{4}{\lambda}v_E \sin\left(\varphi_M + \vartheta_M\right) \approx \frac{4}{\lambda}v_E \sin\varphi_M \tag{3.13}$$

where φ_M and ϑ_M are respectively the Earth's rotational angle during the semi-illumination time and semi-beamwidth, both factors are related as follows:

$$\varphi_M = \frac{D_M}{R_E}\vartheta_M \tag{3.14}$$

From Eqs. (3.13) and (3.14), one comes up with the approximate Doppler bandwidth:

$$B_D \approx \frac{4}{\lambda} v_E \frac{D_M}{R_E} \vartheta_M = \frac{2v_E}{L} \frac{D_M}{R_E} \tag{3.15}$$

It follows that the azimuth resolution is expressed in terms of the above Doppler bandwidth:

$$\Delta_y = \frac{v_E}{B_D} \approx \frac{L}{2} \frac{R_E}{D_M} \tag{3.16}$$

From [16], the azimuthal resolution is augmented by an additional *resolution gain factor*, namely the ratio of Earth's radius to Earth–Moon distance, which in turn leads to an increase in aperture synthesis and results in a boost in azimuth resolution. Therefore, they emphasized that the "Stripmap" mode is essentially a "sliding spotlight mode," providing a justification for the improvement factor in the azimuthal resolution of MBSAR. We note that the derived resolution is the nadir point's azimuthal resolution, which is not attainable by any SAR systems due to technical constraints. Also, the Earth's curvature may render such an azimuthal resolution invalid for positions other than the nadir point.

Guo et al. expanded upon the research conducted by Moccia et al. and Fornaro et al., and introduced a novel formula for the azimuthal resolution of MBSAR [17, 18]. They employed a simplified ECI coordinate system (as depicted in [17]) for the MBSAR to derive their improved expression of the azimuthal resolution, as outlined below.

In the defined MBSAR's geometric model of [17], the simplified coordinate system features the Z-axis pointing toward the North Pole and the X-axis perpendicular to the projection of the Moon on the Earth's equatorial plane. In this regard, the MBSAR's position can be determined:

$$\begin{cases} X_m = 0 \\ Y_m = D_{em} \cos \delta_m \\ Z_m = D_{em} \sin \delta_m \end{cases} \tag{3.17}$$

where D_{em} represents the Earth-Moon distance, δ_m stands for the Moon's declination.

In the defined coordinate system, the shortest slant range can be acquired at zero-azimuth time, given the MBSAR's nadir point and TOI at the same longitude under such a condition. To determine the TOI's location for a given azimuth time, we use the relation:

$$\begin{cases} X_g = R_e \cos \delta_g \cos a_g (t) \\ Y_g = R_e \cos \delta_g \sin a_g (t) \\ Z_g = R_e \sin \delta_g \end{cases} \tag{3.18}$$

with:

$$a_g (\eta) = \omega_e t \tag{3.19}$$

where:

R_e is the Earth's radius,
δ_g is the target's latitude in the ECI coordinate system,
a_g is the target's longitude in the ECI coordinate system,
ω_e is the Earth's rotational angular velocity, and
t is the azimuth time.

It follows that the MBSAR's slant range history under the *stationary Moon assumption* can be expressed as:

$$R(\eta) = \sqrt{M \cos a_g(t) + N} \tag{3.20}$$

with:

$$\begin{cases} M = -2R_e Y_m \cos \delta_g \\ N = D_{em}^2 + R_e^2 - 2R_e Z_m \sin \delta_g \end{cases} \tag{3.21}$$

By applying Eq. (3.20), it is possible to estimate the Doppler frequency at a given moment by:

$$f_D(t) = \frac{-2R'(t)}{\lambda} = \frac{M \sin \alpha_g(t)}{\lambda \sqrt{M \cos \alpha_g(t) + N}} \omega_e \tag{3.22}$$

And the estimated Doppler bandwidth is:

$$B_D = f_{D\_\max} - f_{D\_\min} \approx 2 D_{em} L_a^{-1} \omega_e \cos \delta_m \tag{3.23}$$

The beam-crossing velocity can be approximated by:

$$V_g = R_e \omega_e \cos \delta_g \tag{3.24}$$

Then, the azimuthal resolution of the MBSAR is given by:

$$\rho_a = \frac{V_g}{B_D} = \frac{L_a}{2} \frac{R_e}{D_{em}} \frac{\cos \delta_g}{\cos \delta_m} \tag{3.25}$$

As long as the TOI is positioned directly beneath the MBSAR (i.e., the nadir point), the azimuthal resolution and *resolution gain factor* in Eq. (3.25) simplifies into the form of Eq. (3.16). Eq. (3.25) constitutes a noteworthy improvement in the portrayal of MBSAR's azimuthal resolution when compared to prior preliminary analyses, from which a more distinct correlation between MBSAR's azimuth resolution, Earth–Moon relative position, and real aperture length is discernible [19]. Also, the MBSAR's "equivalent sliding spotlight mode" is far more complicated in contrast to the sliding spotlight mode in the airborne case, as asserted by Eq. (3.25): The beam does not steer toward a virtual focal point at the geocenter. Instead, the beam steering is determined by the relative position between the Moon and Earth in conjunction.

Bear in mind that the derivations mentioned above and associated parameters share the same variables as those in [15–17]. Those preliminary analyses serve as the foundation for exploring the azimuthal resolution of MBSAR. Nevertheless, applying the above results to a universal scenario in the context of MBSAR proves to be a challenging task due to the effects of Earth's curvature. Furthermore, the overlooked MBSAR's inertial motion can affect the accuracy of azimuthal resolution to a certain extent. In the following section, we shall strive to develop a more comprehensive portrayal of MBSAR's azimuthal resolution, considering both the Earth's curvature and MBSAR's inertial motion.

3.2.3 Derivation of MBSAR's Azimuth Resolution

To ensure coherence with the preceding analysis [15–18, 20], we incorporate the *stop-and-go assumption* into the derivation of azimuthal resolution. This assumption induces minimal error on

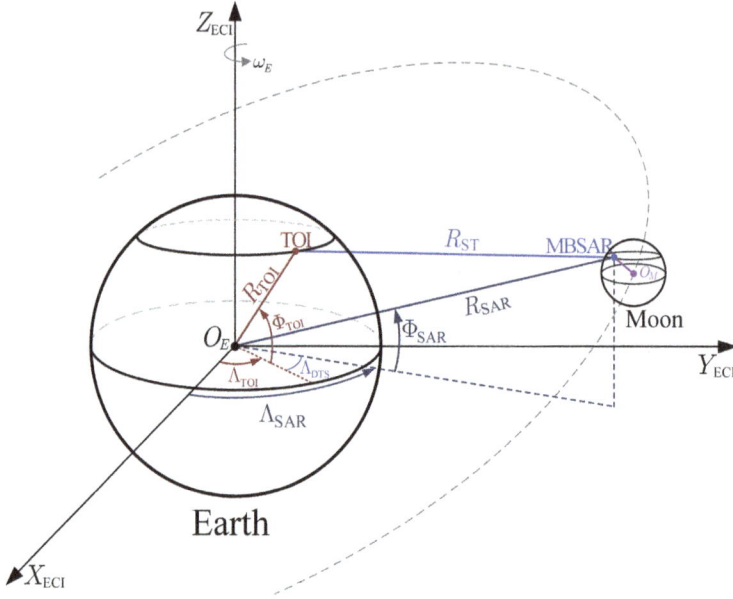

FIGURE 3.1 The geometric model of the MBSAR in the ECI coordinate, where the celestial coordinates of MBSAR and TOI are represented by $(R_{SAR}, \Lambda_{SAR}, \Phi_{SAR})$ and $(R_{TOI}, \Lambda_{TOI}, \Phi_{TOI})$, respectively. This diagram is not to scale for the sake of clarity.

the Doppler bandwidth and hence the associated azimuthal resolution, as will be demonstrated in Chapter 5. Also, this section employs the ECI framework, just as it was employed in the preceding analysis, to delineate the inertial motion of MBSAR. Figure 3.1 describes the geometric model that integrates the TOI and MBSAR in the ECI coordinate system, and thus, all orbital elements in this and subsequent sections are defined in the ECI coordinate system. Knowing that these celestial coordinates differ from those utilized in a geographic coordinate system is essential. Therefore, the symbols used to represent longitude and latitude are distinct from the ones in the ECEF previously defined in Chapter 2.

In the subsequent section, the intention was to ensure a smooth transition from previous research and obtain the MBSAR's azimuthal resolution. To accomplish this, we shall formulate a representation for MBSAR's azimuthal resolution under the stationary Moon assumption, building upon the works of Moccia et al., Fornaro et al., and Guo et al. [15–18, 20]. Subsequently, the influence of MBSAR's inertial motion should be integrated into derivation, thereby enabling the representation of MBSAR's azimuthal resolution with a more sophisticated formula.

3.2.3.1 A Successive Derivation under the Stationary Moon Assumption

Under the stationary Moon assumption, the slant range history of MBSAR relative to the TOI is [19]:

$$R_0\left(\eta\right) = \left\|\mathbf{R}_{\mathbf{SAR},\eta=\eta_c} - \mathbf{R}_{\mathbf{TOI},\eta}\right\|_2 \tag{3.26}$$

where $\| \cdot \|_2$ represents the ℓ^2–norm product, the subscript "0" indicates the scenario where the stationary Moon assumption is made, and the subscript "η" denotes the position vector at the slow time η. Under the stationary Moon assumption, the position of MBSAR refers to that at the beam center crossing time (denoted by η_c).

To simplify the derivation, we focus on Earth's rotation and ignore the other factors like nutation, precession, and polar motion, which exert relatively small impacts on the azimuthal resolution.

Besides, we can set the beam center crossing time to zero (i.e., $\eta_c = 0$), resulting in a streamlined calculation of the slant range:

$$R_0(\eta) = \{R_{SAR}^2 + R_{TOI}^2 - 2R_{SAR}R_{TOI}$$
$$\cdot\left[\sin\Phi_{SAR}\sin\Phi_{TOI} + \cos\Phi_{SAR}\cos\Phi_{TOI}\cos(\Lambda_{DTS} + \omega_E\eta)\right]\}^{0.5} \tag{3.27}$$

where:

R_{SAR} is the distance of MBSAR from Earth's center,

Φ_{SAR} is the MBSAR's declination,

R_{TOI} is the Earth's radius at the TOI,

Φ_{TOI} is the TOI's latitude,

ω_E is the Earth's rotational angular velocity, and

Λ_{DTS} is the meridional deviation of TOI with respect to the MBSAR at the beam center crossing time, defined as:

$$\Lambda_{DTS} = \Lambda_{TOI} - \Lambda_{SAR} \tag{3.28}$$

where:

Λ_{TOI} is the TOI's longitude,

Φ_{SAR} is the right ascension of the MBSAR.

Correspondingly, we define a latitudinal deviation between TOI and MBSAR that facilitates our analysis in subsequent sections:

$$\Phi_{DTS} = \Phi_{TOI} - \Phi_{SAR} \tag{3.29}$$

In Eq. (3.27), the term $\cos(\Lambda_{DTS} + \omega_E\eta)$ can be expanded about $\eta = 0$:

$$\cos(\Lambda_{DTS} + \omega_E\eta) = \cos\Lambda_{DTS} - \omega_E\eta\sin\Lambda_{DTS} - 0.5\omega_E^2\eta^2\cos\Lambda_{DTS} + \mathcal{O}(\eta^3) \tag{3.30}$$

Using Eq. (3.30), the slant range history becomes:

$$R_0(\eta) = \{R_{ST} + 2R_{SAR}R_{TOI}\cos\Phi_{SAR}\cos\Phi_{TOI}\sin\Lambda_{DTS}\cdot\omega_E\eta$$
$$+ R_{SAR}R_{TOI}\cos\Phi_{SAR}\cos\Phi_{TOI}\cos\Lambda_{DTS}\cdot\omega_E^2\eta^2\}^{0.5} + \mathcal{O}(\eta^3) \tag{3.31}$$

with:

$$R_{ST} = \left[R_{SAR}^2 + R_{TOI}^2\right.$$
$$\left. - 2R_{SAR}R_{TOI}\left(\sin\Phi_{SAR}\sin\Phi_{TOI} + \cos\Phi_{SAR}\cos\Phi_{TOI}\cos\Lambda_{DTS}\right)\right]^{0.5} \tag{3.32}$$

The derivation that follows can be accomplished by employing the Hyperbolic Range Equation (HRE), which is commonly used to describe the trajectory of a SAR system [21]. To recall the definition of HRE, it can be expressed as follows:

$$R_{HRE}(\eta) = \sqrt{R_c^2 - 2R_cV_r\eta\sin\theta_r + V_r^2\eta^2} + \mathcal{O}(\eta^3) \tag{3.33}$$

where R_c is the slant range of the beam centerline, identical to R_{ST} that given in Eq. (3.32). It is noteworthy that HRE cannot be employed directly in the signal processing of MBSAR, given its inclusion of higher-order phase errors that necessitate compensation during imaging formation [22]. Nevertheless, those phase errors do not impede the Doppler centroid and Doppler frequency modulation rate (DFMR), thus allowing for the utilization of HRE in determining the Doppler bandwidth and azimuthal resolution.

Upon comparing Eqs. (3.31) and (3.33), one can derive expressions for equivalent velocity and squint angle, assuming a stationary Moon:

$$V_{r0} = \omega_E \sqrt{R_{SAR} R_{TOI} \cos \Phi_{SAR} \cos \Phi_{TOI} \cos \Lambda_{DTS}} \tag{3.34}$$

$$\theta_{r0} = -\sin^{-1} \left\{ \frac{\sqrt{R_{SAR} R_{TOI} \cos \Phi_{SAR} \cos \Phi_{TOI}}}{R_c \cdot \sqrt{\cos \Lambda_{DTS}}} \sin \Lambda_{DTS} \right\} \tag{3.35}$$

Hence, we can express the DFMR of the MBSAR by:

$$f_{dr0} = 2\lambda^{-1} R_c^{-1} V_{r0}^2 \cos^2 \theta_{r0} \tag{3.36}$$

As in Eq. (3.27), the subscript "0" indicates the corresponding parameters under the stationary Moon assumption.

Next, the synthetic aperture time (SAT) is considered. In the preceding analysis, the SAT model assumes a linear trajectory and a level Earth surface [16, 17]. To maintain congruity with these assumptions, we express the SAT in the following form:

$$T_{SAR} \approx 0.886 \frac{\lambda R_c}{\ell_a V_{g0} \cos \theta_{r0}} \tag{3.37}$$

where a 3-dB bandwidth-related factor of 0.886 is included. V_{g0} is the beam-crossing velocity when employing the stationary Moon assumption; in such a scenario, it is solely attributed to Earth's rotation and can be expressed as:

$$V_{g0} = R_{TOI} \omega_E \cos \Phi_{TOI} \tag{3.38}$$

Keep in mind that the above assumption affects the SAT's accuracy and, consequently, the MBSAR's azimuthal resolution. A more precise approach for determining the SAT is required, which will be discussed further later when we account for the effects of MBSAR's inertial motion on azimuthal resolution.

The Doppler bandwidth of the MBSAR under the stationary Moon assumption becomes:

$$B_D = \left| f_{dr0} \right| T_{SAR} = 0.886 \frac{2 V_{r0}^2 \cos \theta_{r0}}{\ell_a V_{g0}} \tag{3.39}$$

Consequently, the azimuthal resolution is:

$$\rho_a = \frac{0.886 V_{g0} \cos \theta_{r0}}{B_D} = \frac{\ell_a V_{g0}^2}{2 V_{r0}^2} \tag{3.40}$$

By substituting Eqs. (3.38) and (3.34) into Eq. (3.40), the MBSAR's azimuthal resolution in the context of the stationary Moon assumption is modified to [19, 22]:

$$\rho_a = \frac{\ell_a}{2} \cdot \frac{R_{TOI}}{R_{SAR}} \cdot \frac{\cos \Phi_{TOI}}{\cos \Phi_{SAR}} \cdot \frac{1}{\cos \Lambda_{DTS}} \tag{3.41}$$

where we come up with a *resolution gain factor* taking account of meridional deviation Λ_{DTS}. The results reveal that the MBSAR's azimuthal resolution depends not just on the real aperture length but also on the Earth's curvature and MBSAR's position relative to TOI.

3.2.3.2 The Azimuthal Resolution Considering MBSAR's Inertial Motion

Under the stationary Moon assumption, we have derived the azimuth resolution of the MBSAR given by Eq. (3.41), which includes the effects of spatial variability induced by Earth's curvature. However, the azimuth resolution in Eq. (3.41) may be inadequate or inapplicable at the high Earth latitudes. As an illustrative example, let us consider observing the Earth's poles by the MBSAR, where the linear velocity of Earth's rotation approaches zero, leading to a null azimuth resolution. Thus, the poles of the Earth cannot be observed in the view of the stationary Moon assumption. The rationale behind this phenomenon is threefold [4, 22]:

1) The impacts of MBSAR's inertial motion have been neglected in the Doppler shift;
2) The impacts of the Earth's curvature and MBSAR's curved trajectory have been neglected in determining SAT;
3) The impact of Earth's ellipsoid has not been considered (though such an influence is relatively minor).

To deal with the above issues, we shall derive the MBSAR's azimuthal resolution by introducing both effects of MBSAR's inertial motion and Earth's curvature. In this context, the range history is:

$$R(\eta) = \left\| \mathbf{R}_{\mathrm{SAR},\eta} - \mathbf{R}_{\mathrm{TOI},\eta} \right\|_2 \tag{3.42}$$

The three components of the MBSAR position vector are:

$$\begin{cases} X_{\mathrm{SAR}}(\eta) = R_{\mathrm{SAR}} \cos \Phi_{\mathrm{SAR}} \cos \Lambda_{\mathrm{SAR}} + V_{\mathrm{SX}} \cdot \eta + A_{\mathrm{SX}} \cdot \eta^2 \\ Y_{\mathrm{SAR}}(\eta) = R_{\mathrm{SAR}} \cos \Phi_{\mathrm{SAR}} \sin \Lambda_{\mathrm{SAR}} + V_{\mathrm{SY}} \cdot \eta + A_{\mathrm{SY}} \cdot \eta^2 \\ Z_{\mathrm{SAR}}(\eta) = R_{\mathrm{SAR}} \sin \Phi_{\mathrm{SAR}} + V_{\mathrm{SZ}} \cdot \eta + A_{\mathrm{SZ}} \cdot \eta^2 \end{cases} \tag{3.43}$$

where $\mathbf{V}_{\mathbf{SAR}} = [V_{\mathrm{SX}}, V_{\mathrm{SY}}, V_{\mathrm{SY}}]^T$ and $\mathbf{A}_{\mathbf{SAR}} = [A_{\mathrm{SX}}, A_{\mathrm{SY}}, A_{\mathrm{SY}}]^T$ are MBSAR's velocity and acceleration vectors in the ECI coordinate system, respectively. Note that the inertial motion of MBSAR is a consequence of both the lunar revolution and rotational motion of the Moon [23]. Determining position and velocity vectors of MBSAR requires the orbit and attitude of the Moon, which can be known by lunar ephemeris. The acceleration vector is typically not available but can be derived from the MBSAR's velocity vector.

The ensuing parameters that pertain to equivalent velocity and squint angle can be derived as follows:

$$\begin{aligned} \left. \frac{dR(\eta)}{d\eta} \right|_{\eta=0} = R_c^{-1} \Big\{ & \omega_E \cdot R_{\mathrm{SAR}} R_{\mathrm{TOI}} \cos \Phi_{\mathrm{SAR}} \cos \Phi_{\mathrm{TOI}} \sin \Lambda_{\mathrm{DTS}} \\ & + V_{\mathrm{SZ}}(R_{\mathrm{SAR}} \sin \Phi_{\mathrm{SAR}} - R_{\mathrm{TOI}} \sin \Phi_{\mathrm{TOI}}) \\ & - R_{\mathrm{TOI}} \cos \Phi_{\mathrm{TOI}}(V_{\mathrm{SY}} \sin \Lambda_{\mathrm{TOI}} + V_{\mathrm{SX}} \cos \Lambda_{\mathrm{TOI}}) \\ & + R_{\mathrm{SAR}} \cos \Phi_{\mathrm{SAR}}(V_{\mathrm{SX}} \cos \Lambda_{\mathrm{SAR}} + V_{\mathrm{SY}} \sin \Lambda_{\mathrm{SAR}}) \Big\} \end{aligned} \tag{3.44}$$

and:

$$\begin{aligned} \left. \frac{d^2 R(\eta)}{d\eta^2} \right|_{\eta=0} = R_c^{-1} \Big\{ & V_{\mathrm{SAR}}^2 + \omega_E^2 R_{\mathrm{SAR}} R_{\mathrm{TOI}} \cos \Phi_{\mathrm{SAR}} \cos \Phi_{\mathrm{TOI}} \cos \Lambda_{\mathrm{DTS}} \\ & + 2A_{\mathrm{SZ}}(R_{\mathrm{SAR}} \sin \Phi_{\mathrm{SAR}} - R_{\mathrm{TOI}} \sin \Phi_{\mathrm{TOI}}) + 2R_{\mathrm{SAR}} \cos \Phi_{\mathrm{SAR}}(A_{\mathrm{SY}} \sin \Lambda_{\mathrm{SAR}} + A_{\mathrm{SX}} \cos \Lambda_{\mathrm{SAR}}) \\ & + 2R_{\mathrm{TOI}} \cos \Phi_{\mathrm{TOI}} \Big[\sin \Lambda_{\mathrm{TOI}}(\omega_E V_{\mathrm{SX}} - A_{\mathrm{SY}}) - \cos \Lambda_{\mathrm{TOI}}(\omega_E V_{\mathrm{SY}} + A_{\mathrm{SX}}) \Big] \Big\} - R_c^{-1} \left[\frac{dR(\eta)}{d\eta} \right]^2 \end{aligned} \tag{3.45}$$

It follows that the equivalent velocity and squint angle of the MBSAR are given by:

$$V_r = \sqrt{\left(\frac{dR(\eta)}{d\eta}\right)^2 + R_c \frac{d^2R(\eta)}{d\eta^2}} = V_{r0}\sqrt{\kappa_{vr}} \tag{3.46}$$

$$\theta_r = \sin^{-1}\left(-V_r^{-1}\frac{dR(\eta)}{d\eta}\right) = \sin^{-1}\left(\frac{\kappa_s \sin\theta_{r0}}{\sqrt{\kappa_{vr}}}\right) \tag{3.47}$$

with:

$$\begin{aligned}
\kappa_{vr} = 1 + &\frac{1}{\omega_E^2 R_{\text{SAR}} R_{\text{TOI}} \cos\Phi_{\text{SAR}} \cos\Phi_{\text{TOI}} \cos\Lambda_{\text{DTS}}} \\
&\cdot \Big\{ V_{\text{SAR}}^2 + 2A_{\text{SZ}} \cdot (R_{\text{SAR}} \sin\Phi_{\text{SAR}} - R_{\text{TOI}} \sin\Phi_{\text{TOI}}) \\
&+ 2R_{\text{SAR}} \cos\Phi_{\text{SAR}} (A_{\text{SY}} \sin\Lambda_{\text{SAR}} + A_{\text{SX}} \cos\Lambda_{\text{SAR}}) \\
&+ 2R_{\text{TOI}} \cos\Phi_{\text{TOI}} \Big[\sin\Lambda_{\text{TOI}}(\omega_E V_{\text{SX}} - A_{\text{SY}}) - \cos\Lambda_{\text{TOI}}(\omega_E V_{\text{SY}} + A_{\text{SX}}) \Big] \Big\}
\end{aligned} \tag{3.48}$$

$$\begin{aligned}
\kappa_s = 1 + &\frac{1}{R_{\text{SAR}} R_{\text{TOI}} \omega_E \cos\Phi_{\text{SAR}} \cos\Phi_{\text{TOI}} \sin\Lambda_{\text{DTS}}} \cdot \Big[V_{\text{SZ}}(R_{\text{SAR}} \sin\Phi_{\text{SAR}} - R_{\text{TOI}} \sin\Phi_{\text{TOI}}) \\
&- R_{\text{TOI}} \cos\Phi_{\text{TOI}}(V_{\text{SY}} \sin\Lambda_{\text{TOI}} + V_{\text{SX}} \cos\Lambda_{\text{TOI}}) + R_{\text{SAR}} \cos\Phi_{\text{SAR}}(V_{\text{SX}} \cos\Lambda_{\text{SAR}} + V_{\text{SY}} \sin\Lambda_{\text{SAR}}) \Big]
\end{aligned} \tag{3.49}$$

When addressing azimuthal resolution, we must take the beam-crossing velocity into account. In the MBSAR, the Earth's rotation predominantly affects beam-crossing velocity, whereas the contribution from the SAR inertial motion is comparatively small [24]. We can estimate the beam-crossing velocity of MBSAR as follows:

$$V_g = \kappa_{vg} \cdot V_{g0} \tag{3.50}$$

$$\kappa_{vg} \approx \frac{1}{\omega_E R_{\text{TOI}} \cos\Phi_{\text{TOI}}} \left\| \mathbf{w_E} \times \mathbf{R_{TOI}} - \frac{\mathbf{R_{SAR}} \cdot \mathbf{R_{TOI}}}{\|\mathbf{R_{SAR}}\|_2} \cdot \frac{\mathbf{V_{SAR}}}{\|\mathbf{R_{SAR}}\|_2} \right\|_2 \tag{3.51}$$

Finally, let us turn our attention to the SAT in the context of MBSAR's inertial motion and Earth's curvature. A generalized form of synthetic aperture time can be accepted [22, 25]:

$$T_{\text{SAR}} = \eta_{\text{end}} - \eta_{\text{start}} \tag{3.52}$$

with:

$$\begin{cases}
\cos^{-1}\left(\dfrac{\mathbf{R}_{\text{ST},\eta_{\text{start}}} \cdot \mathbf{R}_{\text{ST},\eta_c}}{\|\mathbf{R}_{\text{ST},\eta_{\text{start}}}\|_2 \cdot \|\mathbf{R}_{\text{ST},\eta_c}\|_2}\right) = -0.886\dfrac{\lambda}{2\ell_a} \\[4mm]
\cos^{-1}\left(\dfrac{\mathbf{R}_{\text{ST},\eta_{\text{end}}} \cdot \mathbf{R}_{\text{ST},\eta_c}}{\|\mathbf{R}_{\text{ST},\eta_{\text{end}}}\|_2 \cdot \|\mathbf{R}_{\text{ST},\eta_c}\|_2}\right) = 0.886\dfrac{\lambda}{2\ell_a}
\end{cases} \tag{3.53}$$

where η_{start} and η_{end} refer to the start and end times, respectively, during which the TOI is illuminated by the MBSAR.

By comparing Eqs. (3.52) and (3.53) with Eq. (3.37), it is possible to restate the SAT as follows:

$$T_{SAR} = 0.886\kappa_t \frac{\lambda R_c}{\ell_a V_g \cos\theta_r} \tag{3.54}$$

with:

$$\kappa_t = \cos\theta_r \frac{V_g \ell_a (\eta_{end} - \eta_{start})}{0.886\lambda R_c} \tag{3.55}$$

As can be discerned from Eqs. (3.47) and (3.55), the factor κ_s is inherently embedded within the factor κ_t.

Taking MBSAR's inertial motion into the Doppler bandwidth, we have:

$$B_D = |f_{dr}| T_{SAR} = 0.886 \cdot \kappa_t \frac{2V_r^2 \cos\theta_r}{\ell_a V_g} \tag{3.56}$$

Recall that the azimuthal resolution is of the form:

$$\rho_a = \frac{0.886V_g \cos\theta_r}{B_D} = \frac{\ell_a V_g^2}{2V_r^2} \tag{3.57}$$

Upon substituting Eqs. (3.46) and (3.50) into Eq. (3.57), one can readily obtain:

$$\rho_a = \frac{\ell_a}{2} \frac{\kappa_{vg}^2}{\kappa_t \kappa_{vr}} \frac{V_{g0}^2}{V_{r0}^2} \tag{3.58}$$

Comparing Eqs. (3.58) to (3.40) and (3.41), one arrives at MBSAR's azimuthal resolution considering both MBSAR's inertial motion and Earth's curvature:

$$\rho_a = \frac{\ell_a}{2} \frac{\kappa_{vg}^2}{\kappa_t \kappa_{vr}} \frac{R_{TOI}}{R_{SAR}} \frac{\cos\Phi_{TOI}}{\cos\Phi_{SAR}} \frac{1}{\cos\Lambda_{DTS}} \tag{3.59}$$

In Table 3.1, we compare the versions of azimuth resolution expression of SAR systems on airborne, spaceborne, and Moon-based platforms. It can be observed that MBSAR surpasses azimuth resolution constraints of half real aperture length, a fact that we can perceive through the 'pseudo-spotlight' perspective, wherein the synthetic aperture length is extended due to the antenna's steering rotation when the nadir point is directed toward the Earth's center during MBSAR orbiting the Earth.

3.2.3.3 The Azimuthal Resolution Relating to Antenna Beam Pointing

Eq. (3.57) provides a fundamental understanding of the MBSAR's azimuthal resolution as it establishes a correlation with the real aperture length and the position and movement of MBSAR relative to TOI. However, this equation does not include the contributions from the orbital elements, beam pointing, and side-looking direction of MBSAR. We shall further revise the expression for MBSAR's azimuthal resolution, as depicted below.

Consider that the MBSAR illuminates the TOI while moving along the azimuth direction during the SAT. The MBSAR effectively rotates with respect to TOI, forming a synthetic angle θ_{syn} (i.e., the angle between the start and end of synthetic aperture processing):

$$\theta_{syn} = \omega_{syn} T_{SAR}, \quad \omega_{syn} = \frac{\Re_{syn}}{0.886 \cdot V_g \cos\theta_r} \tag{3.60}$$

TABLE 3.1

A Comparison of Azimuthal Resolutions in SAR Systems Deployed on Various Platforms

Scenarios	Representations of Azimuthal Resolutions			
Airborne SAR [26]	$\dfrac{\ell_a}{2}$			
Spaceborne SAR [11]	$\dfrac{\ell_a}{2}\dfrac{V_g}{V_{SAR}}$			
MBSAR	Moccia & Renga [15]	Fornaro et al. [16]	Guo et al. [17]	Eq. (3.61)
	$\dfrac{\ell_a}{2}$	$\dfrac{\ell_a}{2}\dfrac{R_{TOI}}{R_{SAR}}$	$\dfrac{\ell_a}{2}\dfrac{R_{TOI}}{R_{SAR}}\dfrac{\cos\Phi_{TOI}}{\cos\Phi_{SAR}}$	$\dfrac{\ell_a}{2}\dfrac{\kappa_{vg}^2}{\kappa_i\kappa_{vr}}\dfrac{R_{TOI}}{R_{SAR}}\dfrac{\cos\Phi_{TOI}}{\cos\Phi_{SAR}}\dfrac{1}{\cos\Lambda_{DTS}}$

Note: For ease of comparison, the azimuthal resolutions of all SAR systems are expressed in unified representations, and the aperture illumination factor of MBSAR's azimuthal resolution in [15] is neglected.

with $\mathfrak{R}_{syn}$ being a function of antenna azimuth angle ϕ and look angle θ_l, which can be decomposed into the following form:

$$\mathfrak{R}_{syn} = \mathfrak{R}_1\left(\xi_{sld};\theta_l,\phi\right) + \mathfrak{R}_2\left(\xi_{sld};\theta_l,\phi\right) + \mathfrak{R}_3\left(\xi_{sld};\theta_l,\phi\right) + \mathfrak{R}_4\left(\xi_{sld};\theta_l,\phi\right) \tag{3.61}$$

where ξ_{sld} signifies side-looking direction ($\xi_{sld} = 1$, left-looking; $\xi_{sld} = -1$, right-looking). The first three terms ($\mathfrak{R}_1 \sim \mathfrak{R}_3$) of Eq. (3.61) are correlated to the DFMR only, while the last term $\mathfrak{R}_4$ ($\mathfrak{R}_4 = \mathfrak{R}_{4S} + \mathfrak{R}_{4E} + \mathfrak{R}_{4SE}$) is connected with the Doppler centroid and DFMR. By identifying contributions from Earth-MBSAR motion dynamics, Eq. (3.61) can be rewritten as follows:

$$\mathfrak{R}_{syn} = \underbrace{\mathfrak{R}_1(\xi_{sld};\theta_l,\phi) + \mathfrak{R}_{4S}(\xi_{sld};\theta_l,\phi)}_{\text{SAR's inertial motion}}$$
$$+ \underbrace{\mathfrak{R}_2(\xi_{sld};\theta_l,\phi) + \mathfrak{R}_{4SE}(\xi_{sld};\theta_l,\phi)}_{\text{Coupling term}} + \underbrace{\mathfrak{R}_1(\xi_{sld};\theta_l,\phi) + \mathfrak{R}_{4E}(\xi_{sld};\theta_l,\phi)}_{\text{Earth's rotational motion}} \tag{3.62}$$

A more in-depth account of each term of Eq. (3.62) will be detailed in Chapter 5.

The beam-crossing velocity, a function of the beam pointing and orbital elements, is given by:

$$V_g = \left\| \mathbf{V}_{TOI(ACS)} - \mathbf{V}_{SAR(ACS)} \cdot (1 - R_{ST}R_{SAR}^{-1}\cos\theta_l) \right\|_2 \tag{3.63}$$

with $\mathbf{V}_{TOI(ACS)} = [V_{TOI, ACS}, V_{TOI, ACS}, V_{TOI, ACS}]^T$ and $\mathbf{V}_{SAR(ACS)} = [V_{SX, ACS}, V_{SY, ACS}, V_{SY, ACS}]^T$ in the ACS respectively expressed by:

$$\begin{cases} V_{TOI,ACS} = \omega_E\left[(R_{SAR} - R_{ST}\cos\theta_l)\cos i - R_{ST}\sin\theta_l\sin\phi\sin i\sin u\right] \\ V_{TOI,ACS} = \omega_E\left[(R_{SAR} - R_{ST}\cos\theta_l)\sin i\cos u - R_{ST}\sin\theta_l\cos\phi\sin i\sin u\right] \\ V_{YOI,ACS} = \omega_E\left[-R_{ST}\sin\theta_l\sin\phi\sin i\cos u + R_{ST}\sin\theta_l\cos\phi\cos i\right] \end{cases} \tag{3.64}$$

$$\begin{cases} V_{SX,ACS} = -V_{SX}\cdot(\sin u\cos\Omega + \cos i\cos u\sin\Omega) \\ \qquad\qquad -V_{SY}\cdot(\sin u\sin\Omega - \cos i\cos u\cos\Omega) + V_{SZ}\cdot\sin i\cos u \\ V_{SY,ACS} = 0 \\ V_{SZ,ACS} = -V_{SX}\cdot(\cos u\cos\Omega - \cos i\sin u\sin\Omega) \\ \qquad\qquad -V_{SY}\cdot(\cos u\sin\Omega + \cos i\sin u\cos\Omega) - V_{SZ}\cdot\sin i\sin u \end{cases} \tag{3.65}$$

where:

i is the orbital inclination,

Ω is the RAAN, and

u is the AOL, all these orbital elements are given in ECI coordinate system.

Note that the contribution of Earth's ellipsoid is implicitly incorporated in the variable R_{ST}, the expression of which is already given in Eq. (2.60). Besides, the explicit derivations of both $\mathbf{V}_{TOI(ACS)}$ and $\mathbf{V}_{SAR(ACS)}$ will be discussed in Chapter 5.

The synthetic angle is confined by the SAT, depending on the imaging modes [11]. The imaging mode resembles sliding spotlight mode for a particular aperture length, albeit lacking a well-defined focal point. In this mode, the SAT is given by Eq. (3.52). On the other hand, the coherent processing of MBSAR can be dissociated from the real aperture length, sharing similarities with the spotlight mode, as reported in [4, 27]. As a matter of fact, the spotlight mode is frequently employed in planetary radar imaging or inverse SAR [28–30], whose imaging mechanism demonstrates a highly resemblance to that of the MBSAR.

With Eq. (3.60), the Doppler bandwidth is given by [4]:

$$B_D = \frac{2}{\lambda} \Re_{syn} T_{SAR} \tag{3.66}$$

It follows that the azimuthal resolution is expressed as:

$$\rho_a = \frac{0.886 V_g \cos\theta_r}{B_D} = 0.886 \frac{\lambda}{2\theta_{syn}} \tag{3.67}$$

As long as we ignore the MBSAR's inertial motion, Eq. (3.68) is reduced to Eq. (3.5). More about the MBSAR's azimuth resolution is given in Section 3.4.

3.3   OTHER GENERIC PARAMETERS

Two essential factors to be considered when evaluating the performance of the MBSAR system are the ambiguity constraint and the power restriction [31–33]. The former dictates the PRF associated with signal sampling, while the power restriction determines the SNR, or NESZ (radiometric resolution), as elaborated below.

3.3.1   THE CHOICE OF PRF (AMBIGUITY CONSTRAINTS)

For a pulse radar system, the PRF selection is primarily driven by preventing range and azimuth ambiguities [11]. The Nyquist-Shannon sampling theorem dictates that the range aliasing arises when the PRF of the SAR system is excessively high, causing a "ghost image" after signal processing [34]. Conversely, If the PRF is insufficiently high, it may not be able to accurately sample the Doppler frequency within the Doppler bandwidth, leading to a phenomenon known as azimuth aliasing [35]. Hence, these competing requirements pose significant challenges in balancing the appropriate PRF, a dilemma of range-Doppler [7]. We summarize some properties of selecting PRF, as shown in Table 3.2.

Figure 3.2 provides a graphical explanation of the slant and ground swaths of MBSAR. The determination of PRT is contingent upon the prerequisite that the received signal from the slant swath should not be subject to concurrent influence from two pulses. The rationale for this requirement is that the received signal must be generated by two pulses emanating from distinct locations on the ground, which brings about differences in their respective slant ranges [36, 37]. Mathematically, this requirement can be expressed as a function of PRF (f_{PRF}):

$$\frac{c}{2} \frac{1}{f_{PRF}} > W_{sr} \tag{3.68}$$

TABLE 3.2

A Summary of Selecting PRFs

	High PRF	Low PRF
Limitations	Narrower unambiguous swath width	Coarser azimuthal resolution
Advantages	1) Reduced range ambiguity (more pulses per second); 2) Enhanced SNR (increased transmitted energy); 3) Improved azimuth resolution.	1) Increased unambiguous swath width; 2) Decreased signal-to-noise ratio; 3) Inferior azimuth resolution.

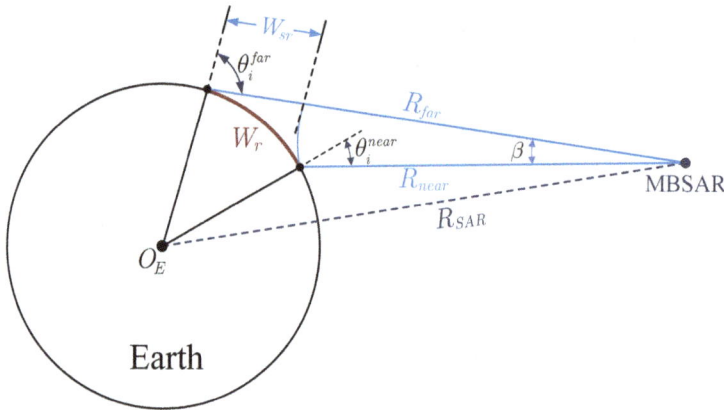

FIGURE 3.2 The schematic diagram of range ambiguity in side view, where W_{sr} represents the slant swath, W_r stands for ground swath (typical swath width); R_{near} and R_{far} denote the near and far ranges that determine the extent of swath width, θ_i^{far} and θ_i^{near} are the far and near incident angles, respectively.

Therefore, the slant swath imposes an upper bound on the PRF:

$$f_{PRF} < \frac{c}{2W_{sr}} \tag{3.69}$$

In conventional SAR (e.g., airborne SAR), the slant swath and ground swath widths are related by [38]:

$$W_{sr} \approx W_r \sin \theta_i \tag{3.70}$$

For MBSAR, Eq. (3.70) is no longer applicable due to the Earth's curvature, and thus, it cannot be used to determine the swath width. We may determine the slant swath by:

$$W_{sr} = R_{ST}\left(\theta_l^{far}\right) - R_{ST}\left(\theta_l^{near}\right) \tag{3.71}$$

where:
  θ_l^{near} is the near look angle
  θ_l^{far} is the far look angle that is defined by: $\theta_l^{far} = \theta_l^{near} + \beta_r$, and
  β_r is the beamwidth along range direction.

To avoid azimuth ambiguity, we shall ensure that the PRF range encompasses the entire Doppler frequency spectrum within the image scene [39]. In the zero-Doppler steering mode, this requirement can be mathematically expressed as:

$$f_{PRF} > B_D \tag{3.72}$$

Note that the DFMR f_{dr} and associated Doppler bandwidth B_D typically decreases with the increasing look angle (to be discussed in Chapter 5). Therefore, when determining the bounds of PRF, it is essential to refer to the Doppler bandwidth corresponding to the near-look angle. Using Eqs. (3.66), (3.69) and (3.72), we have the lower and upper bound of PRF:

$$\frac{2}{\lambda} \Re_{syn} T_{SAR} < f_{PRF} < \frac{c}{2W_{sr}} \tag{3.73}$$

We may apply safety factors a_{az}, a_{rg} given rise from the antenna illuminating edge effects to the range and azimuth ambiguities [6, 40]. Subsequently, the constraints on the PRF are modified to:

$$\frac{2a_{az}}{\lambda} \Re_{syn} T_{SAR} < f_{PRF} < \frac{c}{2W_{sr}a_{rg}}, \quad a_{az} \geq 1, a_{rg} \geq 1 \tag{3.74}$$

Both the Doppler bandwidth and slant swath vary with the MBSAR position. To assess the tolerance for selecting the PRF within one specific cycle T_{lc}, we define the ensuing parameters:

$$\Delta f_{PRF} = \left(\frac{c}{2a_{rg}W_{sr}(T_{lc})} \right)_{min} - \left(\frac{2a_{az}}{\lambda} \cdot \Re_{syn}(T_{lc}) \cdot T_{SAR} \right)_{max} \tag{3.75}$$

Further, Eq. (3.74) also places constraints on the slant swath with respect to the SAT. That is to say, it imposes a cap on the maximum slant swath in relation to the SAT, as specified by:

$$W_{sr} < \frac{c\lambda}{4a_{rg}a_{az}\Re_{syn}T_{SAR}} \tag{3.76}$$

Equivalently, we may set the SAT threshold from Eq. (3.74) if the maximum slant swath is predetermined:

$$T_{SAR} < \frac{c\lambda}{4a_{rg}a_{az}\Re_{syn}W_{rs}} \tag{3.77}$$

3.3.2   SNR and NESZ (Power Considerations)

Now, we assess the power-related factors, i.e., the SNR and NESZ, in the MBSAR. The SNR in the SAR system is given by [41, 42]:

$$SNR = \frac{c}{256\pi^3} \frac{P_{ave}G^2\lambda^3}{kT_0B_rF} \frac{V_g}{R_{ST}^3V_r^2\sin\theta_i} \sigma^0 \tag{3.78}$$

with the average transmitted power given by [43]:

$$P_{\text{ave}} = P_T f_{\text{PRF}} T_p \tag{3.79}$$

where:
  P_T is the peak power of the transmitted signal;
  G is the antenna gain;
  k is Boltzmann's constant;
  T_0 is the reference temperature of the receiver;
  F is the receiver noise figure;
  θ_i is the incident angle; and
  σ^0 is the scattering coefficient.

In the SAR system, the radiometric resolution is frequently defined by NESZ [44]:

$$\text{SNR}\left(\text{NESZ}\right) = 1 \rightarrow \frac{c}{256\pi^3} \frac{P_{\text{ave}} G^2 \lambda^3 \sigma^0}{kT_0 B_r F} \frac{V_g}{R_{\text{ST}}^3 V_r^2 \sin\theta_i} \text{NESZ} = 1 \tag{3.80}$$

Upon solving Eq. (3.80), we have:

$$\text{NESZ} = \frac{256\pi^3}{c} \frac{kT_0 BF}{P_{\text{ave}} G^2 \lambda^3} \frac{R_{\text{ST}}^3 V_r^2 \sin\theta_i}{V_g} \tag{3.81}$$

If the TOI has an σ^0 value below NESZ, it is impossible to differentiate it from the background or the noise floor [45, 46].

When the observation regions are restricted to the low-to-mid latitudes, the following approximation can be utilized:

$$R_{\text{ST}}^{-3} V_r^{-2} V_g \approx \omega_E^{-1} \cdot R_{\text{ST}}^{-3} R_{\text{SAR}}^{-1} \cdot \cos^{-1}\Phi_{\text{SAR}} \cdot \cos^{-1}\Lambda_{\text{DTS}} \tag{3.82}$$

Then, we can express the SNR and NESZ to a reasonable accuracy level:

$$\begin{cases} \text{SNR} \approx \dfrac{1}{\omega_E} \dfrac{c}{256\pi^3} \dfrac{P_{\text{ave}} G^2 \lambda^3}{kT_0 B_r F} \dfrac{1}{R_{\text{ST}}^3 R_{\text{SAR}} \cos\Phi_{\text{SAR}} \cos\Lambda_{\text{DTS}} \sin\theta_i} \cdot \sigma^0 \\[4mm] \text{NESZ} \approx \omega_E \dfrac{256\pi^3}{c} \dfrac{kT_0 BF \sin\theta_i}{P_{\text{ave}} G^2 \lambda^3} R_{\text{ST}}^3 R_{\text{SAR}} \cos\Phi_{\text{SAR}} \cos\Lambda_{\text{DTS}} \sin\theta_i \end{cases} \tag{3.83}$$

The validity of Eq. (3.83) can be attributed to the following rationale: In regions with low-to-mid latitudes, the impact of MBSAR's inertial motion is relatively insignificant. Therefore, we can make estimations or approximations in the following manner: $\kappa_{vg}\kappa_{vr}^{-1} \approx 1$. In this event, the parameters related to velocity can be replaced with those based on the stationary Moon assumption.

Eq. (3.81) states that the NESZ is influenced by the central slant range, beam-crossing and equivalent velocities, and incident angle, all of which exhibit spatial variability within the image scene and temporal variability under MBSAR's inertial motion. Consequently, the NESZ demonstrates spatiotemporal variability in the MBSAR. To explicate the inherent properties of the NESZ, we introduce a factor:

$$\Delta\text{NESZ} = \frac{\text{NESZ}}{\text{NESZ}_{\text{ref}}} \tag{3.84}$$

where NESZ_{ref} denotes the NESZ at a specified location under a particular beam-pointing direction. Eq. (3.84) can be also expressed in dB:

$$\Delta\text{NESZ}(\text{dB}) = 10\log_{10}\left(\frac{\text{NESZ}}{\text{NESZ}_{\text{ref}}}\right) = \text{NESZ}(\text{dB}) - \text{NESZ}_{\text{ref}}(\text{dB}) \tag{3.85}$$

3.4 NUMERICAL ILLUSTRATIONS

In this section, we shall employ the lunar cycles provided in Table 2.4 to examine the MBSAR's system parameters. We start with the azimuthal resolution, PRF, and radiometric resolution, assuming the MBSAR is located at $(0°,0°)$ on the lunar surface while it looks from the left-hand side, unless otherwise specified.

3.4.1 ILLUSTRATIONS OF THE AZIMUTHAL RESOLUTION

We show how the stationary Moon assumption affects the azimuthal resolution of MBSAR in what follows. Given that the influence of MBSAR's inertial motion is closely intertwined with factors κ_{vr} and κ_{vg}, while the factor κ_t is given rise by Earth's curvature. We examine the fluctuations in κ_{vr} and κ_{vg} versus Φ_{DTS} while considering varying Λ_{DTS} values. As the manifestations of MBSAR's inertial motion on azimuthal resolution can be perceived as a function of $\kappa_{vr}^{-1}\kappa_{vg}^{2}$, the corresponding variations are also examined. Figure 3.3 shows results at different locations in cycle T_{lc1}. The selected positions were the AOL of $90°$, $180°$, and $270°$, corresponding to the nadir point's latitude approaches $28.57°$, $0°$, and $-28.61°$, respectively.

According to Figure 3.3, a conspicuous correlation exists between the impact of MBSAR's inertial motion and Earth-MBSAR's relative position. In this case, the influence of TOI's latitude is particularly noteworthy, given its more substantial variations on the impact of MBSAR's inertial motion. In the low-to-mid latitudes, where the Earth's rotational motion is dominant in determining the Doppler shift, the magnitudes of all those factors are above 0.9, suggesting that the corresponding variation in azimuthal resolution is relatively small. As the MBSAR beam illuminates high-latitude regions (such as when the MBSAR is located at the AOL of $90°$), each factor undergoes significant variations with the increase Φ_{DTS}. In this particular instance, the inertial motion of MBSAR could lead to a notable fluctuation in azimuthal resolution. This phenomenon is attributed to the reduction in Earth's linear rotational velocity at higher latitudes, while the MBSAR's inertial motion becomes increasingly significant in generating the Doppler shift effect. Meanwhile, we note that the impact of MBSAR's inertial motion on the azimuthal resolution varies across various meridian deviations, as illustrated in the following.

We further inspect the variation pattern of the factors κ_{vr}, κ_{vg}, and $\kappa_{vr}^{-1}\kappa_{vg}^{2}$ with respect to Λ_{DTS} (the meridian deviation) under the above conditions, as presented in Figure 3.4.

The results in Figure 3.4 suggest that the meridian deviation Λ_{DTS} in the azimuthal resolution strongly correlates to the relative positions between the MBSAR and TOI. The following properties are discovered:

1) When the MBSAR is placed at the descending node (AOL of $180°$), the Earth's latitude does not have a notable effect on the factor κ_{vg}. At this juncture, the factor κ_{vg} only displays an approximate symmetry around $\Lambda_{\text{DTS}} = 0°$. Conversely, if the nadir point is situated elsewhere, the influence of latitude on this factor becomes more conspicuous: The factor κ_{vg} undergoes more significant variations and tends to be asymmetric, and more so as the latitude increases.

2) The factor κ_{vr} is more significantly impacted by the latitude of the Earth compared to the factor κ_{vg}: the factor κ_{vr} is asymmetric regardless of the MBSAR's location. As the nadir point of MBSAR moves toward higher latitudes, the degree of asymmetry becomes more pronounced.

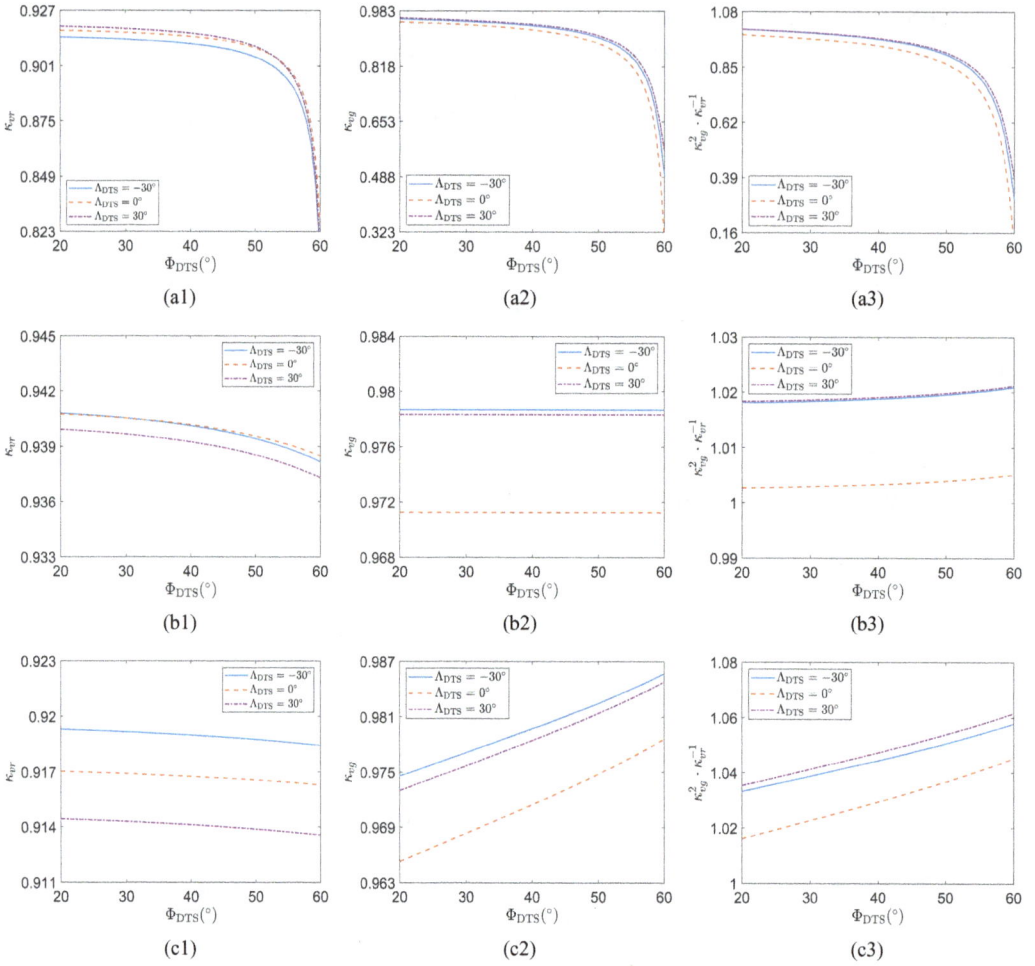

FIGURE 3.3 The factor associated with the effect of MBSAR's inertial motion versus Φ_{DTS}, where the 1st (a1–d1), 2nd (a2–d2) and 3rd (a3–d3) columns respectively represent factors κ_{vr}, κ_{vg}, and $\kappa_{vr}^{-1}\kappa_{vg}^{2}$, while the 1st (a1–a3), 2nd (b1–b3), and 3rd (c1–c3) rows stand for the AOLs of 90°, 180°, and 270°.

3) The interplay of κ_{vg} and κ_{vr} affects the MBSAR's azimuth resolution by working in tandem through the coupling factor $\kappa_{vr}^{-1}\kappa_{vg}^{2}$. Indeed, the coupling effect is less pronounced when observing at low-to-mid Earth latitudes, while it is significantly magnified in high-latitude regions. The inertial motion's influence on MBSAR also correlates with its nadir point's latitude: raising the nadir point's latitude has a more significant impact on MBSAR's azimuth resolution.

The impact of MBSAR's inertial motion on azimuth resolution may vary over time as the perturbation effects can affect its orbital elements. Consequently, it is crucial to inspect the variations in the factor $\kappa_{vr}^{-1}\kappa_{vg}^{2}$ versus Φ_{DTS} and Λ_{DTS} across various cycles, as respectively depicted in Figures 3.5 and 3.6, with the MBSAR located at the AOL of 180°.

Figures 3.5 and 3.6 show that when the MBSAR is located at the descending node, the factors related to the effects of MBSAR's inertial motion on the azimuthal resolution varies across different cycles. Such a variability is highly correlated to the relative position between the MBSAR and TOI, and is relatively small when observing the low-to-mid latitudes of Earth. However, once the MBSAR

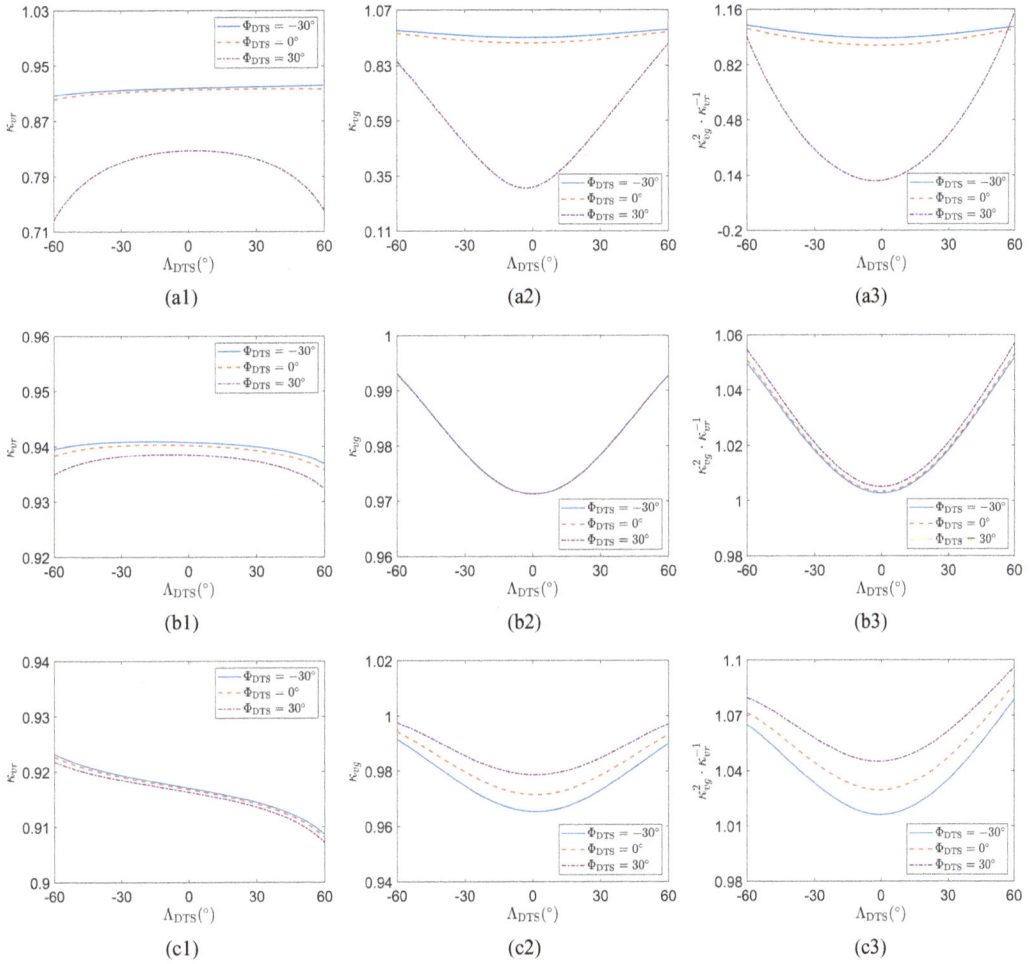

FIGURE 3.4 The factor associated with the effect of MBSAR's inertial motion versus Λ_{DTS}, where the 1st (a1–d1), 2nd (a2–d2) and 3rd (a3–d3) columns respectively represent factors κ_{vr}, κ_{vg}, and $\kappa_{vr}^{-1}\kappa_{vg}^{2}$, while 1st (a1–a3), 2nd (b1–b3), and 3rd (c1–c3) rows stand for the AOLs of 90°, 180°, and 270°.

is positioned near its maximal declination, it can be reasonably inferred that the azimuthal resolution variability would be relatively large.

Hence, we see that the stationary Moon assumption limits the applicability of Eqs. (3.25) and (3.41) to give the azimuthal resolution at high latitudes. Furthermore, it demands a tolerance for spatial resolution errors even at low-to-mid latitudes. In cases where precision is crucial, the azimuthal resolution that relies on the stationary Moon assumption might be called into question and, in some instances, may be deemed impossible. Eqs. (3.59) and (3.67) provide more reliable alternatives for determining the azimuthal resolution by taking account of MBSAR's inertial motion in conjunction with the Earth's curvature.

In the subsequent discussions, we shall expound upon the distinctive attributes of MBSAR's azimuthal resolution in the "spotlight" mode, wherein a relatively consistent azimuthal resolution is achieved across the entire image scene [4]. To provide a systematical analysis, we take account of the azimuthal resolution of the MBSAR in relation to the orbital elements, beam pointing, and side-looking direction, as given in Eq. (3.67).

The use of Eq. (3.67) necessitates configuring the SAT based on Eq. (3.77). To this end, we shall refer to Figure 3.7, which shows the SAT relating to the bound of the slant swath in the zero-Doppler

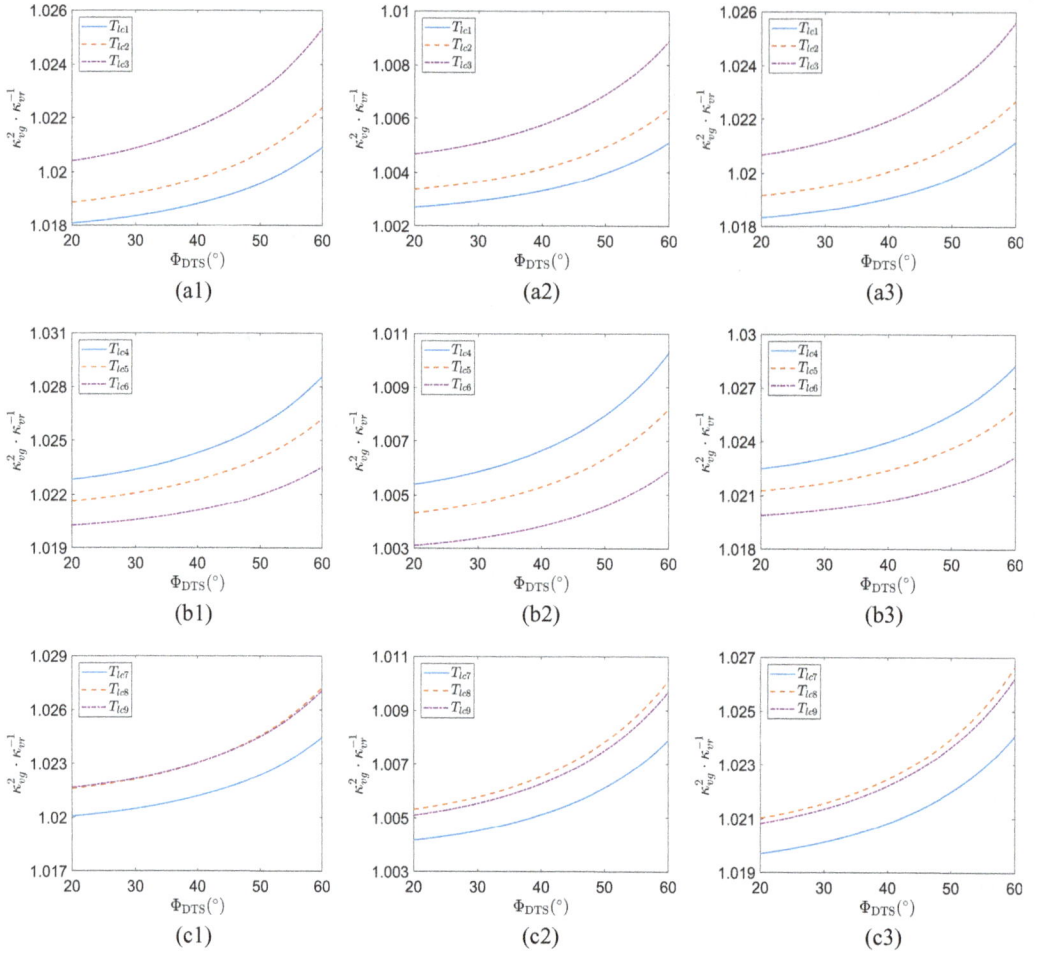

FIGURE 3.5 At the descending node, the factor $\kappa_{vr}^{-1}\kappa_{vg}^{2}$ versus Φ_{DTS} across various cycles, where the 1st (a1–c1), 2nd (a2–c2), and 3rd (a3–c3) columns represent the cases: $\Lambda_{DTS} = -30°$, $0°$, $30°$, while 1st (a1–a3), 2nd (b1–b3), 3rd (c1–c3) rows stand for epochs of $T_{lc1} \sim T_{lc3}$, $T_{lc4} \sim T_{lc6}$, and $T_{lc7} \sim T_{lc9}$, respectively.

steering mode. In this case, the carrier frequency of MBSAR was set to 1.2 GHz at the zero-azimuth time of 00:00:00 on Mar. 20, 2024, TDB.

Figure 3.7 shows that the MBSAR possesses a unique advantage in accomplishing a wide slant swath associated with the coverage performance. Notably, a relatively short SAT offers higher flexibility for achieving extensive coverage; hence, with a carrier frequency at L-band, an SAT ranging from 100 to 200 s is deemed suitable for Earth observation by MBSAR. Further, such an SAT effectively mitigates defocusing phenomena arising from atmospheric or orbital perturbation effects [47–50]. Given these considerations, we shall employ a carrier frequency of 1.2 GHz and an SAT of 200 s in the subsequent analysis regarding the azimuthal resolution of MBSAR.

In the context of MBSAR, the orbital elements experience continuous variations over time due to the significant orbital perturbation effects, distinguishing it from spaceborne SAR. We inspect fluctuations in the azimuthal resolution throughout various cycles. Figure 3.8 depicts the azimuthal resolutions of the MBSAR at different look angles (i.e., 0.2°, 0.5°, and 0.8°) with respect to the AOL across cycles from T_{lc1} to T_{lc9}, where the antenna beam points toward the zero-Doppler plane.

From Figure 3.8, a general pattern of variation in the MBSAR's azimuthal resolution is shown as: The magnitude of azimuthal resolution oscillates, showing trough and crest at the AOLs of 0°

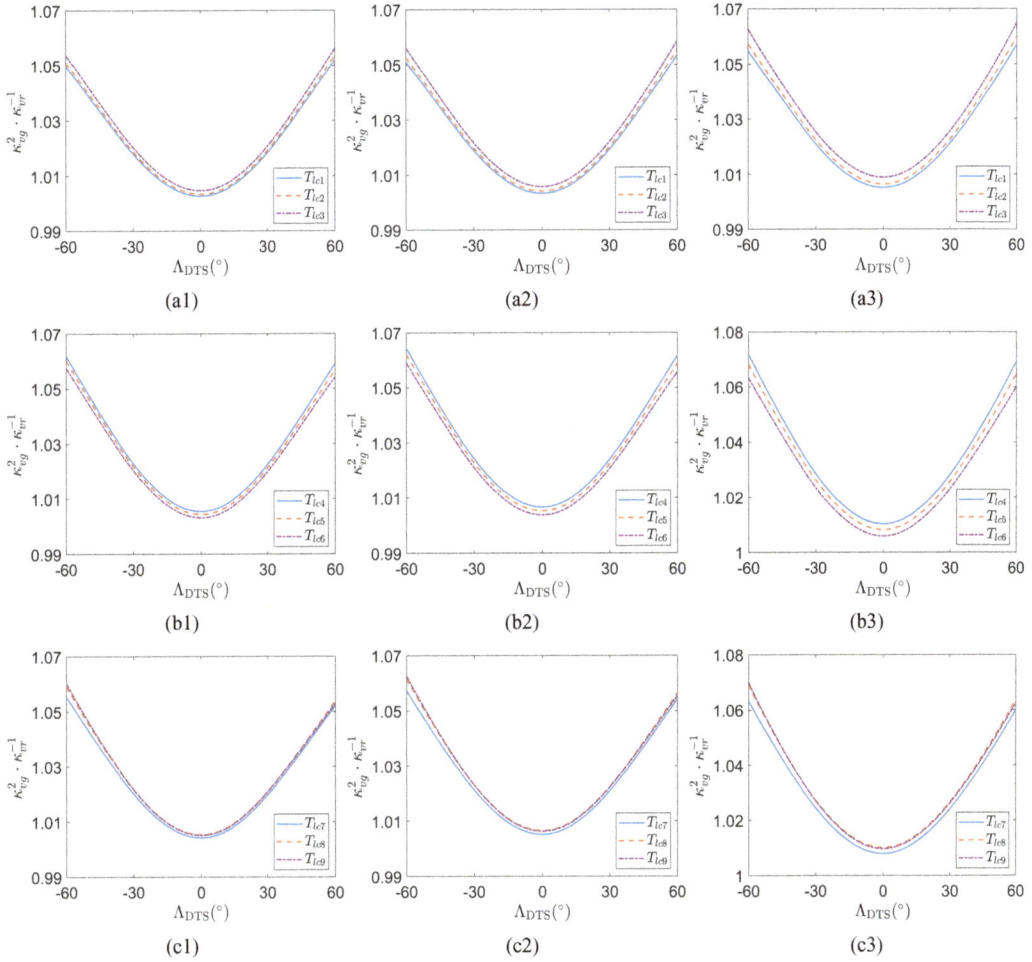

FIGURE 3.6 At the descending node, the factor $\kappa_{vr}^{-1}\kappa_{vg}^{2}$ versus Λ_{DTS} across various cycles, where the 1st (a1–c1), 2nd (a2–c2), and 3rd (a3–c3) columns respectively represent the cases: $\Phi_{DTS} = 20°, 40°, 60°$, while 1st (a1–a3), 2nd (b1–b3), 3rd (c1–c3) rows stand for epochs of $T_{lc1} \sim T_{lc3}$, $T_{lc4} \sim T_{lc6}$, and $T_{lc7} \sim T_{lc9}$.

and approximately 90°. Subsequently, the azimuthal resolution manifests a declining tendency in magnitude concerning the escalating AOL. At approximately 180° of the AOL, the azimuthal resolution's magnitude attains its nadir. This recurrence occurs in the AOL range of 180° to 360°. Further, the magnitude of azimuthal resolution and its variation strongly correlate with MBSAR's look angle regardless of cycles, with the far look angle (i.e., 0.8° in this case) exerting a more significant impact.

Upon considering the diverse orbital inclinations throughout these cycles, a more intricate pattern of variation regarding MBSAR's azimuthal resolution can be discerned, further explained below.

1) Within the cycles of $T_{lc1} \sim T_{lc3}$, the fluctuation in azimuthal resolution correlates to the MBSAR's location. For the AOL spanning from 0° to 180°, a significant fluctuation in azimuthal resolution can be observed across different cycles when a look angle of 0.8° is utilized. On the other hand, a comparison of azimuthal resolution at look angles of 0.2° and 0.5° reveals that the corresponding differences remain consistently insignificant across cycles. When the MBSAR is located at AOL from 180° to 360°, the relative variation of azimuthal resolution across each cycle is comparatively small regardless of the look angle.

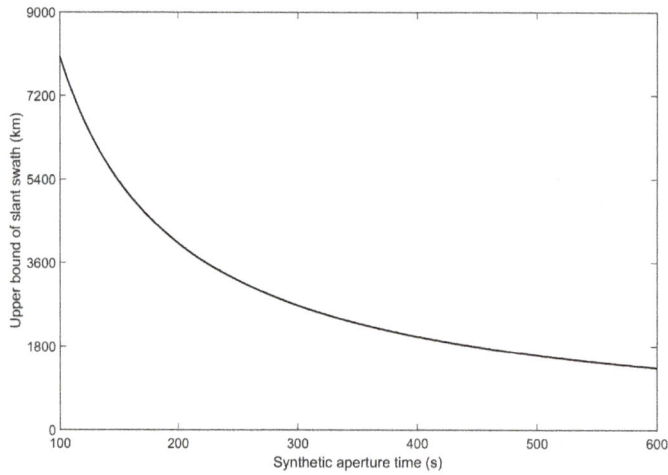

FIGURE 3.7 The relationship between the SAT and upper bound of the slant swath in the MBSAR.

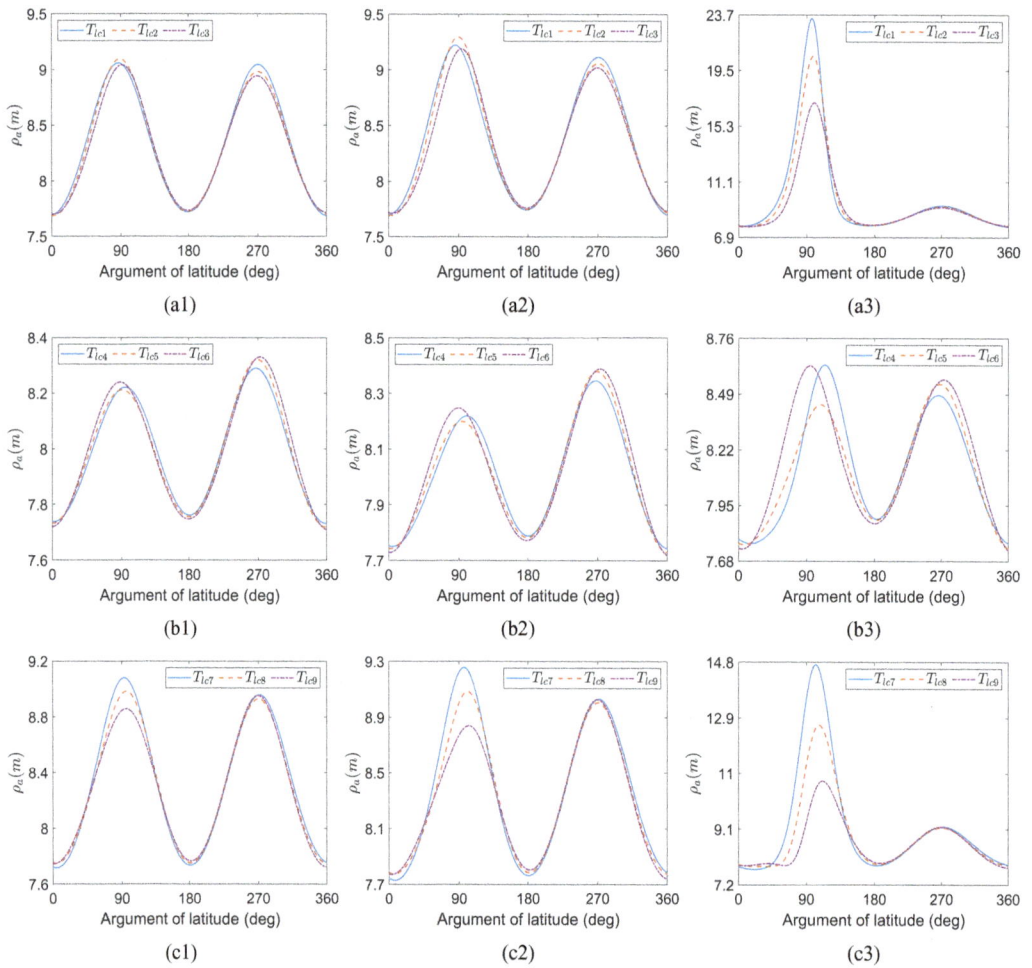

FIGURE 3.8 The azimuthal resolution of the MBSAR versus the AOL across various cycles, where the 1st (a1–c1), 2nd (a2–c2), and 3rd (a3–c3) columns respectively represent the look angles of 0.2°, 0.5°, and 0.8°, while 1st (a1–a3), 2nd (b1–b3), 3rd (c1–c3) rows stand for epochs of $T_{lc1} \sim T_{lc3}$, $T_{lc4} \sim T_{lc6}$, and $T_{lc7} \sim T_{lc9}$.

2) In the epochs from T_{lc4} to T_{lc6}, the first peak of the azimuthal resolution's variation curve exhibits a significant magnitude reduction with 0.8° of look angle, as compared to cycles of $T_{lc1} \sim T_{lc3}$. This suggests the MBSAR's azimuthal resolution improves as the orbital inclination decreases. Also, the nonuniform characteristics of azimuthal resolution among cycles are minimized. It can be inferred that the MBSAR has the potential to produce a more consistent and refined azimuthal resolution in cases where the orbital inclination is minimal. Another finding pertains to the migration of peaks on curves that depict alterations in azimuthal resolution across different cycles. This occurrence is particularly conspicuous when the look angle is large. Therefore, orbital perturbations may impact the distribution of MBSAR's azimuthal resolution along its orbit, particularly under a small orbital inclination condition.

3) Upon examination of Figure 3.8(c1)–(c3), it is clear that after a period of 18.6 years, the variation of azimuthal resolution with respect to AOL deviates from that depicted in Figure 3.8 (a1)–(a3), despite there exists comparable regularity. Hence, the fluctuation in azimuthal resolution under orbital perturbations lacks a definitive cycle. The reason behind the irregular fluctuations lies in the existence of distinct periods corresponding to various orbital elements. Nonetheless, a noticeable trend can be observed where the fluctuation curve of azimuthal resolution for the MBSAR displays a prominent peak, especially when the look angle is large.

The MBSAR's azimuthal resolution shown in Figure 3.8 differs slightly from that in [4]. The reason is that DE 440 was used in plotting Figure 3.8 and considered an antenna factor of 0.886 and lunar physical librations, while DE 430 was used in [4] only.

We find that the azimuthal resolution of MBSAR is also influenced by the side-looking direction (left-looking or right-looking). Henceforth, we shall delve into the consequences of this impact on the azimuthal resolution in the MBSAR. For this purpose, we employ identical parameters illustrated in Figure 3.8 in the following analysis. Figures 3.9–3.11 show variation in azimuthal resolution of left-looking or right-looking MBSAR across various cycles, as a function of AOL, under look angles of 0.2°, 0.5°, and 0.8°, respectively.

The effect of side-looking direction on the azimuthal resolution is connected to the MBSAR's position, orbiting cycle, and look angle, as shown in Figures 3.9–3.11. The discrepancy in azimuthal resolution between the left-looking and right-looking MBSAR is insignificant at a small look angle, e.g., 0.2°. As the look angle increases, the impact of the side-looking direction on the MBSAR's azimuthal resolution increases. A considerable divergence in azimuthal resolutions exists at a look angle of 0.8° between the left-looking and right-looking MBSARs. Throughout various cycles, the left-looking azimuthal resolution shows different tendencies compared to the right-looking one. The MBSAR's azimuthal resolution is subject to various sources of influence, rendering it challenging to achieve a consistent azimuthal resolution per cycle that is typically observed by spaceborne SAR.

We have delved into the MBSAR's azimuthal resolution variation in the zero-Doppler steering mode. However, in practical scenarios, MBSAR may operate in the squint-looking mode, particularly when a vast spatial coverage is required in a short duration [12, 27]. Therefore, we shall examine the fluctuation pattern of azimuthal resolution over distinct cycles when the MBSAR works in the squint-looking mode, as presented in Figures 3.12–3.14.

Figures 3.12–3.14 show that in the squint-looking mode, the azimuthal resolution is subject to variation with the changing AOL within one cycle of MBSAR. Besides, each cycle correlates with distinctive fluctuation regularity in the azimuthal resolution. There are notable discrepancies in fluctuation regularities of azimuthal resolution when the orbital inclination approaches its minimum and maximum values. Further, a discernible fluctuation pattern, which is highly dependent on the look angle, can be observed. For example, at a look angle of 0.2°, the variation pattern of azimuthal resolution with respect to AOL exhibits a resemblance to that under the zero-Doppler steering mode. As the look angle increases, the variation pattern of azimuthal resolution with respect to AOL

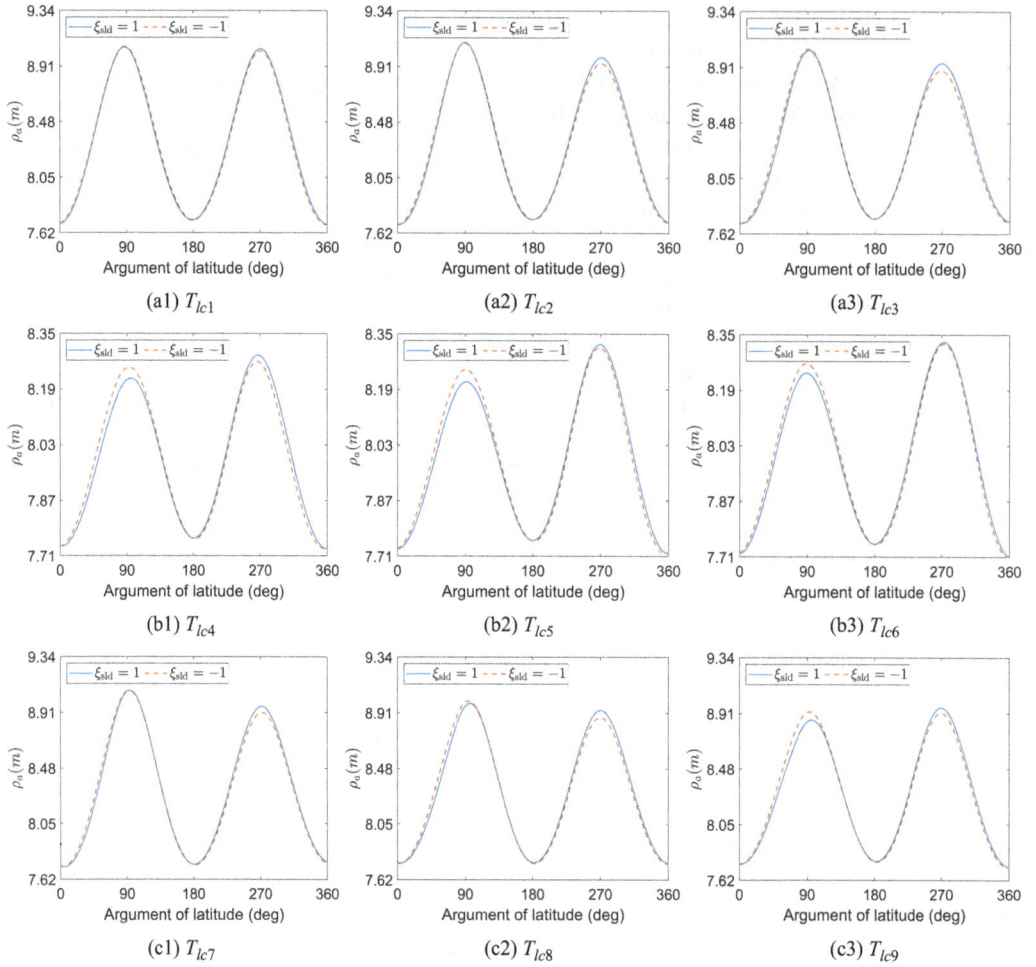

FIGURE 3.9 At a look angle of 0.2°, the comparison of left-looking and right-looking azimuthal resolutions with respect to the AOL in cycles of $T_{lc1} \sim T_{lc9}$, where $\xi_{sld} = 1$ and $\xi_{sld} = -1$ denotes the MBSAR is left-looking and right-looking, respectively.

exhibits a markedly distinct pattern compared to that observed in Figures 3.8–3.11. The most prominent distinction is noticeable at a look angle of 0.8°, where the variation trend of azimuthal resolution deviates entirely from that under the zero-Doppler steering mode. At this juncture, a solitary peak is discernible in the variation curve of the azimuthal resolution, as opposed to the two peaks visible at look angles of 0.2° and 0.5°. Moreover, in contrast to the zero-Doppler steering mode, the azimuthal resolution of MBSAR is not exclusively influenced by the look angle; it is also affected by the antenna azimuth angle under the squint-looking mode. Consequently, the phenomenon of squint-looking not only results in image distortion (as will be expounded upon in Chapter 6), but also disturbs the azimuthal resolution, ultimately deteriorating the MBSAR's imaging performance.

3.4.2 ILLUSTRATIONS OF THE SNR AND NESZ

As previously in Section 3.4.1, the squint-looking mode could exert adverse effects on the MBSAR's imaging performance. Henceforth, the subsequent analysis of the PRF and NESZ is carried out under the zero-Doppler steering mode.

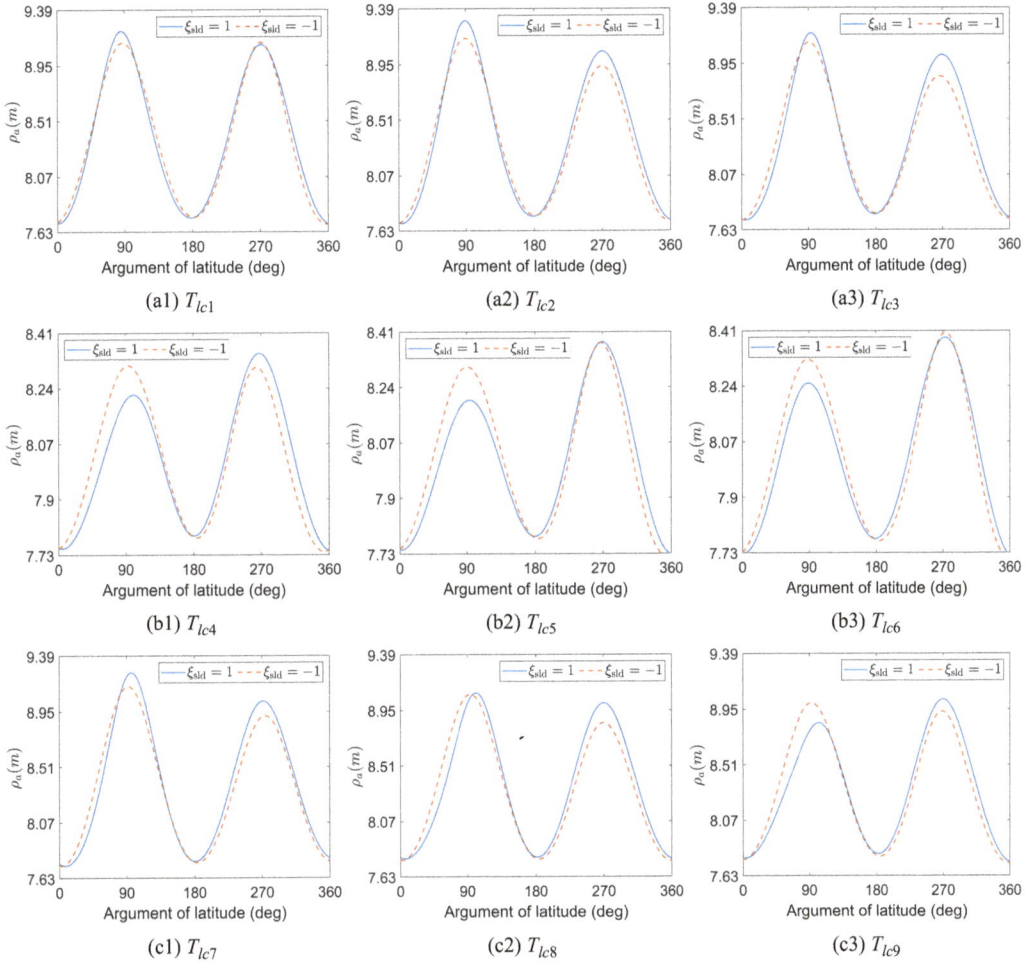

FIGURE 3.10 At a look angle of 0.5°, the comparison of left-looking and right-looking azimuthal resolutions with respect to the AOL in cycles of $T_{lc1} \sim T_{lc9}$, where $\xi_{sld} = 1$ and $\xi_{sld} = -1$ denotes the MBSAR is left-looking and right-looking, respectively.

3.4.2.1 On Selection of PRF in the MBSAR

When it comes to selecting the PRF for the MBSAR, one should bear in mind that both slant swath and Doppler bandwidth are determined by SAR configurations. One may select the near-look angle of 0.2° and beamwidth of 0.6°, along with SAT of 200 s, to examine the lower and upper bounds for the MBSAR's PRF across different cycles. The corresponding results are displayed in Figure 3.15 as a function of AOL. In this case, neither the azimuth safety factor nor the range safety factor is considered.

It is evident that the MBSAR has varying upper and lower bounds for its PRF at different AOLs within one cycle, and both bounds exhibit various regularities across different cycles. When considering the selection of PRF, it is important to contemplate its suitability throughout the entire cycle or even the entire lifespan of the MBSAR system. Hence, the tolerant range for selecting the MBSAR's PRF during a specific epoch must be the constrained range between the largest lower and smallest upper bounds within the time span. This certainly curtails the PRF options for MBSAR. In this instance, Δf_{PRF}, the acceptable range for PRF during each cycle is limited to a narrow frequency band

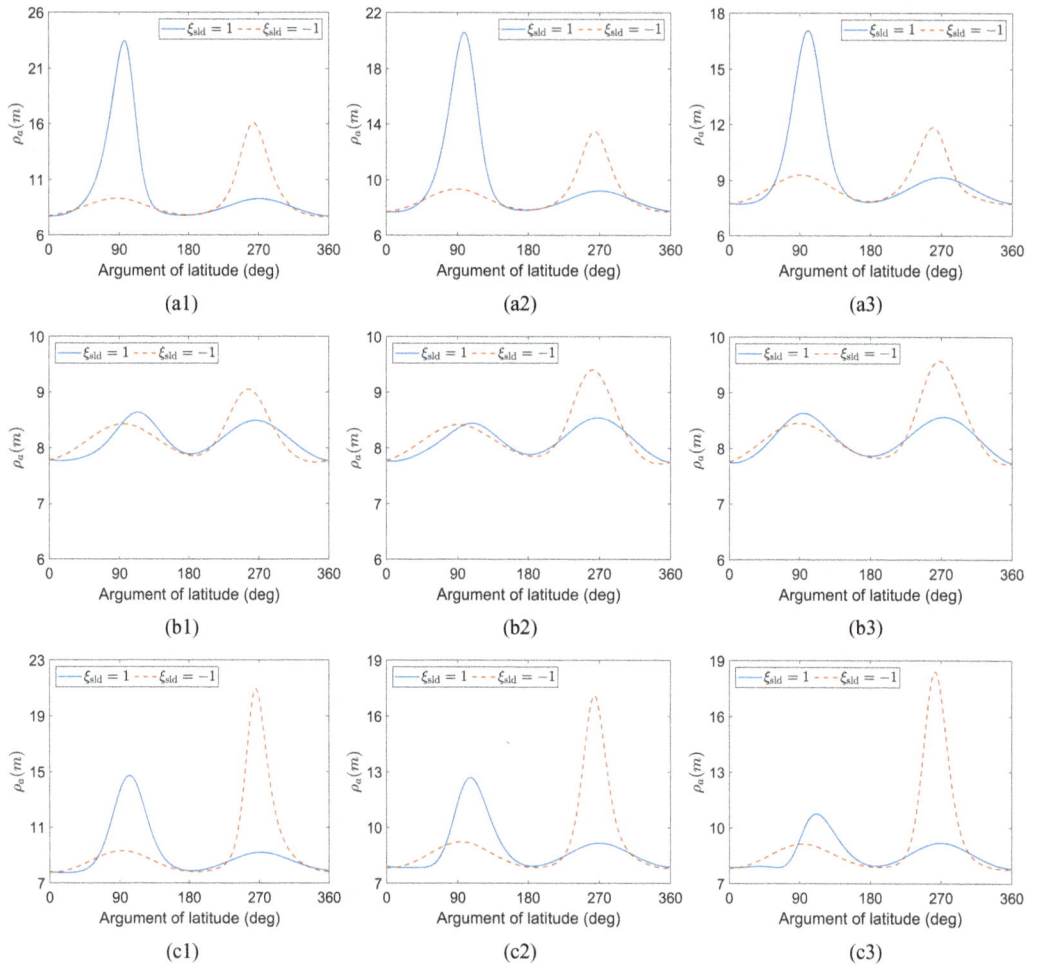

FIGURE 3.11 At a look angle of 0.8°, the comparison of left-looking and right-looking azimuthal resolutions with respect to the AOL in cycles of $T_{lc1} \sim T_{lc9}$, where $\xi_{sld} = 1$ and $\xi_{sld} = -1$ denotes the MBSAR is left-looking and right-looking, respectively.

ranging from 41.81 Hz to 43.81 Hz. This acceptable range becomes even narrower once the antenna beam edges are considered.

As the antenna beam typically shows an edge effect in practice, the safety factors a_{az} and a_{rg}, both greater than 1 in general, may be included. Consequently, both upper and lower bounds for the PRF are adjusted accordingly. Given both safety factors of 1.3, Figure 3.16 illustrates the modified upper and lower bounds for the MBSAR's PRF across multiple cycles. As shown in Figure 3.16, the upper threshold of PRF significantly decreases compared to that in Figure 3.15. In contrast, the lower bound of PRF increases considerably. As a result, the acceptable range of PRF, Δf_{PRF}, is further restricted in the MBSAR, within the range of 5.03–6.76 Hz.

We further examine the variation of Δf_{PRF} with a near-look angle of 0.2° to understand the tolerance range of MBSAR's PRF in various cycles as a function of SAT, as shown in Figure 3.17. It is apparent that despite slight variations in the specific details of the tolerance range of PRF over various cycles, their general trend remains consistent versus the SAT under the specified antenna beamwidth. When the beamwidth is narrower, which corresponds to a smaller coverage width, the tolerance range of PRF widens for a specified SAT. Moreover, this narrower

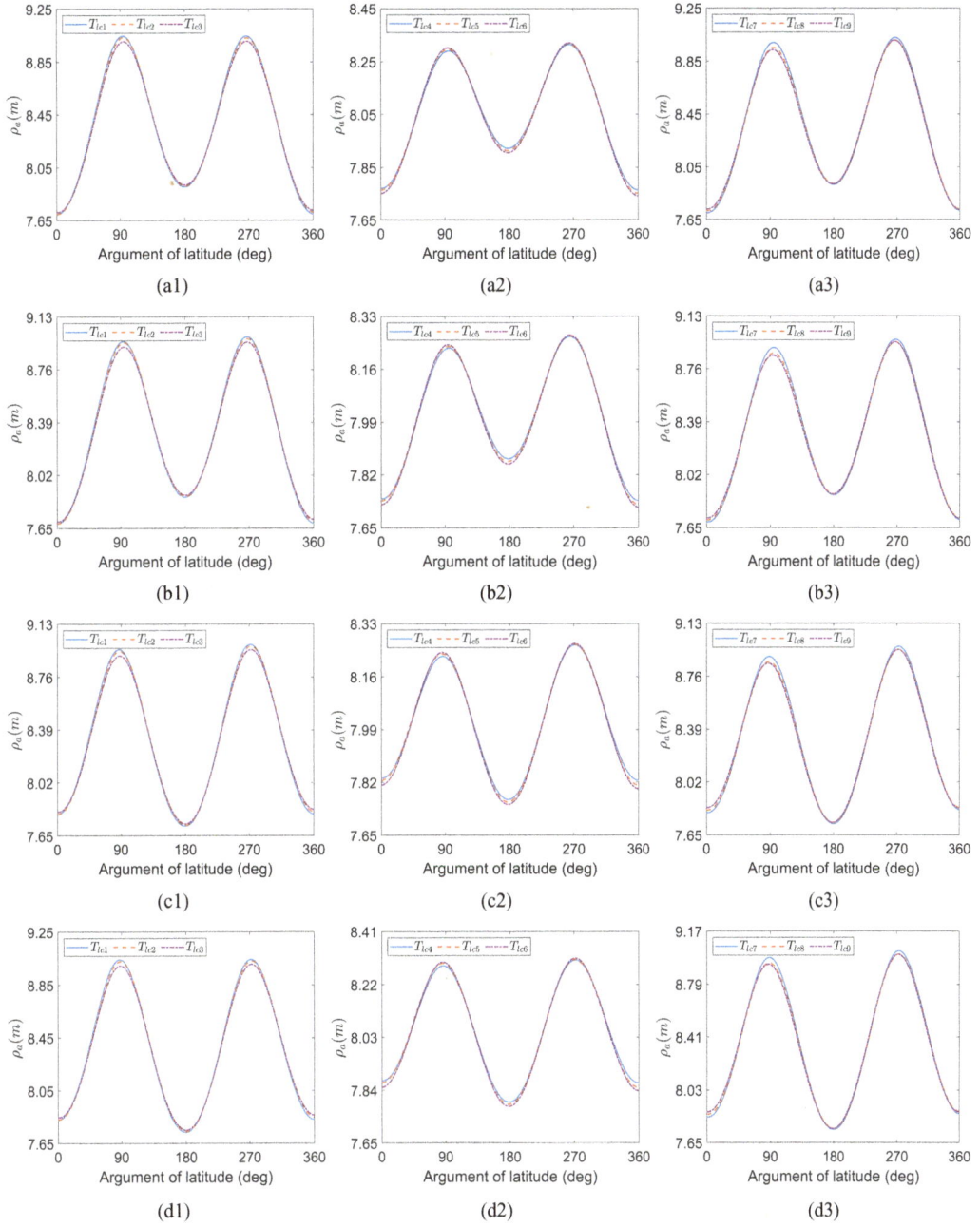

FIGURE 3.12 At a look angle of $0.2°$, the azimuthal resolutions across various cycles with respect to the AOL under various antenna azimuth angles, where the 1st (a1–a3), 2nd (b1–b3), 3rd (c1–c3), and 4th (d1–d3) rows represent the antenna azimuth angles of $45°$, $60°$, $120°$, and $135°$, while 1st (a1–d1), 2nd (a2–d2) and 3rd (a3–d3) columns stand for epochs of $T_{lc1} \sim T_{lc3}$, $T_{lc4} \sim T_{lc6}$, and $T_{lc7} \sim T_{lc9}$.

beamwidth allows greater flexibility in designing the SAT (or the azimuth resolution). The coverage width expands as the beamwidth broadens at the expense of reducing the usable range of PRF and the adaptability in designing SAT (azimuthal resolution). The augmentation in swath width is accompanied by a concomitant limitation in the range for selecting PRF and degradation in azimuth resolution.

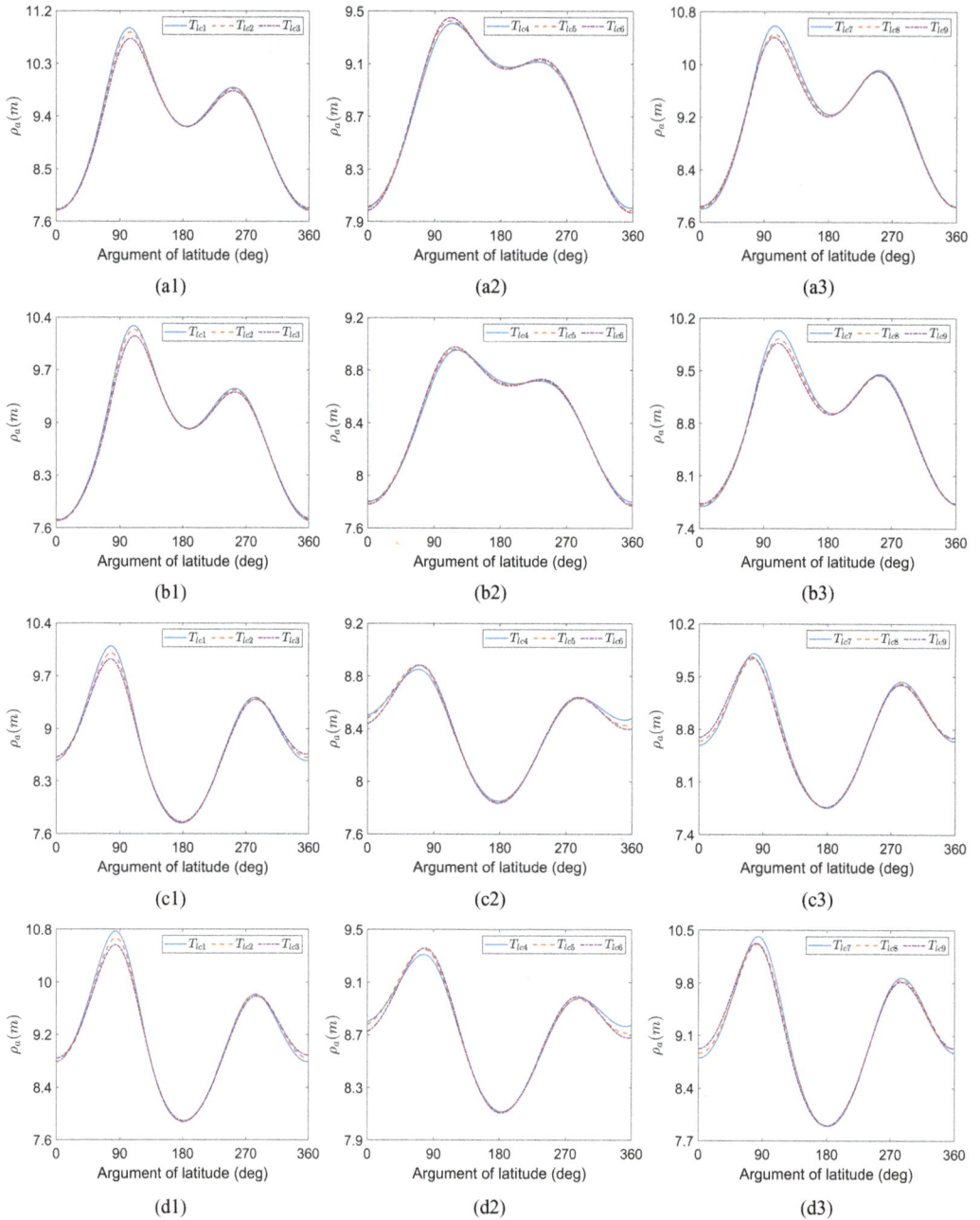

FIGURE 3.13 At a look angle of 0.5°, the azimuthal resolutions across various cycles with respect to the AOL under various antenna azimuth angles, where the 1st (a1–a3), 2nd (b1–b3), 3rd (c1–c3), and 4th (d1–d3) rows represent the antenna azimuth angles of 45°, 60°, 120°, and 135°, while 1st (a1–d1), 2nd (a2–d2) and 3rd (a3–d3) columns stand for epochs of $T_{lc1} \sim T_{lc3}$, $T_{lc4} \sim T_{lc6}$, and $T_{lc7} \sim T_{lc9}$.

3.4.2.2 Spatiotemporal Variation of the MBSAR's NESZ

Given the extensive image scene of the MBSAR, it is conceivable for the radiometric resolution to exhibit spatial variability within the swath width. Furthermore, considering the orbital perturbation effects, the parameters linked to MBSAR's orbit in radiometric resolution may also experience

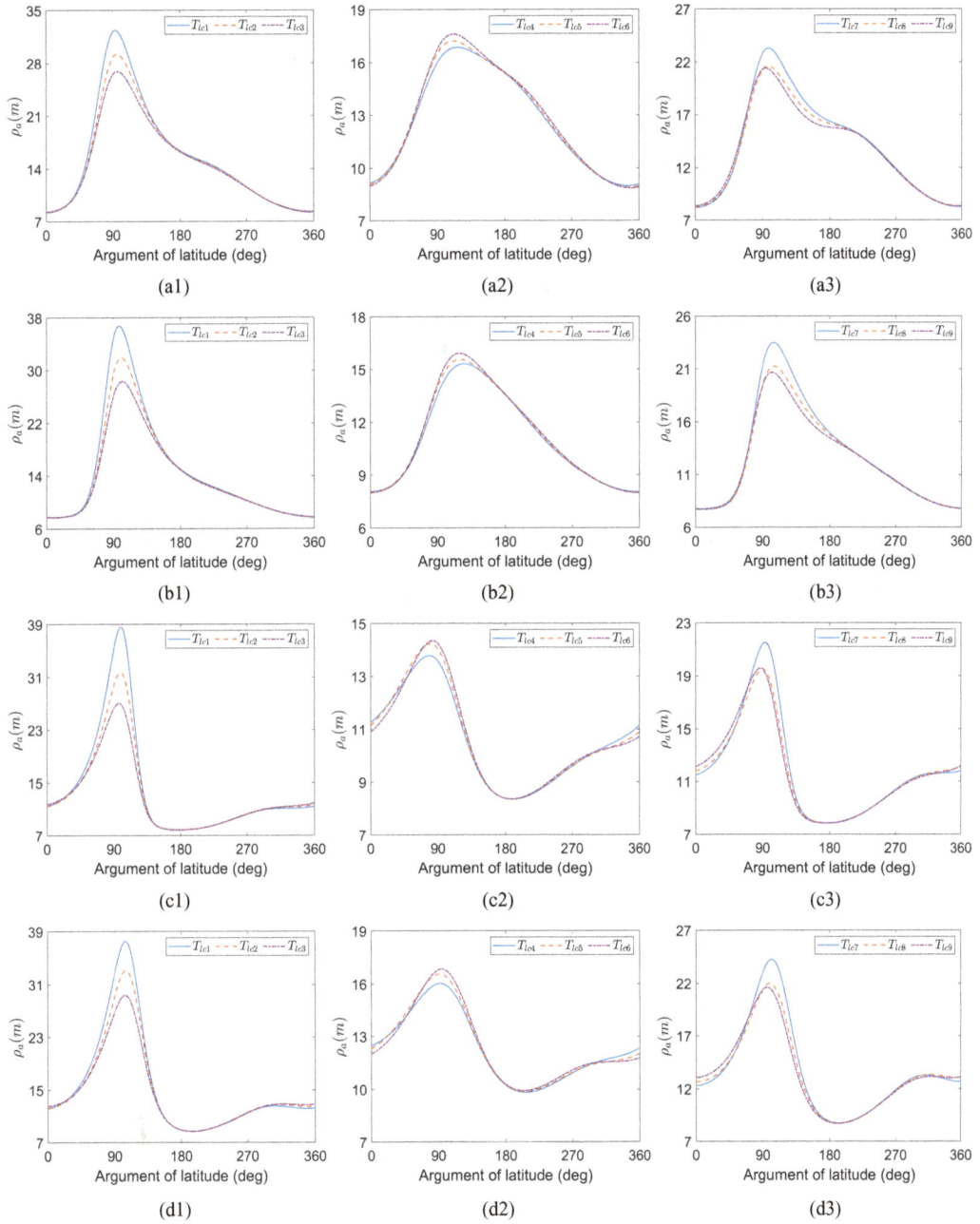

FIGURE 3.14 At a look angle of $0.8°$, the azimuthal resolutions across various cycles with respect to the AOL under various antenna azimuth angles, where the 1st (a1–a3), 2nd (b1–b3), 3rd (c1–c3), and 4th (d1–d3) rows represent the antenna azimuth angles of $45°$, $60°$, $120°$, and $135°$, while 1st (a1–d1), 2nd (a2–d2) and 3rd (a3–d3) columns stand for epochs of $T_{lc1} \sim T_{lc3}$, $T_{lc4} \sim T_{lc6}$, and $T_{lc7} \sim T_{lc9}$.

temporal fluctuations. As a result, there is spatiotemporal variation in the radiometric resolution of the MBSAR. Let us investigate the correlation between ΔNESZ and the look angle and AOL, shown in Figure 3.18, where ΔNESZ is the difference between the actual NESZ and reference NESZ for a look angle of $0.2°$ at the onset of cycle T_{lc1}.

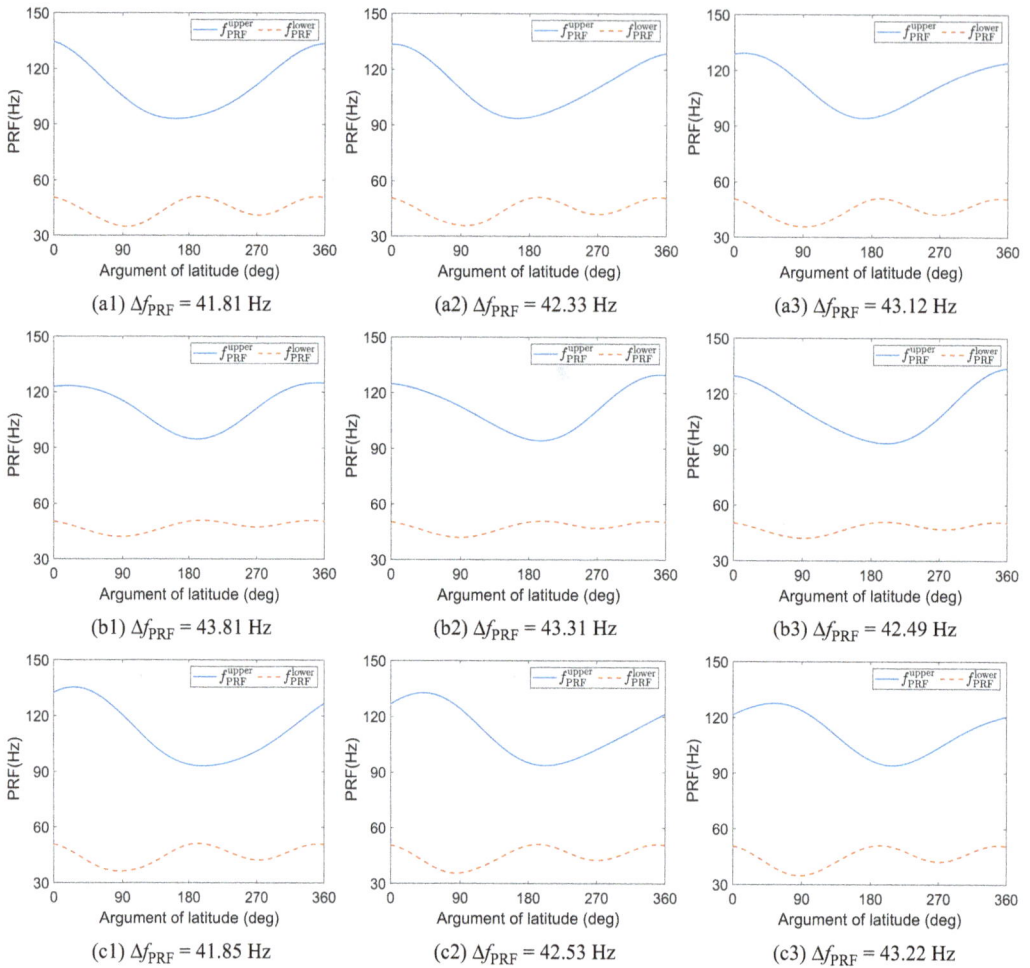

FIGURE 3.15 When disregarding azimuth and range safety factors, the lower and upper bounds of PRF versus the AOL across cycles from T_{lc1} to T_{lc9}, where f_{PRF}^{lower} and f_{PRF}^{upper} represents the lower and upper bounds, respectively.

As depicted in Figure 3.18, the radiometric resolution fluctuates with changes in both the look angle and AOL during a given cycle. For most of MBSAR's locations, an increase in the look angle leads to an increased magnitude of ΔNESZ, implying that the radiometric resolution is essentially decreasing. However, at an AOL of around 90°, there is a discernible decrease in ΔNESZ as the look angle surpasses a particular threshold.

We shall see the variation in MBSAR's radiometric resolution across various cycles. Employing the same reference NESZ as depicted in Figure 3.18, Figure 3.19 displays ΔNESZ at various AOLs with respect to look angles across different cycles. Further, Figure 3.20 illustrates the fluctuation of ΔNESZ for the AOLs under various look angles and across multiple cycles.

As revealed in Figures 3.19 and 3.20, the magnitude of ΔNESZ tends to increase with the increasing look angle at varying extents across different cycles, although there are a few cases where this trend does not hold. Hence, it can be inferred that the radiometric resolution tends to diminish as the look angle increases in most cases. The fluctuation observed within each cycle indicates that the radiometric resolution is indeed influenced by orbital perturbations. In the context of large orbital inclination, when the MBSAR is situated around the AOL of 90°, it can be noticed that initially, there is an increase in ΔNESZ with the increasing look angle. However, the radar beam is directed

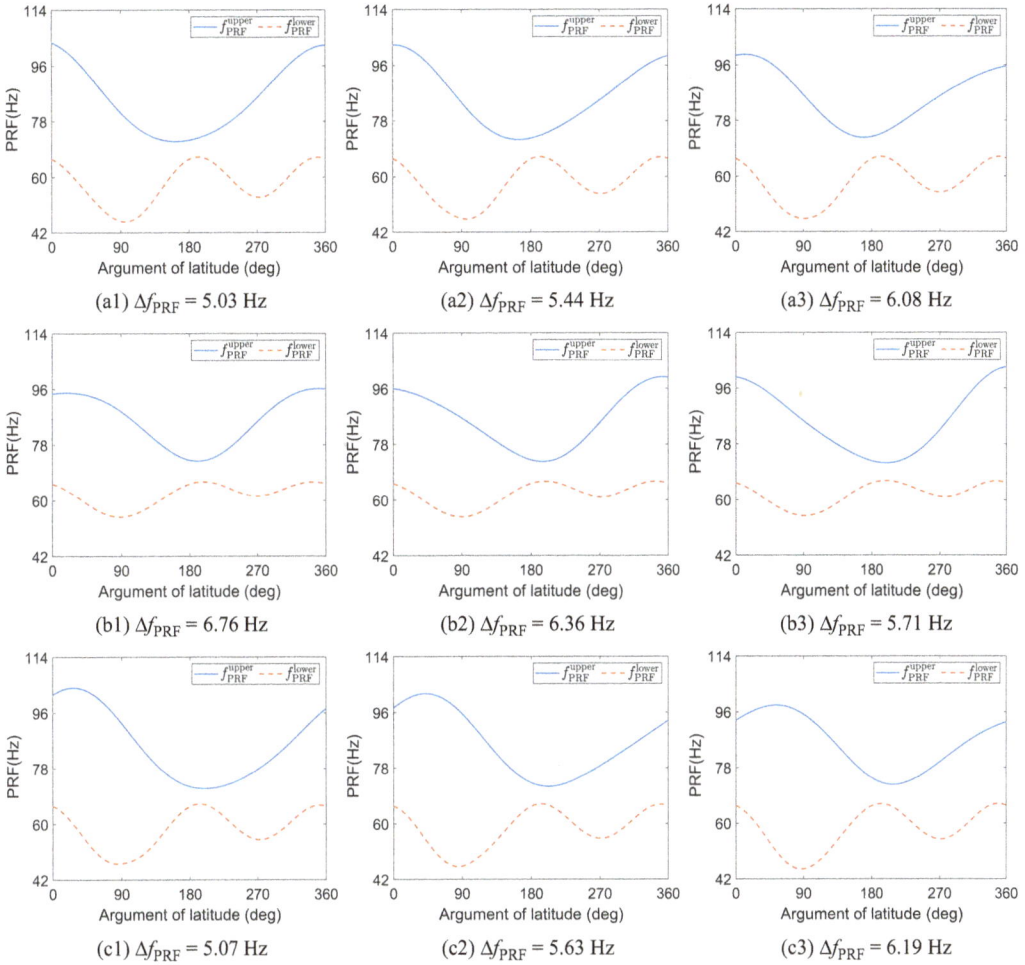

FIGURE 3.16 With the range and azimuth safety factors of 1.3, the lower and upper bounds of PRF versus the AOL across cycles from T_{lc1} to T_{lc9}, where f_{PRF}^{lower} and f_{PRF}^{upper} represents the lower and upper bounds, respectively.

toward higher latitudes (polar regions) at larger look angles, leading to a noticeable decrease in ΔNESZ. This suggests that the radiometric resolution of the MBSAR improves in this circumstance. After conducting a more thorough analysis, it is discovered that while different cycles may result in variations in radiometric resolution, their influence is relatively minor compared to that of the look angle: The reduction in radiometric resolution due to an increase in look angle is considerably more substantial. Based on this analysis, we conclude that the influence of orbital perturbations on MBSAR's radiometric resolution is comparatively insignificant. By contrast, the observation geometry and coverage width are the principal factors determining the radiometric resolution of the MBSAR.

Compared to the spaceborne SAR, the MBSAR stands out for its exceptional altitude and unique orbit. While the Earth's rotation is the primary factor contributing to Doppler effects in the MBSAR, the Moon's inertial motion must also be taken into account, leading to more complex imaging modes. Additionally, the MBSAR's configuration design must consider both Earth's rotation and MBSAR's inertial motions when determining spatial resolution, power budget, and PRF selection. This chapter offers a comprehensive discussion of these parameters in the context of MBSAR, thus providing an informative reference for the future design and implementation of the MBSAR system. It is imperative to additionally consider the spatiotemporal coverage of MBSAR when discussing its configuration design. We shall explore this matter in-depth through quantitative analysis in the next chapter.

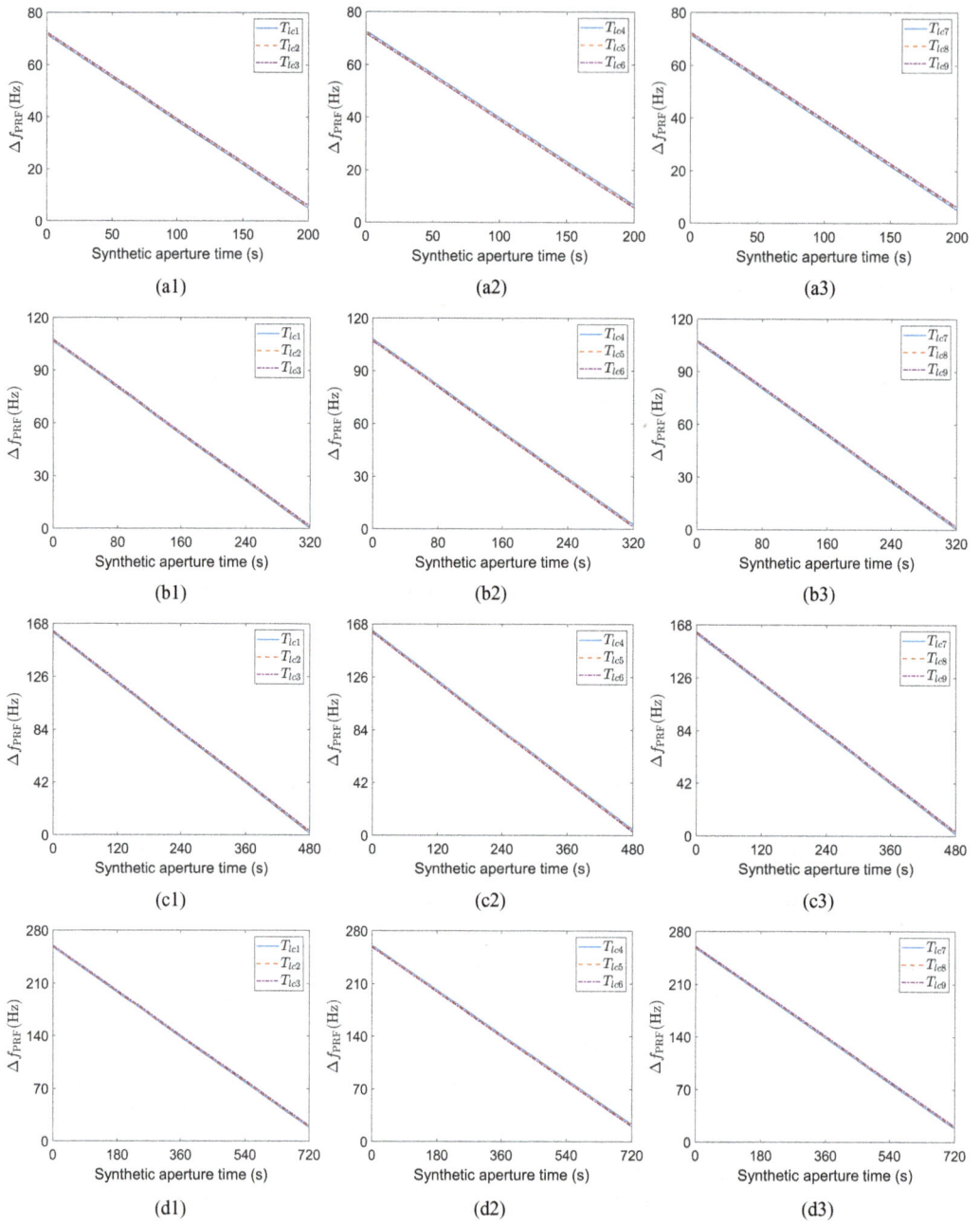

FIGURE 3.17 With the range and azimuth safety factors of 1.3, the Δf_{PRF} versus SAT across various cycles under different beamwidths, where the 1st (a1–d1), 2nd (a2–d2), and 3rd (a3–d3) columns respectively represent epochs of $T_{lc1} \sim T_{lc3}$, $T_{lc4} \sim T_{lc6}$, and $T_{lc7} \sim T_{lc9}$, while 1st (a1–a3), 2nd (b1–b3), 3rd (c1–c3) and 4th (d1–d3) rows stand for the beamwidths of 0.6°, 0.5°, 0.4°, and 0.3°, respectively.

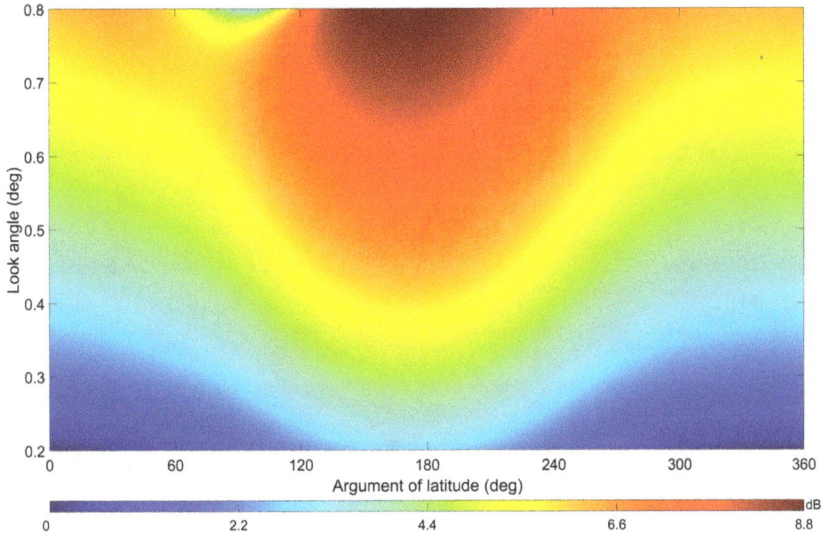

FIGURE 3.18   In cycle T_{lc1}, ΔNESZ versus AOL and look angle in the MBSAR.

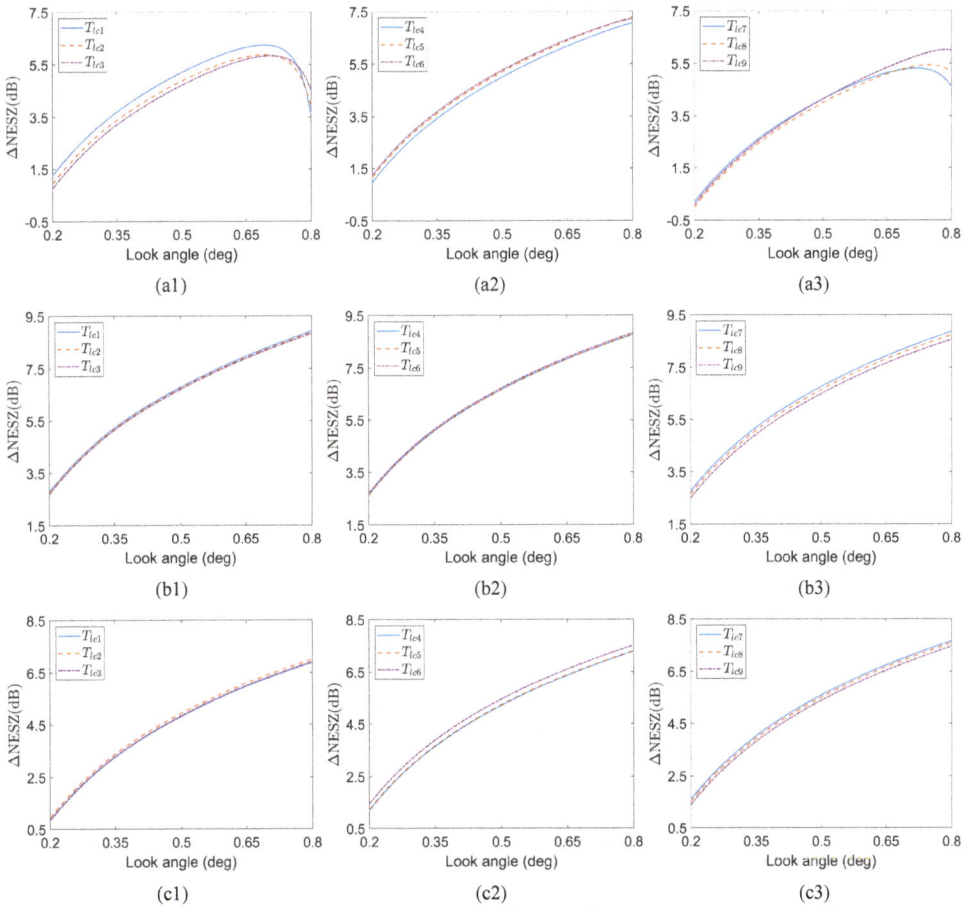

FIGURE 3.19   At different locations of MBSAR, ΔNESZ versus beamwidth across various cycles, where the 1st (a1–d1), 2nd (a2–d2), and 3rd (a3–d3) columns respectively represent epochs of $T_{lc1} \sim T_{lc3}$, $T_{lc4} \sim T_{lc6}$, and $T_{lc7} \sim T_{lc9}$, while 1st (a1–a3), 2nd (b1–b3), and 3rd (c1–c3) rows stand for the AOLs of 90°, 180°, and 270°, respectively.

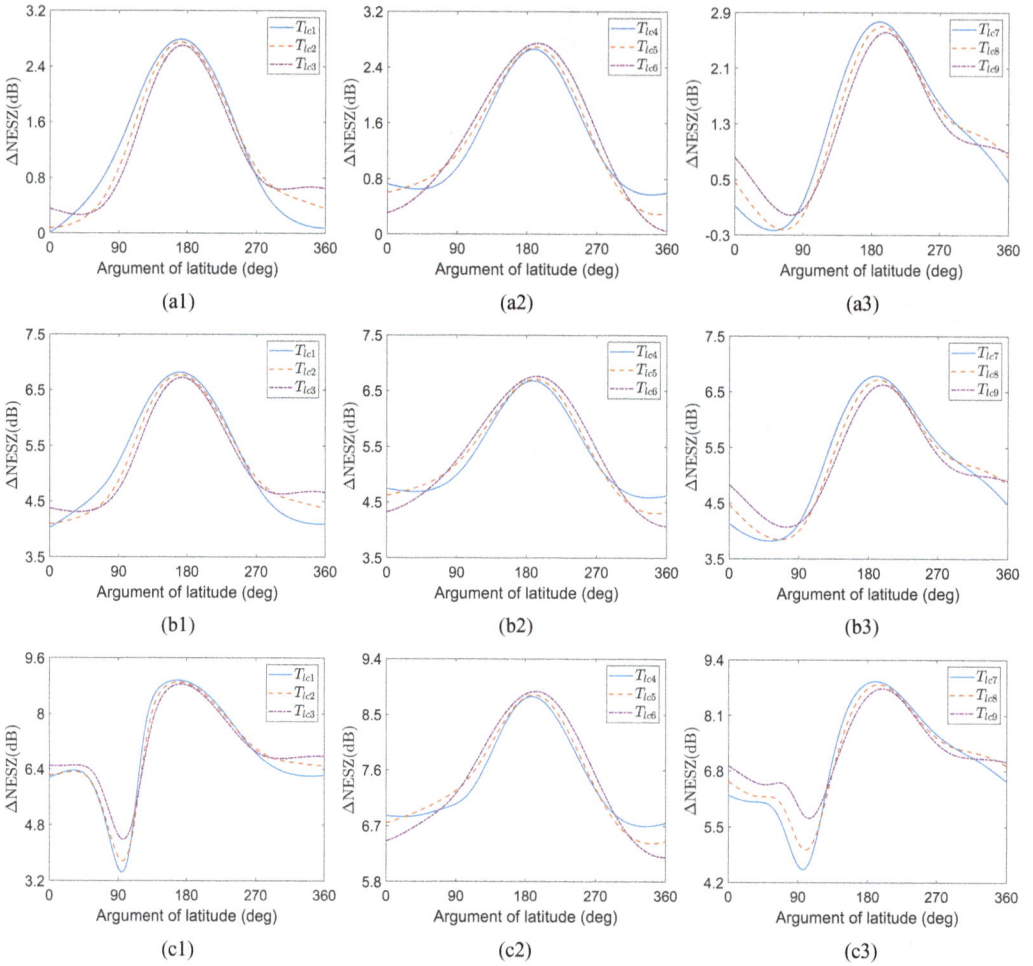

FIGURE 3.20 At various look angles, ΔNESZ versus AOL over various cycles, where the 1st (a1–c1), 2nd (a2–c2), and 3rd (a3–c3) columns respectively represent epochs of $T_{lc1} \sim T_{lc3}$, $T_{lc4} \sim T_{lc6}$, and $T_{lc7} \sim T_{lc9}$, while 1st (a1–a3), 2nd (b1–b3), and 3rd (c1–c3) rows stand for look angles of 0.2°, 0.5°, 0.8°, respectively.

REFERENCES

[1] M. Ramsey, "ESS Science Planning and Lunar Workshop Overview," presented at the *NASA Advisory Council Workshop on Science Associated with the Lunar Exploration Architecture*, Feb. 27–Mar. 2, 2007. [Online]. Available at: http://www.lpi.usra.edu/meetings/LEA/presentations/OpeningPlenary/Ramsey_ESS.pdf

[2] M. Ramsey, "ESS Findings: Lunar Science Planning and Workshop Overview," presented at the *NASA Advisory Council Workshop on Science Associated with the Lunar Exploration Architecture*, Feb. 27–Mar. 2, 2007. [Online]. Available at: https://www.lpi.usra.edu/meetings/LEA/presentations/closing_plenary/Ramsey_ESS_summary_20070302.pdf

[3] J. B. Campbell and R. H. Wynne, *Introduction to Remote Sensing*, 5th ed. New York: Guilford Press, 2011.

[4] Z. Xu, K. S. Chen, and H. Guo, "On Azimuthal Resolution of the Lunar-Based SAR under the Orbital Perturbation Effects," *IEEE Transactions on Geoscience and Remote Sensing*, vol. 61, Apr. 2023, doi: 10.1109/TGRS.2023.3266548

[5] N. Levanon and E. Mozeson, *Radar Signals*. Hoboken: Wiley, 2004.

[6] F. T. Ulaby and D. G. Long, *Microwave Radar and Radiometric Remote Sensing*. Ann Arbor: Univ. Michigan Press, 2014.

[7] A. Moreira, et al., "A Tutorial on Synthetic Aperture Radar," *IEEE Geoscience and Remote Sensing Magazine*, vol. 1, no. 1, pp. 6–43, Mar. 2013.

[8] M. I. Skolnik, *Radar Handbook*, 3rd ed. New York: McGraw-Hill Education, 2008.

[9] K. S. Chen, *Principles of Synthetic Aperture Radar: A System Simulation Approach*. Boca Raton: CRC Press, 2015.

[10] J. C. Curlander and R. N. McDonough, *Synthetic Aperture Radar: Systems and Signal Processing*. New York: Wiley, 1991.

[11] I. G. Cumming and F. H. Wong, *Digital Signal Processing of Synthetic Aperture Radar Data: Algorithms and Implementation*. Boston: Artech House, 2005.

[12] Z. Xu, K. S. Chen, and G. Q. Zhou, "Zero-Doppler Centroid Steering for the Moon-Based Synthetic Aperture Radar: A Theoretical Analysis," *IEEE Geoscience and Remote Sensing Letters*, vol. 17, no. 7, Jul. 2020.

[13] A. Renga and A. Moccia, "Preliminary Analysis of a Moon-based Interferometric SAR System for Very High-Resolution Earth Remote Sensing," *Proc. 9th ILEWG International Conference on Exploration and Utilisation of the Moon*, Sorrento, Italy, pp. 22–26, Oct. 2007.

[14] A. Renga, "Configurations and Performance of Moon-based SAR Systems for Very High-Resolution Earth Remote Sensing," *Proc. AIAA Pegasus Aerospace Conference*, Naples, Italy, pp. 12–13, Apr. 2007.

[15] A. Moccia and A. Renga, "Synthetic Aperture Radar for Earth Observation from a Lunar Base: Performance and Potential Applications," *IEEE Transactions on Aerospace and Electronic Systems*, vol. 46, no. 3, pp. 1034–1051, Jul. 2010.

[16] G. Fornaro, et al., "Potentials and Limitations of Moon-Borne SAR Imaging," *IEEE Transactions on Geoscience and Remote Sensing*, vol. 48, no. 7, pp. 3009–3019, Apr. 2010.

[17] H. Guo, et al., "Conceptual Study of Lunar-Based SAR for Global Change Monitoring," *Science China Earth Sciences*, vol. 57, no. 8, pp. 1771–1779, Aug. 2014.

[18] H. Guo, G. Liu, and Y. Ding, "Moon-based Earth Observation: Scientific Concept and Potential Applications," *International Journal of Digital Earth*, vol. 11, no. 6, pp. 546–557, Jul. 2017.

[19] Z. Xu and K. S. Chen, "On Signal Modeling of Moon-Based Synthetic Aperture Radar (SAR) Imaging of Earth," *Remote Sensing*, vol. 10, no. 3, p. 486, Mar. 2018.

[20] A. Renga and A. Moccia, "Moon-Based Synthetic Aperture Radar: Review and Challenges," *Proc. IEEE International Geoscience and Remote Sensing Symposium*, pp. 3708–3711, Jul. 2016.

[21] L. Huang, X. Qiu, D. Hu, and C. Ding, "Focusing of Medium-Earth-Orbit SAR with Advanced Nonlinear Chirp Scaling Algorithm," *IEEE Transactions on Geoscience and Remote Sensing*, vol. 49, no. 1, pp. 500–508, Jan. 2011.

[22] Z. Xu and K. S. Chen, "Effects of the Earth's Curvature and Lunar Revolution on the Imaging Performance of the Moon-Based Synthetic Aperture Radar," *IEEE Transactions on Geoscience and Remote Sensing*, vol. 57, no. 8, pp. 5868–5882, Mar. 2019.

[23] H. Ye, H. Guo, G. Liu, and Y. Ren, "Observation Duration Analysis for Earth Surface Features from a Moon-based Platform," *Advances in Space Research*, vol. 62, no. 2, pp. 274–287, Jul. 2018.

[24] Z. Xu, K. S. Chen, and G. Q. Zhou, "Effects of the Earth's Irregular Rotation on the Moon-Based Synthetic Aperture Radar Imaging," *IEEE ACCESS*, vol. 7, pp. 155014–155027, Oct. 2019.

[25] Z. Xu and K. S. Chen, "Temporal-Spatial Varying Background Ionospheric Effects on the Moon-Based Synthetic Aperture Radar Imaging: A Theoretical Analysis," *IEEE ACCESS*, vol. 6, pp. 66767–66786, Jul. 2018.

[26] M. Shimada, *Imaging from Spaceborne and Airborne SARs, Calibration, and Applications*. Boca Raton: CRC Press, 2018.

[27] Z. Xu, K. S. Chen, G. Liu, and H. Guo, "Spatiotemporal Coverage of a Moon-Based Synthetic Aperture Radar: Theoretical Analyses and Numerical Simulations," *IEEE Transactions on Geoscience and Remote Sensing*, vol. 58, no. 12, pp. 8735–8750, 2020.

[28] S. J. Ostro, "Planetary Radar Astronomy," *Reviews of Modern Physics*, vol. 65, no. 4, p. 1235, Oct. 1993.

[29] V. C. Chen and H. Ling, *Time-Frequency Transforms for Radar Imaging and Signal Analysis*. Norwood: Artech House, 2002.

[30] C. Ozdemir, *Inverse Synthetic Aperture Radar Imaging with MATLAB Algorithms*. Hoboken, NJ: Wiley, 2021.

[31] G. Franceschetti and R. Lanari, *Synthetic Aperture Radar Processing*. Boca Raton: CRC Press, 1999.

[32] H. Maître, *Processing of Synthetic Aperture Radar Images*. Hoboken, NJ: Wiley, 2013.

[33] C. Oliver and S. Quegan, *Understanding Synthetic Aperture Radar Images*. Raleigh: SciTech Publishing, 2004.

[34] M. A. Richards, *Fundamentals of Radar Signal Processing*. New York: McGraw-Hill Education, 2014.

[35] H. Nyquist, "Regeneration Theory," *The Bell System Technical Journal*, vol. 11, no. 1, pp. 126–147, Jan. 1932.

[36] J. Li and P. Stoica, *MIMO Radar Signal Processing*. Hoboken, NJ: Wiley, 2008.

[37] N. Gebert, G. Krieger, and A. Moreira, "Digital Beamforming on Receive: Techniques and Optimization Strategies for High-Resolution Wide-Swath SAR Imaging," *IEEE Transactions on Aerospace and Electronic Systems*, vol. 45, no. 2, pp. 564–592, Apr. 2009.

[38] G. W. Stimson, H. D. Griffiths, C. J. Baker, and A. Dave, *Introduction to Airborne Radar*, 2nd ed. Raleigh: SciTech Publishing, 1998.

[39] R. Wang and Y. Deng, *Bistatic SAR System and Signal Processing Technology*. Singapore: Springer, 2018.

[40] R. K. Moore, J. P. Claassen, and Y. H. Lin, "Scanning Spaceborne Synthetic Aperture Radar with Integrated Radiometer," *IEEE Transactions on Aerospace and Electronic Systems*, vol. AES-17, no. 3, pp. 410–421, May. 1981.

[41] F. Ulaby, M. C. Dobson, and J. L. Álvarez-Pérez, *Handbook of Radar Scattering Statistics for Terrain*. Norwood, MA: Artech House, 2019.

[42] F. T. Ulaby, R. K. Moore, and A. K. Fung, *Microwave Remote Sensing: Active and Passive*. Norwood, MA: Artech House, 1981.

[43] C. V. Jakowatz, et al., *Spotlight-Mode Synthetic Aperture Radar: A Signal Processing Approach*. New York, NY: Springer, 2012.

[44] J. Marquez-Martinez, J. Mittermayer, and M. Rodriguez-Cassola, "Radiometric Resolution Optimization for Future SAR Systems," *Proc. IEEE International Geoscience and Remote Sensing Symposium*, Anchorage, AK, USA, vol. 3, pp. 1738–1741, Sep. 2004.

[45] B. J. Döring and M. Schwerdt, "The Radiometric Measurement Quantity for SAR Images," *IEEE Transactions on Geoscience and Remote Sensing*, vol. 51, no. 12, pp. 5307–5314, Feb. 2013.

[46] M. M. Espeseth, et al., "The Impact of System Noise in Polarimetric SAR Imagery on Oil Spill Observations," *IEEE Transactions on Geoscience and Remote Sensing*, vol. 58, no. 6, pp. 4194–4214, Jun. 2020.

[47] Z. Xu, K. S. Chen, P. Xu, and H. Guo, "Ionospheric Effects on the Lunar-Based Radar Imaging," *Proc. 2017 IEEE International Geoscience and Remote Sensing Symposium*, Fort Worth, USA, pp. 5390–5393, Jul. 2017.

[48] Z. Xu, K. S. Chen, and H. Guo, "Effects of Temporally-Varying Tropospheric Path Delay on the Imaging Performance of Moon-based SAR," *Proc. 2019 Photonics and Electromagnetics Research Symposium - Fall*, Xiamen, China, pp. 617–623, 17–20 Dec. 2019.

[49] Z. Xu, K. S. Chen, Z. L. Li, and G. Y. Du, "Apsidal Precession Effects on the Lunar-Based Synthetic Aperture Radar Imaging Performance," *IEEE Geoscience and Remote Sensing Letters*, vol. 18, no. 6, pp. 1079–1083, Jun. 2021.

[50] Z. Xu, K. S. Chen, and G. Liu, "On Evaluating the Imaging Performance and Orbital Determination under Perturbations of Orbital Inclination and RAAN in the Lunar-Based SAR," *IEEE Transactions on Geoscience and Remote Sensing*, vol. 60, Jul. 2022, doi: 10.1109/TGRS.2022.3188294

4 Spatiotemporal Coverage in the MBSAR System

4.1 SPATIOTEMPORAL COVERAGE OF MBSAR WITH SINGLE ANTENNA

The coverage performance holds immense significance for Earth observation; a higher orbital altitude can provide extensive spatial coverage within a shorter revisit period [1–3]. In the case of Moon-Based SAR (MBSAR) with a single antenna, the spatial coverage is determined by the swath width that awards the extent of the image scene and the trajectory of the beam footprint on Earth [4]. Therefore, evaluating spatial coverage for a given time is dependent on the swath width for the MBSAR in this event.

To accurately determine the location of the beam footprint on Earth and to map the swath width, it is necessary to employ the geometric model in the ECEF coordinate system, as illustrated in Figure 4.1. Notably, all orbital elements discussed in this chapter are defined in the ECEF coordinate system. For the sake of brevity, we shall exclude the additional subscript "ECEF" from symbols associated with orbital elements previously outlined in Chapter 2.

From the MBSAR's observation geometry shown in Figure 4.1, we see that it is possible to obtain the latitude and longitude of the nadir point at a specified time in the ECEF coordinate system, both referring to the geocenter, by [5]:

$$\Phi_{SAR} = \tan^{-1}\left(\frac{Z_{SAR}}{\sqrt{X_{SAR}^2 + Y_{SAR}^2}}\right) \tag{4.1}$$

$$\Lambda_{SAR} = \sin^{-1}\left(\frac{Y_{SAR}}{\sqrt{X_{SAR}^2 + Y_{SAR}^2}}\right) \tag{4.2}$$

where $\mathbf{R}_{SAR} = [X_{SAR}, Y_{SAR}, Z_{SAR}]^T$ is the position of MBSAR in the ECEF coordinate system.

Referring to the nadir point, the far and near-look angles of MBSAR that correspond to far and near slant ranges are related by:

$$\theta_l^{far} = \theta_l^{near} + \beta_r \tag{4.3}$$

where β_r is the beamwidth along the range direction that takes the form of:

$$\beta_r = \lambda k_r \ell_r^{-1} \tag{4.4}$$

where:
- λ is the radar wavelength,
- ℓ_r refers to the aperture length along the range direction,
- k_r is the taper factor related to antenna beam steepness. The taper factor $k_r = 0.88$ in a uniformly distributed current density. When an antenna has a highly tapered distribution, it would have a taper factor of $k_r = 2.0$ [6].

DOI: 10.1201/9781003308430-4

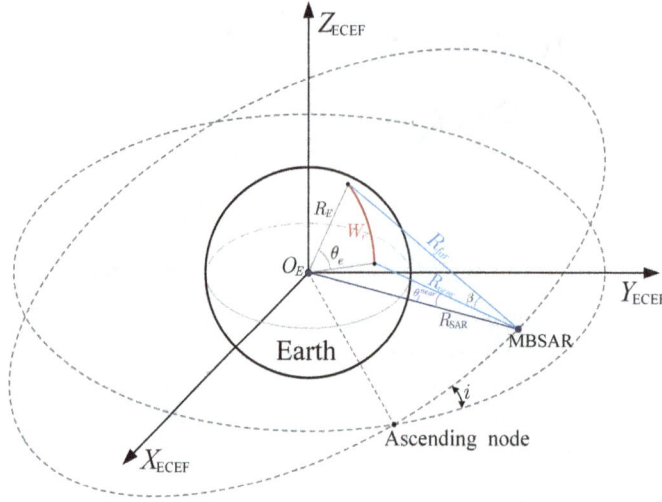

FIGURE 4.1   The schematic diagram of MBSAR's observation geometry in the ECEF coordinate.

The lower and upper bounds of swath width can be derived through the near and far look angles of the MBSAR by:

$$\mathbf{R}_{\mathbf{LBS}} = \mathbf{C}_{\mathbf{ECEF}}^{\mathbf{ACS}} \mathbf{M}_{\mathbf{AE}} \left[\mathbf{R}_{\mathbf{SAR(ACS)}} - \mathbf{R}_{\mathbf{ST(ACS)}}^{\mathbf{near}} \right] \tag{4.5}$$

and:

$$\mathbf{R}_{\mathbf{UBS}} = \mathbf{C}_{\mathbf{ECEF}}^{\mathbf{ACS}} \mathbf{M}_{\mathbf{AE}} \left[\mathbf{R}_{\mathbf{SAR(ACS)}} - \mathbf{R}_{\mathbf{ST(ACS)}}^{\mathbf{far}} \right] \tag{4.6}$$

where $\mathbf{C}_{\mathbf{ECEF}}^{\mathbf{ACS}}$ and $\mathbf{M}_{\mathbf{AE}}$ are the coordinate transformation matrices, and $\mathbf{R}_{\mathbf{SAR(ACS)}}$, $\mathbf{R}_{\mathbf{ST(ACS)}}^{\mathbf{near}}$ and $\mathbf{R}_{\mathbf{ST(ACS)}}^{\mathbf{far}}$ are the SAR's position, near slant range, and far slant range vectors in the ACS, respectively. More details of each term are given in Chapter 2.

Note that the slant range highly depends on the beam pointing in terms of the antenna azimuth and look angles. In the context of the MBSAR, employing the zero-Doppler centroid beam pointing confers an appreciable advantage in optimizing signal processing, attaining high-quality images, and mitigating the complexity of system design [7–9]. When a solitary antenna is utilized in the MBSAR, it is desirable to devise zero-Doppler steering to align the swath width with the zero-Doppler plane [10]. Details are treated in Chapter 5.

Then, the following expression is given for the geocentric angle:

$$\theta_e = \frac{\mathbf{R}_{\mathbf{LBS}} \mathbf{R}_{\mathbf{UBS}}}{\left\| \mathbf{R}_{\mathbf{LBS}} \right\|_2 \left\| \mathbf{R}_{\mathbf{UBS}} \right\|_2} \tag{4.7}$$

Assuming a spherical Earth model, we can approximate the swath width with an average Earth's radius as follows:

$$W_{r(\mathrm{sph})} = R_E \theta_e \tag{4.8}$$

The above MBSAR's swath width has not accounted for the spatial variability of Earth's radius. For an ellipsoidal Earth, it is difficult, if not impossible, to provide a closed-form solution for the swath width, and thus, a numerical solution is sought. In this event, we divide the geocentric angle into n sub-intervals to determine the swath width on the Earth's ellipsoidal surface as:

$$
\begin{cases}
W_r = \dfrac{\theta_e}{n}\left(R_{\text{TOI}}^{\text{far}} + R_{\text{TOI}}^{\text{near}} + \displaystyle\sum_{k=1}^{n-1} R_{\text{TOI}}(\vartheta_k) \right) \\
\vartheta_k = k \cdot n^{-1}\theta_e, k = 1, 2, \cdots, n-1
\end{cases}
\tag{4.9}
$$

where:
  $R_{\text{TOI}}^{\text{far}}$ is the Earth's radius corresponding to the far look angle,
  $R_{\text{TOI}}^{\text{near}}$ is the Earth's radius corresponding to the near look angle,
  $R_{\text{TOI}}(\vartheta_k)$ represents the spatially varying Earth's radius corresponding to ϑ_k.

When $R_{\text{TOI}}^{\text{far}}$, $R_{\text{TOI}}^{\text{near}}$, and $R_{\text{TOI}}(\vartheta_k)$ is replaced by the average radius of Earth, Eq. (4.9) becomes the numerical solution for the MBSAR's swath width on the spherical Earth.

To consider the effects of Earth's ellipsoid on the swath width, we establish a relative error metric to quantify the relative fluctuations in the swath width resulting from the spatially varying Earth's radius:

$$
\delta W_r = W_{r(\text{epd})}^{-1} \cdot \left[W_{r(\text{epd})} - W_{r(\text{sph})} \right]
\tag{4.10}
$$

To demonstrate how the spatial variability of Earth's radius affects the swath width, we plot Figure 4.2 to illustrate the swath width of MBSAR based on spherical and ellipsoidal Earth models at various beamwidths versus the AOL in cycle T_{lc1}, as well as the relative errors resulting from the spatially varying Earth's radius.

It is observed that the magnitude of swath width derived from a spherical Earth model approximates that obtained by an ellipsoidal Earth model. Besides, the relative error δW_r varies temporally and depends on antenna beamwidth. Its value is typically below 4×10^{-3} within one cycle, regardless of the MBSAR's location. Assuming a constant Earth's radius is only applicable if δW_r no more than 4×10^{-3} is accepted.

Next, we are concerned with the spatiotemporal variation in the MBSAR's swath width. Due to orbital perturbations encountered in the MBSAR, uncertainty in spatiotemporal swath width is induced [11–13]. Hence, it is important to analyze the variations in swath width along the orbit during different epochs. To this end, the swath widths with various near-look angles in the epochs from T_{lc1} to T_{lc9} are inspected, and the corresponding results are presented in Figure 4.3 as a function of AOL. To maintain simplicity while ensuring generality, a beamwidth of $0.573°$ at a center frequency of 1.2 GHz, corresponding to an aperture length along the range of 50 m and a taper factor of 2.0, is applied in this case.

Figure 4.3 reveals that varying cycles result in distinct swath widths at the same AOL, indicating that fluctuations in swath width display different regular patterns in various cycles. Changes in near-look angles disturb variation patterns in the swath width with respect to the AOL in one certain cycle. Despite the differences, the swath widths in these cycles share some similarities: When employing identical MBSAR configurations, the magnitude of swath width and the associated fluctuations persist at similar levels. Specifically, for a near-look angle of $0.2°$, the swath width can be up to thousands of kilometers. Further, the difference between the maximum and minimum swath widths within one cycle is over several hundred kilometers, and this difference can be exacerbated and exceed 1000 km as the near-look angle increases. Consequently, the MBSAR with a higher near-look angle enables a wider swath width.

By analyzing the variations in swath width across multiple cycles, we find that between two gapped cycles, the fluctuations in swath width are more pronounced than those between two

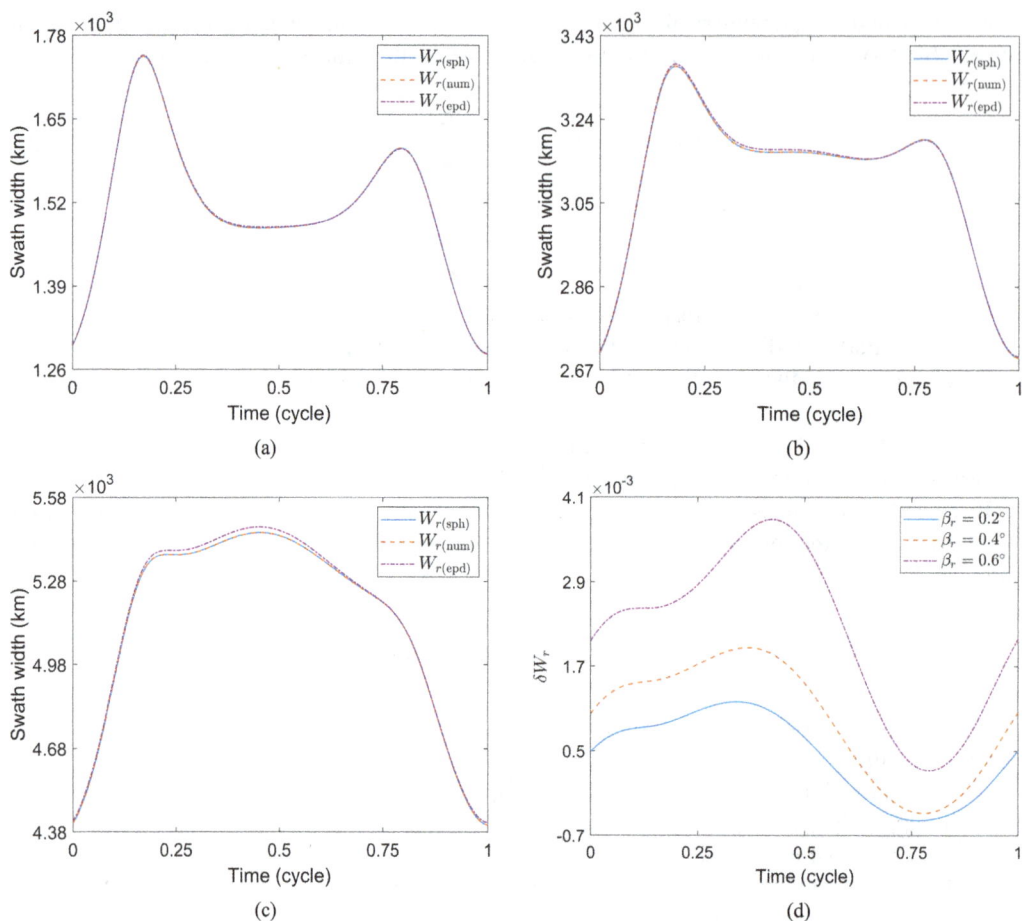

FIGURE 4.2 Referring to spherical and ellipsoidal Earth models, the swath width of MBSAR within cycle T_{lc1} under the beamwidth of (a) 0.20°, (b) 0.40°, (c) 0.60°, where $W_{r(sph)}$ and $W_{r(num)}$ demotes the analytical and numerical swath widths on a spherical Earth, while $W_{r(epd)}$ represents the swath width on an ellipsoidal Earth. (d) The relative error in swath width with various beamwidths.

consecutive cycles, particularly at a larger near-look angle. The reason behind this phenomenon is given rise by the aperiodic variation of MBSAR's orbital element induced by orbital perturbations. Moreover, the longer the interval between two inconsecutive cycles, the stronger the fluctuation of swath width. This suggests that the accumulation of orbital perturbations impacts more on spatio-temporal swath width. Hence, it can be concluded that the orbital perturbation induces irregular fluctuations in the spatiotemporal swath widths of the MBSAR.

The MBSAR's spatiotemporal coverage is determined by its swath width and temporally varying beam footprint. Figures 4.4 and 4.5 illustrate the temporal changes in longitude and latitude of the nadir point in ECEF coordinate system across different cycles of MBSAR.

Figures 4.4 and 4.5 show that the latitude and longitude of the MBSAR's nadir point are tem-porally varying. Reminding that the determination of the latitude of the nadir point is mainly influ-enced by the MBSAR's inertial motion, while the longitude of the nadir point is predominantly determined by the Earth's rotational motion. After one cycle of MBSAR, the longitude of the nadir point deviates from its original position due to the Earth's rotational motion and the MBSAR's inertial motion. The temporal variation of the nadir point's geographical coordinates is continu-ous. However, the temporal variation of MBSAR's coverage does not fully align with that of the nadir point. In order to generate a high-quality SAR image of the coverage region, it is essential

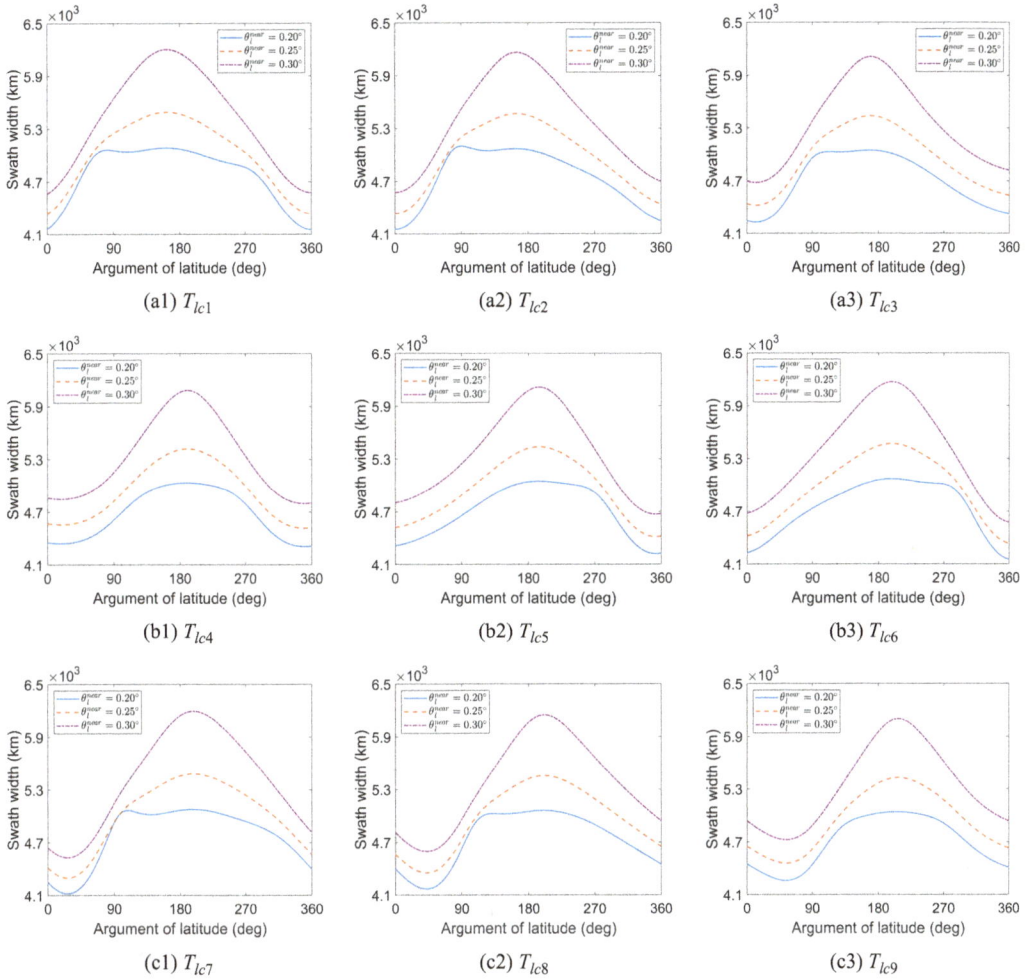

FIGURE 4.3 The swath width of the MBSAR under various near-look angles as a function of AOL over various cycles from $T_{lc1} \sim T_{lc9}$.

to consider the time interval associated with the synthetic aperture, known as synthetic aperture time (SAT) [14], when assessing the spatiotemporal coverage of the MBSAR. As discussed in Chapter 3, an SAT of 200 s can engender a relatively uniform azimuthal resolution, amounting to approximately 10 m. Consequently, this option proves to be a viable choice for Earth observation by the MBSAR.

Figure 4.6 depicts the trajectory of MBSAR's nadir point over one week. It is observed that there is a slight daily shift in latitude (around 3°); the longitudinal variance of the nadir point over one day may surpass 340°. As a result of this dynamic nadir point, in conjunction with the wide swath width, the revisit time for the overlap imaging regions covered by the MBSAR is approximately one day. This revisit time could be reduced further if a multi-antenna system is implemented.

In the subsequent analysis, we shall examine the spatiotemporal imaging regions of the MBSAR using a single antenna with the MBSAR's near-look angle setting at 0.3°. In Figures 4.7 and 4.8, we showcase MBSAR's coverage during two distinct cycles, T_{lc1} and T_{lc4}, when the orbital inclination is around its maximum and minimum. To evaluate the MBSAR's repeated observation performance, we also identify three specific locations along the 113°E longitude, denoted as TOI_1, TOI_2, and TOI_3, each located at latitudes of 15°N, 45°N, and 70°N, respectively.

From Figure 4.7, it is evident that the left-looking MBSAR has the capacity to provide coverage for a vast majority of the Northern Hemisphere of the Earth in cycle T_{lc1}. However, its ability to observe

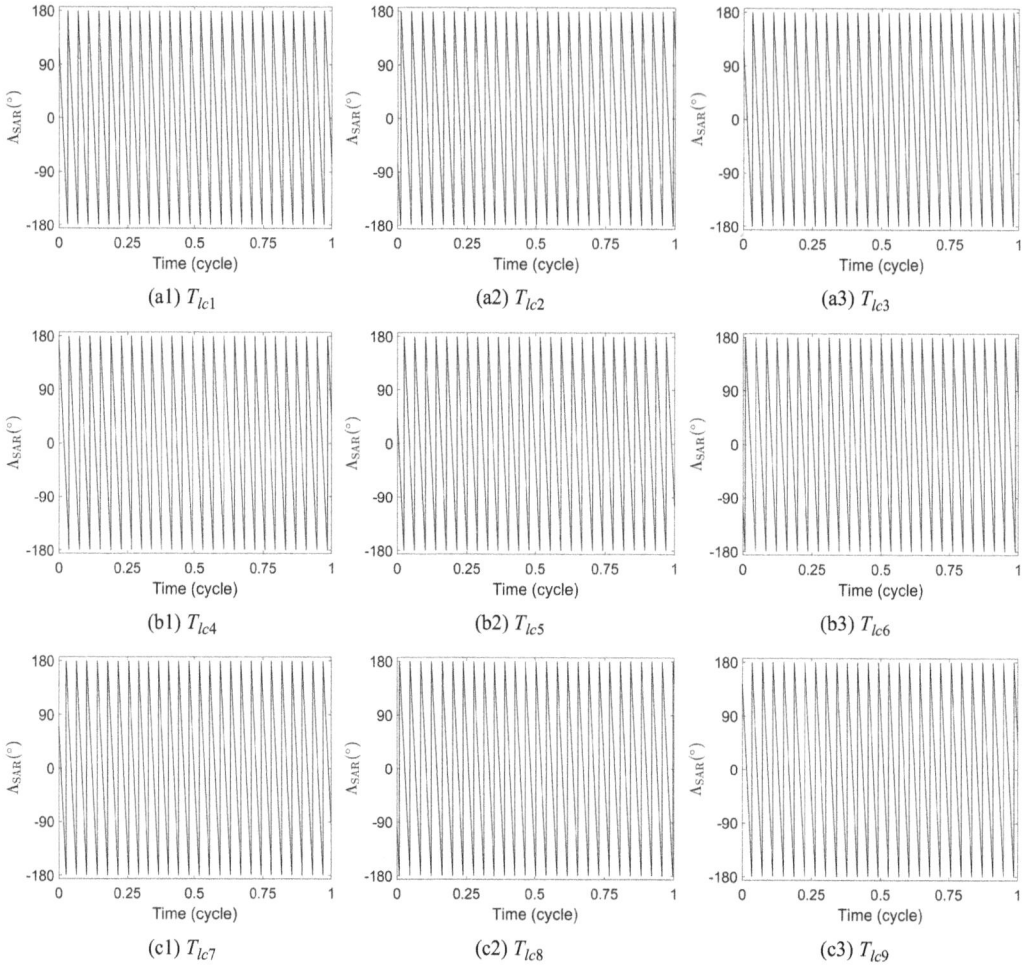

FIGURE 4.4 The longitude of the MBSAR's nadir point in the ECEF over various cycles from $T_{lc1} \sim T_{lc9}$.

areas in the Southern Hemisphere is limited to a specific region. Besides, the frequency of MBSAR's repeated observations varies irregularly with the Earth's latitude. As an example, TOI_1, located at low latitude, will be observable by the MBSAR from March 26, 2024, to April 06, 2024. However, during other parts of the cycle, it is unfeasible to detect TOI_1. On the other hand, TOI_3 can be observed approximately once a day from March 14, 2024, to March 26, 2024. This indicates that at high latitudes, particularly in Polar Regions, the MBSAR observation frequency decreases as the Earth's latitude increases. It is worth noting that TOI_2 can be observed up to 20 times, indicating that mid-latitudes, including the major inhabited regions, can be observed with high frequency when the orbital inclination is large.

Regarding the coverage performance in cycle T_{lc4}, it can be observed from Figure 4.8 that the left-looking MBSAR cannot capture certain parts of the Arctic region. Moreover, the MBSAR's observation of the region in the Southern Hemisphere experiences a sharp decline. As a result, under the small orbital inclination, the MBSAR covers a narrower range of latitudes within a single cycle. Repetitive observation displays disparate trends compared to those observed under a high orbital inclination. As an illustration, the quantitative assessment reveals that TOI_1 and TOI_3 are each observed 11 and 13 times, whereas TOI_2 is observed repetitively 26 times during the cycle T_{lc4}. Thus, with a smaller orbital inclination, the repetitive observation of low and high latitudes decreases compared to those in the cycle T_{lc1}. Interestingly, the MBSAR provides the most repeated observations for the mid-latitude in both cycles T_{lc1} and T_{lc4}.

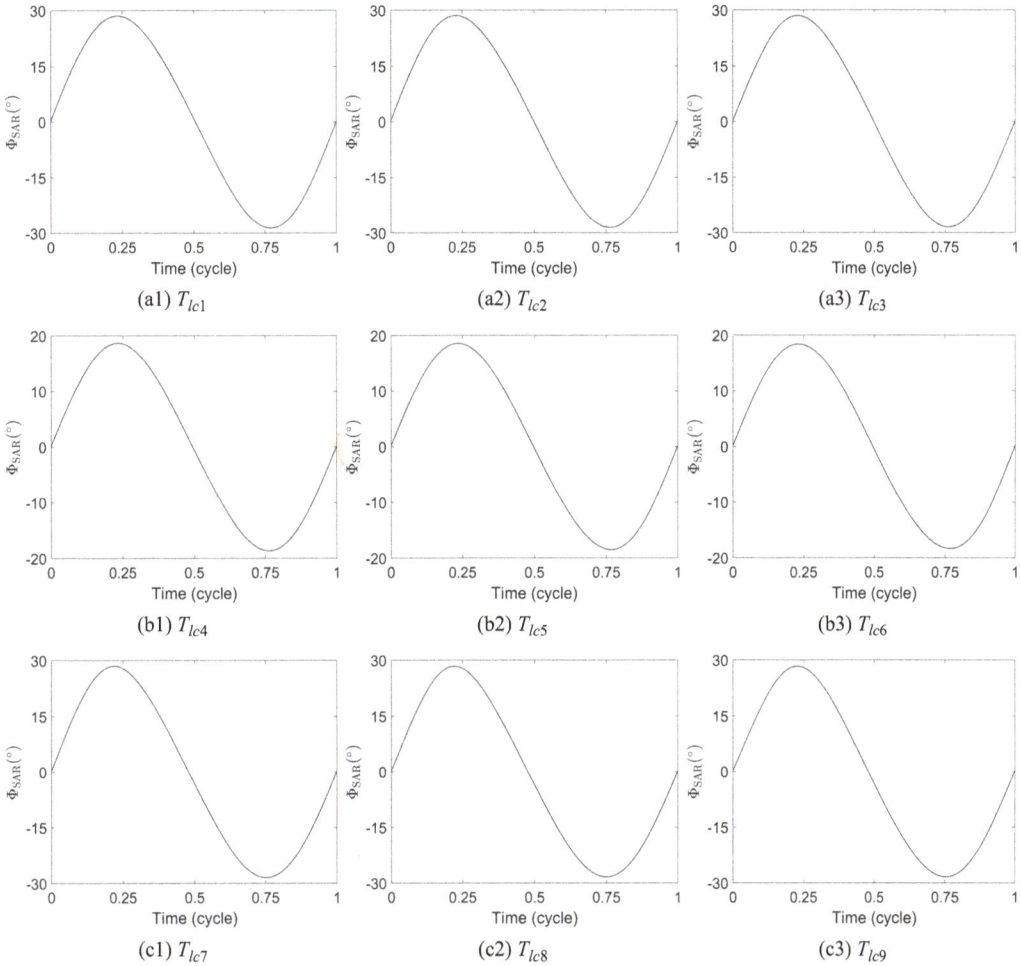

FIGURE 4.5   The latitude of the MBSAR's nadir point in the ECEF over various cycles from $T_{lc1} \sim T_{lc9}$.

We can deduce that a single-antenna MBSAR system can continuously observe the Earth on a hemispheric scale within one cycle. However, the repetitive observation of a particular region on Earth, depending on geographical location and orbital perturbations, may be interrupted and even erratic in various cycles. As a result, it becomes arduous, if not impossible, to calculate a generic repeat cycle for a certain region in an Earth-observing MBSAR system, be it left-looking or right-looking, though a revisit time for the overlap covered region could be available.

To perform more thorough evaluations, we simulate the cumulative imaging regions for one cycle of MBSAR, as shown in Figures 4.9 and 4.10, with near-look angles of 0.2° and 0.3° over the cycles $T_{lc1} \sim T_{lc3}$. Similarly, Figures 4.11 and 4.12 show the same cumulative imaging regions across the cycles $T_{lc4} \sim T_{lc6}$. Both the left-looking and right-looking modes are considered here.

As depicted in Figures 4.9–4.12, the cumulative imaging region of the MBSAR is strongly influenced by the orbital inclination. Specifically, a smaller orbital inclination leads to a restricted range of latitudes covered by the MBSAR. Conversely, a larger inclination angle enables the MBSAR to extend its coverage to a broader range of latitudes over one lunar cycle. The left-looking and right-looking MBSAR systems have comparable coverage performance in general. However, a noticeable dissimilarity exists between their imaging regions; to be precise, the left-looking MBSAR is primarily imaging the Northern Hemisphere, whereas the right-looking one is observing the Southern Hemisphere of Earth.

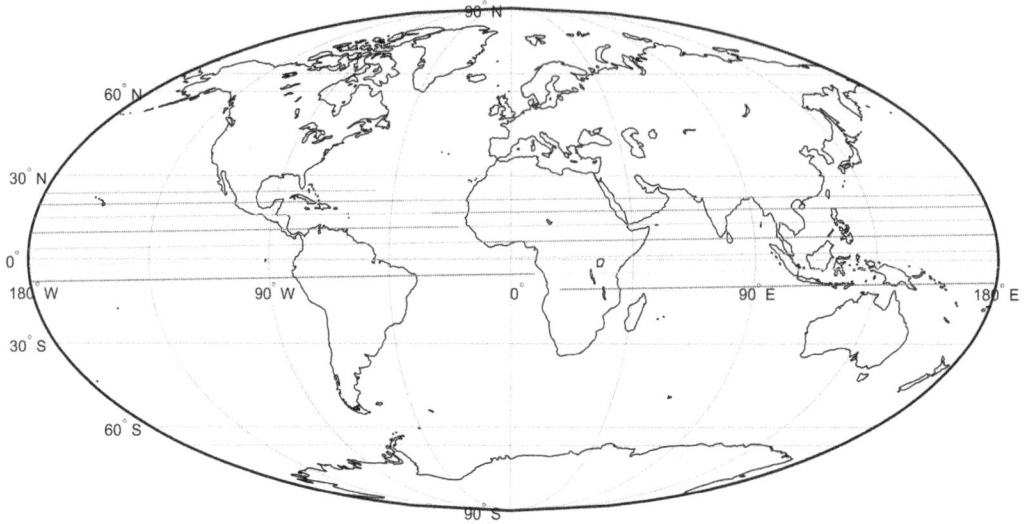

FIGURE 4.6 The track of the MBSAR's nadir point from March 20, 2024, to March 26, 2024, during which the latitude of the nadir point shifts from 24.60°N to 10.69°S. Each color means a daily track of nadir point, with a time interval of 200 s.

Further, the near-look angle can also affect the cumulative imaging region: a larger near-look angle enables the MBSAR to effectively measure the high latitudes of the Northern Hemisphere when operating in left-looking mode, albeit at the cost of a reduction in coverage of the Southern Hemisphere. In contrast, a comparable phenomenon arises in the high latitudes of the Southern Hemisphere when the MBSAR is configured in a right-looking mode.

In this section, we have evaluated the spatiotemporal swath width and the corresponding imaging region of the MBSAR that is equipped with a solitary antenna. It is found that the MBSAR has the potential to cover a vast swath width, spanning thousands of kilometers. When a larger near-look angle is employed, it becomes more probable for the MBSAR to observe high latitudes. Such a phenomenon is particularly noteworthy under the condition of a larger orbital inclination. What's more, the imaging region is also subject to the influence of the side-looking direction, which renders the left-looking MBSAR an optimal instrument for monitoring the Northern Hemisphere, and conversely for the right-looking MBSAR to monitor the Southern Hemisphere. In addition, it is of utmost importance to acknowledge that the MBSAR's trajectory is highly susceptible to orbital perturbations, potentially leading to irregular fluctuations in both the spatiotemporal swath width and the corresponding imaging region.

A single antenna MBSAR, in conjunction with zero-Doppler steering, can achieve superior imaging quality. Meanwhile, a well-balanced coverage performance can also be maintained. However, this method falls short of delivering the optimal spatial coverage that can be achieved through MBSAR. In the scenario where the goal is to optimize spatial coverage within a given time frame, the observation by multi-antennas can be employed in MBSAR, as discussed below.

4.2   SPATIOTEMPORAL COVERAGE WITH MULTI-ANTENNAS

This section evaluates the spatiotemporal coverage performance of MBSAR when employing a collaborative deployment of SAR antennas. In the scenario of multiple antennas, the MBSAR's coverage performance cannot be simply conveyed through the swath width [15]. Rather, it can be explicated by projecting the spatial coverage onto the Earth's surface. Such a projection is affected by several factors, including the geocentric angle, look angle, grazing angle, and observation azimuth

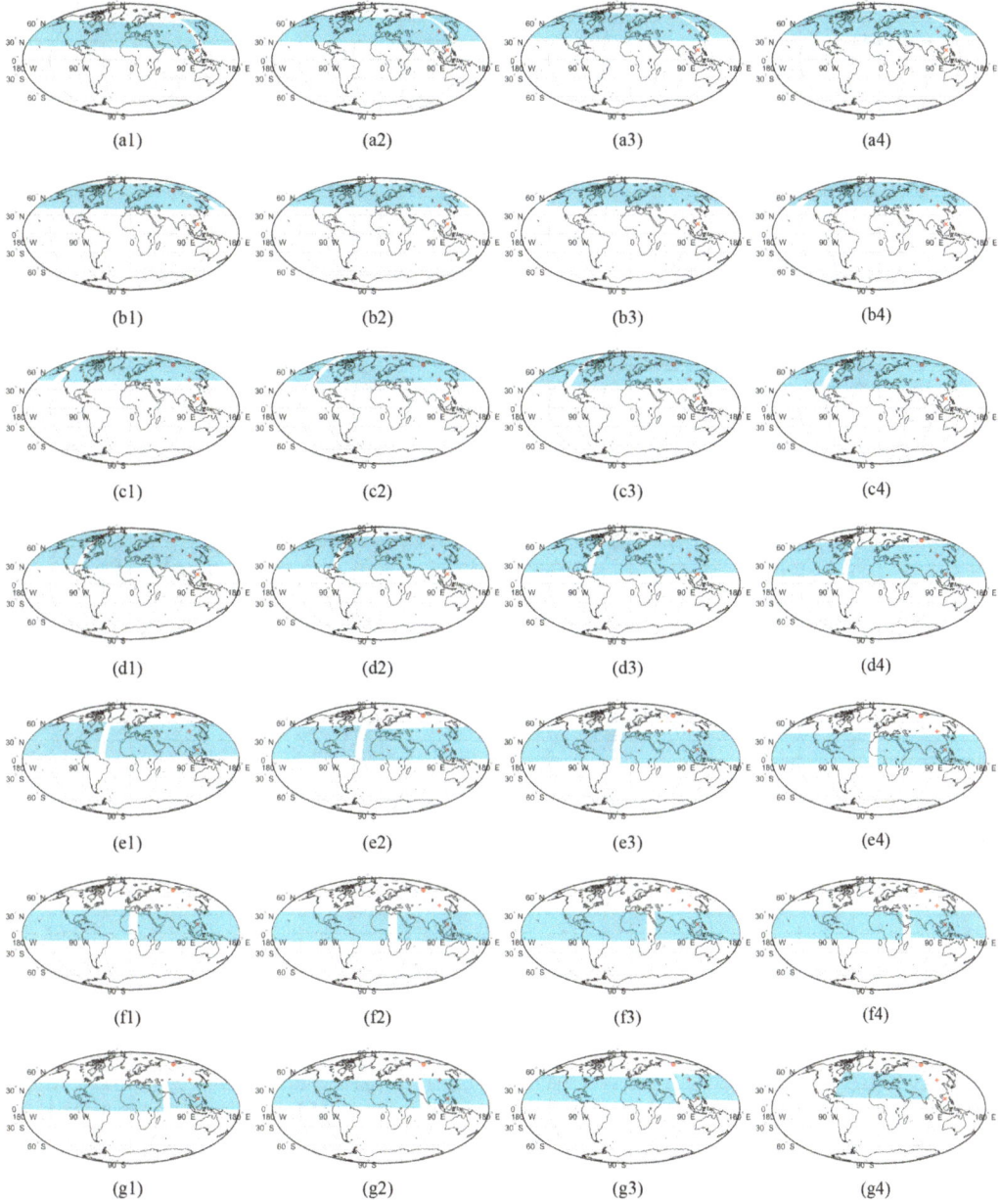

FIGURE 4.7 The imaging region of the left-looking MBSAR with a near-look angle of 0.3° within the cycle T_{lc1}, where the symbols "×", "+" and "⊕" stand for the TOIs located at low, middle, and high latitudes, respectively. Each graph represents a daily coverage except for the last one, which stands for the coverage from 05:05:29, Apr. 07, 2024, to 16:13:50, Apr. 07, 2024.

angle. To be more precise, the MBSAR's coverage is determined by the observation parameters illustrated in Figure 4.13, whose definitions and expressions are given below:

The geocentric angle, defined as the angle between the nadir point of the MBSAR and the ground target referring to the Earth's center, is:

$$\theta_e = \cos^{-1}\left(\cos\Phi_{\text{SAR}}\cos\Phi_{\text{TOI}}\cos\Lambda_{\text{DTS}} + \sin\Phi_{\text{SAR}}\sin\Phi_{\text{TOI}}\right) \tag{4.11}$$

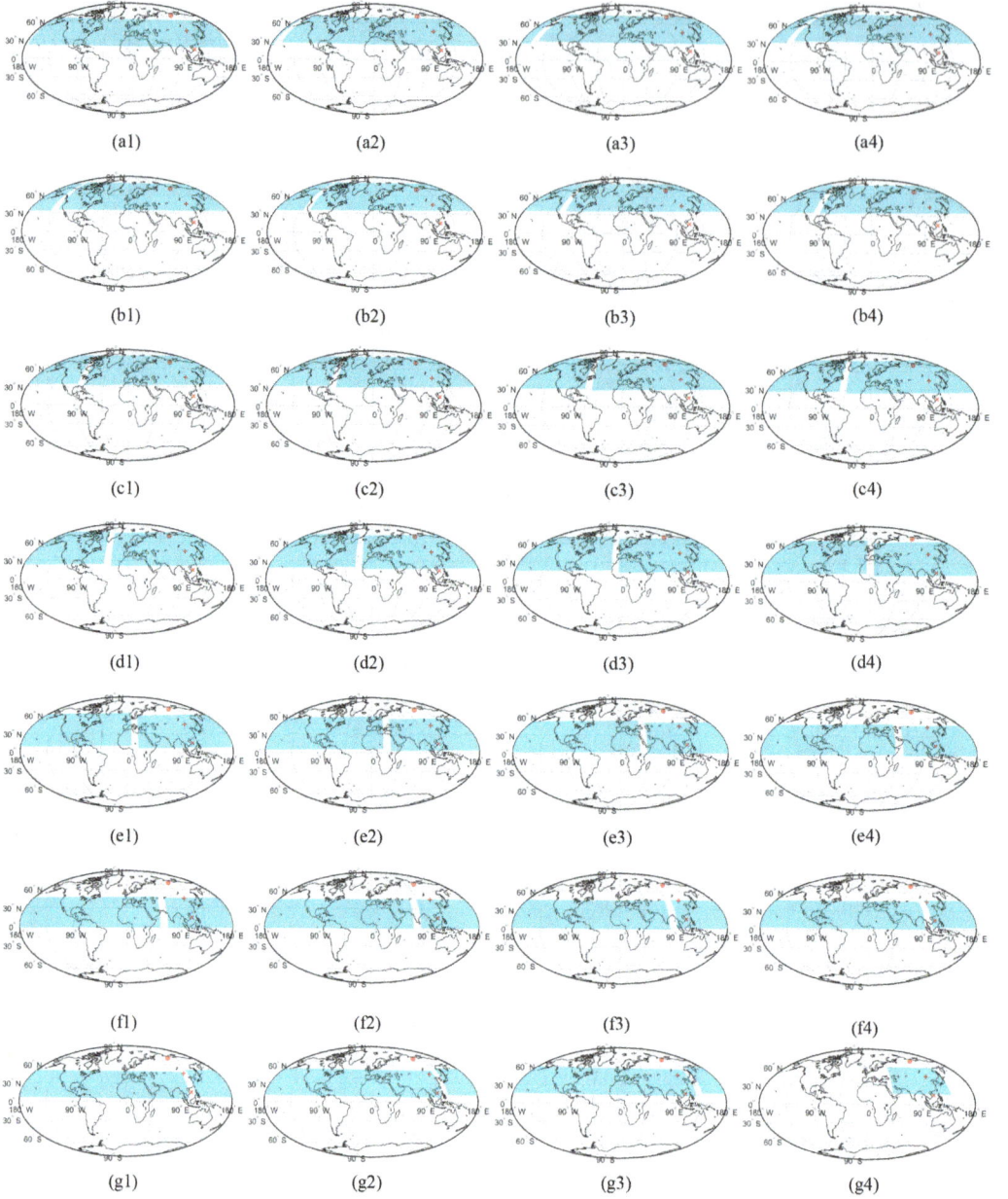

FIGURE 4.8 The imaging region of the left-looking MBSAR with a near-look angle of 0.3° within the cycle T_{lc4}, where the symbols "×", "+" and "⊕" stand for the TOIs located at low, middle, and high latitudes, respectively. Each graph represents a daily coverage except for the last one, which stands for the coverage from 18:22:19, Jul. 16, 2033, to 01:06:30.68, Jul. 17, 2033.

The look angle refers to the angle between the position vector and the slant range vector, which is specified by:

$$\theta_l = \cos^{-1}\left\{\frac{R_{\mathrm{SAR}} - R_{\mathrm{TOI}}\left(\cos\Phi_{\mathrm{SAR}}\cos\Phi_{\mathrm{TOI}}\cos\Lambda_{\mathrm{DTS}} + \sin\Phi_{\mathrm{SAR}}\sin\Phi_{\mathrm{TOI}}\right)}{R_{\mathrm{ST}}}\right\} \quad (4.12)$$

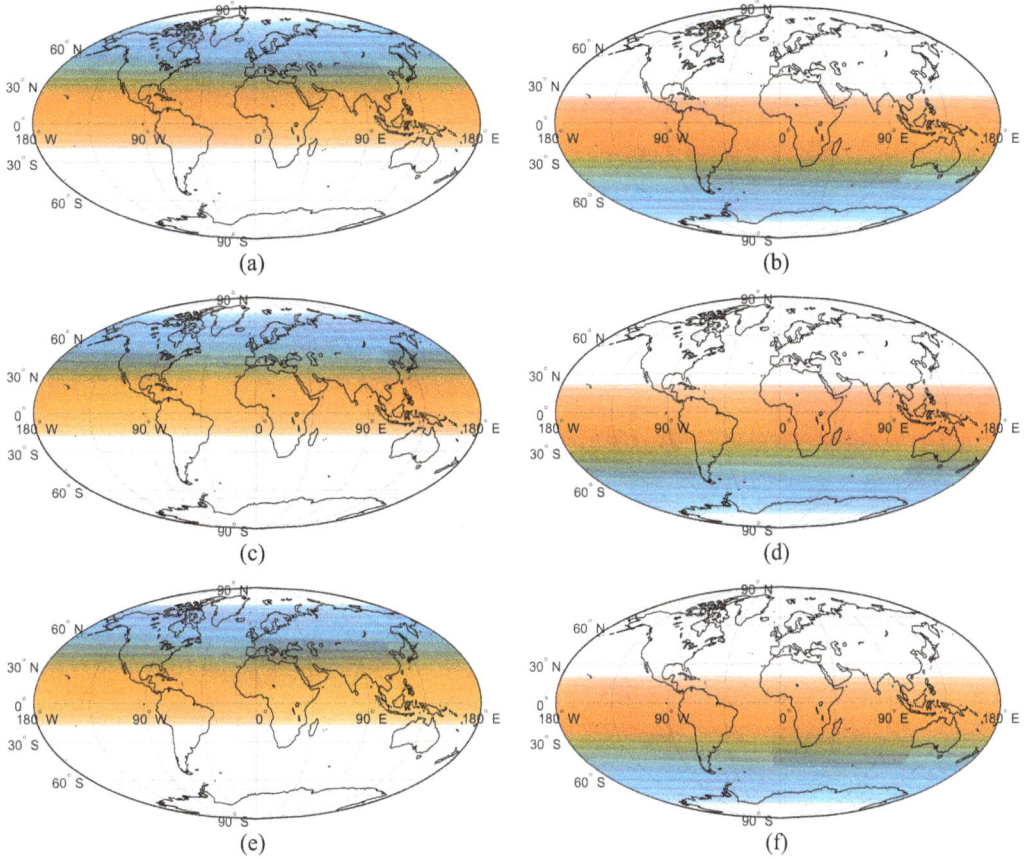

FIGURE 4.9 The cumulative imaging region with a near-look angle of 0.2° within various cycles when the orbital inclination approaches its maximum, where the 1st (a, c, e) and 2nd (b, d, e) columns represent the left-looking and right-looking MBSAR systems, while the 1st (a, b), 2nd (c, d), and 3rd (e, f) rows indicate cycles T_{lc1}, T_{lc2}, and T_{lc3}.

The grazing angle, defined as the angle between the slant range and the ground range plane, is mathematically expressed by:

$$\psi = \sin^{-1}\left\{\frac{R_{SAR}\left(\cos\Phi_{SAR}\cos\Phi_{TOI}\cos\Lambda_{DTS} + \sin\Phi_{SAR}\sin\Phi_{TOI}\right) - R_{TOI}}{R_{ST}}\right\} \quad (4.13)$$

The observation azimuth angle, denoting the angle between the slant range vector in the ground plane and the east–west direction of the Earth's rotation, can be expressed as:

$$\phi_{oa} = \cos^{-1}\left\{\frac{\cos\Phi_{SAR}\sin\Lambda_{DTS}}{\sqrt{1-\left(\cos\Phi_{TOI}\cos\Phi_{SAR}\cos\Lambda_{DTS} + \sin\Phi_{TOI}\sin\Phi_{SAR}\right)^2}}\right\} \quad (4.14)$$

The observation azimuth angle should not be confused with the antenna azimuth angle. The detailed derivation for the observation parameters associated with MBSAR's coverage capability can be found in Appendix A.

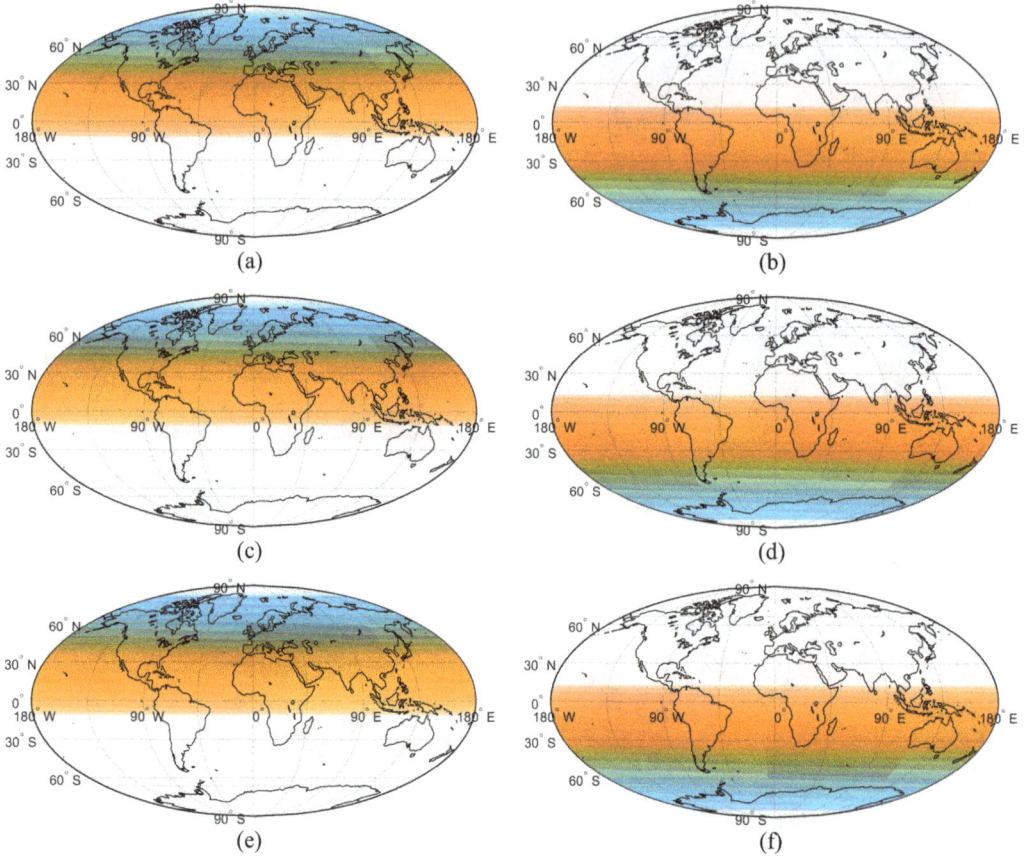

FIGURE 4.10 The cumulative imaging region with a near-look angle of 0.3° within various cycles when the orbital inclination approaches its maximum, where the 1st (a, c, e) and 2nd (b, d, e) columns represent the left-looking and right-looking MBSAR systems, while the 1st (a, b), 2nd (c, d), and 3rd (e, f) rows indicate cycles T_{lc1}, T_{lc2}, and T_{lc3}.

Either the grazing or look angle imposes the MBSAR's ground coverage as both angles are related to each other. For illustration, we use the grazing angle to define the MBSAR's spatiotemporal coverage in what follows. From the geometrical arrangement depicted in Figure 4.14, we may infer a relationship between the grazing angle within coverage and its bounds for MBSAR, which can be expressed as:

$$\psi_{\text{far}} \leq \psi \leq \psi_{\text{near}} \tag{4.15}$$

where ψ_{far} and ψ_{near} denote the minimum and maximum limits of grazing angle, respectively, to be elaborated in the following sections.

Afterward, the geocentric angle can be equivalently expressed as a function of grazing angle, as:

$$\theta_e = \cos^{-1}\left(\sin\psi \frac{R_{\text{ST}}}{R_{\text{SAR}}} + \frac{R_{\text{TOI}}}{R_{\text{SAR}}}\right) \tag{4.16}$$

Regarding the MBSAR's spatial coverage, it is possible to estimate the slant range as:

$$R_{\text{ST}} \approx R_{\text{STA}} \tag{4.17}$$

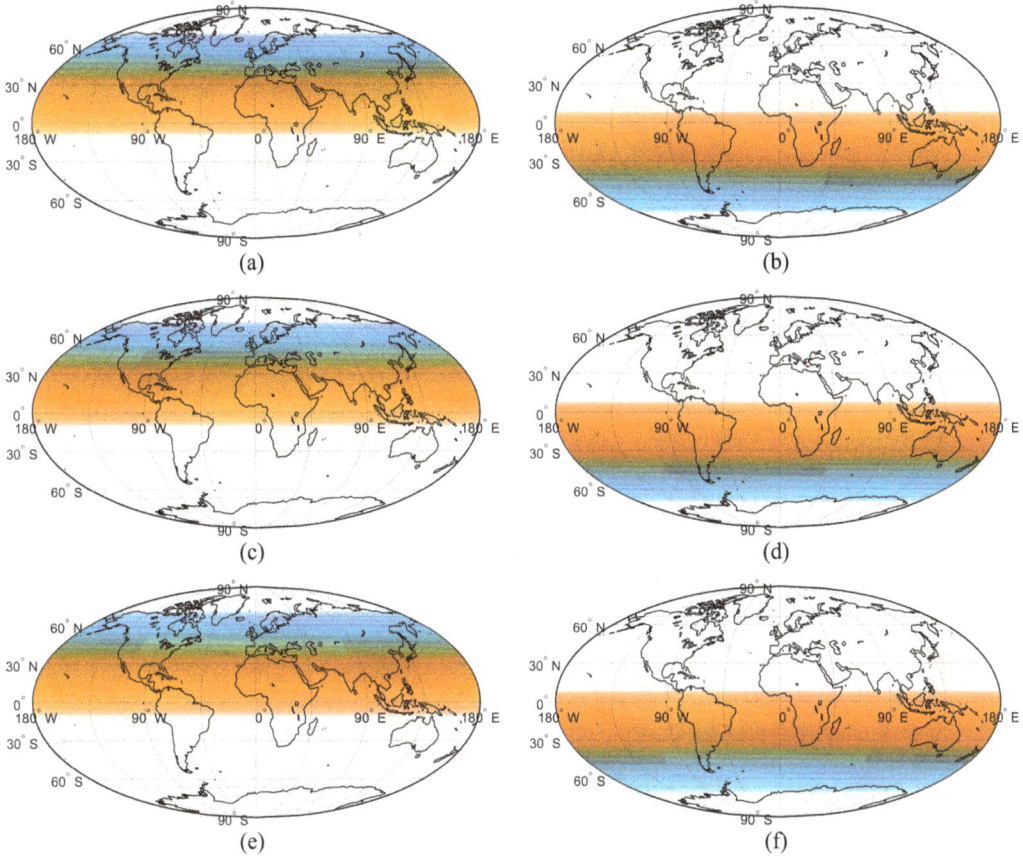

FIGURE 4.11 The cumulative imaging region with a near-look angle of 0.2° within various cycles when the orbital inclination approaches its minimum, where the 1st (a, c, e) and 2nd (b, d, e) columns represent the left-looking and right-looking MBSAR systems, while the 1st (a, b), 2nd (c, d), and 3rd (e, f) rows indicate cycles T_{lc4}, T_{lc5}, and T_{lc6}.

with:

$$R_{\text{STA}} = R_{\text{SAR}} \left(1 + \frac{0.5 R_{\text{TOI}}^2 - R_{\text{SAR}} R_{\text{TOI}} \cos \theta_e}{R_{\text{SAR}}^2} \right) \tag{4.18}$$

To verify the accuracy of Eq. (4.17) in evaluating the MBSAR's spatial coverage, we define a relative error as:

$$\delta R_{\text{ST}} = \frac{R_{\text{ST}} - R_{\text{STA}}}{R_{\text{SAR}}} \tag{4.19}$$

Figure 4.15 illustrates the relative errors δR_{ST} in the $\Lambda_{\text{DTS}} - \Phi_{\text{TOI}}$ domain at perigee and apogee of the MBSAR. For the sake of simplicity but not at the expense of generality, we assume the latitude of the MBSAR's nadir point to be 0°, and the grazing angle to range from 0° to 90°.

As indicated by Figure 4.15, the maximum value of the relative error δR_{ST} is far below 10^{-3}, regardless of the Earth–MBSAR distance. This finding suggests that the approximation of the slant range in has a negligible impact on the dimensionless quantity, δR_{ST}. As a consequence, we can

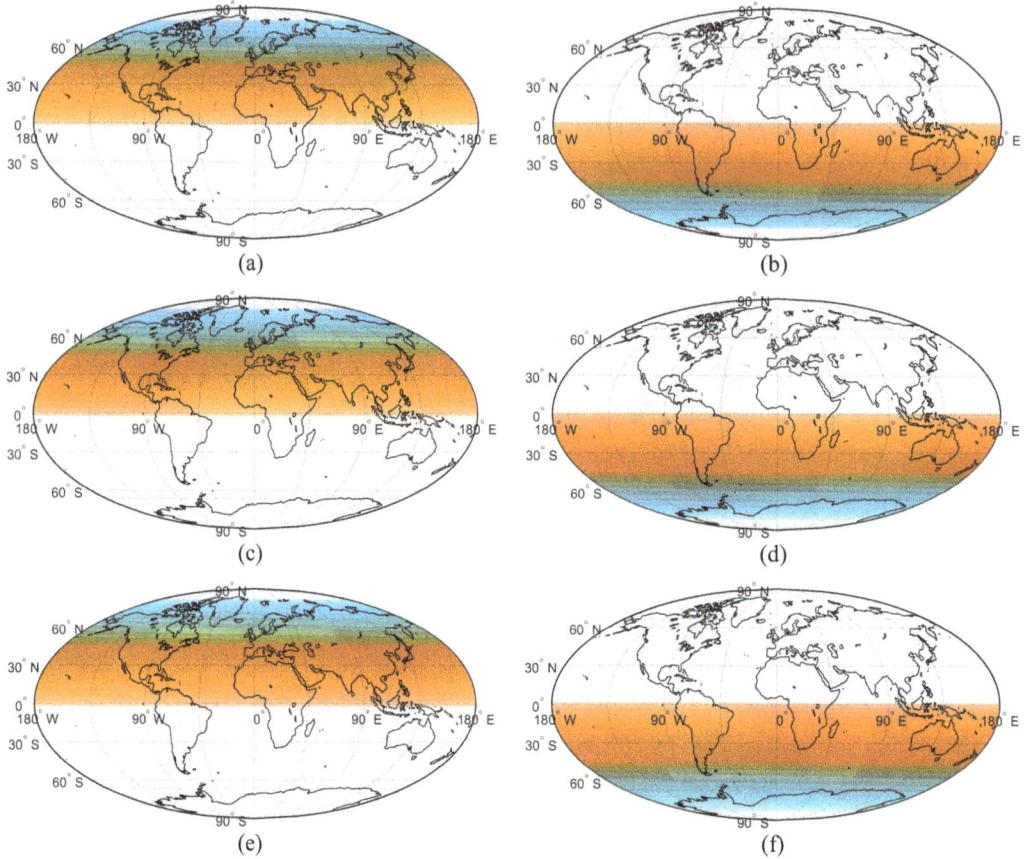

FIGURE 4.12 The cumulative imaging region with a near-look angle of 0.3° within various cycles when the orbital inclination approaches its maximum, where the 1st (a, c, e) and 2nd (b, d, e) columns represent the left-looking and right-looking MBSAR systems, while the 1st (a, b), 2nd (c, d), and 3rd (e, f) rows indicate cycles T_{lc4}, T_{lc5}, and T_{lc6}.

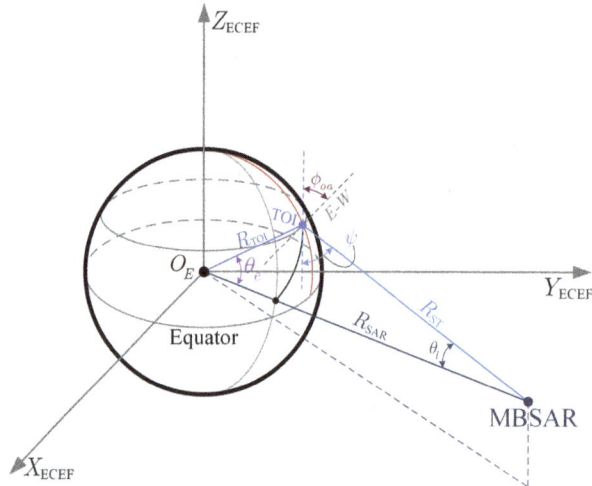

FIGURE 4.13 The schematic diagram of parameters associated with the MBSAR's coverage in the ECEF coordinate, where "E–W" means the east–west direction.

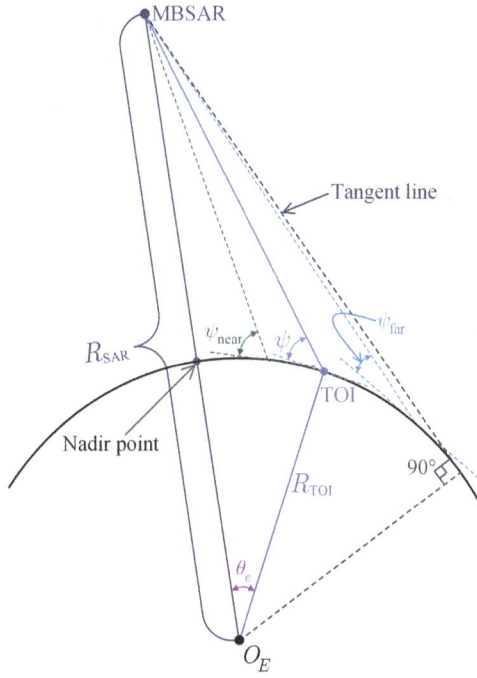

FIGURE 4.14   The diagram of the near and far grazing angles, namely the lower and upper bounds of the grazing angle in the MBSAR.

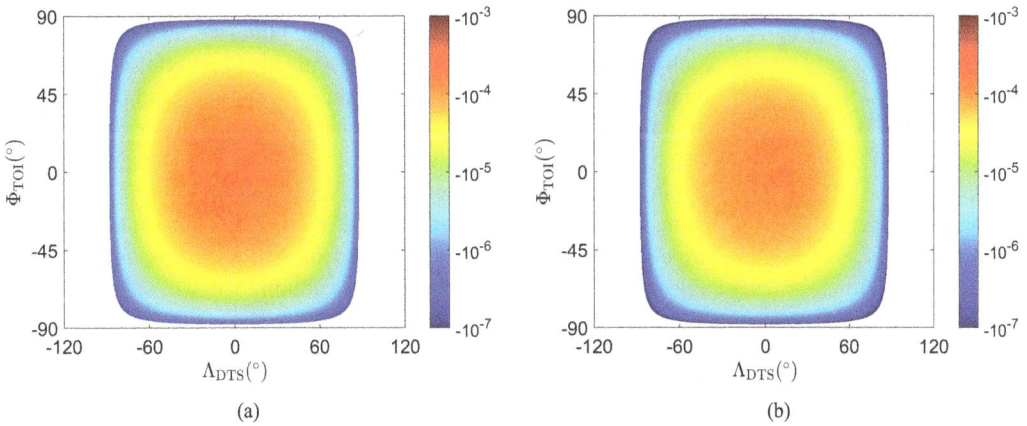

FIGURE 4.15   The relative errors δR_{ST} in the $\Lambda_{DTS} - \Phi_{TOI}$ domain at the (a) perigee, (b) apogee of the MBSAR in cycle T_{lc1}.

reasonably disregard the error caused by the approximation of the slant range when determining the spatial coverage of an MBSAR with multiple antennas.

Therefore, it is possible to approximate Eq. (4.16) as follows:

$$\cos\theta_e \approx \kappa_s\left(\psi, R_{SAR}\right) \tag{4.20}$$

with:

$$\kappa_s\left(\psi, R_{SAR}\right) = \frac{\sin\psi + 0.5R_{TOI}R_{SAR}^{-1}}{\sin\psi R_{TOI}R_{SAR}^{-1} + 1} + 0.5R_{TOI}R_{SAR}^{-1} \tag{4.21}$$

Take note that the ratio $R_{TOI}R_{SAR}^{-1}$ varies spatiotemporally according to the following:

$$\frac{R_{TOI}}{R_{SAR}} \approx \frac{R_E}{R_{SARA}}\left(1 - \Delta R_{SART}R_{SARA}^{-1} + \Delta R_{ES}R_E^{-1}\right) \tag{4.22}$$

where ΔR_{ES} is the spatially varying component of Earth's radius relative to the average Earth's radius, R_E; ΔR_{SART} is the temporally varying component of Earth–MBSAR distance pertaining to the mean Earth–MBSAR distance, R_{SARA}. It is estimable that $\Delta R_{SART}R_{SARA}^{-1}$ exceeds $\Delta R_{ES}R_E^{-1}$ by a ratio of about 30 times during one cycle of MBSAR. As such, the spatially varying component of Earth's radius is negligible. Therefore, it is reasonable to assume a constant Earth radius.

From the discussions above, we lay out the observation parameters that are relevant to the coverage performance of MBSAR with multiple antennas. Here, to balance the imaging performance of the SAR system within spatial coverage, we shall first determine the effective scope for spatial coverage.

4.2.1 Effective Range of the MBSAR's Spatial Coverage

The Moon, a stable and enduring celestial body, is an ideal platform where a set of antennas or an array can be installed to observe the Earth [16–18]. A single antenna SAR allows for observation of a limited portion of the Earth's surface within a given time frame, as demonstrated in Section 4.1. On the other hand, the entire Earth's disk appears smaller than a view field of $2°$ when observed from the Moon-based viewpoint [19, 20]; multiple antennas or an antenna array allow for optimal spatial coverage by steering the pointing direction of each antenna [15, 21]. However, the potential issues of left–right ambiguity, image degradation, and system noise may arise when multiple antennas are utilized.

The left–right ambiguity stems from the fact that TOIs situated symmetrically about the radar's moving trace exhibit identical range histories [22, 23]. Such an issue could be resolved by using two sets of antenna systems, each comprising multiple antennas with identical configurations, to conduct simultaneous right- and left-looking operations and disentangle ambiguous echoes. Further, the MBSAR's spatial coverage is contingent upon the imaging geometry, given that the SAR system is incapable of nadir imaging and may experience suboptimal performance at large incident angles [24–26]. Hence, it is crucial to constrain the grazing angle within a valid range for effective imaging. Typically, spaceborne SAR systems observe areas within a range of $20°$ to $60°$ of incident angles or, equivalently, from $70°$ to $30°$ of grazing angles. In the context of MBSAR, it is imperative to limit the near and far grazing angles to a practical range that enables comprehensive global coverage within each cycle of MBSAR.

To ensure the global coverage capabilities of MBSAR, we select the far grazing angle so that both the North and South Poles of Earth are observable by satisfying the following conditions:

$$\psi_{far} \leq \sin^{-1}\left\{\frac{\sin|\Phi_{SAR}| - R_{TOI}R_{SAR}^{-1}}{(1 + R_{TOI}^2 R_{SAR}^{-2} - 2\sin|\Phi_{SAR}| \cdot R_{TOI}R_{SAR}^{-1})^{0.5}}\right\} \tag{4.23}$$

The MBSAR fails to cover the North or South Pole if the far grazing angle surpasses its upper limit, irrespective of the nadir point's location. Keep in mind that the MBSAR still is not necessarily able to observe the North or South Pole even when the far grazing angle is restricted. The observation of these regions is contingent upon the latitude of the nadir point, as demonstrated in the following. Similarly, the minimum near grazing angle must conform to cover the equatorial regions:

$$\psi_{near} \geq \sin^{-1}\left\{\frac{R_{TOI}R_{SAR}^{-1} - \cos\Phi_{SAR}}{\cos\Phi_{SAR} \cdot R_{TOI}R_{SAR}^{-1} - 0.5 \cdot R_{TOI}^2 R_{SAR}^{-2} - 1}\right\}. \tag{4.24}$$

The inclination of the lunar orbit to the Earth's equatorial plane ranges between 18.3° and 28.6° [27], which translates to a maximum scale of the MBSAR's nadir point's latitude within a similar range. Figure 4.16 depicts the bounds of far and near grazing angles at apogee and perigee of the lunar orbit against the nadir point's latitude.

Figure 4.16 shows that the upper limit of the far grazing angle at the apogee of the lunar orbit can reach 27.80° when the orbital inclination is at its maximum. However, to achieve continuous global observation during one MBSAR cycle with minimum orbital inclination, we must limit the upper bound of the far grazing angle to 17.32°. Failure to do so could result in the North Pole (or the South Pole) revisiting periods of several years, which is highly undesirable. Similarly, to cover the equatorial regions during one cycle of MBSAR with the minimum orbital inclination, the near grazing angle must exceed 71.40°. As a result, the optimal range for the grazing angle is [15°, 75°].

Although the MBSAR system has a limited spatial coverage range due to its near and far grazing angles, ensuring high-quality MBSAR imagery remains a significant challenge. Pixel skewing on the ground is common when the system operates in squint-looking mode to observe bounded regions [28]. In such cases, imaging is only possible when the iso-range and iso-Doppler profiles are considerably non-parallel [29]. Typically, parallel regions of iso-range and iso-Doppler contours occur when the observation azimuth angles are close to 0° or 180°. Hence, it is practical to limit the observation azimuth angle to:

$$\left|\cos \phi_{oa}\right| \leq \cos \phi_b \tag{4.25}$$

where ϕ_b is the bound of the azimuth angle. Here we set it to 30° to ensure a good image quality.

Figure 4.17 displays the iso-Doppler and iso-range profiles of the MBSAR in the $\Lambda_{\mathrm{DTS}} - \Phi_{\mathrm{TOI}}$ domain at 00:00:00 on March 20, March 25, April 01, and April 07, 2024. One can observe that the iso-Doppler and iso-range contours intersect with one another at all orbital locations of the MBSAR system. This finding suggests that the image quality remains satisfactory even when the MBSAR operates in a squint-looking mode. Hence, the skewing effect has little impact on the spatial coverage of the MBSAR within the defined scope.

Finally, by applying the previously mentioned bounds, we can establish limits of the geocentric angle that determine the MBSAR's spatial coverage as follows:

$$\theta_e^{\mathrm{low}} \leq \theta_e \leq \theta_e^{\mathrm{up}} \tag{4.26}$$

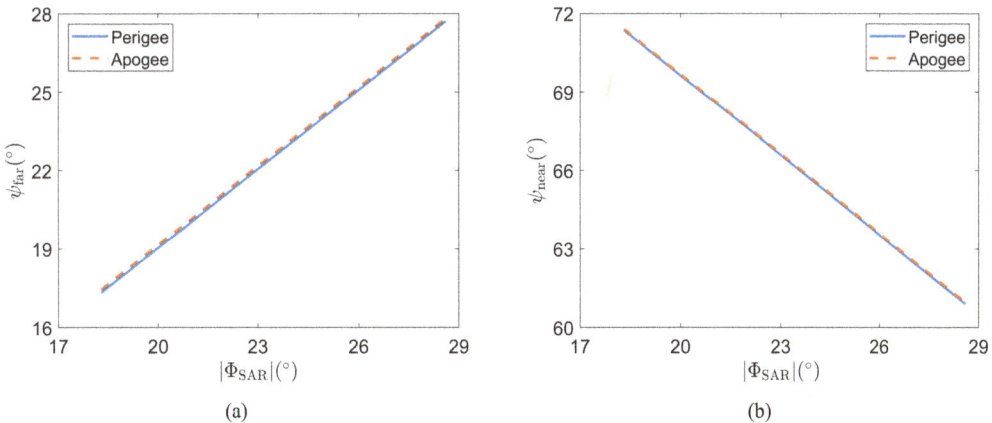

FIGURE 4.16   The Moon situated at the apogee or perigee: (a) the bound of far grazing angle (b) the bound of near grazing angle.

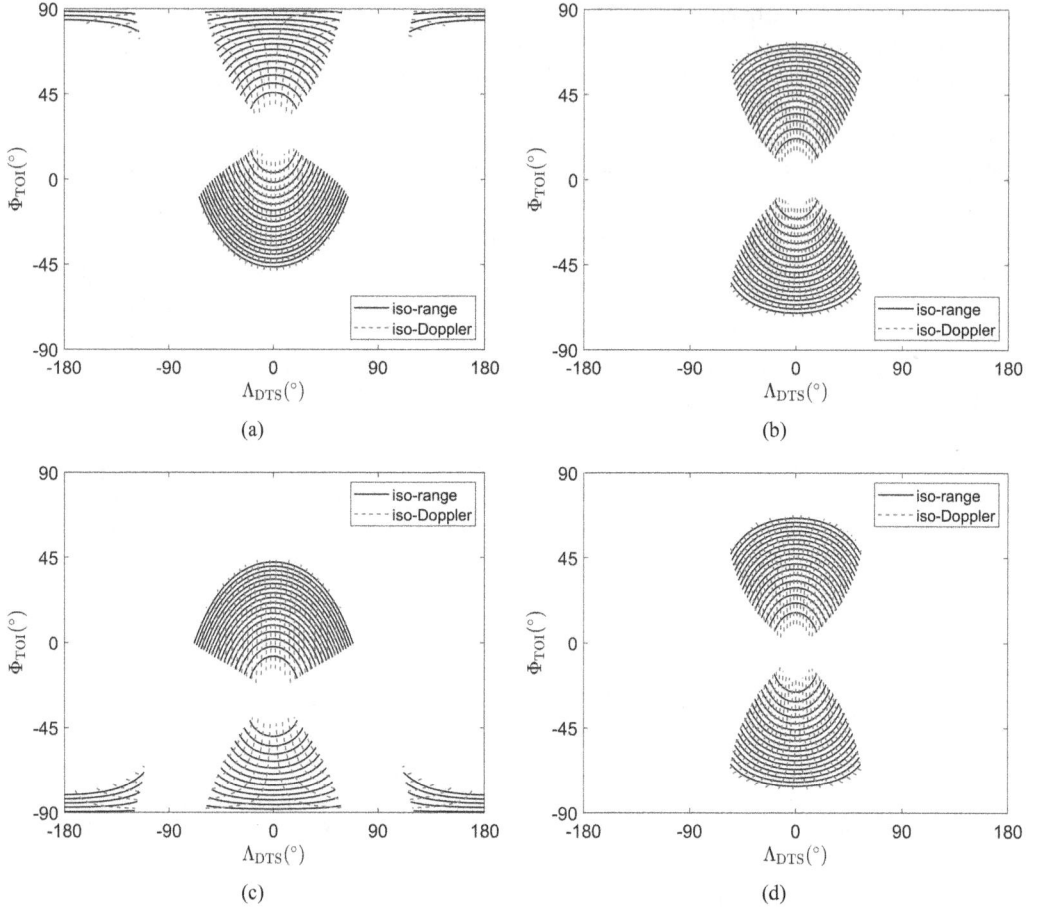

FIGURE 4.17   The iso-Doppler and iso-range profiles in the $\Lambda_{\mathrm{DTS}} - \Phi_{\mathrm{TOI}}$ domain at 00:00:00 on the following dates: (a) March 20, 2024, (b) March 25, 2024, (c) April 01, 2024, and (d) April 07, 2024.

where θ_e^{up} is the upper limit of the geocentric angle:

$$\theta_e^{\mathrm{up}} = \cos^{-1}\left[\kappa_s\left(\psi_{\mathrm{far}}, R_{\mathrm{SAR}}\right)\right] \tag{4.27}$$

The geocentric angle's lower bound is given:

$$\theta_e^{\mathrm{low}} = \begin{cases} \cos^{-1}\left[\kappa_s(\psi_{\mathrm{near}}, R_{\mathrm{SAR}})\right], & |\Lambda_{\mathrm{DTS}}| \leq \alpha_{\mathrm{th}} \\ \cos^{-1}\left[\cos\phi_b(\cos^2\phi_b - \cos^2\Phi_{\mathrm{SAR}}\sin^2\alpha)^{0.5}\right], & |\Lambda_{\mathrm{DTS}}| \geq \alpha_{\mathrm{th}} \end{cases} \tag{4.28}$$

with:

$$\alpha_{\mathrm{th}} = \sin^{-1}\left\{\cos\phi_b \cdot \cos^{-1}(\Phi_{\mathrm{SAR}}) \cdot \left[1 - \kappa_s^2(\psi_{\mathrm{near}}, R_{\mathrm{SAR}})\right]^{0.5}\right\} \tag{4.29}$$

We note that the outer limit of the MBSAR's spatial coverage is determined by the maximum geocentric angle, which is associated with the far grazing angle. On the other hand, the minimum geocentric angle, linked to the near grazing angle and observation azimuth angle, sets the boundary for the blind region that the MBSAR cannot observe. The regions covered by MBSAR lie between the upper and lower bounds of the geocentric angle.

4.2.2 Examination of the MBSAR's Spatial Coverage

The spatial coverage of the MBSAR can be evaluated with regard to its coverage area and imaging region. We shall explore further the coverage area of MBSAR within the bounds of Eq. (4.26). Under the model of a spherical Earth, the MBSAR's coverage area on the Earth's surface can be estimated by making use of:

$$
\begin{aligned}
S_{EC} = 4R_E^2 \cdot \Big\{ & \cos^{-1}\Big[\xi_s(\phi_b) \cdot \kappa_s(\psi_{\text{far}}, R_{\text{SAR}}) \cdot \zeta_s(\psi_{\text{far}}, R_{\text{SAR}}) \Big] \\
& - \cos^{-1}\Big[\xi_s(\phi_b) \cdot \kappa_s(\psi_{\text{far}}, R_{\text{SAR}}) \cdot \zeta_s(\psi_{\text{near}}, R_{\text{SAR}}) \Big] \\
& + \sin^{-1}\Big[\zeta_s(\psi_{\text{far}}, R_{\text{SAR}}) \Big] - \sin^{-1}\Big[\zeta_s(\psi_{\text{near}}, R_{\text{SAR}}) \Big] \Big\}
\end{aligned}
\tag{4.30}
$$

with:

$$
\begin{cases}
\xi_s(\phi_b) = \cos\phi_b - \sin\phi_b \\
\zeta_s(\psi_{\text{far}}, R_{\text{SAR}}) = \Big[1 + \kappa_s^2(\psi_{\text{far}}, R_{\text{SAR}}) \Big]^{-0.5} \\
\zeta_s(\psi_{\text{near}}, R_{\text{SAR}}) = \Big[1 + \kappa_s^2(\psi_{\text{near}}, R_{\text{SAR}}) \Big]^{-0.5}
\end{cases}
\tag{4.31}
$$

The percentage of the MBSAR's coverage area to Earth's overall surface area can be approximately given by:

$$
\begin{aligned}
r_{EC} = \frac{1}{\pi} \cdot \Big\{ & \cos^{-1}\Big[\xi_s(\phi_b) \cdot \kappa_s(\psi_{\text{far}}, R_{\text{SAR}}) \cdot \zeta_s(\psi_{\text{far}}, R_{\text{SAR}}) \Big] \\
& - \cos^{-1}\Big[\xi_s(\phi_b) \cdot \kappa_s(\psi_{\text{far}}, R_{\text{SAR}}) \cdot \zeta_s(\psi_{\text{near}}, R_{\text{SAR}}) \Big] \\
& + \sin^{-1}\Big[\zeta_s(\psi_{\text{far}}, R_{\text{SAR}}) \Big] - \sin^{-1}\Big[\zeta_s(\psi_{\text{near}}, R_{\text{SAR}}) \Big] \Big\} \times 100\%
\end{aligned}
\tag{4.32}
$$

A detailed derivation of the above expression is given in Appendix B.

Following Eqs. (4.30) and (4.32), the coverage area of the MBSAR on Earth's surface depends on the Earth's radius, Earth–MBSAR distance, near and far grazing angles, and bounds of azimuth angle. Given that the grazing and observation azimuth angle limitations have already been determined, the coverage area of the MBSAR is determined by the ratio of the Earth's radius to the Earth–MBSAR distance. To comprehensively analyze the coverage area, we establish the relationship between the Earth–MBSAR distance and MBSAR's coverage area as a proportion of the total global area in Figure 4.18.

By inspecting Figure 4.18, there is a positive correlation between the Earth–MBSAR distance and MBSAR's coverage area. Additionally, the surface coverage area exhibits minor variations ranging from 21.01% to 21.08% of the global area. Considering the average eccentricity of the lunar orbit to be 0.0549, the Earth–MBSAR distance experiences fluctuations on a magnitude of millions of meters during a single MBSAR cycle [30]. Therefore, we can conclude that the determination of MBSAR's coverage area suffers a relatively small impact by the variations in Earth–MBSAR distance.

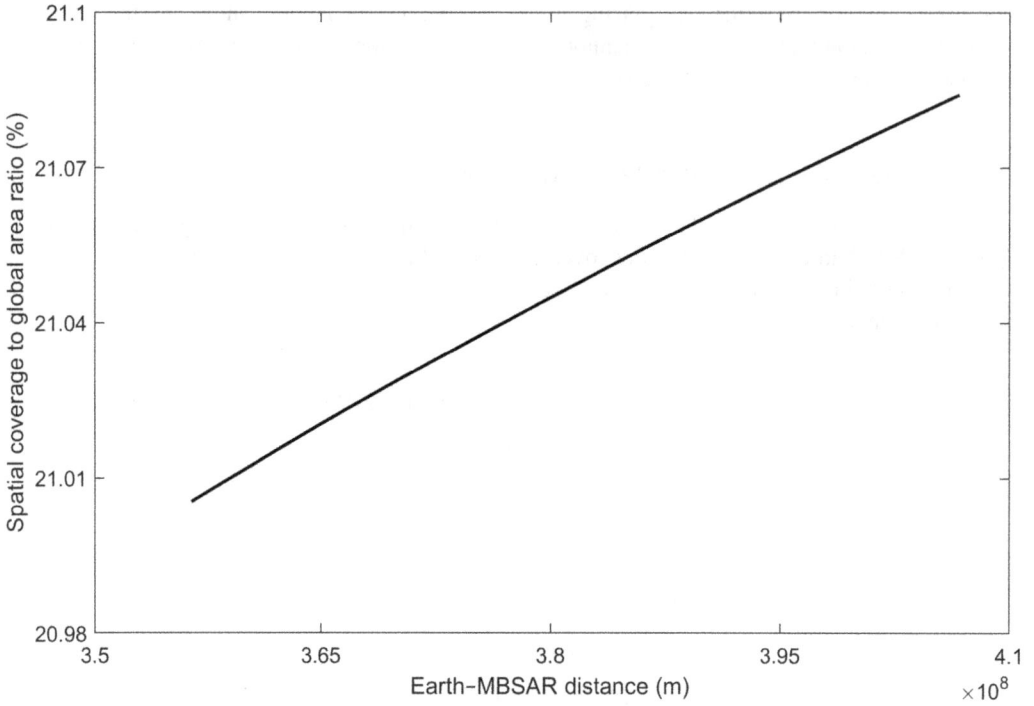

FIGURE 4.18 The ratio of MBSAR's spatial coverage area to total global area versus Earth–MBSAR distance.

4.2.3 LATITUDINAL AND LONGITUDINAL GROUND COVERAGES

Consider that the latitude of maximum ground coverage is attained at a longitudinal deviation of 0°, namely $\Lambda_{DTS} = 0°$. In this situation, the maximum and minimum latitudes of the MBSAR's coverage can be determined as:

$$\Phi_{TOI} = \pm\cos^{-1}\left[\kappa_s(\psi_{far}, R_{SAR})\right] + \Phi_{SAR} \tag{4.33}$$

Given that the maximum value for the Earth's latitude is ±90°, the angle threshold for the latitude of the nadir point is:

$$\Phi_{STH} = 90° - \cos^{-1}\left[\kappa_s(\psi_{far}, R_{SAR})\right] \tag{4.34}$$

When the MBSAR's grazing angle is constrained by the far grazing angle, the coverage of the Earth's North and South Poles depends on the nadir point's latitude, along with the threshold value Φ_{STH} and its inverse value $-\Phi_{STH}$. To this end, we present a graph in Figure 4.19 depicting the variation in the latitude of the nadir point over a single Julian year (365.25 days), along with the corresponding thresholds Φ_{STH} and $-\Phi_{STH}$. For numerical analysis, we set the time beginning at 00:00:00 on March 20, 2024.

The MBSAR would be able to cover the North Pole and South Pole when the latitude of the nadir point exceeds the threshold Φ_{STH} and $-\Phi_{STH}$, respectively. In either scenario, the MBSAR experiences a maximum longitudinal deviation of up to 180°. In the case of the MBSAR with a nadir point's latitude between $-\Phi_{STH}$ and Φ_{STH}, neither the North nor South Pole would be within its

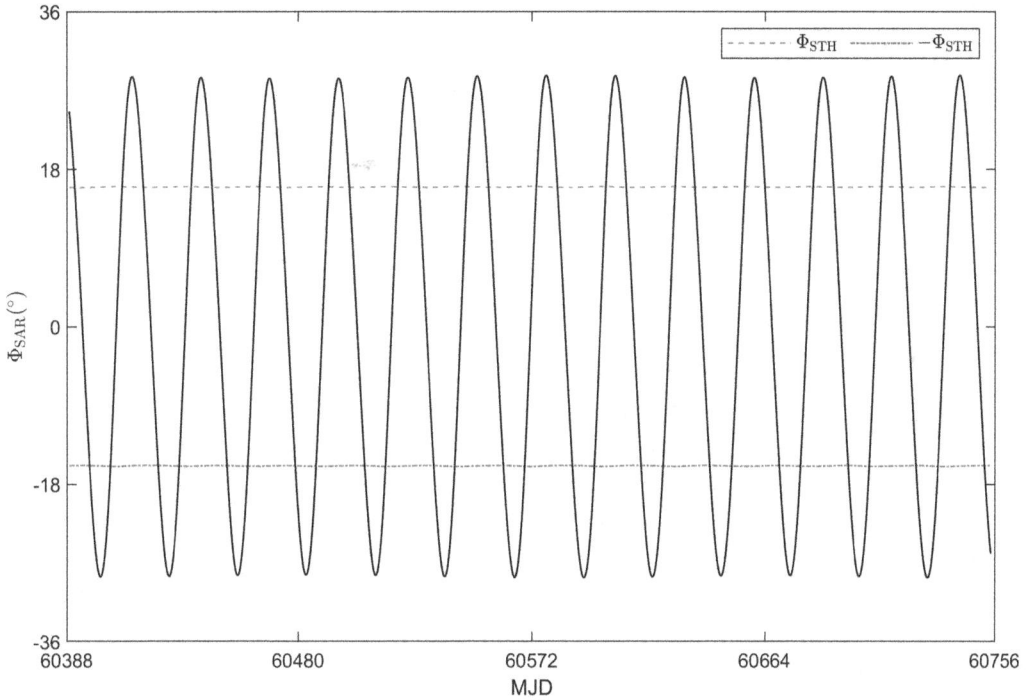

FIGURE 4.19 The latitude of MBSAR's nadir point varies with time over one year, with a corresponding threshold for its latitude.

observation range. Meanwhile, the maximum longitudinal deviation would always be less than 90°, whose magnitude is given by Eq. (4.35).

$$\Lambda_{\mathrm{dmax}} = \sin^{-1}\left\{ \frac{\cos\phi_b \cdot \cos^{-1}\left(\Phi_{\mathrm{SAR}}\right)}{\sqrt{1 - \kappa_s^2\left(\psi_{\mathrm{far}}, R_{\mathrm{SAR}}\right)}} \right\} \tag{4.35}$$

Figure 4.20 presents a graphical representation of the ground coverage of the MBSAR under different nadir point's latitudes (−28°, −10°, 0°, and 18°) in the $\Lambda_{\mathrm{DTS}} - \Phi_{\mathrm{TOI}}$ domain. Notably, the distance between Earth and MBSAR utilized in this analysis has been set at 385,000 km. In this case, the threshold value of Φ_{STH} is 15.916°.

As depicted in Figure 4.20, the ground coverage is susceptible to blind regions confined by both far and near grazing angles and bound of observation azimuth angle. Moreover, the longitudinal ground coverage exhibits rotational symmetry about the axis of $\alpha = 0°$, whereas the latitudinal ground coverage is asymmetrical, barring the nadir point located at the equator of the Earth. If the nadir point's latitude falls into the range between $-\Phi_{\mathrm{STH}}$ and Φ_{STH}, it becomes impossible for MBSAR to cover either of the Earth's poles. Meanwhile, the longitudinal deviation can never exceed 90° in magnitude. In the case where the latitude of the MBSAR's nadir point is higher than Φ_{STH}, as depicted in Figure 4.20(c), the North Pole is covered, but the high latitudes in the Earth's Southern Hemisphere cannot be observed. On the other hand, if the latitude of the nadir point is lower than $-\Phi_{\mathrm{STH}}$, the MBSAR can cover the South Pole, but observing the high latitudes in the Northern Hemisphere becomes impossible. In both instances, the maximum longitudinal deviation can span 180°.

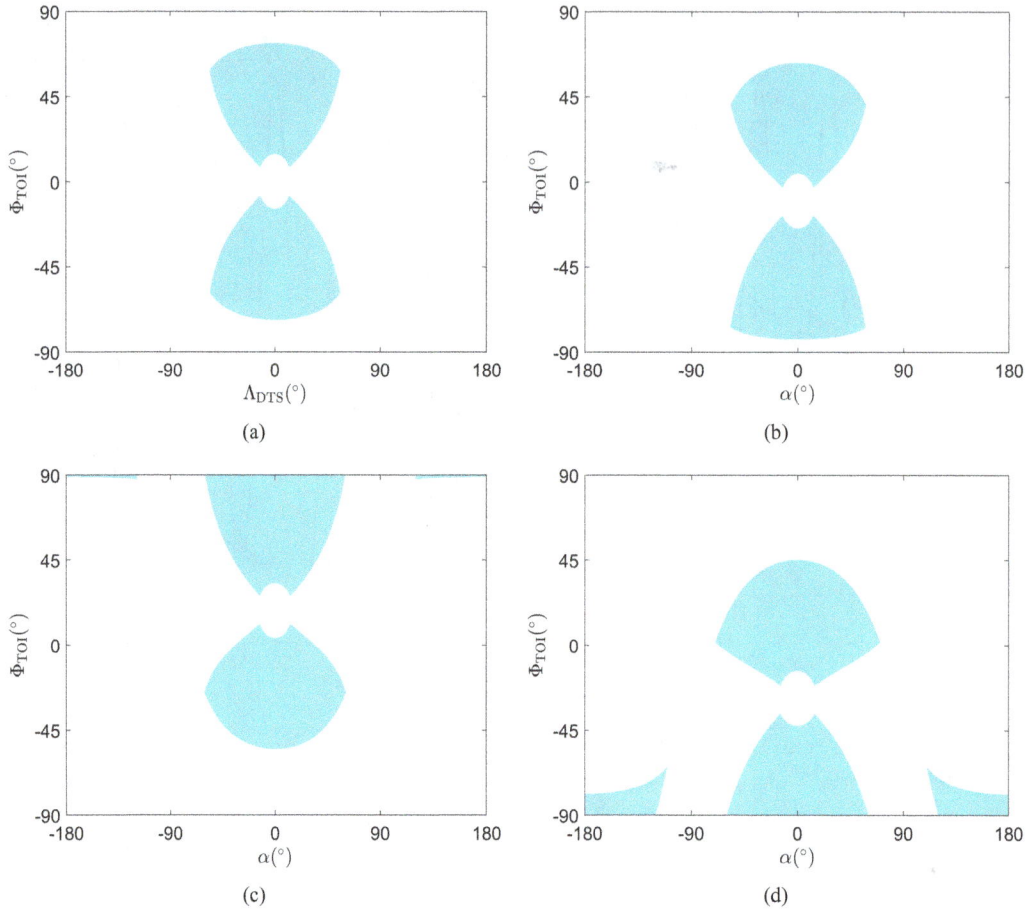

FIGURE 4.20 The ground coverage of MBSAR in the $\Lambda_{\mathrm{DTS}} - \Phi_{\mathrm{TOI}}$ domain at different nadir point's latitudes of (a) 0°, (b) −10°, (c) 18°, and (d) −28°.

4.2.4   Numerical Simulations of Spatiotemporal Coverage

In this section, we present numerical simulations to observe the spatiotemporal coverage of the MBSAR with multiple antennas. To begin with, we shall examine the hourly variations in the spatial coverage within an Earth day.

4.2.4.1   Hourly Variations in the Spatial Coverage

Considering the visit time to be specifically on March 20, 2024, when the MBSAR's nadir point has a relatively higher latitude, we plot the hourly variation in spatial coverage of the MBSAR within an Earth Day, as shown in Figure 4.21, with 2-hour time intervals.

    Figure 4.21 illustrates that the MBSAR's spatial coverage progresses from east to west, and the time interval for revisiting the previously covered area is typically less than a day. This indicates that the Earth's rotation plays a significant role in determining the temporal variations in the MBSAR's spatial coverage. Comparing Figure 4.21(a1) with (e1), we note that the spatial coverage has shifted toward the east after a lapse of one day, this occurrence is mainly attributed to the MBSAR's inertial motion around the Earth. Further, it is worth mentioning that the MBSAR captures a substantial portion of the Earth's surface. For instance, South America is entirely encompassed by the MBSAR at 00:00:00. Although the far grazing angle provides extensive coverage, certain areas still remain unobserved, creating blind spots. This highlights the impacts of near grazing angle and observation

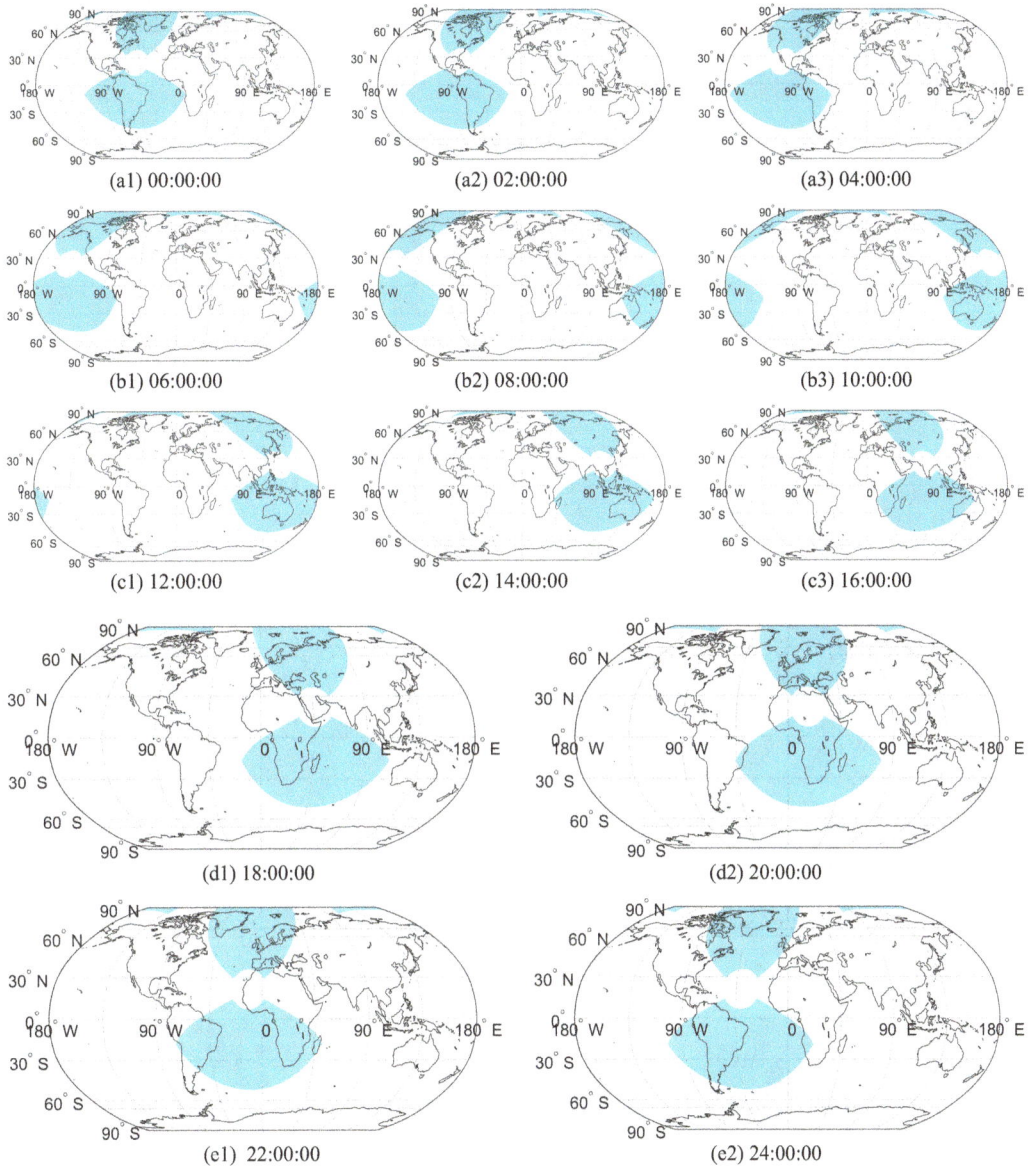

FIGURE 4.21 The MBSAR's spatial coverage within one Earth Day on March 20, 2024.

azimuth angle's bound on the spatial coverage. In this regard, the spatial coverage is influenced by the combination of near and far grazing angles and the bound of observation azimuth angle.

To provide a visual representation, Figure 4.22 shows the temporal changes in spatial coverage of the MBSAR on March 25, 2024, with a time interval of 2 hours, when the nadir point's latitude is relatively small. Figure 4.22 illustrates that the MBSAR maintains noteworthy spatial coverage over an Earth Day even when the nadir point's latitude is small. Besides, owing to the vast spatial coverage and Earth's rotational motion, the observed region experiences a revisit time of less than one day. In addition, the MBSAR's inertial motion causes the spatial coverage of the MBSAR to move eastward after one Earth Day. However, it is not feasible for the MBSAR to observe either the North Pole or the South Pole under the prevailing circumstances. Moreover, a significant blind region emerges on the area enclosed by the far grazing angle. Correspondingly, the influence of near

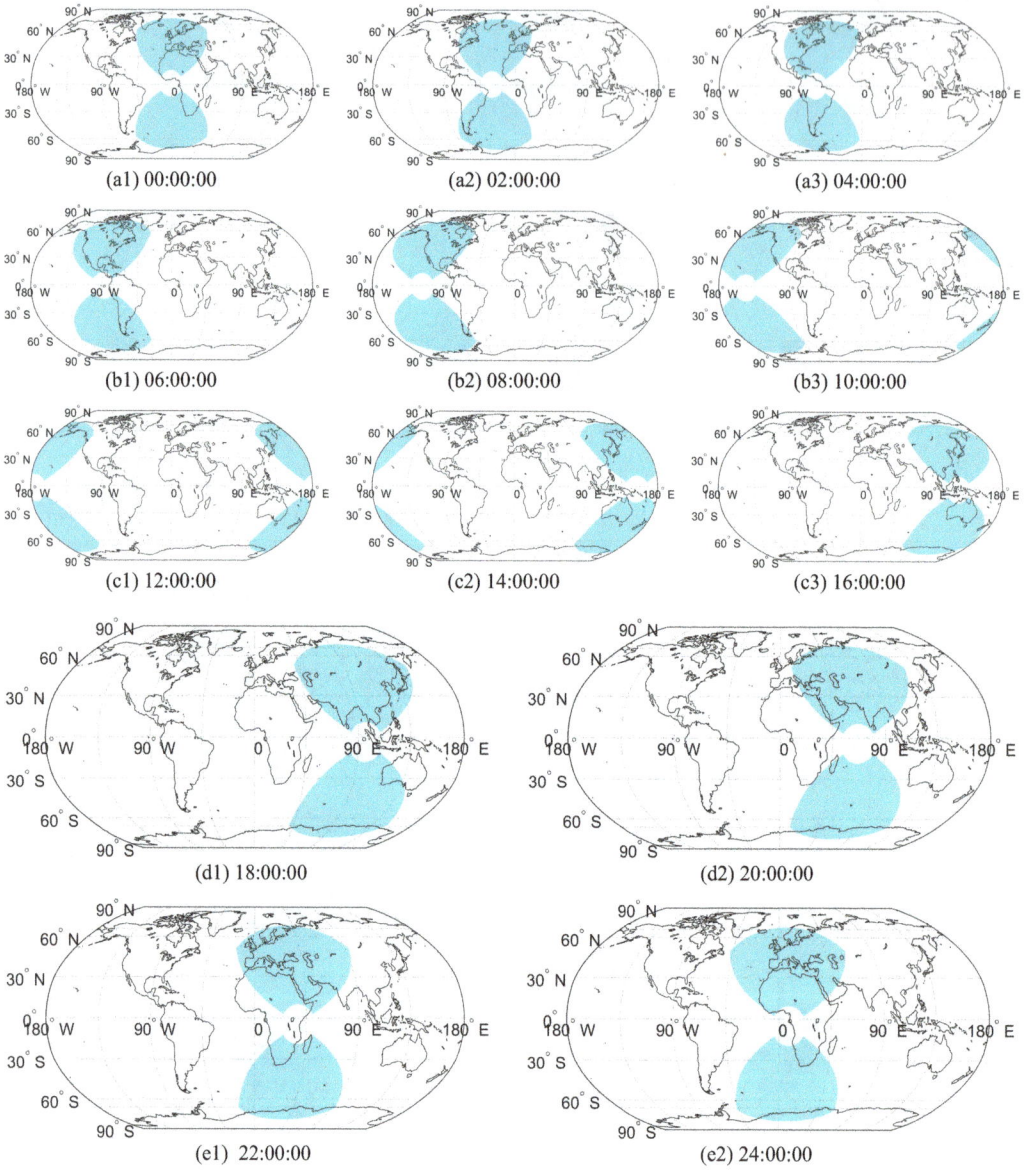

FIGURE 4.22 The MBSAR's spatial coverage within one Earth Day on March 25, 2024.

grazing angle and observation azimuth angle's bound cannot be ignored, even when the nadir point's latitude is small.

As a part of the assessment regarding spatiotemporal coverage, we conducted simulations to scrutinize the spatial coverage of the MBSAR on April 1, 2024, when the nadir point was located in the Southern Hemisphere at a relatively high latitude. The simulation outcomes, reflecting hourly variations of spatial coverage, are depicted in Figure 4.23. The spatial coverage of MBSAR presents a comparable pattern to what is observed when the nadir point is placed in the Northern Hemisphere. Nevertheless, there is a crucial differentiation in this specific scenario; namely, the MBSAR can cover a majority of areas in the Southern Hemisphere. By contrast, the MBSAR is deficient in its capacity to observe high-latitude regions in the Northern Hemisphere.

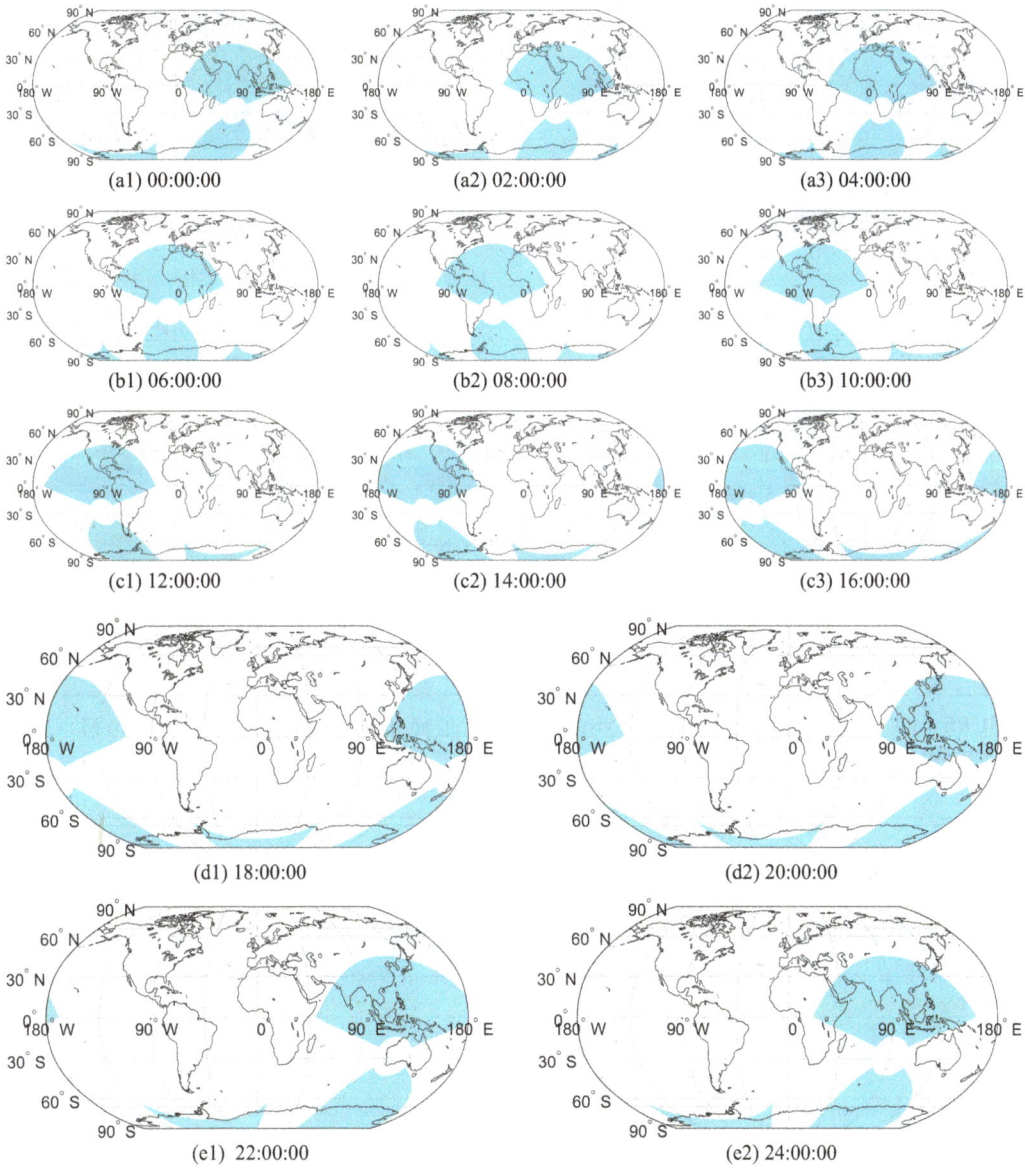

FIGURE 4.23 The MBSAR's spatial coverage within one Earth Day on April 01, 2024.

4.2.4.2 Global Cumulative Visible Time within Different Periods

To examine the spatiotemporal variations in the ground coverage of MBSAR over a given time frame, we introduce the concept of cumulative visible time, referring to the total duration for which MBSAR has covered a specific area within the specified time frame. Figure 4.24 illustrates the global daily-cumulative visible time of MBSAR on four distinct dates: March 20, March 25, April 1, and April 7 in the year 2024, the results of which are consistent with the preceding analysis.

It is essential to ascertain the complete span of Earth's visibility acquired by MBSAR throughout the nodical months. To accomplish this goal, we depict the MBSAR's cumulative visible time over diverse cycles in T_{lc1}, T_{lc2}, and T_{lc3}, as displayed in Figure 4.25. It is noteworthy that the MBSAR's orbital inclination is around its maximum within those cycles.

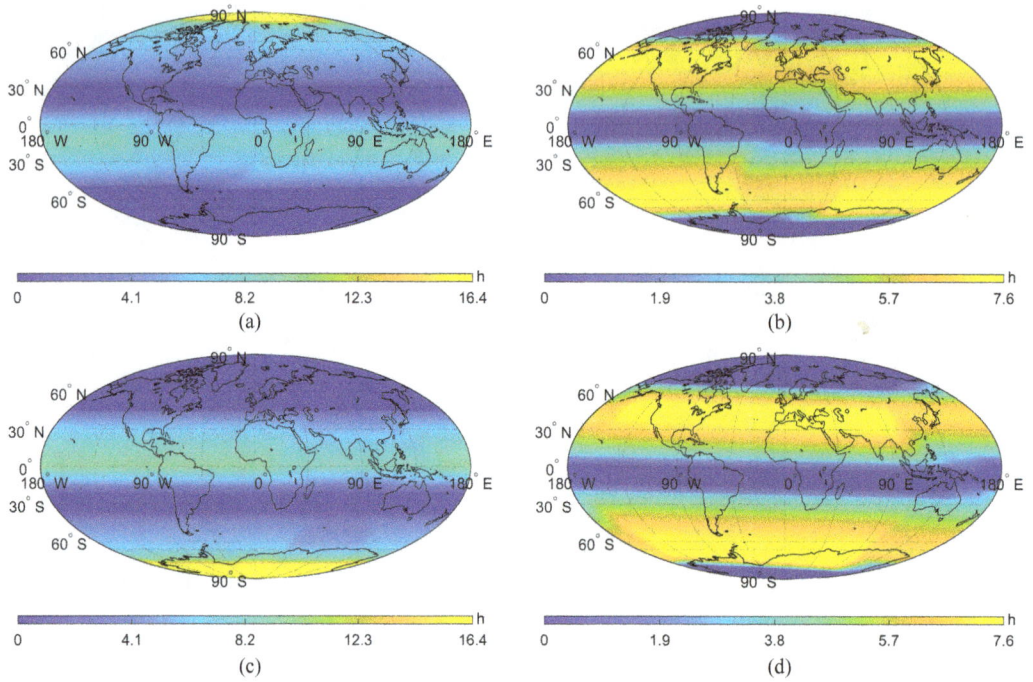

FIGURE 4.24 The MBSAR's cumulative visible time within an Earth Day on (a) March 20, (b) March 25, (c) April 01, and (d) April 06, 2024.

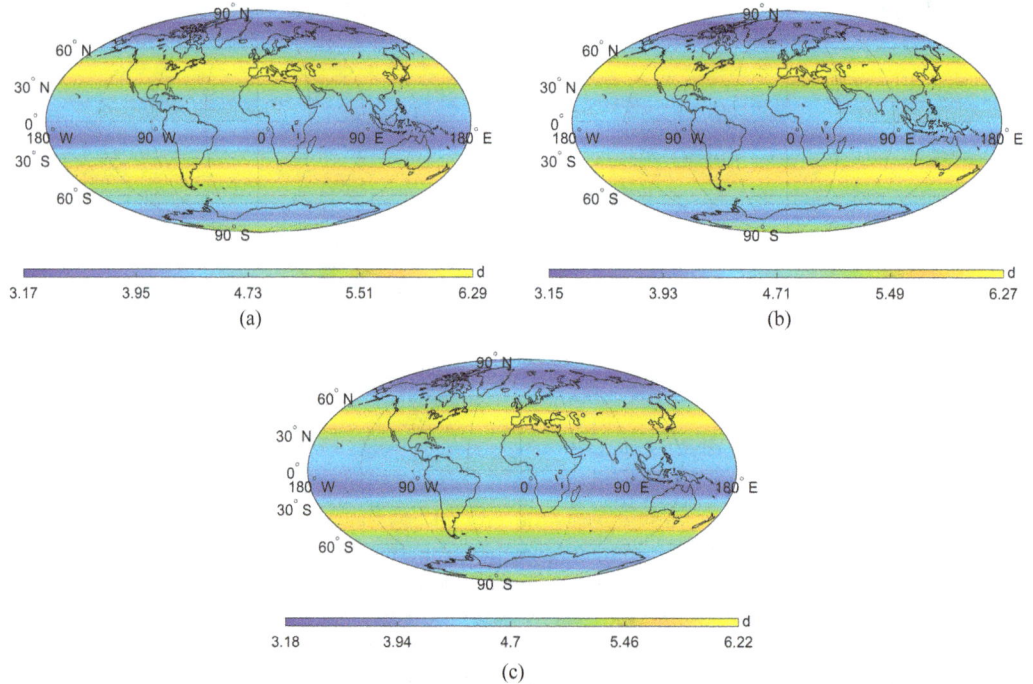

FIGURE 4.25 The MBSAR's cumulative visible time within cycles (a) T_{lc1}, (b) T_{lc2}, (c) T_{lc3}. The symbol 'd' indicates days.

Upon examining Figure 4.25, it becomes apparent that the global cumulative visible time spans a duration of several days within an MBSAR cycle. It is also relevant that the peak value of latitudinal cumulative visible time is observed at latitudes of ±40°, irrespective of the orbital cycles. It is worth mentioning that the Northern Hemisphere experiences a greater amount of cumulative visible time. Interestingly, at latitudes of ±20°, two stripes can be observed, both of which correspond to the local minimum of latitudinal cumulative visible time; this valley, which is more prominent in the Southern Hemisphere, is also independent of the orbiting cycles. Such a phenomenon can be attributed to the near grazing angle. In the high latitudes, yet another valley of latitudinal cumulative visible time given rise by the far grazing angle comes into view. What is intriguing here is that this valley is particularly prominent in the Northern Hemisphere despite the orbiting cycles. Regarding the longitudinal cumulative visible time, the position and amplitude of its peak and valley fluctuate with each cycle. Regarding this, there is a disparity in MBSAR's global cumulative visible time between each cycle, implying that the nodical monthly cumulative visible time is subject to orbital perturbations in the MBSAR.

To delve deeper into the spatiotemporal coverage of the MBSAR during various nodical months under the condition of small orbital inclination, we shall conduct simulations on the global cumulative visible time of MBSAR in cycles from T_{lc4} to T_{lc6}, as illustrated in Figure 4.26. Notably, the MBSAR's inclination angle approaches its minimum during those cycles.

Figure 4.26 shows that a decrease in the orbital inclination leads to a narrower range of accessible global nodical monthly cumulative visible time in the MBSAR. Additionally, the distribution of the latitudinal cumulative visible time is distinct from that under larger orbital inclination. In this case, the peak of latitudinal cumulative visible time is apparent at the equator and latitudes of ±45°. Concerning the local minimum of latitudinal cumulative visible time, comparing Figures 4.25 and 4.26 shows that the one located at a lower latitude tends to move toward higher latitudes, while another located at higher latitudes shifts in the opposite direction, toward lower latitudes.

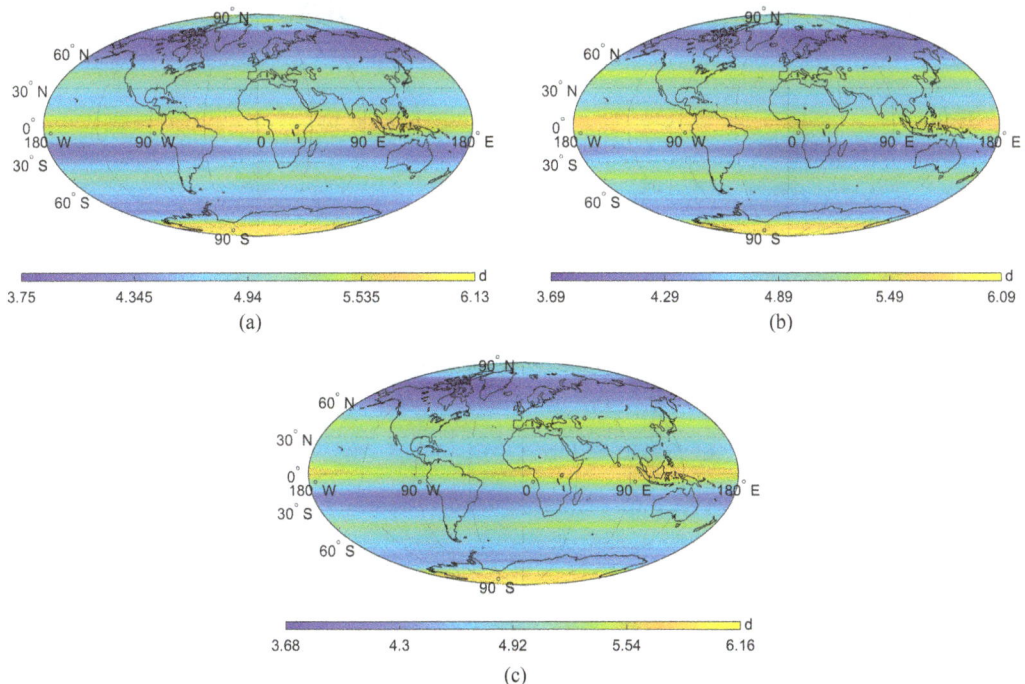

FIGURE 4.26   The MBSAR's cumulative visible time within cycles (a) T_{lc4}, (b) T_{lc5}, (c) T_{lc6}. The symbol "d" indicates days.

The regularity of the variation in latitudinal cumulative visible time is not contingent upon the orbiting cycle. In contrast, the longitudinal cumulative visible time strongly correlates with the orbiting cycle of the MBSAR. Correspondingly, the cumulative visible time across the globe conforms to an unevenly distributed pattern. Additionally, there are discrepancies in the global cumulative visible time over three consecutive cycles. From this, it can be inferred that the orbital perturbation also exerts a marked influence on the MBSAR's spatiotemporal coverage in scenarios where the inclination angle is small.

In order to gain a more comprehensive understanding of the MBSAR's spatiotemporal coverage, it is also necessary to inspect the nodical monthly cumulative visible time of the MBSAR after a period of 18.6 years with respect to those depicted in Figure 4.25. To this end, we simulate the global cumulative visible time across cycles T_{lc7} to T_{lc9}. The accompanying outcomes presented in Figure 4.27 demonstrate that the trend of the cumulative visible time versus Earth's location after a span of 18.6 years resembles that illustrated in Figure 4.25. For a particular geographical location, the nodical monthly cumulative visible time does not entirely align with that after a span of 18.6 years. This suggests that the MBSAR's performance in spatiotemporal coverage is still subject to aperiodic changes when considering the cumulative visible time on a nodical monthly basis. Consequently, the MBSAR's spatiotemporal coverage undergoes fluctuations due to the influence of orbital perturbations, resulting in irregular variations across different nodical months.

The MBSAR's inertial motion does not follow a predictable pattern within a single Julian year. Thus, the yearly cumulative visible time does not necessarily signify any discernible patterns in the spatiotemporal coverage of MBSAR, which is not elucidated in this context. On the other hand, the MBSAR experiences cyclical changes in its orbital inclination every 18.6 years, leading to the highest or lowest range of the nadir point's latitude during such a time interval. Hence, it is intriguing to depict the cumulative visible time of MBSAR during an 18.6-year cycle; one can accomplish this

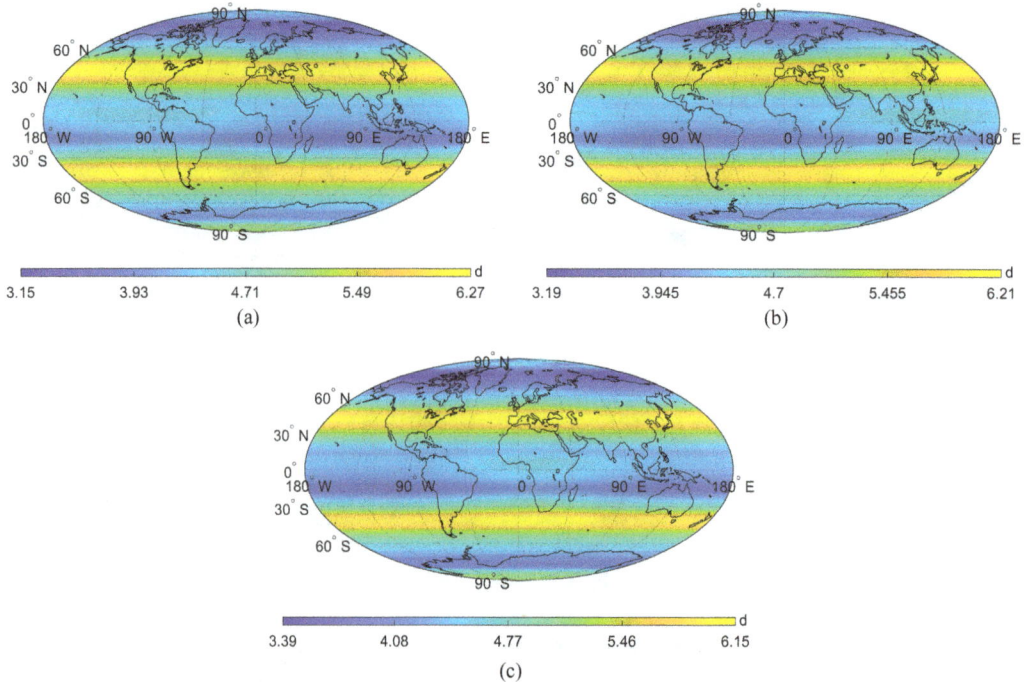

FIGURE 4.27   The MBSAR's cumulative visible time within cycles (a) T_{lc7}, (b) T_{lc8}, (c) T_{lc9}. The symbol "d" indicates days.

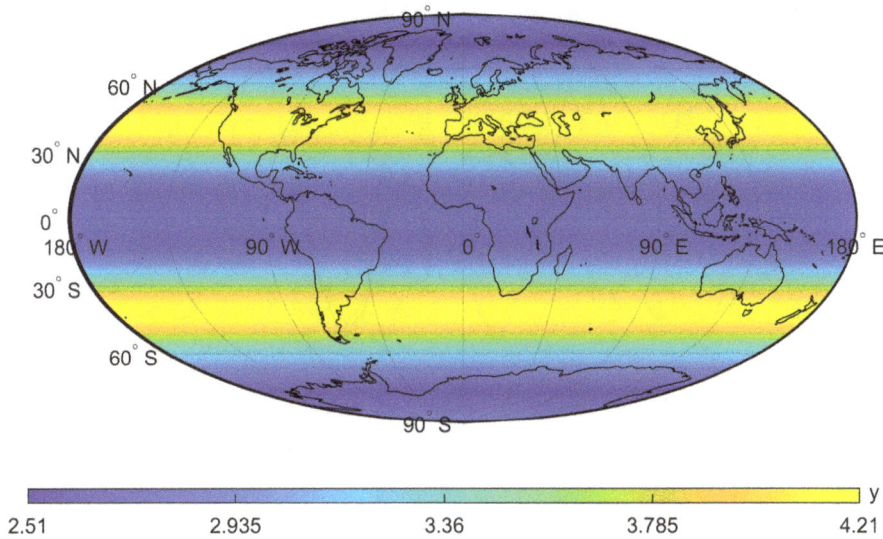

FIGURE 4.28 The cumulative visible time of the MBSAR over 18.6 years. The symbol "y" represents the Julian year (365.25 days).

by initiating the visit time at 00:00:00 on March 20, 2024, TDB. The resulting MBSAR's global cumulative visible time during a period of 18.6 years is depicted in Figure 4.28.

As observed from Figure 4.28, over a period of 18.6 years, the global cumulative visible time of the MBSAR can be several years. The pattern of cumulative visible time across longitude exhibits a uniform distribution. By contrast, the latitudinal cumulative visible time exhibits a symmetrical distribution relative to the equator. Moreover, the latitudinal cumulative visible time reaches its zenith at the equator and latitudes of ±40°, while the local minimum values can be observed at low latitudes of ±15° and high latitudes of ±80°. Further, there are discernible stripes in the cumulative visible time of the MBSAR in Earth's northern and southern hemispheres. This phenomenon is a consequence of the Earth's rotation, the inertial motion of MBSAR, and observation parameters regarding the near and far grazing angles and bound of observation azimuth angle.

The inclination of the MBSAR's orbit experiences a periodic fluctuation with an average period of 18.6 years. However, there is no observable regularity in the instantaneous or short-term cumulative spatiotemporal coverage of the MBSAR within such an epoch, as demonstrated previously. Notwithstanding, it is still worthwhile to analyze the MBSAR's spatiotemporal coverage for a duration exceeding 18.6 years to explore prospective variation trends in the coverage performance. As such, Figure 4.29 portrays the MBSAR's cumulative visible time at various Earth's longitudes as a function of versus Earth's latitude throughout three consecutive 18.6-year cycles.

Figure 4.29 indicates that there seems to be negligible variance in the 18.6-year cumulative visible time at distinct longitudes for a given latitude. As a result, the 18.6-year cumulative visible time of the MBSAR is uniformly distributed along the longitudinal axis. Additionally, the MBSAR's latitudinal cumulative visible time exhibits negligible discrepancy throughout each 18.6-year cycle. From this perspective, it can be concluded that the cumulative visible time approximately experiences periodic oscillations over an 18.6-year cycle in the MBSAR. Moreover, this cumulative visible time helps select appropriate sites for MBSAR by discerning the long-term variation patterns of MBSAR observations across different locations on the lunar terrain, as expounded below.

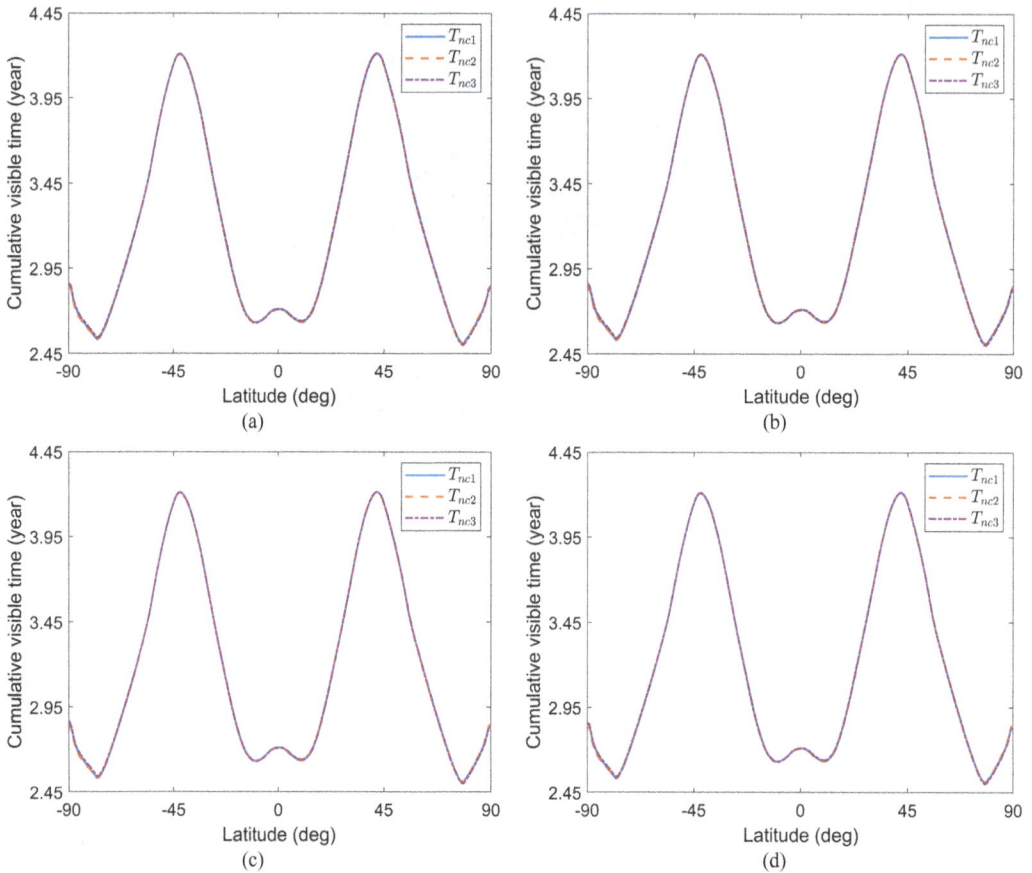

FIGURE 4.29 The 18.6-yearly cumulative visible time versus the Earth's latitudes at four different longitudes on Earth: (a) 0°, (b) 60°E, (c) 120°E, (d) 180°E. The data are simulated spanning three consecutive 18.6-year cycles, namely T_{nc1}, T_{nc2}, and T_{nc3}, commencing at 00:00:00 on March 20, 2024.

4.3   OPTIMAL SITE SELECTION FOR THE MBSAR

To ascertain the most suitable location for the MBSAR system, we have selected five prospective sites that may provide unobstructed spatial coverage for Earth observation. These sites are illustrated in Figure 4.30, and their respective coordinates are tabulated in Table 4.1. It is pertinent to mention that this analysis is simplified by disregarding the physical librations of the Moon and assuming it to be a spherical body.

The reason for choosing these five locations is that the sites M_2 and M_4 are selected to maximize and minimize the nadir point's latitudes, respectively. The sites M_3 and M_5 are chosen to attain the maximum and minimum values of the nadir point's longitude. Furthermore, site M_1 is chosen for comparison with the selected sites. With this particular configuration, the impact of the MBSAR's sites can be categorized into two aspects: longitude and latitude. The effect of the site's longitude can be analyzed by investigating sites M_1, M_3, and M_5, while the influence of the site's latitude can be examined by focusing on sites M_1, M_2, and M_4.

The cumulative visible time within a single nodical month may differ from that within the following nodical month, owing to temporal fluctuations in the MBSAR's orbital elements. Hence, the nodical monthly cumulative visible time is inadequate for determining the optimal site for an MBSAR. On the other hand, the inclination of the lunar orbit experiences cyclic fluctuations every 18.6 years, which would lead to a comparable buildup of cumulative visible time for the MBSAR during the

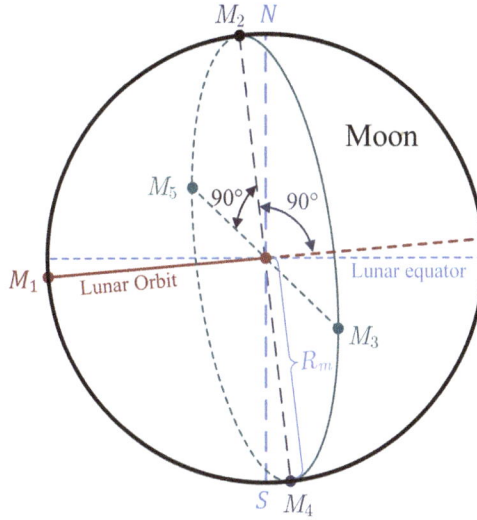

FIGURE 4.30   The schematic diagram of the Moon-based station's sites. R_M is the Moon's radius. N and S are the North and South poles of the Moon, respectively.

TABLE 4.1

Coordinates of the MBSAR's Candidate Sites in the ECEF

Site	Earth–MBSAR Distance	Nadir Point's Longitude	Nadir Point's Latitude
M_1	$R_{MC} - R_m$	Λ_{MC}	Φ_{MC}
M_2	$\sqrt{R_{MC}^2 + R_m^2}$	Λ_{MC}	$\Phi_{MC} + \tan^{-1}\left(R_m R_{MC}^{-1}\right)$
M_3	$\sqrt{R_{MC}^2 + R_m^2}$	$\Lambda_{MC} - \tan^{-1}\left(R_m R_{MC}^{-1}\right)$	Φ_{MC}
M_4	$\sqrt{R_{MC}^2 + R_m^2}$	Λ_{MC}	$\Phi_{MC} - \tan^{-1}\left(R_m R_{MC}^{-1}\right)$
M_5	$\sqrt{R_{MC}^2 + R_m^2}$	$\Lambda_{MC} + \tan^{-1}\left(R_m R_{MC}^{-1}\right)$	Φ_{MC}

Note: R_{MC}, Λ_{MC}, and Φ_{MC} are given by $\mathbf{R_{MC}}$, the position vector of Moon's barycenter. The lunar librations are ignored in this case.

next 18.6-year cycle, as evidenced in Figure 4.29. Henceforth, we shall assess the aggregation of cumulative visible time for a period of 18.6 years, considering five potential locations to ascertain the optimal site. To this end, Figure 4.31 depicts the longitudinal aggregation of cumulative visible time over 18.6 years at Earth latitudes of 0°, 30°N, 60°N, and 80°N, commencing from 00:00:00 on March 20, 2024.

Figure 4.31 shows that the cumulative visible time along the longitudinal axis is evenly spread across all MBSAR sites, though slight fluctuations exist in the longitudinal cumulative visible time whatever the Earth's latitudes are. At all latitudes on Earth, the cumulative visible time along the longitudinal axis at site M_1 is similar to that at sites M_3 and M_5. The evidence points toward the conclusion that the longitude of the nadir point has an insignificant impact on the cumulative visible time over 18.6 years. On the other hand, the cumulative visible time along the longitudinal axis at site M_1 differs from that at sites M_2 and M_4. Such an observation suggests that the nadir point's latitude could certainly impact the cumulative visible time of the MBSAR.

Next, we focus on ascertaining the influence of the nadir point's latitude (M_1, M_2, and M_4) on the cumulative visible time over an 18.6-year duration. To achieve this, we shall generate a graphical

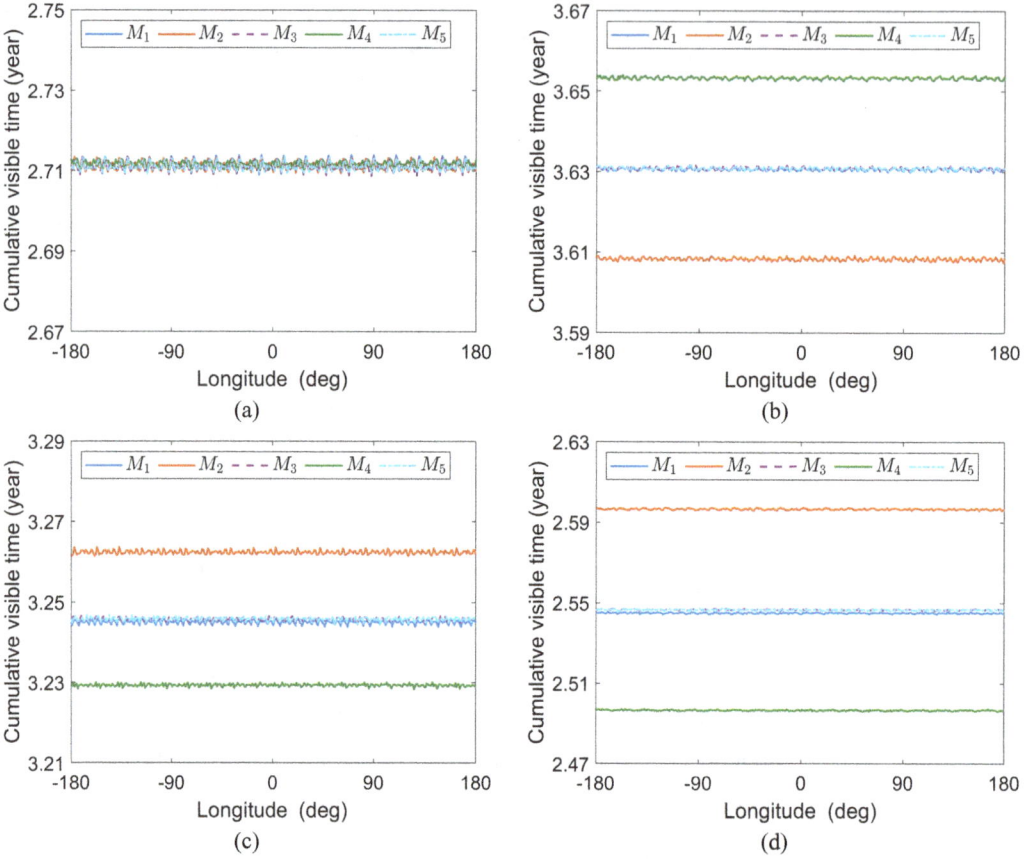

FIGURE 4.31 The 18.6-yearly longitudinal cumulative visible time at the Earth's latitude of (a) 0°, (b) 30°N, (c) 60°N, and (d) 80°N.

representation of corresponding cumulative visible time, plotted against the Earth's longitudes at 0°, 60°E, 120°E, and 180°E, as depicted in Figure 4.32, where the commencement time is configured to begin at 00:00:00 on March 20, 2024.

Figure 4.32 illustrates that the latitudinal cumulative visible time exhibits a consistent pattern across varying longitudes. On the other hand, the latitudinal cumulative visible time displays variability dependent on the Earth's latitude. As for an MBSAR situated at the M_1 site, the latitudinal cumulative visible time reaches its peak at the Earth's latitude of 42° (both North and South), whereas the latitudinal cumulative visible time dwindles to its minimum at latitudes of 78°N and 78°S. Furthermore, the MBSAR's location significantly impacts the cumulative visible time at high latitudes along the latitudinal axis. Once the MBSAR is located at M_2, near the Moon's North Pole, the cumulative visible time along the latitudinal axis is higher in Earth's Northern Hemisphere compared to the Southern Hemisphere, particularly at higher latitudes. By contrast, the trend is reversed when MBSAR is placed at M_4, near the Moon's South Pole. If the MBSAR is placed at M_1, the site near the lunar equator, the cumulative visible time along the latitudinal axis is symmetrically distributed with respect to the Earth's equator.

Hence, it may be advisable to position the MBSAR near M_2 to observe the Arctic region. Conversely, the site surrounding M_4 on the Moon could be an ideal location for measuring Antarctica due to its consistently higher latitudinal cumulative visible time. The MBSAR's site location bears no significance for other regions, as the corresponding cumulative visible time over 18.6 years is almost identical for most of Earth's regions.

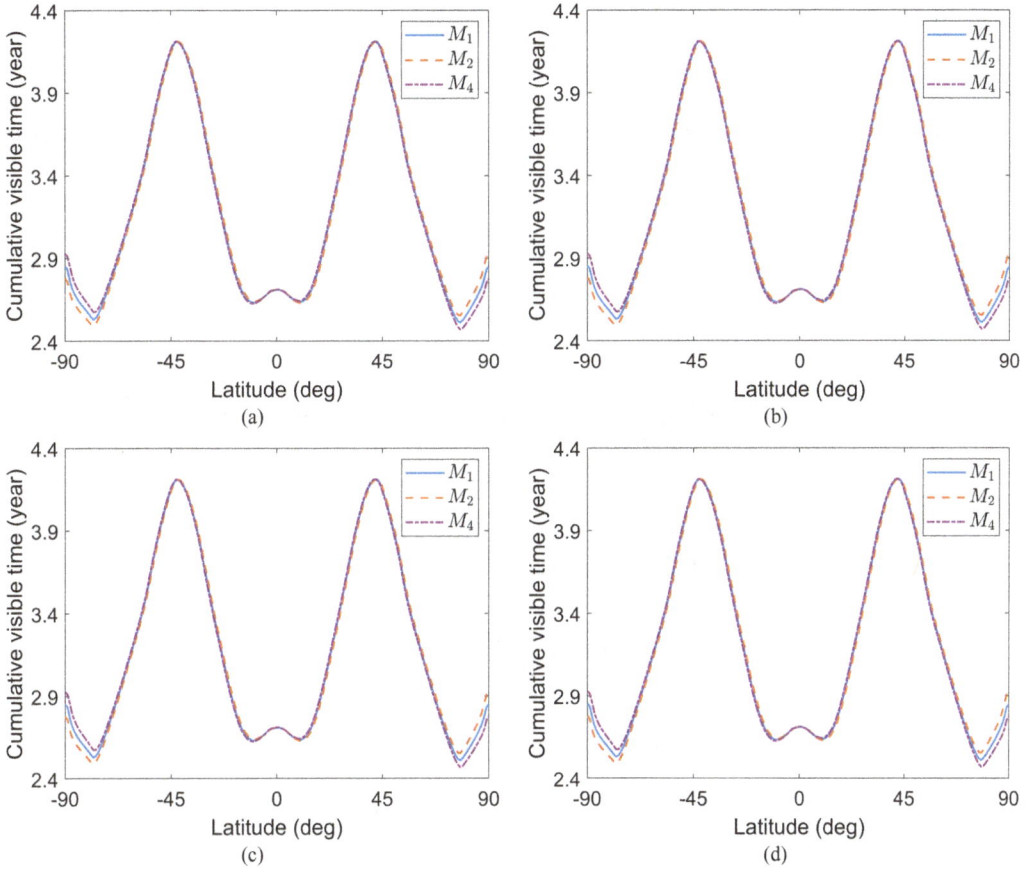

FIGURE 4.32 The 18.6-yearly latitudinal cumulative visible time at the Earth's longitude of (a) 0°, (b) 60°E, (c) 120°E, and (d) 180°E.

APPENDIX A: THE DERIVATIONS OF OBSERVATION PARAMETERS

In the ECEF coordinate system, the position vectors for both MBSAR and TOI can be depicted as follows:

$$\mathbf{R_{SAR}} = \left[R_{SAR} \cos \Lambda_{SAR} \cos \Phi_{SAR}, R_{SAR} \cos \Lambda_{SAR} \sin \Phi_{SAR}, R_{SAR} \sin \Phi_{SAR} \right]^T \tag{A.1}$$

$$\mathbf{R_{TOI}} = \left[R_{TOI} \cos \Phi_{TOI} \cos \Lambda_{TOI}, R_{TOI} \cos \Phi_{TOI} \sin \Lambda_{TOI}, R_{TOI} \sin \Phi_{TOI} \right]^T \tag{A.2}$$

To acquire the slant range vector, one can employ the following relation:

$$\mathbf{R_{ST}} = \mathbf{R_{SAR}} - \mathbf{R_{TOI}} = \begin{bmatrix} R_{SAR} \cos \Lambda_{SAR} \cos \Phi_{SAR} - R_{TOI} \cos \Phi_{TOI} \cos \Lambda_{TOI} \\ R_{SAR} \cos \Lambda_{SAR} \sin \Phi_{SAR} - R_{TOI} \cos \Phi_{TOI} \sin \Lambda_{TOI} \\ R_{SAR} \sin \Phi_{SAR} - R_{TOI} \sin \Phi_{TOI} \end{bmatrix} \tag{A.3}$$

As such, the slant range of the beam centerline can be formulated as:

$$\begin{aligned} R_{ST} &= \left\| \mathbf{R_{ST}} \right\|_2 \\ &= \sqrt{R_{SAR}^2 + R_{TOI}^2 - 2 R_{SAR} R_{TOI} (\cos \Phi_{TOI} \cos \Phi_{SAR} \cos \Lambda_{DTS} + \sin \Phi_{TOI} \sin \Phi_{SAR})} \end{aligned} \tag{A.4}$$

with:

$$\Lambda_{DTS} = \Lambda_{TOI} - \Lambda_{SAR} \tag{A.5}$$

The MBSAR's look angle, the angle between the MBSAR's position and slant range vectors, can be articulated as:

$$\theta_l = \cos^{-1}\left\{\frac{\mathbf{R}_{ST} \cdot \mathbf{R}_{SAR}}{\|\mathbf{R}_{ST}\|_2 \cdot \|\mathbf{R}_{SAR}\|_2}\right\} \tag{A.6}$$

Upon incorporating Eqs. (A.1) and (A.3) into Eq. (A.5), the resultant expression is:

$$\theta_l = \cos^{-1}\left\{\frac{R_{SAR} - R_{TOI}\left(\cos\Phi_{SAR}\cos\Phi_{TOI}\cos\Lambda_{DTS} + \sin\Phi_{SAR}\sin\Phi_{TOI}\right)}{R_{ST}}\right\} \tag{A.7}$$

The grazing angle, the angle formed between the slant range vector and the normal vector to the Earth's surface at the TOI's point, is given by:

$$\psi = \sin^{-1}\left\{\frac{\mathbf{R}_{ST} \cdot \mathbf{R}_{TOI}}{\|\mathbf{R}_{ST}\|_2 \cdot \|\mathbf{R}_{TOI}\|_2}\right\} \tag{A.8}$$

One can determine the grazing angle by incorporating Eqs. (A.2) and (A.3) into Eq. (A.8):

$$\psi = \sin^{-1}\left\{\frac{R_{SAR}\left(\cos\Phi_{SAR}\cos\Phi_{TOI}\cos\Lambda_{DTS} + \sin\Phi_{SAR}\sin\Phi_{TOI}\right) - R_{TOI}}{R_{ST}}\right\} \tag{A.9}$$

The geocentric angle, which represents the angle between the MBSAR and TOI vectors, can be mathematically formulated as:

$$\theta_e = \cos^{-1}\left\{\frac{\mathbf{R}_{SAR} \cdot \mathbf{R}_{TOI}}{\|\mathbf{R}_{SAR}\|_2 \cdot \|\mathbf{R}_{TOI}\|_2}\right\} \tag{A.10}$$

Upon scrutinizing Eqs. (A.1) and (A.2), it is feasible to rephrase Eq. (A.10) as:

$$\theta_e = \cos^{-1}\left(\cos\Phi_{SAR}\cos\Phi_{TOI}\cos\Lambda_{DTS} + \sin\Phi_{SAR}\sin\Phi_{TOI}\right)[0,1] \tag{A.11}$$

Henceforth, we shall probe into the representation for observation azimuth angle. One may ascertain the vector oriented perpendicular to the ground plane through the following process:

$$\mathbf{u}_{pg} = \left[\cos\Phi_{TOI}\cos\Lambda_{TOI}, \cos\Phi_{TOI}\sin\Lambda_{TOI}, \sin\Phi_{TOI}\right]^T \tag{A.12}$$

Upon projection, the vector along the ground range direction can be delineated as:

$$\mathbf{u}_{gr} = \left(\mathbf{I} - \mathbf{u}_{pg}^T\mathbf{u}_{pg}\right) \tag{A.13}$$

where $\mathbf{I}$ is a unit matrix.

When the observation azimuth angle is set to $0°$ in the eastward direction, the corresponding unit vector can be expressed as follows:

$$\mathbf{u}_{ew} = \frac{\mathbf{u}_N \times \mathbf{R}_{TOI}}{\left\|\mathbf{u}_N \times \mathbf{R}_{TOI}\right\|_2}$$

(A.14)

where $\mathbf{u}_N = [0, 0, 1]^T$.

One may acquire the MBSAR's observation azimuth angle via the following method:

$$\phi_{oa} = \cos^{-1}\left\{\frac{\mathbf{u}_{gr} \cdot \mathbf{u}_{ew}}{\left\|\mathbf{u}_{gr}\right\|_2 \cdot \left\|\mathbf{u}_{ew}\right\|_2}\right\}$$

(A.15)

Upon substitution of Eqs. (A.1), (A.2), (A.12), (A.13), and (A.14) into Eq. (A.15), the observation azimuth angle takes on the subsequent expressions:

$$\phi_{oa} = \cos^{-1}\left\{\frac{\cos\Phi_{SAR}\sin\Lambda_{DTS}}{\sqrt{1-\left(\cos\Phi_{TOI}\cos\Phi_{SAR}\cos\Lambda_{DTS}+\sin\Phi_{TOI}\sin\Phi_{SAR}\right)^2}}\right\}$$

(A.16)

APPENDIX B: THE DERIVATION OF MBSAR'S COVERAGE AREA

In order to scrutinize the MBSAR's coverage area, we shall introduce a pair of contiguous spherical triangles denoted by $\triangle ABC$ and $\triangle BCD$, respectively. The triangular shapes are a result of the intersection of three great circles on a unit sphere, as depicted in Figure 4.A1.

The law of cosines asserts that:

$$\cos(c+d) = \cos b \cos e + \sin b \sin e \cos\angle ACD$$

(A.17)

$$\cos\angle CBD = \sin\angle BCD \sin\angle ADC \cos e - \cos\angle BCD \cos\angle ADC.$$

(A.18)

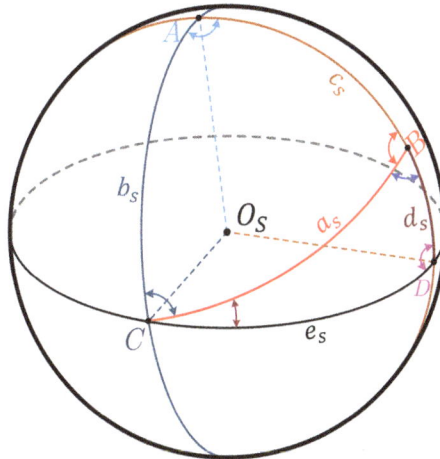

FIGURE 4.A1 Two neighboring unit spherical triangles, where $\triangle ABC$ consists of sides a_s, b_s, and, c_s; $\triangle BCD$ is composed of sides a_s, d_s, and e_s. The side a_s serves as the common side between the two neighboring spherical triangles.

According to the law of sines for spherical triangles, one can obtain:

$$\frac{\sin e}{\sin \angle \text{CAD}} = \frac{\sin b}{\sin \angle \text{ADC}} = \frac{\sin(c+d)}{\sin \angle \text{ACD}} \tag{A.19}$$

In the context of a spherical triangle, employing the law of sines results in the following relation:

$$\frac{\sin e_s}{\sin \angle \text{CAD}} = \frac{\sin b_s}{\sin \angle \text{ADC}} = \frac{\sin(c_s+d_s)}{\sin \angle \text{ACD}} \tag{A.20}$$

where $\angle \text{ACD} = 90°$.

With the assumption that the latitude of the MBSAR's nadir point is 0°, Eq. (A.11) can be simplified to:

$$\theta_e = \cos^{-1}\left(\cos \Phi_{\text{TOI}} \cos \Lambda_{\text{DTS}}\right) \tag{A.21}$$

Therefore, one can approximate the geocentric angle as follows:

$$\cos \Phi_{\text{TOI}} \cos \Lambda_{\text{DTS}} \approx \kappa_s(\psi, R_{\text{SAR}}) \tag{A.22}$$

with:

$$\kappa_s\left(\psi, R_{SAR}\right) = \frac{\sin \psi + 0.5 R_{\text{SAR}}^{-1} R_{\text{TOI}}}{\sin \psi \cdot R_{\text{SAR}}^{-1} R_{\text{TOI}} + 1} + 0.5 R_{\text{SAR}}^{-1} R_{\text{TOI}} \tag{A.23}$$

We can assume that the latitude with a longitudinal deviation of 0° ($\Lambda_{\text{DTS}} = 0°$) is represented by the side b, while let e denote the longitudinal deviation with a latitude of 0° ($\Phi_{\text{TOI}} = 0°$). From Eq. (A.22), it can be deduced that:

$$\begin{cases} \cos b_s \approx \kappa_s\left(\psi, R_{\text{SAR}}\right) \\ \cos e_s \approx \kappa_s\left(\psi, R_{\text{SAR}}\right) \end{cases} \tag{A.24}$$

Also, it is possible to obtain that:

$$\begin{cases} \sin b_s = \sqrt{1 - \kappa_s^2\left(\psi, R_{\text{SAR}}\right)} \\ \sin e_s = \sqrt{1 - \kappa_s^2\left(\psi, R_{\text{SAR}}\right)} \end{cases} \tag{A.25}$$

Drawing from the established conclusions in Eq. (A.17), one can derive the following relation:

$$\cos(c_s+d_s) = \cos b_s \cos e_s + \sin b_s \sin e_s = \kappa_s^2\left(\psi, R_{\text{SAR}}\right) \tag{A.26}$$

Subsequently, we can express the sine of the side $c_s + d_s$ as follows:

$$\sin(c_s+d_s) = \sqrt{1 - \kappa_s^4\left(\psi, R_{\text{SAR}}\right)} \tag{A.27}$$

As the law of sines outlined in Eq. (A.20) is taken into account, one can acquire:

$$\begin{cases} \sin \angle ADC = \left[1 + \kappa_s^2 \left(\psi, R_{SAR}\right)\right]^{-0.5} \\ \sin \angle CAD = \left[1 + \kappa_s^2 \left(\psi, R_{SAR}\right)\right]^{-0.5} \end{cases}$$

(A.28)

and:

$$\cos \angle ADC = \kappa_s \left(\psi, R_{SAR}\right) \cdot \left[1 + \kappa_s^2 \left(\psi, R_{SAR}\right)\right]^{-0.5}$$

(A.29)

By substituting Eq. (A.29) into Eq. (A.18), one arrives at the following expression:

$$\cos \angle CBD = \kappa_s \left(\psi, R_{SAR}\right) \cdot \left(\sin \angle BCD - \cos \angle BCD\right) \cdot \left[1 + \kappa_s^2 \left(\psi, R_{SAR}\right)\right]^{-0.5}$$

(A.30)

$$\begin{aligned} \cos \angle ABC &= \cos\left(180^\circ - \angle CBD\right) \\ &= \kappa_s \left(\psi, R_{SAR}\right) \cdot \left(\cos \angle BCD - \sin \angle BCD\right) \cdot \left[1 + \kappa_s^2 \left(\psi, R_{SAR}\right)\right]^{-0.5} \end{aligned}$$

(A.31)

Through the utilization of the principles of spherical trigonometry, one may derive the ensuing connections:

$$\angle ABC = \cos^{-1}\left\{\kappa_s \left(\psi, R_{SAR}\right) \frac{\cos \angle BCD - \sin \angle BCD}{\sqrt{1 + \kappa_s^2 \left(\psi, R_{SAR}\right)}}\right\}$$

(A.32)

$$\angle BAC = \angle CAD = \sin^{-1}\left\{\left[1 + \kappa_s^2 \left(\psi, R_{SAR}\right)\right]^{-0.5}\right\}$$

(A.33)

and:

$$\angle ACB = \angle ACD - \angle BCD = 90^\circ - \angle BCD$$

(A.34)

Let $\angle BCD$ be ϕ_b, the magnitude of surface area of $\triangle ABC$ can be ascertained as

$$\begin{aligned} S_{\triangle ABC} &= \angle ABC + \angle BAC + \angle ACB - \pi \\ &= \sin^{-1}\left\{\left[1 + \kappa_s^2(\psi, R_{SAR})\right]^{-0.5}\right\} - \phi_b - \pi/2 \\ &+ \cos^{-1}\left\{(\cos\phi_b - \sin\phi_b) \cdot \kappa_s(\psi, R_{SAR}) \cdot \left[1 + \kappa_s^2(\psi, R_{SAR})\right]^{-0.5}\right\} \end{aligned}$$

(A.35)

As the observation azimuth and grazing angles of the MBSAR are limited by

$$\begin{cases} \psi_{far} \leq \psi \leq \psi_{near} \\ \left|\cos\phi\right| \leq \cos\phi_b \end{cases}$$

(A.36)

Given the spatial coverage's symmetry with respect to the nadir point, the MBSAR's coverage area assumes takes the following form:

$$
\begin{aligned}
S_{\mathrm{EC}} = 4R_E^2 \cdot \Bigg\{ & \cos^{-1}\left[(\cos\phi_b - \sin\phi_b) \cdot \kappa_s(\psi_{\mathrm{far}}, R_{\mathrm{SAR}}) \cdot \left(1 + \kappa_s^2(\psi_{\mathrm{far}}, R_{\mathrm{SAR}})\right)^{-0.5} \right] \\
& - \cos^{-1}\left[(\cos\phi_b - \sin\phi_b) \cdot \kappa_s(\psi_{\mathrm{near}}, R_{\mathrm{SAR}}) \cdot \left(1 + \kappa_s^2(\psi_{\mathrm{near}}, R_{\mathrm{SAR}})\right)^{-0.5} \right] \\
& + \sin^{-1}\left[\left(1 + \kappa_s^2(\psi_{\mathrm{far}}, R_{\mathrm{SAR}})\right)^{-0.5} \right] - \sin^{-1}\left[\left(1 + \kappa_s^2(\psi_{\mathrm{near}}, R_{\mathrm{SAR}})\right)^{-0.5} \right] \Bigg\}
\end{aligned}
\tag{A.37}
$$

Ultimately, the proportion of the area covered by the MBSAR to the entirety area of the Earth's surface may be approximated as follows:

$$
\begin{aligned}
r_{\mathrm{EC}} = \frac{1}{\pi} \cdot \Bigg\{ & \cos^{-1}\left[(\cos\phi_b - \sin\phi_b) \cdot \kappa_s(\psi_{\mathrm{far}}, R_{\mathrm{SAR}}) \cdot \left(1 + \kappa_s^2(\psi_{\mathrm{far}}, R_{\mathrm{SAR}})\right)^{-0.5} \right] \\
& - \cos^{-1}\left[(\cos\phi_b - \sin\phi_b) \cdot \kappa_s(\psi_{\mathrm{near}}, R_{\mathrm{SAR}}) \cdot \left(1 + \kappa_s^2(\psi_{\mathrm{near}}, R_{\mathrm{SAR}})\right)^{-0.5} \right] \\
& + \sin^{-1}\left[\left(1 + \kappa_s^2(\psi_{\mathrm{far}}, R_{\mathrm{SAR}})\right)^{-0.5} \right] - \sin^{-1}\left[\left(1 + \kappa_s^2(\psi_{\mathrm{near}}, R_{\mathrm{SAR}})\right)^{-0.5} \right] \Bigg\} \times 100\%
\end{aligned}
\tag{A.38}
$$

REFERENCES

[1] A. Moussessian, et al., "System Concepts and Technologies for High Orbit SAR," *Proc. IEEE MTT-S International Microwave Symposium Digest*, Long Beach, USA, pp. 1623–1626, Jul. 2005.

[2] D. M. Tralli, W. Foxall, and C. Schultz, "Concept for a High MEO InSAR Seismic Monitoring System," *Proc. 2007 IEEE Aerospace Conference*, Big Sky, USA, pp. 1–7, Mar. 2007.

[3] O. Dubovik, et al., "Grand Challenges in Satellite Remote Sensing," *Frontiers in Remote Sensing*, vol. 2, p. 619818, Feb. 2021.

[4] M. Shimada, *Imaging from Spaceborne and Airborne SARs, Calibration, and Applications*. Boca Raton: CRC Press, 2018.

[5] J. L. Awange, E. W. Grafarend, B. Paláncz, and P. Zaletnyik, *Algebraic Geodesy and Geoinformatics*. Berlin: Springer, 2010.

[6] F. T. Ulaby and D. G. Long, *Microwave Radar and Radiometric Remote Sensing*. Ann Arbor: Univ. Michigan Press, 2014.

[7] Z. Xu, K. S. Chen, and G. Q. Zhou, "Zero-Doppler Centroid Steering for the Moon-Based Synthetic Aperture Radar: A Theoretical Analysis," *IEEE Geoscience and Remote Sensing Letters*, vol. 17, no. 7, pp. 1208–1212, Jul. 2020.

[8] H. Fiedler, E. Boerner, J. Mittermayer, and G. Krieger, "TotaL Zero Doppler Steering-A New Method for Minimizing the Doppler Centroid," *IEEE Geoscience and Remote Sensing Letters*, vol. 2, no. 2, pp. 141–145, Apr. 2005.

[9] E. Boerner, H. Fiedler, G. Krieger, and J. Mittermayer, "A New Method for Total Zero Doppler Steering," *Proc. IEEE International Geoscience and Remote Sensing Symposium*, Anchorage, AK, USA, vol. 2, pp. 1526–1529, Sep. 2004.

[10] J. Dong, et al., "Spatio-Temporal Distribution of the Zero-Doppler Line of Lunar-Based SAR," *Remote Sensing Letters*, vol. 12, no. 2, pp. 113–121, Dec. 2021.

[11] Z. Xu, K. S. Chen, Z. L. Li, and G. Y. Du, "Apsidal Precession Effects on the Lunar-Based Synthetic Aperture Radar Imaging Performance," *IEEE Geoscience and Remote Sensing Letters*, vol. 18, no. 6, pp. 1079–1083, Jun. 2021.

[12] Z. Xu, K. S. Chen, and G. Liu, "On Evaluating the Imaging Performance and Orbital Determination under Perturbations of Orbital Inclination and RAAN in the Lunar-Based SAR," *IEEE Transactions on Geoscience and Remote Sensing*, vol. 60, Jul. 2022, doi: 10.1109/TGRS.2022.3188294

[13] Z. Xu, K. S. Chen, and G. Liu, "On Orbital Determination of the Lunar-Based SAR under Apsidal Precession," *IEEE Transactions on Geoscience and Remote Sensing*, vol. 60, May. 2022, doi: 10.1109/TGRS.2022.3176836

[14] I. G. Cumming and F. H. Wong, *Digital Signal Processing of Synthetic Aperture Radar Data: Algorithms and Implementation*. Boston: Artech House, 2005.

[15] Z. Xu, K. S. Chen, G. Liu, and H. Guo, "Spatiotemporal Coverage of a Moon-Based Synthetic Aperture Radar: Theoretical Analyses and Numerical Simulations," *IEEE Transactions on Geoscience and Remote Sensing*, vol. 58, no. 12, pp. 8735–8750, 2020.

[16] M. Ramsey, "ESS Science Planning and Lunar Workshop Overview," presented at the *NASA Advisory Council Workshop on Science Associated with the Lunar Exploration Architecture*, Feb. 27–Mar. 2, 2007. [Online]. Available at: http://www.lpi.usra.edu/meetings/LEA/presentations/OpeningPlenary/Ramsey_ESS.pdf

[17] M. Ramsey, "ESS Findings: Lunar Science Planning and Workshop Overview," presented at the *NASA Advisory Council Workshop on Science Associated with the Lunar Exploration Architecture*, Feb. 27–Mar. 2, 2007. [Online]. Available at: https://www.lpi.usra.edu/meetings/LEA/presentations/closing_plenary/Ramsey_ESS_summary_20070302.pdf

[18] J. Dong, et al., "An Analysis of Spatiotemporal Baseline and Effective Spatial Coverage for Lunar-Based SAR Repeat-Track Interferometry," *IEEE Journal of Selected Topics in Applied Earth Observations and Remote Sensing*, vol. 12, no. 9, pp. 3458–3469, Sep. 2019.

[19] G. H. Heiken, D. T. Vaniman, and B. M. French, *Lunar Sourcebook: A User's Guide to the Moon*. Cambridge, UK: Cambridge Univ. Press, 1991.

[20] H. Guo, G. Liu, and Y. Ding, "Moon-Based Earth Observation: Scientific Concept and Potential Applications," *International Journal of Digital Earth*, vol. 11, no. 6, pp. 546–557, Jul. 2017.

[21] A. Moccia and A. Renga, "Synthetic Aperture Radar for Earth Observation from a Lunar Base: Performance and Potential Applications," *IEEE Transactions on Aerospace and Electronic Systems*, vol. 46, no. 3, pp. 1034–1051, Jul. 2010.

[22] J. C. Curlander and R. N. McDonough, *Synthetic Aperture Radar: Systems and Signal Processing*. New York: Wiley, 1991.

[23] G. Franceschetti and R. Lanari, *Synthetic Aperture Radar Processing*. Boca Raton: CRC Press, 1999.

[24] M. A. Richards, *Fundamentals of Radar Signal Processing*. New York: McGraw-Hill Education, 2014.

[25] K. S. Chen, *Principles of Synthetic Aperture Radar: A System Simulation Approach*. Boca Raton: CRC Press, 2015.

[26] A. Moreira, et al., "A Tutorial on Synthetic Aperture Radar," *IEEE Geoscience and Remote Sensing Magazine*, vol. 1, no. 1, pp. 6–43, Mar. 2013.

[27] M. C. Gutzwiller, "Moon-Earth-Sun: The Oldest Three-Body Problem," *Review of Modern Physics*, vol. 70, no. 2, pp. 589–639, Apr. 1998.

[28] Z. Xu and K. S. Chen, "Effects of the Earth's Curvature and Lunar Revolution on the Imaging Performance of the Moon-Based Synthetic Aperture Radar," *IEEE Transactions on Geoscience and Remote Sensing*, vol. 57, no. 8, pp. 5868–5882, Mar. 2019.

[29] R. Wang and Y. Deng, *Bistatic SAR System and Signal Processing Technology*. Singapore: Springer, 2018.

[30] J. Meeus, *Mathematical Astronomy Morsels*. Richmond, USA: Willmann-Bell, 1997.

5 Signal Model of the MBSAR System

5.1 INTRODUCTION

The signal model is essential for SAR signal processing and image formation; thus, understanding the signal properties provides a reliable key to converting raw data into SAR images [1, 2]. The signal properties are linked to the Doppler shift between the TOI and SAR, making it essential to comprehend Doppler properties in the SAR signal analysis [3]. In the Moon-Based SAR (MBSAR), the Doppler parameters exhibit distinct properties due to the distinctive observation geometry and elliptical perturbed orbit [4–6]. This chapter focuses primarily on the properties of two major Doppler parameters: Doppler frequency and Doppler frequency modulation rate (DFMR). To this end, we derive the mathematical representations for both Doppler parameters and discuss how they vary with the beam pointing direction, side-looking direction, and orbit elements in the MBSAR, highlighting how their properties differ from those in spaceborne SAR.

The combined Earth's rotation and MBSAR's elliptical orbit give rise to a variation of non-zero Doppler centroid over the slant range swath for the broadside (non-squint) looking MBSAR [7]. This, in effect, leads to severe range cell migration, further posing a challenge to image focusing and system performance. As the conventional zero-Doppler steering methods are unsuitable for MBSAR [7], this chapter introduces a general approach based on the phase scan to minimize the Doppler centroid to zero Hz. We also clarify the accuracy requirements for beam pointing direction when employing the zero-Doppler steering to the MBSAR.

Another concern associated with the MBSAR's signal properties is the ultra-long round-trip delay of the signal propagation, invalidating the stop-and-go assumption (also known as the start-stop approximation, stop-and-go approximation) used for signal modeling in the conventional SAR [8]. This chapter, therefore, examines the signal error due to the stop-and-go assumption and establishes a non-stop-and-go echo model for the unique MBSAR.

5.2 DOPPLER PROPERTIES OF THE MBSAR

The Doppler frequency and DFMR are key parameters in image focusing and quality [9, 10]. Due to Earth-MBSAR relative motion and peculiar imaging geometry, the Doppler properties of the MBSAR differ from those of the spaceborne SAR [4]. This section discusses the Doppler frequency and DFMR in line with the orbital elements and SAR configurations, as detailed below.

5.2.1 DOPPLER FREQUENCY AND DFMR

The Doppler frequency is induced by the SAR movement relative to TOI, which can be expressed as:

$$f_d = f_{d(\text{SAR})} + f_{d(\text{ER})} \tag{5.1}$$

where $f_{d(\text{SAR})}$ and $f_{d(\text{ER})}$ respectively denote the Doppler frequency components given rise by the SAR inertial motion and Earth's rotation, both are defined by [1]:

$$f_{d(\text{SAR})} = -\frac{2}{\lambda} \cdot \frac{\mathbf{V}_{\text{SAR}} \cdot \mathbf{R}_{\text{ST}}}{R_{\text{ST}}} \tag{5.2}$$

DOI: 10.1201/9781003308430-5

$$f_{d(\text{ER})} = -\frac{2}{\lambda} \cdot \frac{\omega_E \mathbf{u_N} \cdot \left(\mathbf{R_{SAR}} \times \mathbf{R_{TOI}} \right)}{R_{ST}} \tag{5.3}$$

where:

λ is the wavelength of the probing signal;

ω_E is the Earth's rotational angular velocity;

R_{ST} is the slant range, given in Eq. (2.60);

$\mathbf{u_N}$ represents the unit vector that equals to $[0, 0, 1]^T$;

$\mathbf{V_{SAR}}$ is the velocity vector of the SAR;

$\mathbf{R_{SAR}}$ is the position vector of the SAR;

$\mathbf{R_{TOI}}$ is the position vector of the TOI; and

$\mathbf{R_{ST}}$ is the slant range vector, written by:

$$\mathbf{R_{ST}} = \mathbf{R_{SAR}} - \mathbf{R_{TOI}} \tag{5.4}$$

For reference purposes, the observation geometry of MBSAR is depicted in Figure 5.1, where two inertial frames, i.e., the ACS and ECI coordinate systems, are used. As for this chapter, it is important to note that all orbital elements are defined in the ECI reference frame. For the sake of simplicity, we shall ignore the subscript "ECI" in the symbols associated with the orbital elements presented in Chapter 2.

For a SAR system operating in an elliptical orbit, the Doppler frequency in Eq. (5.1) can be expressed in terms of the orbital elements and antenna beam pointing. There are two components in the Doppler frequency, wherein the component related to the contribution of Earth's rotation takes the following forms:

$$f_{d(\text{ER})} = 2\lambda^{-1}\omega_E \xi_{\text{sld}} R_{SAR} \sin \theta_l \left(\sin i \cos u \sin \phi - \cos \phi \cos i \right) \tag{5.5}$$

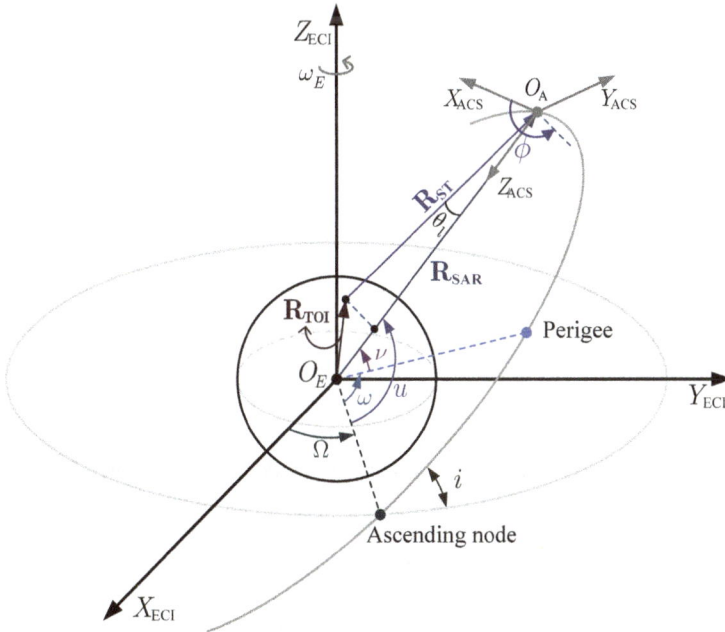

FIGURE 5.1 The observation geometry of the MBSAR in the ACS and ECI coordinate systems.

where:

   R_{SAR} is the distance from Earth's center to MBSAR;
   θ_l is the look angle of the MBSAR;
   ϕ is the antenna azimuth angle, $\phi = 90°$ representing the SAR is broadside looking;
   i is the orbit inclination;
   u is the AOL, the sum of the true anomaly and AOP; and
   ξ_{sld} equals to ± 1, $\xi_{sld} = 1$ denotes the SAR is left-looking, while $\xi_{sld} = -1$ denotes the SAR looks
       from the right-hand side.

The Doppler frequency component induced by SAR's inertial motion is susceptible to the orbital elements. For an elliptical orbit that operates in a Keplerian orbit (non-perturbed orbit), the Doppler component is of the form of $f_{d(SAR)}^{kep}$, and its representation is given by:

$$f_{d(SAR)}^{kep} = 2\lambda^{-1}V_{SAR}\left(\xi_{sld}\cos\gamma\sin\theta_l\cos\phi - \sin\gamma\cos\theta_l\right) \tag{5.6}$$

with:

$$V_{SAR} = \left\|\mathbf{V_{SAR}}\right\|_2 \tag{5.7}$$

$$\gamma = \tan^{-1}\left(\frac{e\sin\nu}{e\cos\nu + 1}\right) \tag{5.8}$$

Then, Eq. (5.6) can be expressed equivalently as:

$$f_{d(SAR)}^{kep} = 2\lambda^{-1}V_{kep}\left[\xi_{sld}\left(e\cos\nu + 1\right)\sin\theta_l\cos\phi - e\sin\nu\cos\theta_l\right] \tag{5.9}$$

with:

$$V_{kep} = \sqrt{\frac{\mu}{a_{kep}\left(1 - e^2\right)}} \tag{5.10}$$

where:

   e is the eccentricity;
   ν is the true anomaly;
   μ is the Earth's gravitational coefficient; and
   a_{kep} is the semi-major axis of the elliptical Keplerian orbit.

The MBSAR's orbit is inherently perturbed, and orbital elements (the eccentricity, orbital inclination, semi-major axis, etc.) are temporally varying, as evidenced in Chapter 2. Thus, it is dubious to calculate the SAR motion-induced Doppler frequency by using Eq. (5.6) or Eq. (5.9). Once the Moon's position and velocity vectors obtained by the numerical ephemeris (like DE440), one can determine the Doppler frequency induced by SAR's inertial motion as:

$$f_{d(SAR)}^{per} = 2\lambda^{-1}\cdot\left(V_{SX,ACS}\cdot\xi_{sld}\sin\theta_l\cos\phi + V_{SZ,ACS}\cos\theta_l\right) \tag{5.11}$$

with:

$$
\begin{cases}
V_{SX,ACS} = -V_{SX} \cdot (\sin u \cos \Omega + \cos i \cos u \sin \Omega) \\
\qquad\quad - V_{SY} \cdot (\sin u \sin \Omega - \cos i \cos u \cos \Omega) + V_{SZ} \cdot \sin i \cos u \\
V_{SY,ACS} = 0 \\
V_{SZ,ACS} = -V_{SX} \cdot (\cos u \cos \Omega - \cos i \sin u \sin \Omega) \\
\qquad\quad - V_{SY} \cdot (\cos u \sin \Omega + \cos i \sin u \cos \Omega) - V_{SZ} \cdot \sin i \sin u
\end{cases}
\tag{5.12}
$$

where Ω is the RAAN in the ECI coordinate system, $\mathbf{V}_{SAR(ACS)} = [V_{SX,ACS}, V_{SY,ACS}, V_{SY,ACS}]^T$ and $\mathbf{V}_{SAR} = [V_{SX}, V_{SY}, V_{SY}]^T$ are the SAR velocity vectors in the ACS and ECI coordinate systems, respectively. Note that the y-axis of the ACS is along the angular momentum direction of the SAR motion, theoretically, the y-component of SAR velocity equals zero in the ACS.

Finally, we arrive at the Doppler frequency of the MBSAR with two components:

$$
\begin{aligned}
f_d = 2\lambda^{-1}\Big[& \left(\xi_{sld} V_{SX,ACS} \sin \theta_l \cos \phi + V_{SZ,ACS} \cos \theta_l \right) \\
& + \xi_{sld} \omega_E R_{SAR} \sin \theta_l \left(\sin i \cos u \sin \phi - \cos \phi \cos i \right) \Big]
\end{aligned}
\tag{5.13}
$$

Now, the DFMR expression is given by [11]:

$$
f_{dr} = 2\lambda^{-1} \cdot \left(\mathfrak{R}_1 + \mathfrak{R}_2 + \mathfrak{R}_3 + \mathfrak{R}_4 \right)
\tag{5.14}
$$

with:

$$
\mathfrak{R}_1 = R_{ST}^{-1} \cdot \left(\mathbf{V}_{SAR}\mathbf{V}_{SAR} + \mathbf{A}_{SAR}\mathbf{R}_{ST} \right)
\tag{5.15}
$$

$$
\mathfrak{R}_2 = -2 R_{ST}^{-1} \cdot \mathbf{V}_{SAR}\mathbf{V}_{TOI}
\tag{5.16}
$$

$$
\mathfrak{R}_3 = R_{ST}^{-1} \cdot \left(\mathbf{V}_{TOI}\mathbf{V}_{TOI} - \mathbf{A}_{TOI}\mathbf{R}_{ST} \right)
\tag{5.17}
$$

$$
\mathfrak{R}_4 = -R_{ST}^{-3} \cdot \left[\left(\mathbf{V}_{SAR} - \mathbf{V}_{TOI} \right) \mathbf{R}_{ST} \right]^2
\tag{5.18}
$$

where $\mathbf{A}_{SAR}$ is the SAR's acceleration vector, $\mathbf{V}_{TOI}$ and $\mathbf{A}_{TOI}$ are velocity and acceleration vectors given rise by Earth's rotation, respectively. The term in Eq. (5.18) is correlated to the Doppler frequency, by:

$$
\mathfrak{R}_4 = -0.25\lambda^{-2} R_{ST}^{-1} f_d^2
\tag{5.19}
$$

By taking account of contributions from the Earth–Moon relative motion dynamics, Eq. (5.18) can be further expressed as:

$$
\mathfrak{R}_4 = \mathfrak{R}_{4S} + \mathfrak{R}_{4E} + \mathfrak{R}_{4SE}
\tag{5.20}
$$

with $\mathfrak{R}_{4S}$ denoting the contribution of the SAR motion, by:

$$
\mathfrak{R}_{4S} = -R_{ST}^{-3} \cdot \left(\mathbf{V}_{SAR}\mathbf{R}_{ST} \right)^2
\tag{5.21}
$$

and $\mathfrak{R}_{4E}$ representing the contribution of Earth's rotation, as:

$$\mathfrak{R}_{4E} = -R_{ST}^{-3} \cdot \left(\mathbf{V_{TOI}} \cdot \mathbf{R_{ST}} \right)^2 \tag{5.22}$$

and $\mathfrak{R}_{4SE}$ being coupling affected by the SAR motion and Earth's rotation:

$$\mathfrak{R}_{4SE} = 2R_{ST}^{-3} \cdot \left(\mathbf{V_{SAR}} \mathbf{R_{ST}} \right) \left(\mathbf{V_{TOI}} \mathbf{R_{ST}} \right) \tag{5.23}$$

It follows that the SAR's DFMR can be rearranged, consisting of three components:

$$
\begin{aligned}
f_{dr} &= f_{dr(SAR)} + f_{dr(SE)} + f_{dr(ER)} \\
&= \underbrace{2\lambda^{-1} \cdot (\mathfrak{R}_1 + \mathfrak{R}_{4S})}_{\text{SAR's motion}} + \underbrace{2\lambda^{-1} \cdot (\mathfrak{R}_2 + \mathfrak{R}_{4SE})}_{\text{Coupling term}} + \underbrace{2\lambda^{-1} \cdot (\mathfrak{R}_3 + \mathfrak{R}_{4E})}_{\text{Earth's rotation}}
\end{aligned}
\tag{5.24}
$$

We find that the DFMR is sensitive to orbital elements as well. For a Keplerian orbit that is elliptical, the DFMR is given by:

$$
\begin{cases}
f_{dr}^{kep} = f_{dr(SAR)}^{kep} + f_{dr(SE)}^{kep} + f_{dr(ER)}^{kep} \\
f_{dr(SAR)}^{kep} = 2\lambda^{-1} \cdot (\mathfrak{R}_1^{kep} + \mathfrak{R}_{4S}^{kep}) \\
f_{dr(SE)}^{kep} = 2\lambda^{-1} \cdot (\mathfrak{R}_2^{kep} + \mathfrak{R}_{4SE}^{kep}) \\
f_{dr(ER)}^{kep} = 2\lambda^{-1} \cdot (\mathfrak{R}_3^{kep} + \mathfrak{R}_{4E}^{kep})
\end{cases}
\tag{5.25}
$$

where the two components due to the SAR motion are:

$$\mathfrak{R}_1^{kep} = -\mu R_{SAR}^{-2} \cos\theta_l + R_{ST}^{-1} V_{kep}^2 \cdot \left(1 + e^2 + 2e\cos v \right) \tag{5.26}$$

$$
\begin{aligned}
\mathfrak{R}_{4S}^{kep} = -R_{ST}^{-1} V_{kep}^2 \cdot \Big[& (1 + e\cos v)^2 \cos^2\theta_l \\
& + 2e(e\cos v + 1) \cdot \xi_{sld} \sin v \sin\theta_l \cos\theta_l \cos\phi + e^2 \sin^2 v \sin^2\theta_l \cos^2\phi \Big]
\end{aligned}
\tag{5.27}
$$

The two components due to the coupling of the SAR motion and Earth's rotation, take the following forms:

$$
\begin{aligned}
\mathfrak{R}_2^{kep} = -2R_{ST}^{-1}\omega_E V_{kep} \Big\{ & \xi_{sld} \cdot e R_{ST} \sin\theta_l \sin v (\sin\phi \sin i \cos u - \cos\phi \cos i) \\
& + (e\cos v + 1) \cdot \big[(R_{SAR} - R_{ST}\cos\theta_l)\cos i - \xi_{sld} \cdot R_{ST} \sin\theta_l \sin\phi \sin i \sin u \big] \Big\}
\end{aligned}
\tag{5.28}
$$

$$
\begin{aligned}
\mathfrak{R}_{4SE}^{kep} = &-2\omega_E R_{SAR} R_{ST}^{-1} V_{kep} \sin\theta_l \\
& \cdot (\sin\phi \sin i \cos u - \cos\phi \cos i) \big[(e\cos v + 1)\sin\theta_l \cos\phi + \xi_{sld} \cdot e \sin v \cos\theta_l \big]
\end{aligned}
\tag{5.29}
$$

Each part from the contribution of Earth's rotation is given:

$$
\begin{aligned}
\mathfrak{R}_3^{kep} = R_{ST}^{-1}\omega_E^2 \Big\{ & R_{SAR}(R_{ST}\cos\theta_l - R_{SAR})\sin^2 i \sin^2 u - R_{SAR}R_{ST}\cos\theta_l \\
& + R_{SAR}^2 - \xi_{sld} \cdot R_{SAR}R_{ST}\sin\theta_l \sin i \sin u (\cos\phi \sin i \cos u + \sin\phi \cos i) \Big\}
\end{aligned}
\tag{5.30}
$$

$$\mathfrak{R}_{4E}^{kep} = \omega_E^2 R_{SAR}^2 R_{ST}^{-1} \sin^2\theta_l \left(2\sin\phi\cos\phi\sin i\cos i\cos u - \cos^2\phi\cos^2 i - \sin^2\phi\sin^2 i\cos^2 u \right) \tag{5.31}$$

Considering the perturbed orbit, the DFMR should be modified to:

$$\begin{cases} f_{dr}^{per} = f_{dr(SAR)}^{per} + f_{dr(SE)}^{per} + f_{dr(ER)}^{per} \\ f_{dr(SAR)}^{per} = 2\lambda^{-1} \cdot (\mathfrak{R}_{1}^{per} + \mathfrak{R}_{4S}^{per}) \\ f_{dr(SE)}^{per} = 2\lambda^{-1} \cdot (\mathfrak{R}_{2}^{per} + \mathfrak{R}_{4SE}^{per}) \\ f_{dr(ER)}^{per} = 2\lambda^{-1} \cdot (\mathfrak{R}_{3}^{per} + \mathfrak{R}_{4E}^{per}) \end{cases} \tag{5.32}$$

where:

$$\mathfrak{R}_{1}^{per} = R_{ST}^{-1} \cdot V_{SAR}^{2} - \xi_{sld} A_{SX,ACS} \sin\theta_l \cos\phi + \xi_{sld} A_{SY,ACS} \sin\theta_l \sin\phi - A_{SZ,ACS} \cos\theta_l \tag{5.33}$$

$$\mathfrak{R}_{4S}^{per} = -R_{ST}^{-1}[V_{SX,ACS}^{2} \sin^{2}\theta_l \cos^{2}\phi + V_{SZ,ACS}^{2} \cos^{2}\theta_l + 2\xi_{sld} V_{SX,ACS} V_{SZ,ACS} \sin\theta_l \cos\theta_l \cos\phi] \tag{5.34}$$

$$\mathfrak{R}_{2}^{per} = -2R_{ST}^{-1}\omega_E \{ -\xi_{sld} V_{SZ,ACS} R_{ST} \sin\theta_l (\cos\phi\cos i - \sin\phi\sin i\cos u) \\ + V_{SX,ACS} \cdot [(R_{SAR} - R_{ST}\cos\theta_l)\cos i - \xi_{sld} R_{ST}\sin\theta_l \sin\phi\sin i\sin u] \} \tag{5.35}$$

$$\mathfrak{R}_{4SE}^{per} = -2\omega_E R_{SAR} R_{ST}^{-1} \cdot \sin\theta_l \\ \cdot (\sin\phi\sin i\cos u - \cos\phi\cos i)(V_{SX,ACS}\sin\theta_l\cos\phi + \xi_{sld}V_{SZ,ACS}\cos\theta_l) \tag{5.36}$$

Note that the terms $\mathfrak{R}_{3}^{per}$ and R_{4E}^{per} depend on Earth's rotation only, thus their forms remain the same with and without lunar orbital perturbations, namely:

$$\mathfrak{R}_{3}^{per} = \mathfrak{R}_{3}^{kep}; \quad \mathfrak{R}_{4E}^{per} = \mathfrak{R}_{4E}^{kep} \tag{5.37}$$

In Eq. (5.33), $\mathbf{A}_{SAR(ACS)} = [A_{SX, ACS}, A_{SY, ACS}, A_{SY, ACS}]^T$ is the SAR acceleration vector in the ACS, the transformation of this vector from the ECI to ACS coordinate systems is given by:

$$\begin{cases} A_{SX,ACS} = -A_{SX} \cdot (\sin u\cos\Omega + \cos i\cos u\sin\Omega) \\ \qquad\qquad - A_{SY} \cdot (\sin u\sin\Omega - \cos i\cos u\cos\Omega) + A_{SZ} \cdot \sin i\cos u \\ A_{SY,ACS} = -A_{SX} \cdot \sin i\sin\Omega + A_{SY} \cdot \sin i\cos\Omega - A_{SZ} \cdot \cos i \\ A_{SZ,ACS} = -A_{SX} \cdot (\cos u\cos\Omega - \cos i\sin u\sin\Omega) \\ \qquad\qquad - A_{SY} \cdot (\cos u\sin\Omega + \cos i\sin u\cos\Omega) - A_{SZ} \cdot \sin i\sin u \end{cases} \tag{5.38}$$

where $\mathbf{A}_{SAR(ECI)} = [A_{SX}, A_{SY}, A_{SY}]^T$ is the SAR acceleration vector in the ECI coordinate system. In numerical lunar ephemerides (e.g., DE 440), the position and velocity are provided, whereas the acceleration vector is not available [12–14]. Thus, the acceleration vector is estimated numerically by the velocity vector in the MBSAR. The detailed derivations of the Doppler frequency and DFMR are given in the Appendix.

5.2.2 Validation of the Framework

The MBSAR's orbit is subject to perturbation effects, differentiating it from Keplerian orbit. We compare the Keplerian and perturbed orbits, using the MBSAR's acceleration as an example. Following Newton's law of gravity in Keplerian orbits, we have [15]:

$$\mathbf{A}_{SAR}^{kep} = -\mu R_{SAR}^{-3} \mathbf{R}_{SAR} \tag{5.39}$$

To clarify the difference between Keplerian and perturbed orbits, we employ the following two error terms:

$$\Delta \mathbf{A}_{SAR} = \mathbf{A}_{SAR}^{per} - \mathbf{A}_{SAR}^{kep} \tag{5.40}$$

$$\delta A_{SAR} = \frac{\left\| \mathbf{A}_{SAR}^{per} \right\|_2 - \left\| \mathbf{A}_{SAR}^{kep} \right\|_2}{\left\| \mathbf{A}_{SAR}^{per} \right\|_2} \times 100\% \tag{5.41}$$

We plot these errors as a function of AOL in the MBSAR within three consecutive cycles $T_{lc1} \sim T_{lc3}$ (their detailed epochs in TDB are given in Chapter 2), as shown in Figure 5.2, where we find that the orbital perturbations undoubtedly impact the acceleration of the MBSAR. Specifically, the acceleration errors in the three components present distinct variation patterns. Therefore, when analyzing system parameters associated with orbital elements, such as Doppler frequency and DFMR, for MBSAR, it is inadequate to assume a Keplerian orbit or an osculating orbit (a Keplerian orbit that has the identical orbital elements as the actual one at a given instant).

We calculate Doppler frequency and DFMR with a carrier frequency of 1.2 GHz, and compare them with the theoretical ones in Eqs. (5.13) and (5.32). For illustration, we use the DE 440 with the

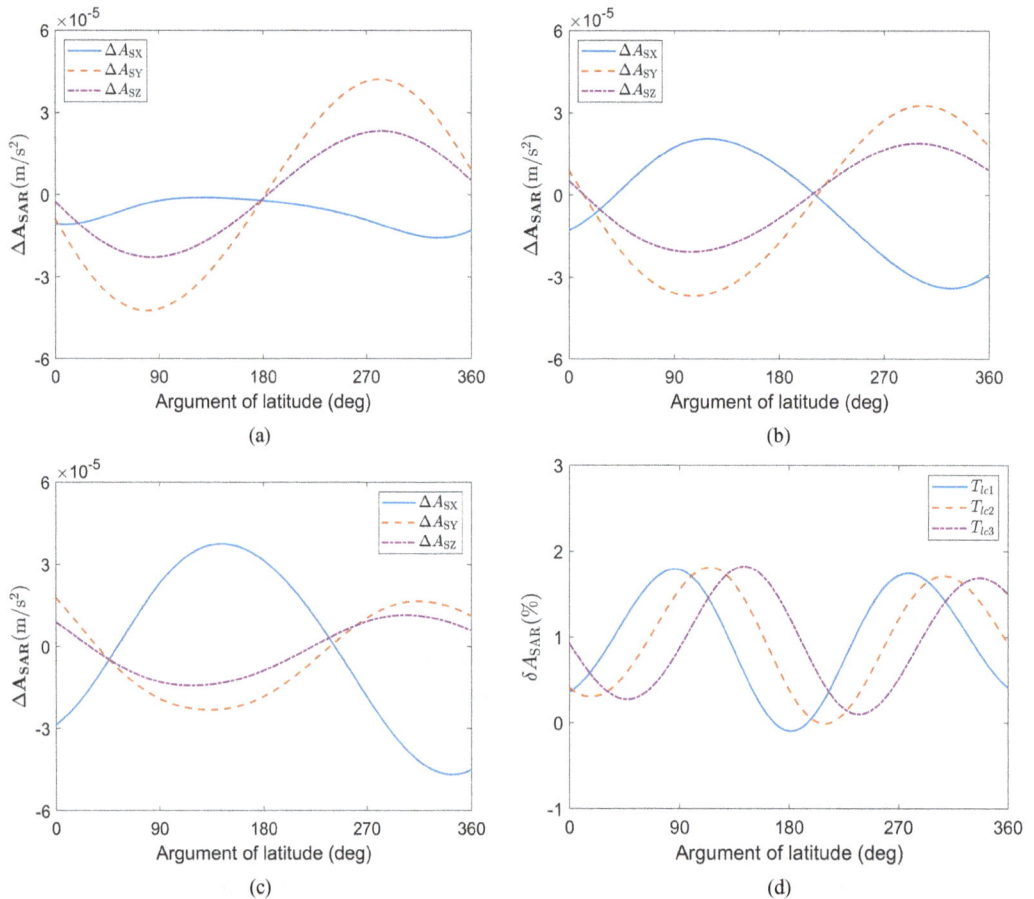

FIGURE 5.2 Within three consecutive cycles, the absolute error in the acceleration vector in cycles (a) T_{lc1}, (b) T_{lc2}, (c) T_{lc3}; (d) the relative error in the SAR acceleration versus AOL.

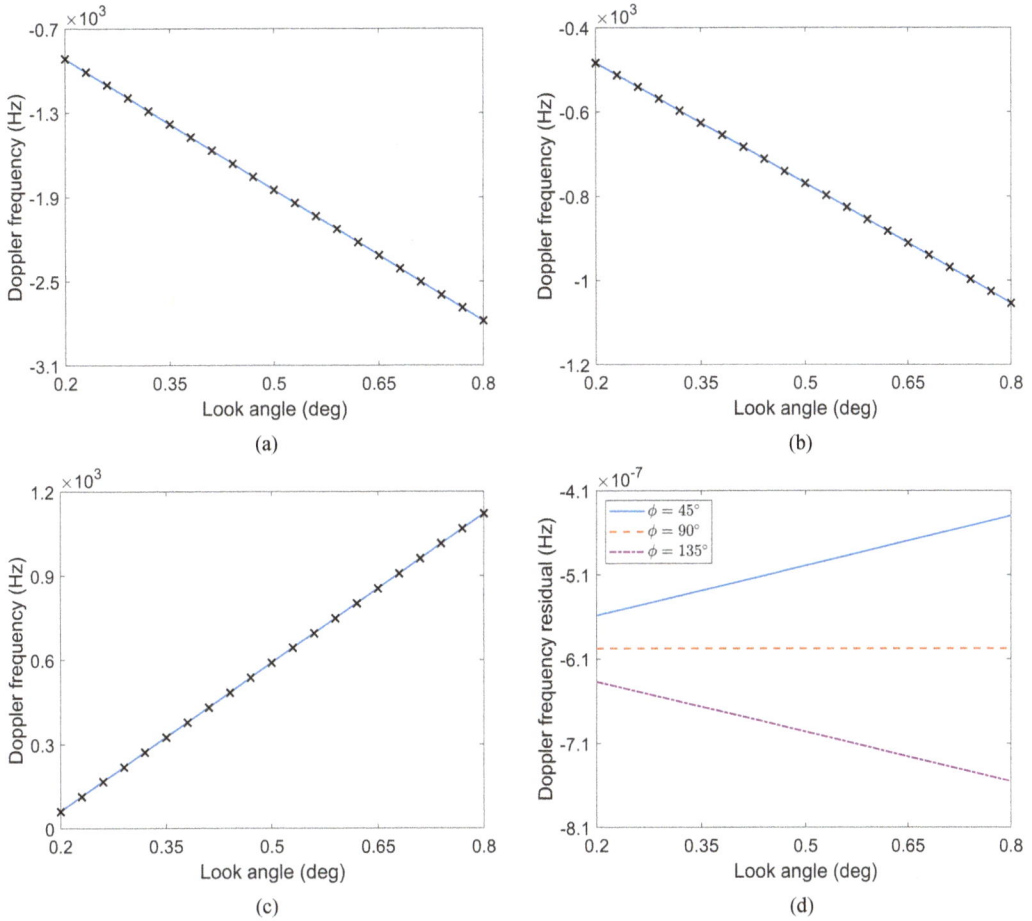

FIGURE 5.3 The Doppler frequency versus look angle under the antenna azimuth angle of (a) 45°, (b) 90°, (c) 135°, where the solid line represents the analytical Doppler frequency and symbol "x" stands for the numerical one. The differences between the analytical and numerical Doppler frequencies are shown in (d).

zero-azimuth time being 00:00:00 on Mar. 20, 2024. We show the numerical and analytical Doppler frequencies, and their differences as a function of the look angle in Figure 5.3 and those as a function of antenna azimuth angle in Figure 5.4.

Upon inspecting the results in Figures 5.3 and 5.4, the analytical Doppler frequency shows good agreement with the numerical values. Quantitative evaluations suggest that the differences between the analytical Doppler frequencies and numerical values, depending on the look angle and antenna azimuth angle, are in the order of 10^{-7} Hz. Therefore, Eq. (5.13) can accurately model the Doppler frequency in the MBSAR.

Next, using the same set of parameters, we compare the numerical DFMR values with analytical ones. Figures 5.5 and 5.6 show the numerical and analytical DFMRs and their differences versus the look angle, and those versus the antenna azimuth angle, respectively.

Figures 5.5 and 5.6 show that the differences between the numerical and analytical DFMRs are consistently below 10^{-10} Hz/s, despite changes in the beam pointing direction. Hence, the analytical DFMR values are consistent with numerical ones in a perturbed orbit. We can explore the Doppler properties of MBSAR using Eqs. (5.13) and (5.32) in the following section.

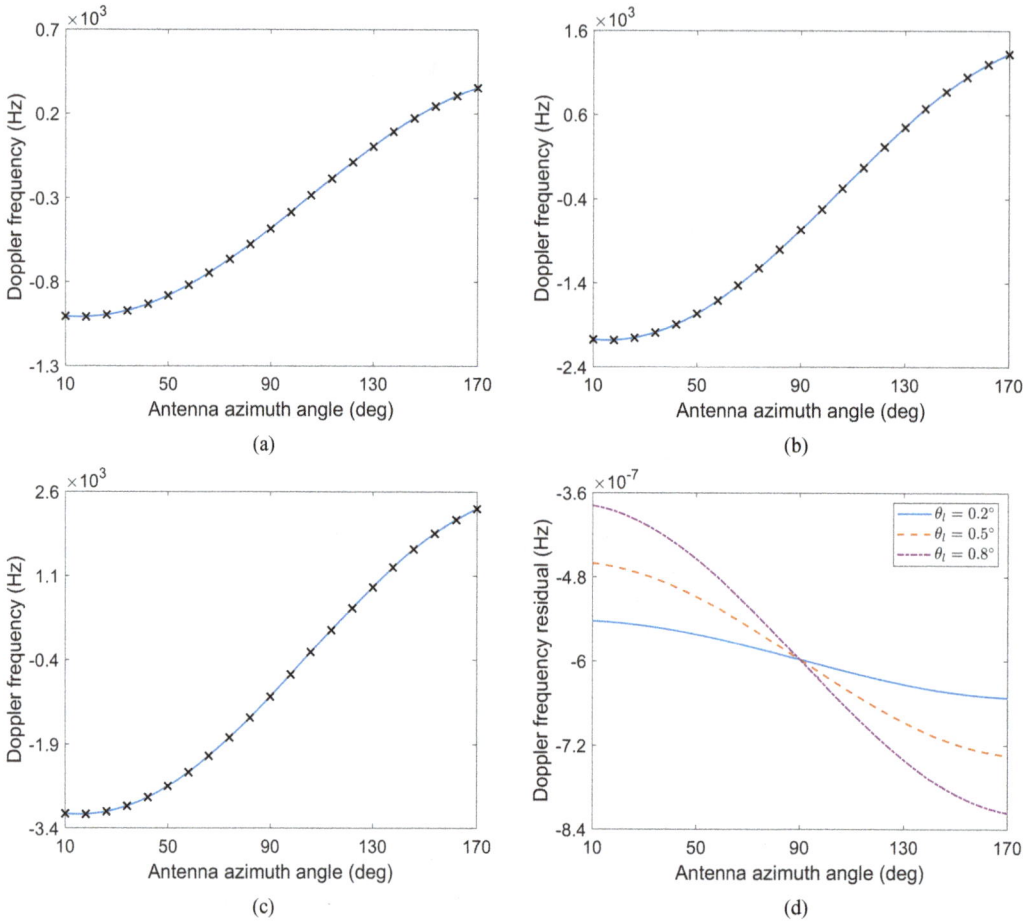

FIGURE 5.4 The Doppler frequency versus antenna azimuth angle under the look angle of (a) 0.2°, (b) 0.5°, (c) 0.8°, where the solid line represents the analytical Doppler frequency, and the symbol "x" stands for the numerical one. The differences between the analytical and numerical Doppler frequencies are shown in (d).

5.2.3 PROPERTIES OF DOPPLER FREQUENCY

In this section, we illustrate the properties of the Doppler frequency and its components under different beam pointing directions and across various cycles in the MBSAR. Besides, we compare the Doppler frequency properties of the MBSAR to that of the spaceborne SAR, showing their differences in detail. For ease of comparison, both the MBSAR and spaceborne SAR are assumed left-looking, and the configurations of the spaceborne SAR listed in Table 5.1 are employed.

Figure 5.7 displays isocontours of the Doppler frequency versus the antenna azimuth and look angles in the MBSAR and spaceborne SAR. The isocontours for each Doppler frequency component are also provided for presenting the contributions from SAR's inertial motion and Earth's rotation.

Figure 5.7 shows that in the MBSAR, the magnitude of Doppler frequency is far smaller than that in the spaceborne SAR. Besides, the Doppler frequency induced by the SAR's inertial motion is more pronounced in the spaceborne SAR, suggesting that the SAR's inertial motion plays a dominant role in the determining Doppler shift in this scenario. By contrast, the Doppler frequency in the MBSAR is more susceptible to Earth's rotation when there are changes in the antenna beam pointing direction. Meanwhile, the MBSAR's inertial motion can affect the Doppler frequency to a certain extent, indicating that Earth's rotation dominates the Doppler effect, whereas the SAR motion induced Doppler shift in MBSAR is not significant but non-negligible.

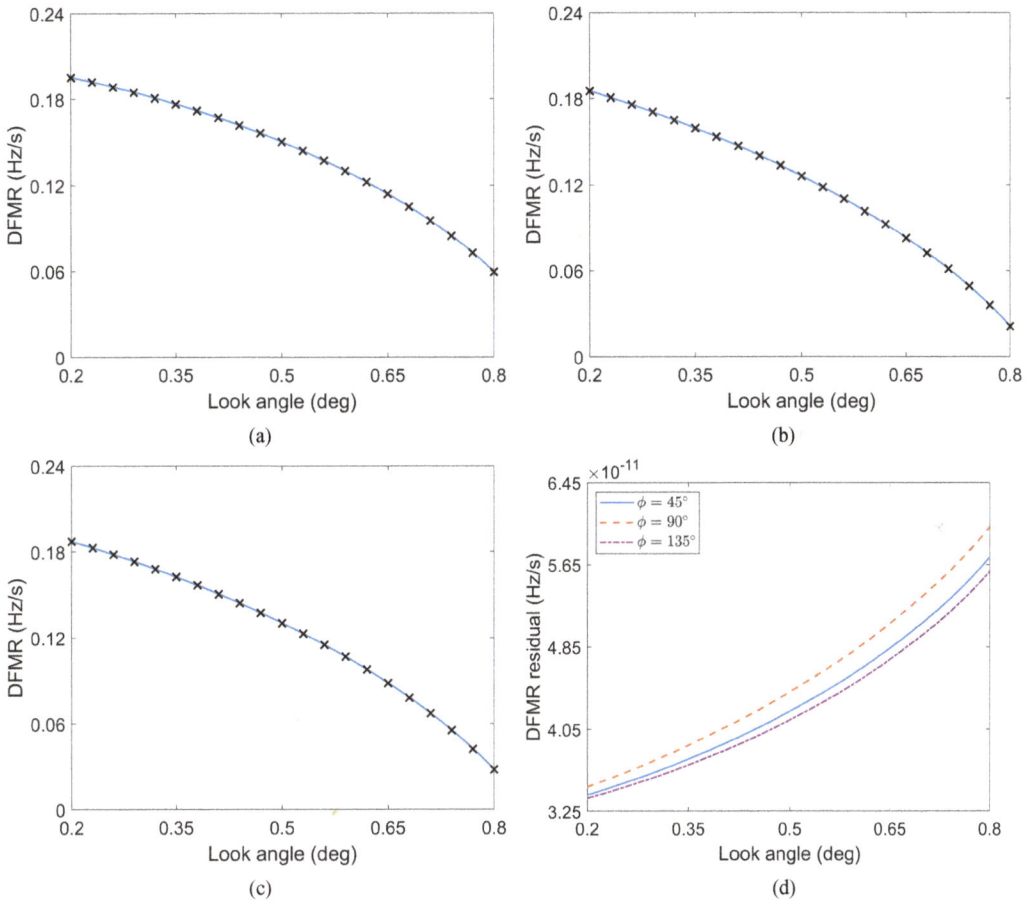

FIGURE 5.5 The DFMR versus the look angle under the antenna azimuth angle of (a) 45°, (b) 90°, (c) 135°, where the solid line analytical DFMR and the symbol "x" stands for the numerical one. The differences between the analytical and numerical Doppler frequencies are shown in (d).

The profiles of each Doppler frequency component of MBSAR and spaceborne SAR in the broadside looking scenario are shown in Figure 5.8 as a function of the look angle.

As shown in Figure 5.8, the Doppler frequency caused by the Earth's rotation increases with the increasing look angle for both the spaceborne SAR and MBSAR. The spaceborne SAR's orbiting-induced Doppler frequency approaches zero Hz when it operates in a circular orbit. However, the Doppler frequency generated by the MBSAR's inertial motion can be up to hundreds of Hz, although it exhibits small variations regarding the look angle. This phenomenon can be attributed to the elliptical orbit of the MBSAR.

To demonstrate the variation of the Doppler frequency for a broadside looking MBSAR as the location and look angle changes, we examine it in the cycle T_{lc1}, and corresponding results versus AOL and look angle are illustrated in Figure 5.9. Additionally, for quantitative and comprehensive analysis, we select three look angles (0.2°, 0.5°, 0.8°), and show their profiles as a function of AOL in Figure 5.10.

Figures 5.9 and 5.10 show that when the MBSAR operates in the broadside looking, the Doppler frequency changes with its look angle and location along the orbit. The Doppler frequency for most locations increases as the look angle increases. Though at certain look angles (e.g., 0.2°, 0.5°, and 0.8°), the Doppler frequency tends to remain constant at specific locations (90° or 270° AOL).

To identify the contributions of Earth's rotation and MBSAR's inertial motion to the variation in Doppler frequency at different locations, we plot the regular pattern of each Doppler component as a function of AOL at various look angles, as shown in Figure 5.11.

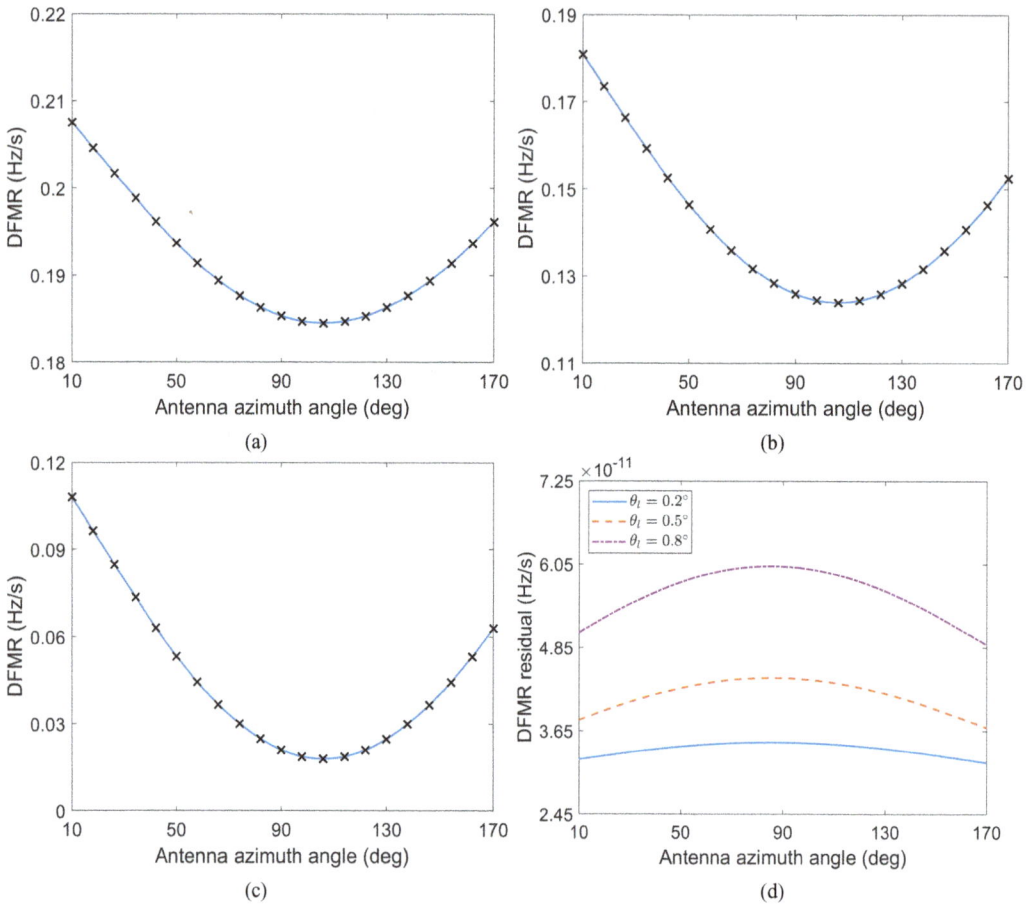

FIGURE 5.6 The DFMR versus the antenna azimuth angle under the look angle of (a) 0.20°, (b) 0.50°, (c) 0.80°, where the solid line analytical DFMR and the symbol "x" stands for the numerical one. The differences between the analytical and numerical Doppler frequencies are shown in (d).

TABLE 5.1

The Sensor Parameters for the Spaceborne SAR

Parameter	Carrier Frequency	Look Angle	Altitude	Inclination	Eccentricity	AOL
Symbol	f_c	θ_l	H_{SAR}	i	e	u
Value	1.27 GHz	20°–60°	694.51 km	97.7°	0	120°

Figure 5.11 shows that the Doppler frequency induced by the MBSAR's inertial motion has a weak dependence on the look angle. In comparison, the Doppler frequency caused by Earth's rotation is more sensitive to the look angle, with the larger look angle accounting for a larger magnitude of this Doppler frequency component. However, when the AOL is close to 90° or 270°, the Earth's rotation-induced Doppler frequency tends to disappear. This explains why the MBSAR's Doppler frequency exhibits the same magnitudes at different look angles at the AOLs of 90° and 270°.

Regarding the variation regularities of each Doppler component versus AOL, the Doppler frequency caused by Earth's rotation decreases with increasing AOL, reaching a valley at an AOL of 180° before positively correlated with the AOL. The SAR motion-induced Doppler frequency decreases with increasing AOL and reaches its local minimum at an AOL of 60° before beginning

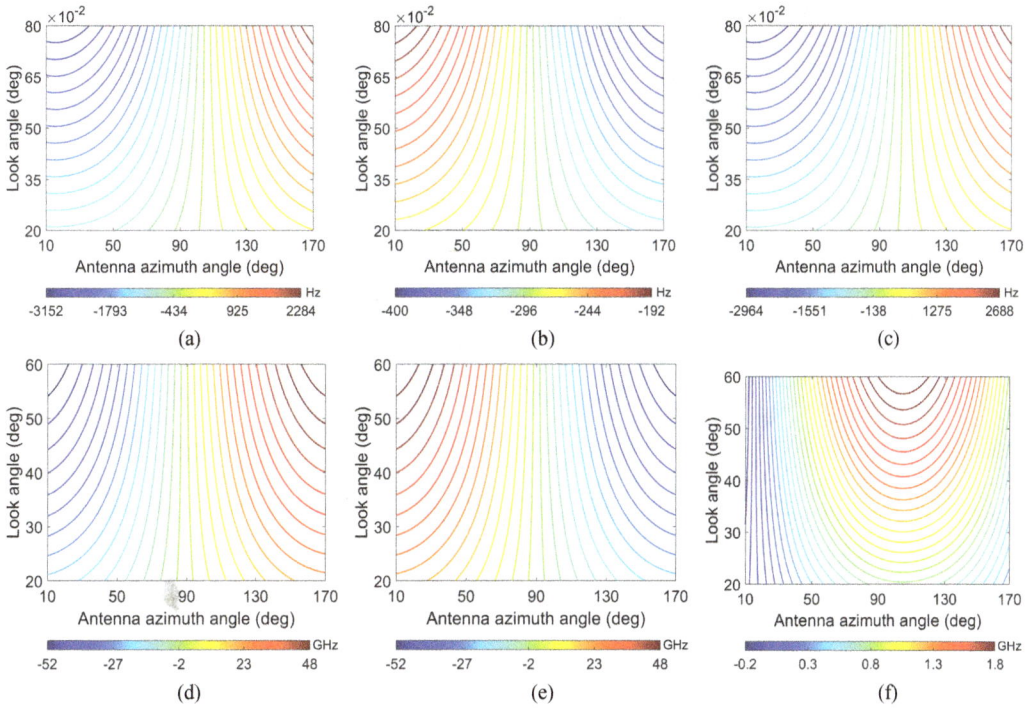

FIGURE 5.7 The isocontours of Doppler frequency and its components versus the antenna azimuth and look angles, where the upper (a, b, c) and lower (d, e, f) rows represent MBSAR and spaceborne SAR scenarios, while the 1st (a, d), 2nd (b, e) and 3rd (c, f) columns stand for f_d, $f_{d(SAR)}$ and $f_{d(ER)}$, respectively.

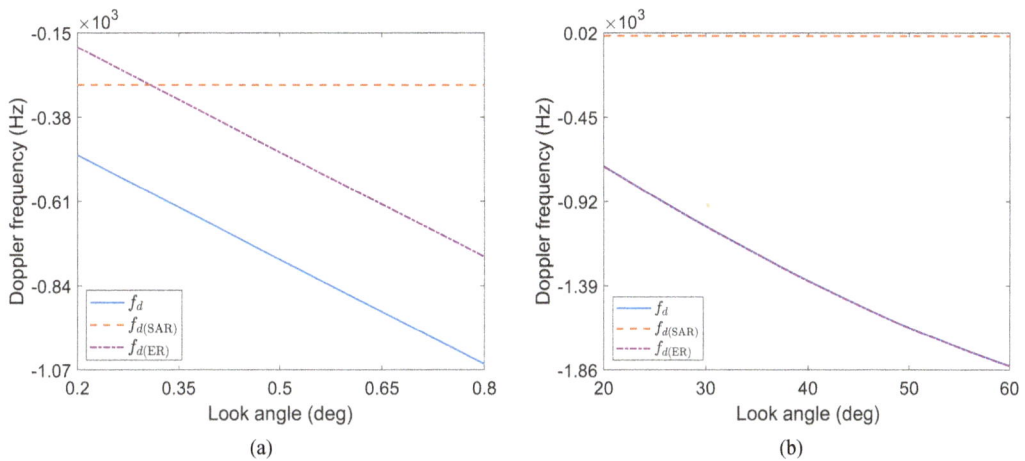

FIGURE 5.8 In the broadside looking scenario, the Doppler frequency and its component versus the look angle in the scenarios of (a) MBSAR, (b) spaceborne SAR.

to ascend and reaching a peak near an AOL of 285°. Afterward, this Doppler frequency component shows a downward trend with respect to AOL again. The Doppler frequency caused by SAR's inertial motion is clearly non-negligible at most positions along the MBSAR's orbit.

In a SAR system, there is a situation where two identical targets are located symmetrically at the right and left sides of the SAR's linear trajectory and induce the same Doppler shift. This phenomenon, known as left-right ambiguity, makes it impossible for the SAR system to distinguish symmetrically-distributed targets. We analyze the influence of side-looking direction on the Doppler

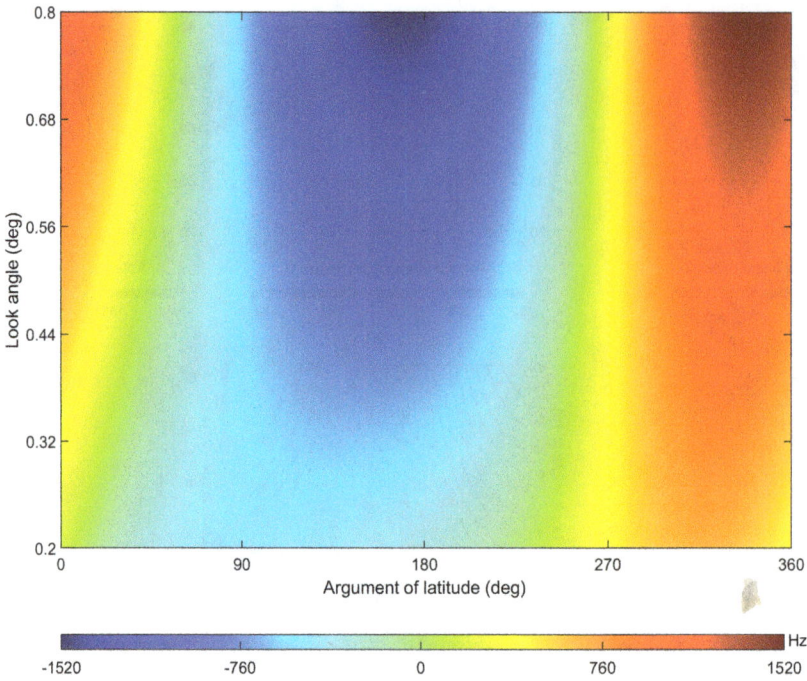

FIGURE 5.9   In one cycle, the Doppler frequency of broadside looking MBSAR versus the look angle and argument of latitude.

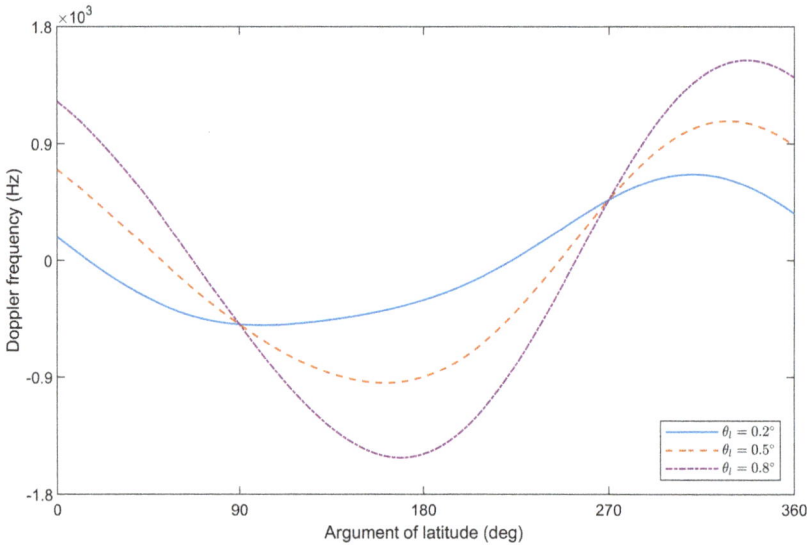

FIGURE 5.10   In one cycle, the Doppler frequency of broadside looking MBSAR with various look angles against the argument of latitude.

frequency in the broadside looking MBSAR, as shown in Figure 5.12. It can be observed that the SAR motion-induced Doppler frequency, which remains approximately constant under different look angles, changes little with the side-looking direction. Interestingly, the Earth's rotation-induced Doppler frequency shows approximately opposite values under different side-looking directions,

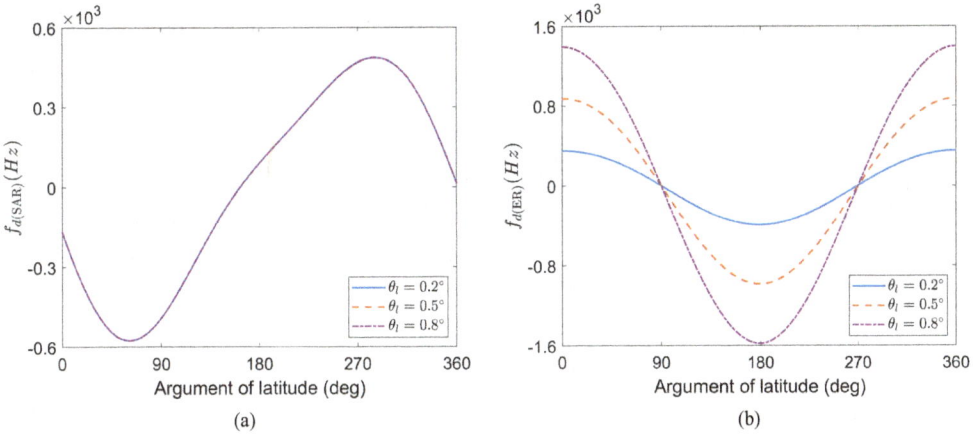

FIGURE 5.11 In the broadside looking MBSAR with various look angles, Doppler frequency component versus the AOL: (a) $f_{d(\text{SAR})}$, (b) $f_{d(\text{ER})}$.

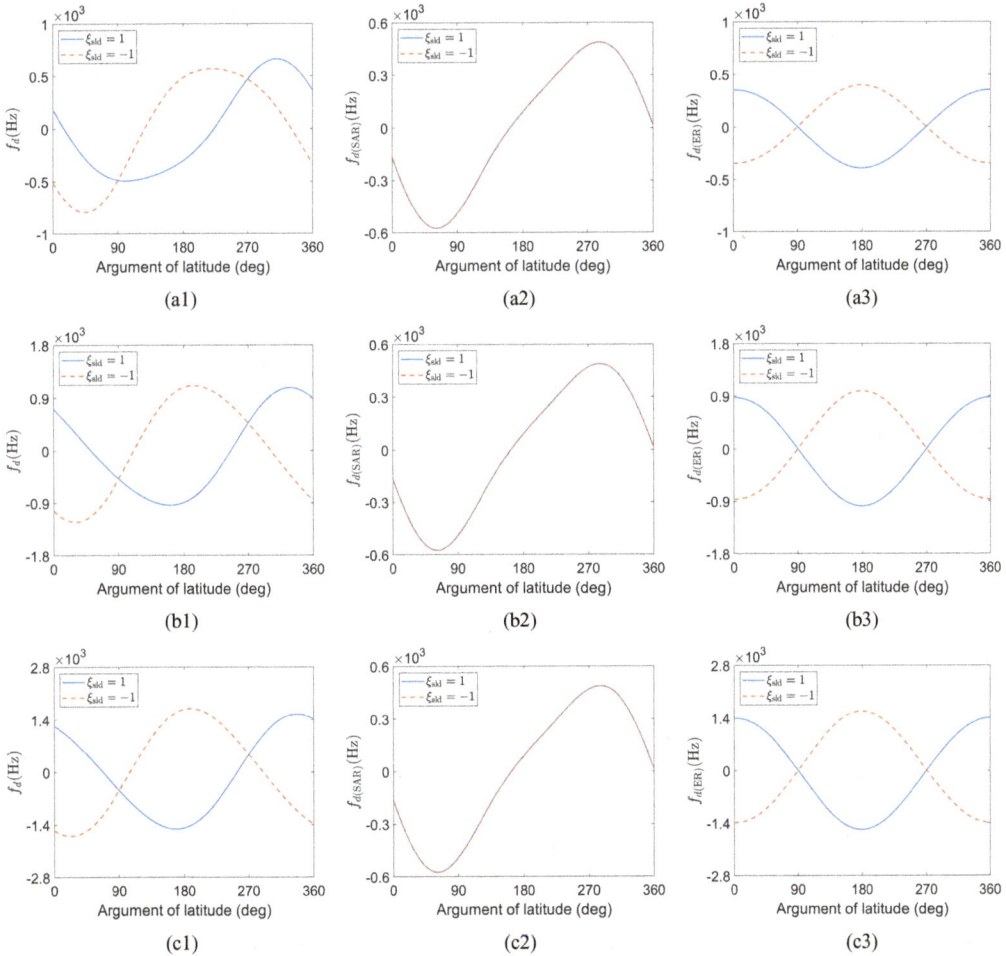

FIGURE 5.12 The Doppler frequency and its components in the left-looking and right-looking MBSARs, where the 1st (a1, b1, c1), 2nd (a2, b2, c2), and 3rd (a3, b3, c3) columns represent the f_d, $f_{d(\text{SAR})}$ and $f_{d(\text{ER})}$, while the 1st (a1, a2, a3), 2nd (b1, b2, b3), and 3rd (c1, c2, c3) rows stand for look angles of 0.2°, 0.5°, and 0.8°, respectively. $\xi_{\text{sld}} = 1$ and $\xi_{\text{sld}} = -1$ denote that the SAR is left-looking and right-looking, respectively.

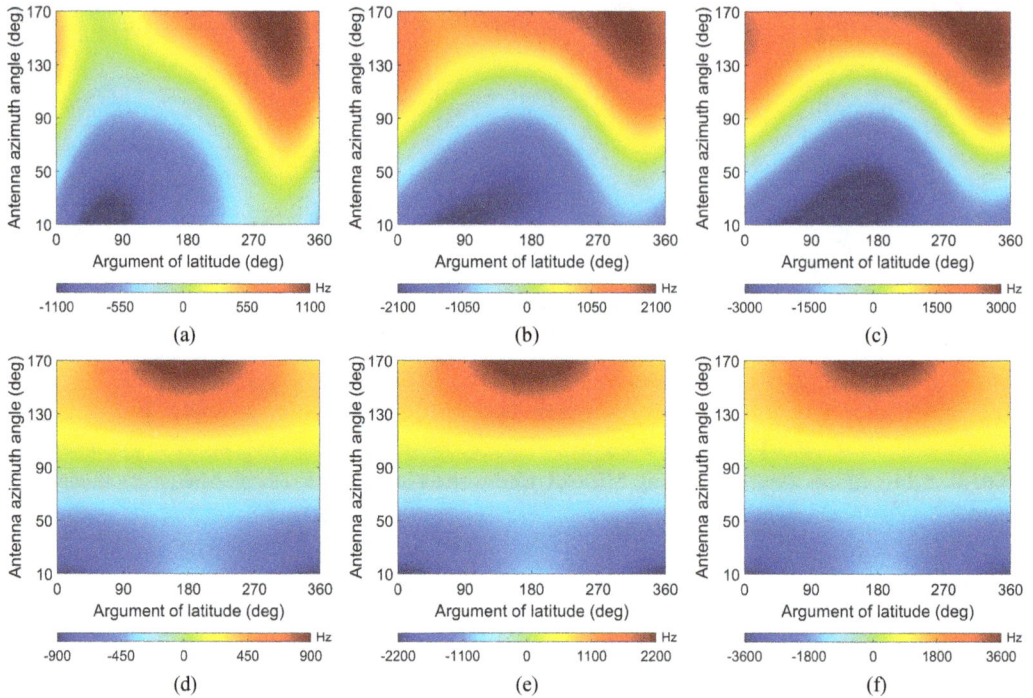

FIGURE 5.13 The Doppler frequencies and their differences (respecting broadside looking ones) versus the antenna azimuth angle and AOL, where the upper (a, b, c) and lower (d, e, f) rows represent the Doppler frequency and Doppler difference, while the 1st (a, d), 2nd (b, e), and 3rd (c, f) columns stand for look angles of 0.20°, 0.50°, and 0.80°, respectively.

with magnitudes increasing with look angle. The entire Doppler frequency results from the coupling effects of Earth's rotation and MBSAR's inertial motion. As a result, the left-looking Doppler frequency differs from the right-looking one at most locations when the MBSAR is broadside looking. Hence, the MBSAR is hardly subject to left–right ambiguity in this event.

Once large-scale spatial coverage is required within a short time span, the MBSAR is more likely to operate in the squint-looking mode [7, 16]. In this scenario, there exists a squint angle that is the angle between the antenna boresight and broadside looking direction. Hence, it is important to analyze the relationship between the Doppler frequency and antenna azimuth angle. To this end, we show the Doppler frequency and their differences with respect to broadside looking ones in Figure 5.13 as functions of the antenna azimuth angle and AOL. For clarity in this illustration, the MBSAR looks from the left-hand side, with three look angles of 0.2°, 0.5°, and 0.8°.

As seen, the Doppler frequency changes differently with the varying antenna azimuth angle at different locations. The variation pattern can be strengthened by a larger look angle, suggesting the look angle could impact the squint effect to a certain extent. As a consequence, the squint effect on the Doppler shift is coupling influenced by the orbital elements and SAR configurations. Notwithstanding, a clear trend can be observed: The larger the squint angle, the greater the change in the Doppler frequency with respect to the look angle.

To explore the rationale behind the reaction of the Doppler frequency to the antenna azimuth angle, it is necessary to examine how changes in the antenna azimuth angle and AOL affect the Doppler frequency components generated by MBSAR's inertial motion and Earth's rotation. Hence, we analyze both Doppler frequency components under different antenna azimuth angles and compared them to those with an antenna azimuth angle of 90°. The results are shown in Figure 5.14 as functions of the antenna azimuth angle and AOL.

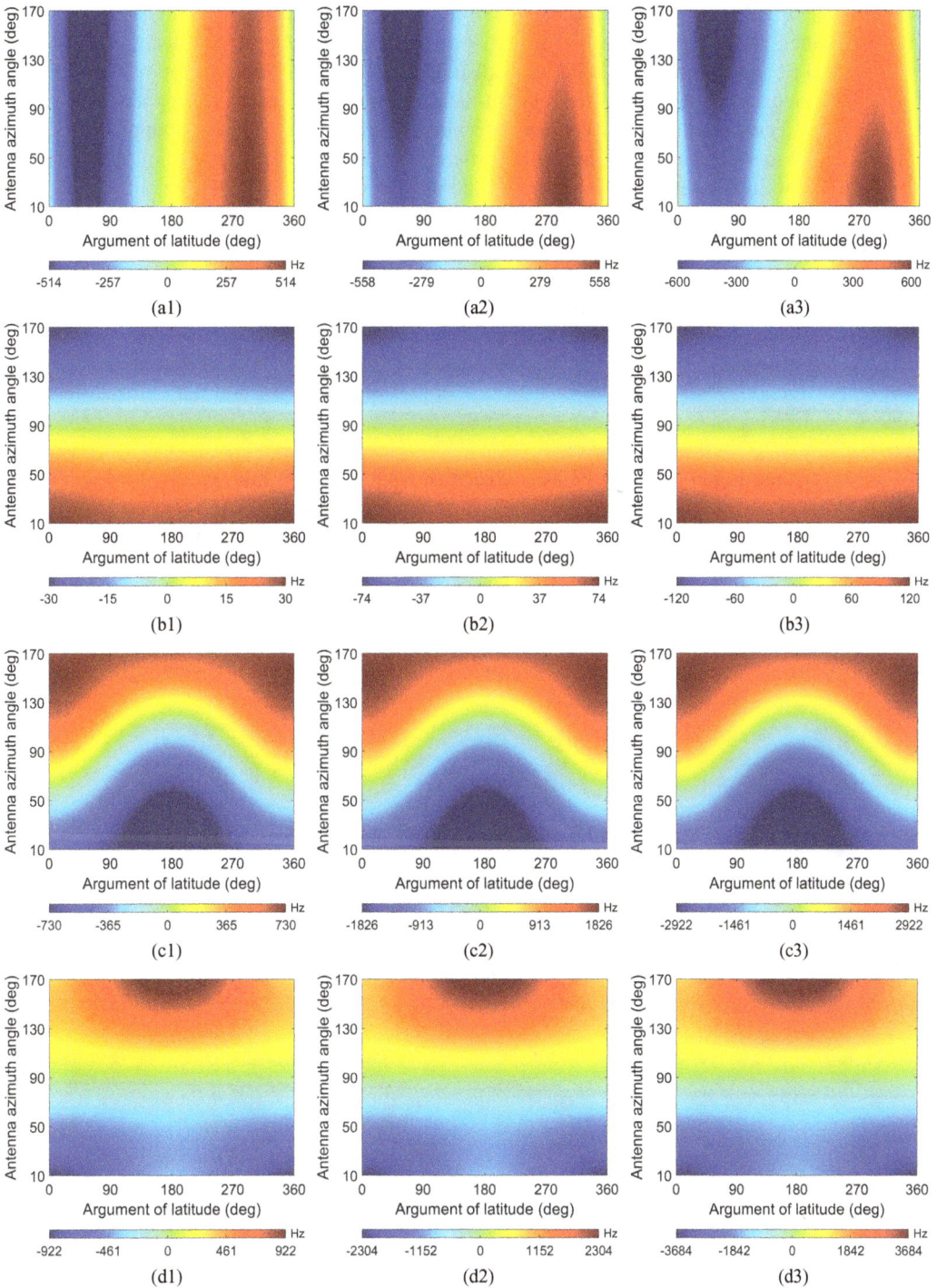

FIGURE 5.14 The Doppler frequency components and their differences (respecting broadside looking ones) versus the antenna azimuth angle and AOL, where the 1st (a1–a3) and 2nd (b1–b3) rows represent the SAR motion-induced Doppler frequency and corresponding Doppler difference; the 3rd (c1–c3) and 4th (d1–d3) rows indicate the Earth's rotation-induced Doppler frequency and corresponding Doppler difference; the 1st (a1–d1), 2nd (a2–d2), and 3rd (a3–d3) columns stand for look angles of 0.20°, 0.50°, and 0.80°, respectively.

Comparing Figure 5.14(a1)–(b3) to Figure 5.14(c1)–(d3) shows that the antenna azimuth angle affects both the Earth's rotation-induced and SAR motion-induced Doppler frequency components. It is a general trend that the large squint and look angles enhance the squint effect on both Doppler frequency components, with the Earth's rotation-induced Doppler frequency component being more significantly affected. The squint effect can influence the SAR motion-induced Doppler frequency component, but to a smaller extent, and with a different variation regularity regarding AOL. Results also suggest that Earth's rotation produces a smaller Doppler frequency with the same squint angle when the antenna azimuth angle is below 90°, regardless of the MBSAR's location. On the other hand, there is an opposite trend in the SAR motion-induced Doppler frequency, still, with a much smaller degree. All the variation patterns regarding the Doppler frequency and its components versus the antenna azimuth and look angles are highly correlated to the position of MBSAR.

To examine the left–right ambiguity issue under squint effects, we have included Figures 5.15–5.17 to display the Doppler frequency and its components versus the AOL and antenna azimuth angles for a right-looking MBSAR at look angles of 0.2°, 0.5°, and 0.8°. The differences between left-looking and right-looking Doppler frequencies, signified as "left–right difference" for short notation, are also depicted in Figures 5.15–5.17.

Upon analyzing Figures 5.15–5.17, it is clear that SAR's inertial motion and Earth's rotation have different effects on the Doppler frequency in squint-looking mode. The SAR motion-induced Doppler frequency creates a symmetrical left–right difference with respect to the broadside looking direction, while Earth's rotation-induced Doppler frequency produces an asymmetrical left–right difference. Furthermore, the magnitude and variability of the Earth's rotation-induced Doppler frequency are greater than those of SAR motion-induced Doppler frequency; an increase in the look angle exacerbates such effects. Consequently, the Earth's rotation predominantly affects the

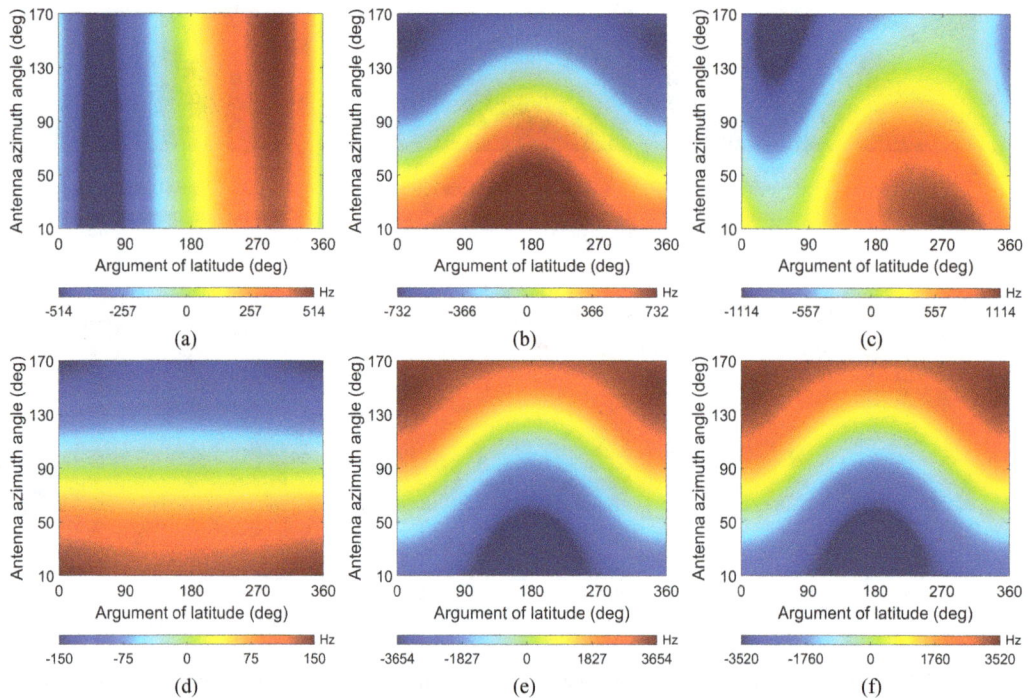

FIGURE 5.15 In the right-looking MBSAR with a look angle of 0.2°, the Doppler frequency (and its components) and left–right Doppler difference versus the AOL and antenna azimuth angle, where the upper (a, b, c) row represent $f_{d(SAR)}, f_{d(ER)}$, and f_d; while the lower (d, e, f) rows stand for the left–right Doppler difference corresponding to $f_{d(SAR)}, f_{d(ER)}$, and f_d.

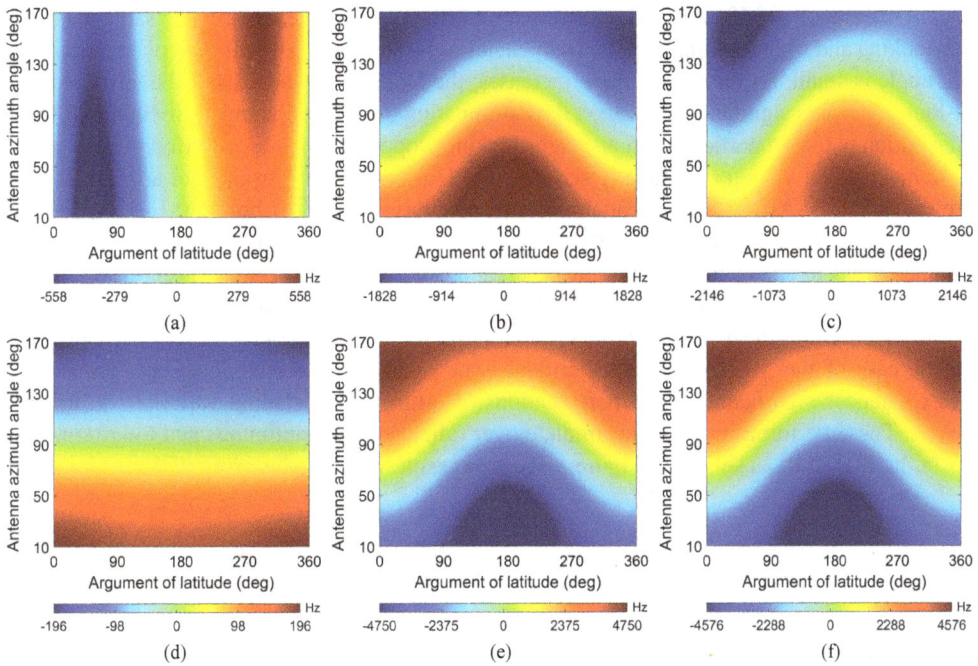

FIGURE 5.16 In the right-looking MBSAR with a look angle of $0.5°$, the Doppler frequency (and its components) and left–right Doppler difference versus the AOL and antenna azimuth angle, where the upper (a, b, c) row represent $f_{d(\text{SAR})}, f_{d(\text{ER})}$, and f_d; while the lower (d, e, f) rows stand for the left–right Doppler difference corresponding to $f_{d(\text{SAR})}, f_{d(\text{ER})}$, and f_d.

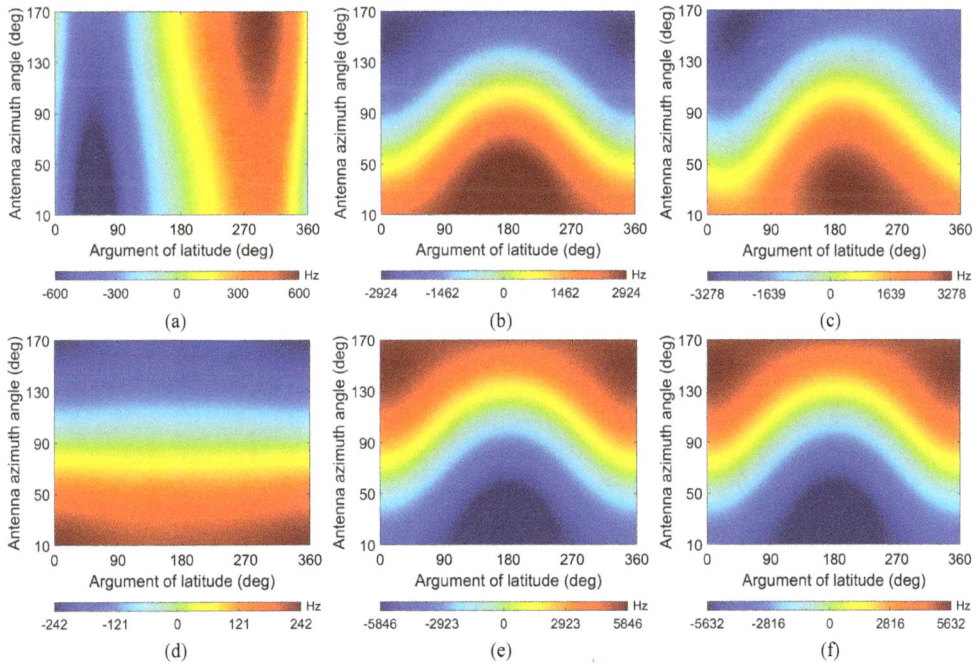

FIGURE 5.17 In the right-looking MBSAR with a look angle of $0.8°$, the Doppler frequency (and its components) and left–right Doppler difference versus the AOL and antenna azimuth angle, where the upper (a, b, c) row represent $f_{d(\text{SAR})}, f_{d(\text{ER})}$, and f_d; while the lower (d, e, f) rows stand for the left–right Doppler difference corresponding to $f_{d(\text{SAR})}, f_{d(\text{ER})}$, and f_d.

left–right difference in MBSAR. Additional quantitative analysis revealed that the same magnitudes of left-looking and right looking Doppler frequencies only occur in a few MBSAR positions when the squint angle is small. As the squint angle increases, the Doppler frequencies in the left-looking and right-looking modes significantly diverge, thus eliminating the left–right ambiguity in the MBSAR.

In addition to the properties of left and right-looking Doppler frequency, the Moon-based platform offers a much wider view for amounting sensors, which distinguishes it from other artificial satellites. Therefore, for MBSAR, it is easily attaining the capability to conduct both left-looking and right-looking observations at the same time. By placing two sets of antennas on the Moon, we can achieve simultaneous left-looking and right-looking observations for monitoring Earth. This not only addresses the left–right ambiguity at certain positions, but also overcomes geometrical observation limitations and power restrictions, resulting in optimal spatial coverage for Earth observation (see Chapter 4).

In the case of the MBSAR, another critical issue related to the Doppler properties is that its orbit experiences significant orbital perturbations, resulting in substantial variations in its orbital elements. These perturbations can accumulate over time, causing continuous and long-term orbital drifts in the MBSAR, as shown in Figure 5.18. As a result, the Doppler properties may be affected

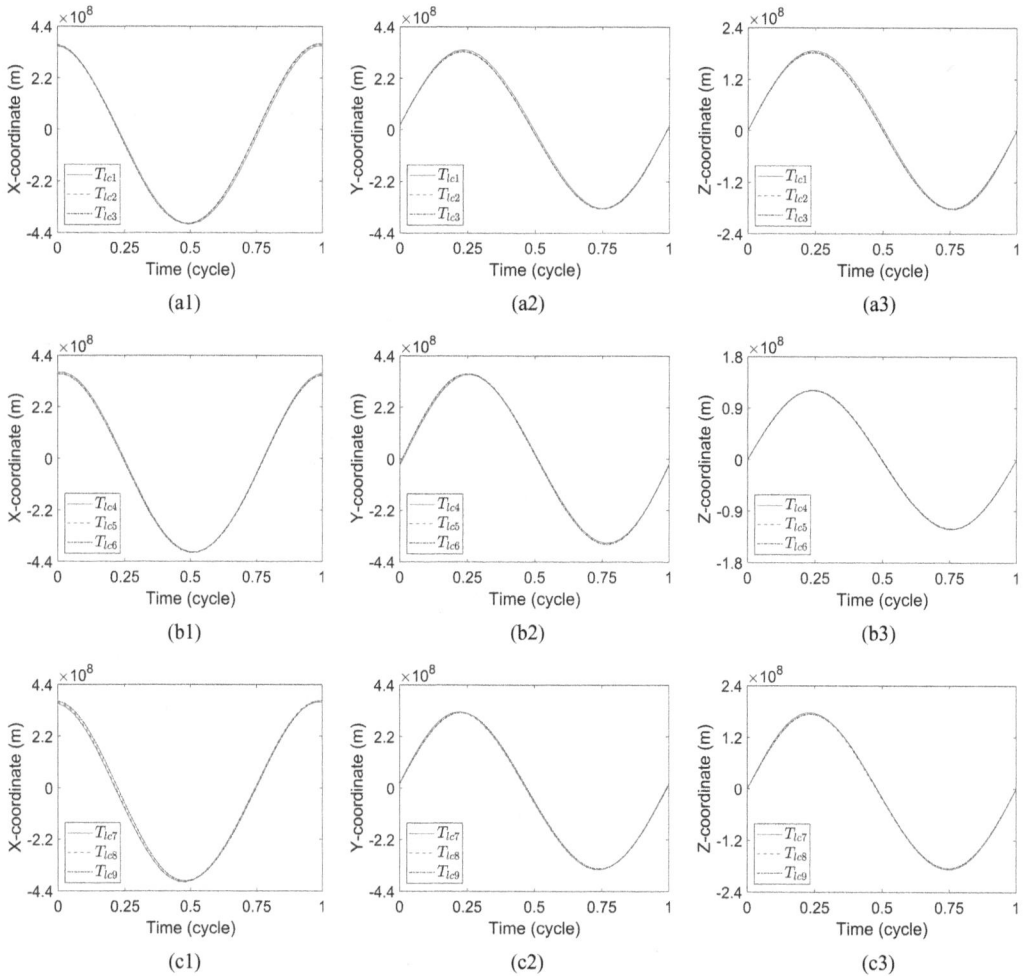

FIGURE 5.18   The variations of MBSAR's coordinates across various cycles, where the 1st (a1–c1), 2nd (a2–c2), and 3rd (a3–c3) columns represent X, Y, and Z components of MBSAR's position vectors, while the 1st (a1–a3), 2nd (b1–b3), and 3rd (c1–c3) rows stand for epochs of $T_{lc1} \sim T_{lc3}$, $T_{lc4} \sim T_{lc6}$, and $T_{lc7} \sim T_{lc9}$, respectively.

by orbital perturbation effects across different cycles. Given that this influence persists throughout the mission lifetime of the MBSAR, it is crucial to take special care with respect to Doppler frequency under such effects.

In Figures 5.19–5.21, by selecting look angles of 0.2°, 0.5°, and 0.8°, we present plots detailing the Doppler frequency and its components in the broadside looking MBSAR across cycles from $T_{lc1} \sim T_{lc9}$. Without loss of generality, the MBSAR looks from the left-hand side in those cases.

The results demonstrate that orbital perturbations exhibit significant impacts on the Doppler frequency. Specifically, Earth's rotation-induced Doppler frequency variation remains relatively small and consistent across three consecutive cycles. By contrast, the spatiotemporal variations in the SAR motion-induced Doppler frequency are much more significant. This implies that the variations in Doppler frequency across a short period (e.g., three consecutive cycles) are primarily given rise by the MBSAR's inertial motion. Over a time interval of 9.3 years, when the orbital inclination angle approaches minimal, significant changes occur in the Doppler frequency component related to Earth's rotation. Therefore, Earth's rotation produces a larger Doppler frequency change across a long time interval; this phenomenon is attributed to cumulative orbital perturbations. After 18.6

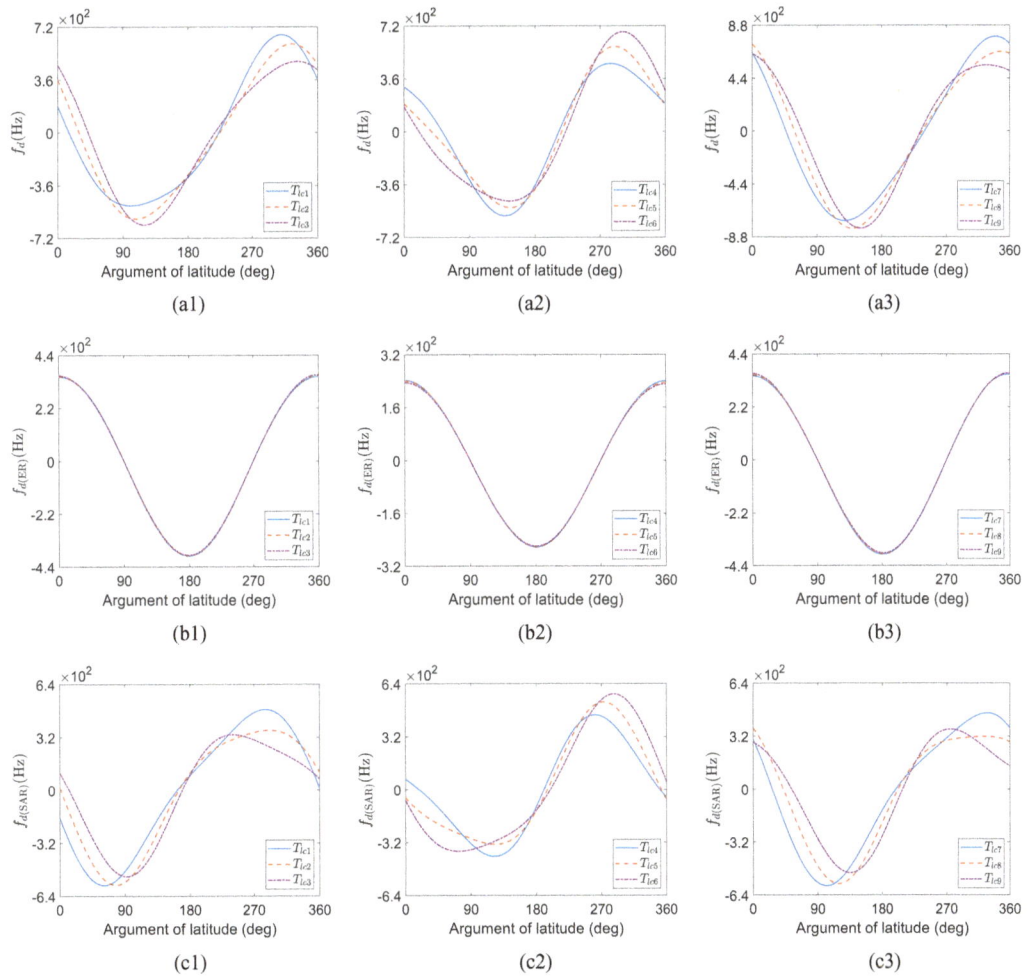

FIGURE 5.19 At a look angle of 0.2°, the Doppler frequency and its component versus AOL across various cycles, where the 1st (a1–a3), 2nd (b1–b3), and 3rd (c1–c3) rows stand for f_d, $f_{d(ER)}$, and $f_{d(SAR)}$, while the 1st (a1–c1), 2nd (a2–c2), and 3rd (a3–c3) columns represent epochs of $T_{lc1} \sim T_{lc3}$, $T_{lc4} \sim T_{lc6}$, and $T_{lc7} \sim T_{lc9}$, respectively.

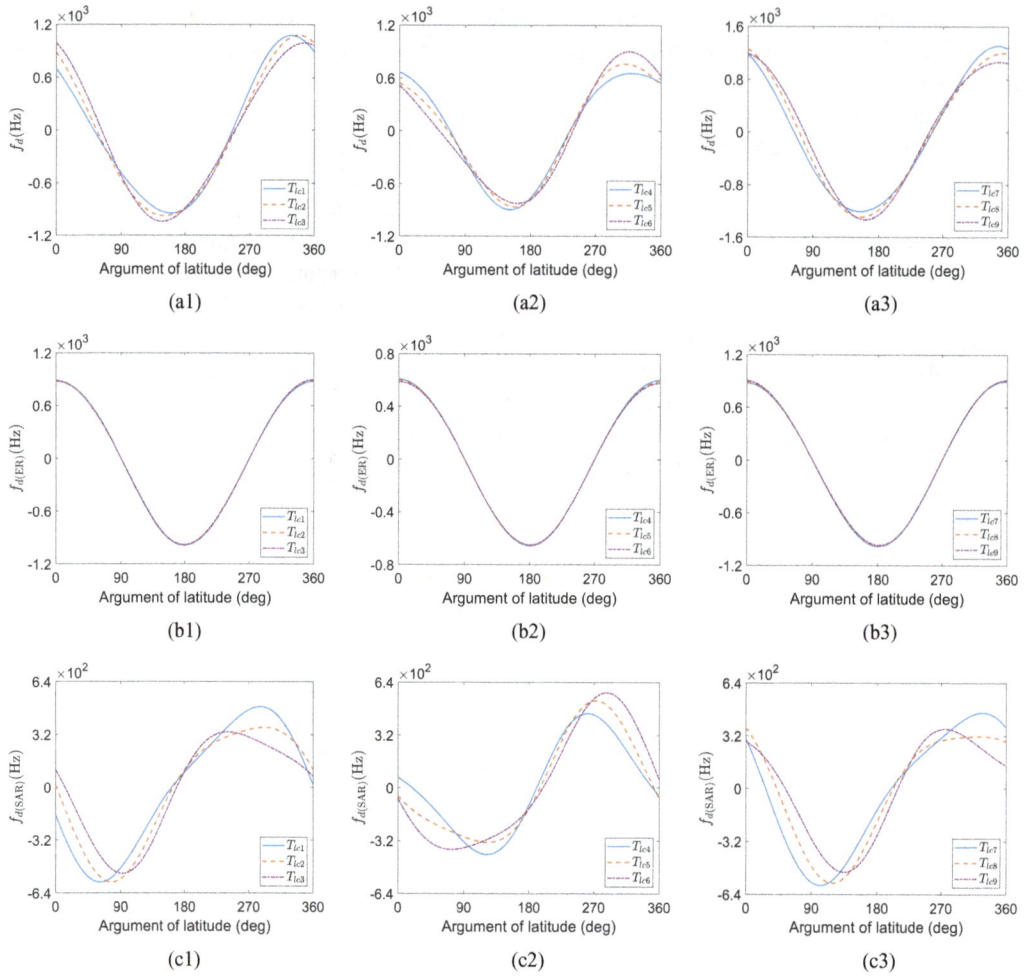

FIGURE 5.20 At a look angle of 0.5°, the Doppler frequency and its component versus AOL across various cycles, where the 1st (a1–a3), 2nd (b1–b3), and 3rd (c1–c3) rows stand for f_d, $f_{d(ER)}$, and $f_{d(SAR)}$, while the 1st (a1–c1), 2nd (a2–c2), and 3rd (a3–c3) columns represent epochs of $T_{lc1} \sim T_{lc3}$, $T_{lc4} \sim T_{lc6}$, and $T_{lc7} \sim T_{lc9}$, respectively.

years, a period of orbital inclination variation, the Doppler frequencies in cycles $T_{lc7} \sim T_{lc9}$ exhibit a certain degree of similarity to those in cycles $T_{lc1} \sim T_{lc3}$, but they are not entirely identical in practice, and further, the variations of both Doppler frequency components are interdependent, resulting in irregular and non-periodic spatiotemporal variations in the Doppler frequency. Notably, the look angle has an impact on the perturbation effects: a larger look angle results in a more pronounced magnitude and variation in Doppler frequency under orbital perturbation effects.

The following analysis explores how changes in cycles affect the Doppler frequency and its components under squint effects in response to the orbital perturbations. Specifically, two cases, where the antenna azimuth angles of 45° and 135° are selected. We have simulated the Doppler frequency and its components across epochs of $T_{lc1} \sim T_{lc6}$ for both cases at three look angles, the results are respectively shown in Figures 5.22–5.24 as a function of the AOL.

Figures 5.22–5.24 demonstrate that orbital perturbation effects significantly impact the Doppler frequency, depending on the MBSAR's location, orbiting cycle, and beam pointing direction. Interestingly, different Doppler frequency components are affected to varying degrees: the antenna azimuth angle could shift in Earth's rotation-induced Doppler frequency, the degree of which

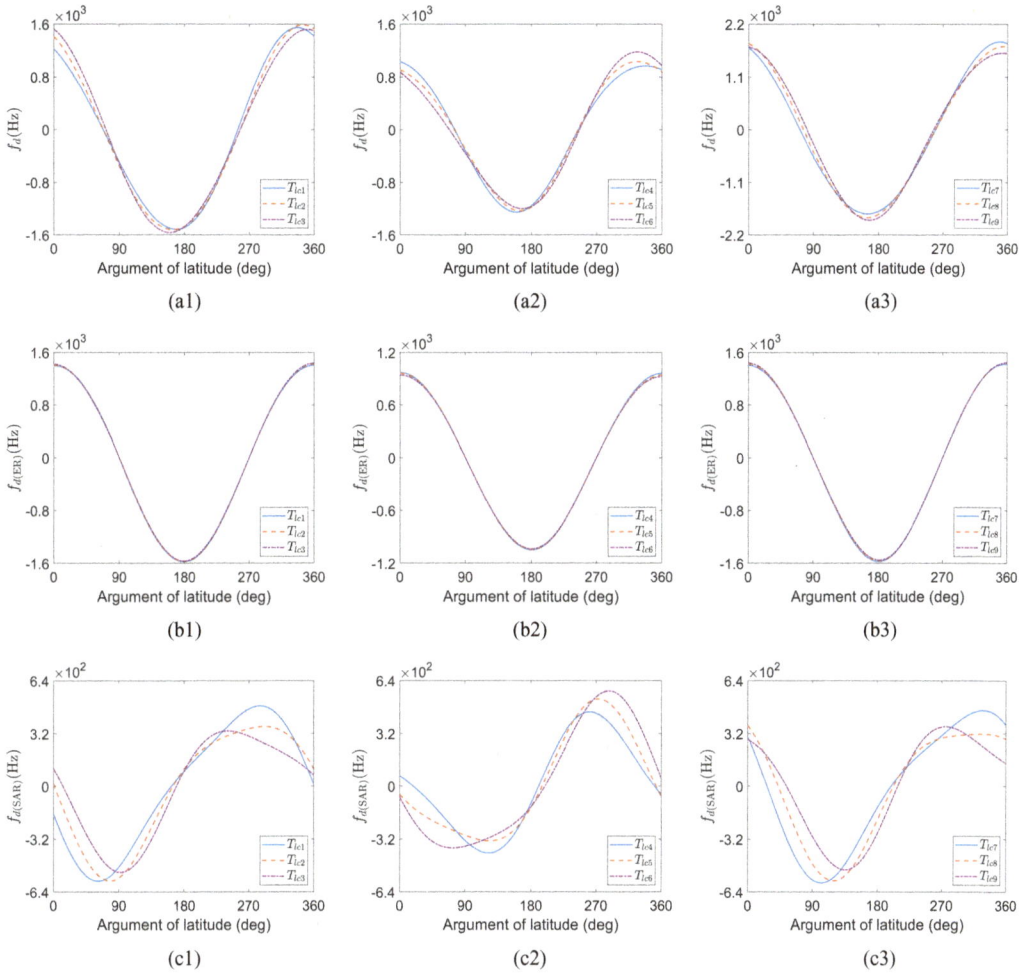

FIGURE 5.21 At a look angle of 0.8°, the Doppler frequency and its component versus AOL across various cycles, where the 1st (a1–a3), 2nd (b1–b3), and 3rd (c1–c3) rows stand for f_d, $f_{d(ER)}$, and $f_{d(SAR)}$, while the 1st (a1–c1), 2nd (a2–c2), and 3rd (a3–c3) columns represent epochs of $T_{lc1} \sim T_{lc3}$, $T_{lc4} \sim T_{lc6}$, and $T_{lc7} \sim T_{lc9}$, respectively.

increases with increasing look angle. Conversely, the antenna azimuth angle shows a greater impact on the SAR motion-induced Doppler frequency component under orbital perturbations. This component, however, is substantially insensitive to the look angle wherever the MBSAR is located.

Regarding the variation of each Doppler frequency component over different cycles, the Earth's rotation contributes to a uniformly spatiotemporal distributed Doppler frequency component with small changes across continuous cycles but more significant changes over longer time intervals. On the other hand, orbital perturbations can significantly affect the Doppler frequency component resulting from SAR's inertial motion, leading to distinct fluctuations across different cycles. Combining both Doppler components brings about irregular variation patterns in the Doppler frequency under perturbation effects. These findings suggest that careful consideration must be given to orbital perturbations when designing the MBSAR system.

We have examined the properties of Doppler frequency in the MBSAR across diverse scenarios. Specifically, we scrutinize how beam pointing directions, side-looking directions, and orbital elements affect the Doppler frequency and its two components generated by Earth's rotation and SAR inertial motion. The general framework presented in this chapter can be straightforwardly applied to analyze Doppler frequency under different scenarios, but this is not presented here.

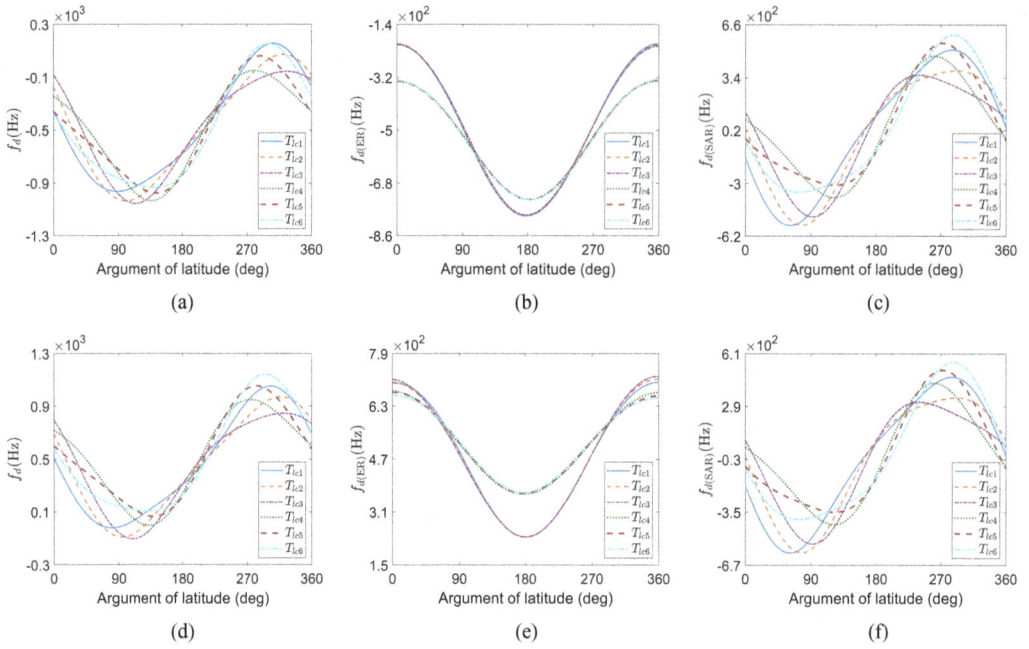

FIGURE 5.22   At a look angle of 0.2°, the Doppler frequency and its component versus AOL across various cycles, where the upper and lower rows stand for the antenna azimuth angles of 45° and 135°, while the 1st (a1–a3), 2nd (b1–b3), and 3rd (c1–c3) columns stand for f_d, $f_{d(ER)}$, and $f_{d(SAR)}$, respectively.

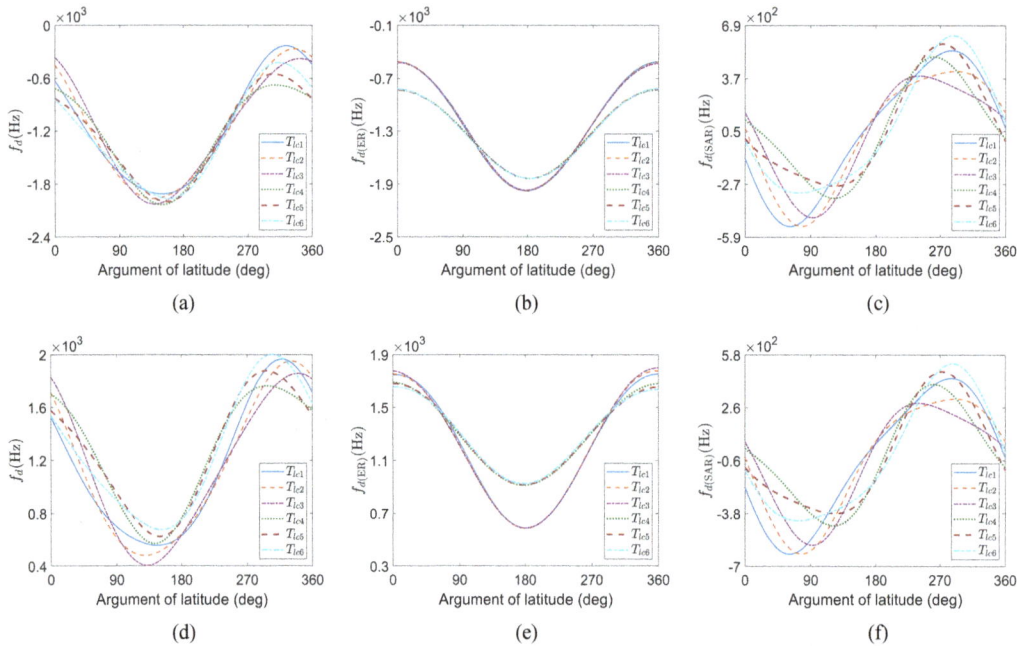

FIGURE 5.23   At a look angle of 0.5°, the Doppler frequency and its component versus AOL across various cycles, where the upper and lower rows stand for the antenna azimuth angles of 45° and 135°, while the 1st (a1–a3), 2nd (b1–b3), and 3rd (c1–c3) columns stand for f_d, $f_{d(ER)}$, and $f_{d(SAR)}$, respectively.

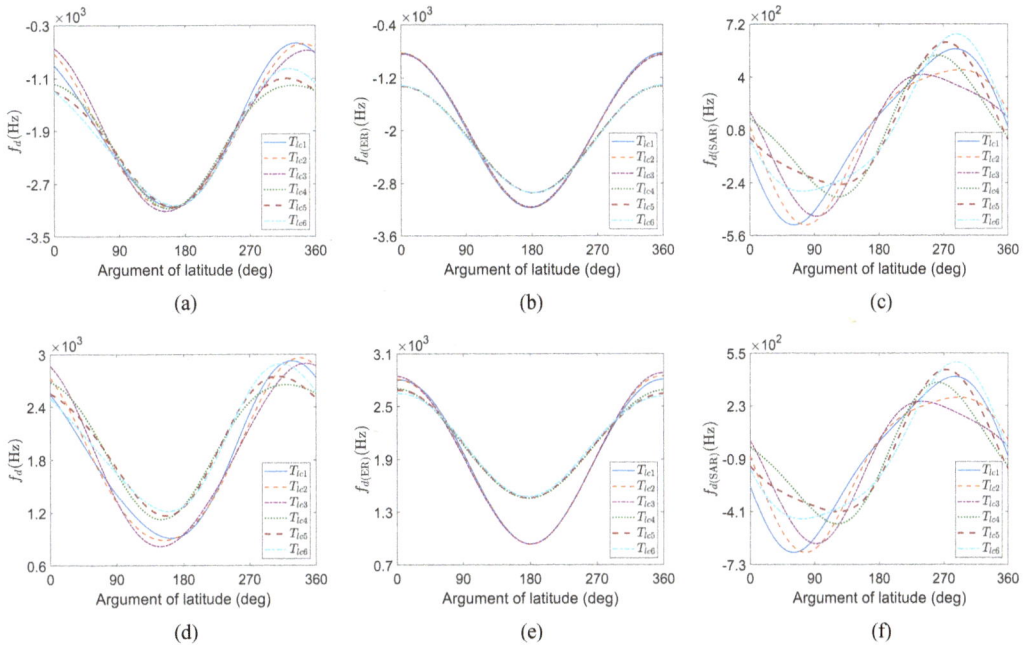

FIGURE 5.24 At a look angle of 0.8°, the Doppler frequency and its component versus AOL across various cycles, where the upper and lower rows stand for the antenna azimuth angles of 45° and 135°, while the 1st (a1–a3), 2nd (b1–b3), and 3rd (c1–c3) columns stand for f_d, $f_{d(ER)}$, and $f_{d(SAR)}$, respectively.

5.2.4 PROPERTIES OF DFMR

In addition to the Doppler frequency, the DFMR is another crucial Doppler parameter linked to the signal properties and system parameters, pivotal in MBSAR's system design, signal modeling, and imaging formation [17, 18]. During the early stages of MBSAR research, DFMR analysis typically ignores the MBSAR's inertial motion to simplify the analysis [19–22]. However, this approach can result in substantial discrepancies between the resulting DFMR properties and their actual values. Therefore, exploring and assessing the impacts of associated factors on DFMR properties in MBSAR is essential. In this section, we shall describe the properties of the DFMR and highlight the role of MBSAR's inertial motion and Earth's rotation in determining the DFMR under different MBSAR configurations in detail.

Figure 5.25 depicts the isocontours of DFMR concerning both antenna azimuth and look angles in MBSAR. Also, it contains a comparison of the DFMR values between MBSAR and the spaceborne SAR system with configurations presented in Table 5.1 to present the unique features of the DFMR in the MBSAR. In the following analysis, both the MBSAR and spaceborne SAR are left-looking unless otherwise specified.

As seen, the MBSAR and spaceborne SAR show a decrease in the DFMR with increasing look angle, albeit with a much greater magnitude in spaceborne SAR. The DFMR is approximately symmetrical at about $\phi = 90°$ in the spaceborne SAR. By comparison, the MBSAR's DFMR shows no such symmetry. These findings indicate notable differences in DFMRs between MBSAR and spaceborne SAR, particularly when considering the antenna azimuth angle.

To quantify the impacts of SAR inertial motion and Earth's rotation on the DFMR, we analyze how each component of DFMR varies with antenna azimuth and look angles. Figure 5.26 displays isocontours for each component in the DFMR versus the antenna azimuth and look angle in MBSAR and spaceborne SAR scenarios.

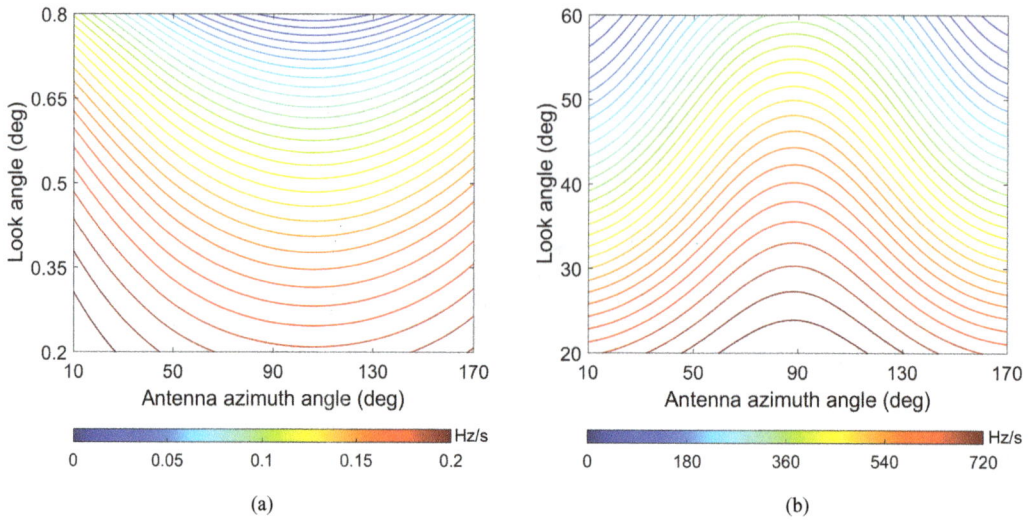

FIGURE 5.25 The isocontours of DFMR versus the antenna azimuth and look angles, (a) in the MBSAR, (b) in the spaceborne SAR.

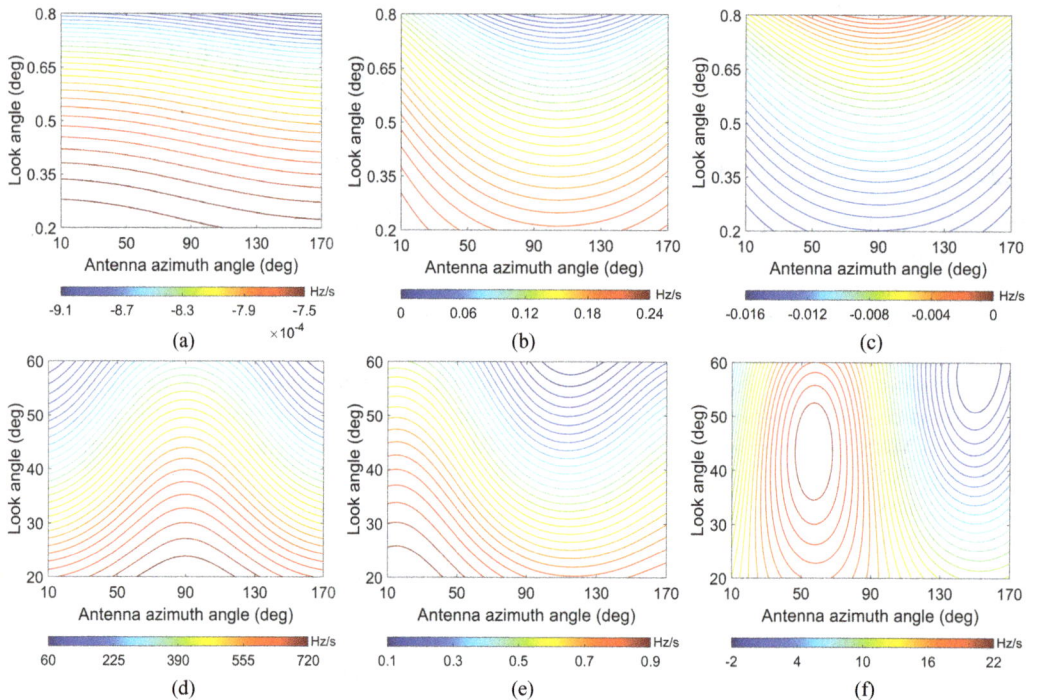

FIGURE 5.26 The isocontours of DFMR's components versus the antenna azimuth and look angles in the MBSAR and spaceborne SAR, where the left (a, c, e) and right (b, d, e) columns represent MBSAR and spaceborne SAR, while the 1st (a, b), 2nd (c, d), and 3rd (e, f) rows stand for $f_{dr(SAR)}, f_{dr(ER)}, f_{dr(SE)}$, respectively.

Figure 5.26 shows significant distinctions in DFMR components and their sensitivity to antenna azimuth and look angles between spaceborne SAR and MBSAR. In spaceborne SAR, the DFMR component generated by SAR motion is nearly axis-symmetric with respect to the $\phi = 90°$, while the remaining two components lack this property. Furthermore, the DFMR resulting from SAR's inertial motion is notably higher than that arising from the other two factors, suggesting that DFMR's magnitude and axis symmetry depend mainly on the SAR's inertial motion in the spaceborne SAR. Conversely, in MBSAR, the DFMR component that is coupling affected by SAR's inertial motion and Earth's rotation displays axis symmetry, with the remaining two components exhibiting distinct variations with varying antenna azimuth and look angles. Further, the DFMR component generated by Earth's rotation has the maximal magnitude. In this regard, Earth's rotation plays a more dominant role in determining the DFMR in MBSAR, underscoring its critical importance in determining the Doppler effect. Although the SAR's inertial motion has the least contribution to DFMR, its magnitude is still significant enough to impact MBSAR imaging even at coarser azimuth resolution, necessitating the inclusion of the MBSAR's inertial motion effect.

To illustrate the characteristics of DFMR and measure the individual contributions of each component to its properties quantitatively, three ratios can be defined as follows:

$$C_{\mathrm{SAR}} = \frac{f_{\mathrm{dr(SAR)}}}{f_{\mathrm{dr}}} \times 100\%, \quad C_{\mathrm{ER}} = \frac{f_{\mathrm{dr(ER)}}}{f_{\mathrm{dr}}} \times 100\%, \quad C_{\mathrm{SE}} = \frac{f_{\mathrm{dr(SE)}}}{f_{\mathrm{dr}}} \times 100\% \qquad (5.42)$$

Figures 5.27 and 5.28 display the defined ratios under various look angles versus antenna azimuth angle and those ratios at various azimuth angles versus look angle in both the MBSAR and spaceborne SAR scenarios.

The effects of antenna azimuth and look angles are mutually dependent, and their contributions to DFMR exhibit no clear trend. Notwithstanding, the magnitudes of those ratios are different in MBSAR compared to those in spaceborne SAR, regardless of the antenna azimuth and look angles. Specifically, in spaceborne SAR, the DFMR is primarily influenced by SAR's inertial motion, with minimal impact from Earth's rotation. In the MBSAR, however, the Earth's rotation plays a dominant role in determining the DFMR, while quite a small contribution from the SAR's inertial motion.

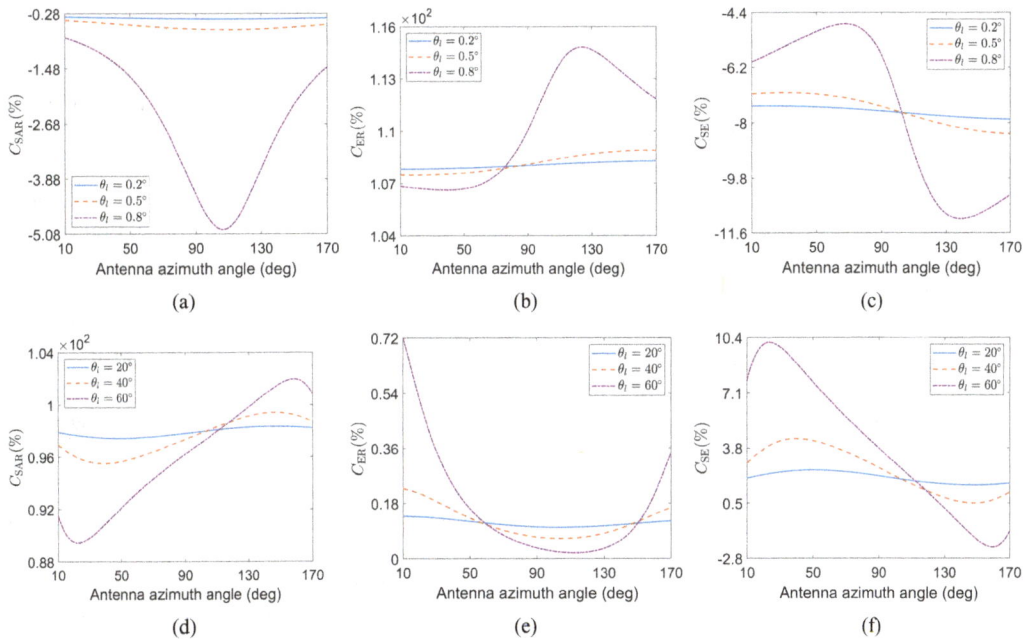

FIGURE 5.27 The contribution of each component to the DFMR versus antenna azimuth angle under various look angles, where the upper (a, b, c) and lower (b, d, e) rows represent MBSAR and spaceborne SAR, while the 1st (a, d), 2nd (b, e), and 3rd (c, f) columns stand for C_{SAR}, C_{ER}, C_{SE}, respectively.

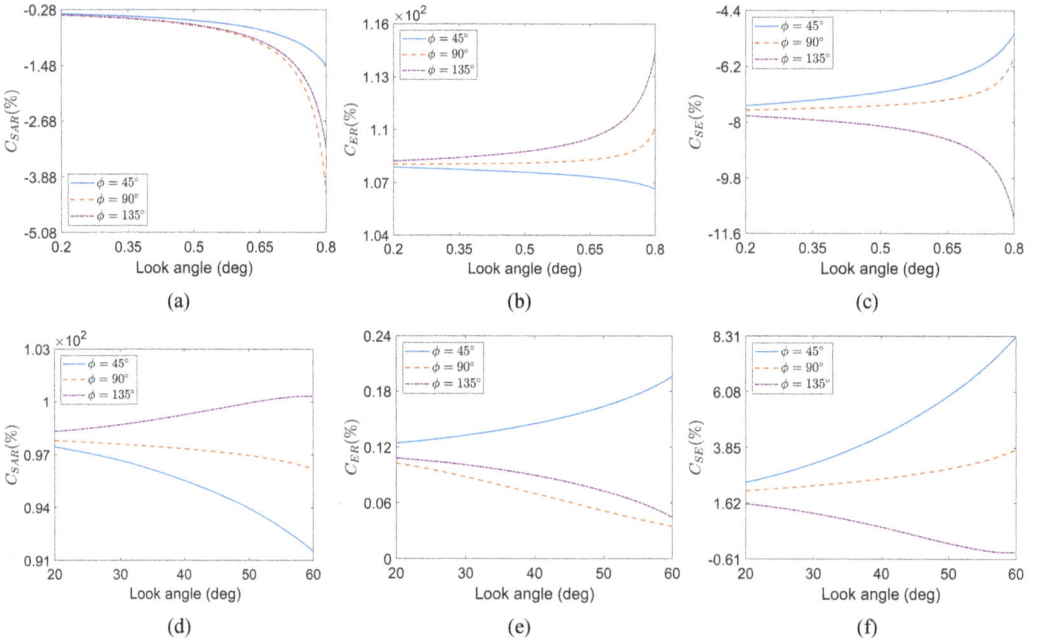

FIGURE 5.28 The contribution of each component to the DFMR versus look angle under various antenna azimuth angles, where the upper (a, b, c) and lower (b, d, e) rows represent MBSAR and spaceborne SAR, while the 1st (a, d), 2nd (b, e), and 3rd (c, f) columns stand for C_{SAR}, C_{ER}, C_{SE}, respectively.

Interestingly, regardless of platform, the coupling effect of SAR's inertial motion and Earth's rotation has a secondary impact on the DFMR compared to the other two factors. Moreover, in MBSAR, both C_{SAR} and C_{SE} have negative values, suggesting that SAR's inertial motion and the coupling effect of SAR's inertial motion and Earth's rotation tend to decrease the DFMR. By contrast, all factors positively impact the DFMR in spaceborne SAR. These findings confirm that MBSAR has a unique Doppler property characterized by the dominance of Earth's rotation over SAR's inertial motion despite the latter's non-negligible influence.

Next, we demonstrate how the DFMR changes as the MBSAR system's location varies in the cycle T_{lc1}. Figure 5.29 presents the DFMR versus AOL and look angle. Additionally, we select three look angles of $0.2°$, $0.5°$, and $0.8°$, and plot the DFMR and its associated components in Figure 5.30 as a function of AOL to provide a quantitative analysis. Throughout the following analysis, we focus specifically on the case when the antenna beam center is steered toward the zero-Doppler plane. This is preferred for both MBSAR and spaceborne SAR since directing the antenna beam center toward the zero-Doppler plane simplifies the signal processing, refrains from image skewing, and allows for high focusing quality in imagery [7, 23, 24].

The results from Figures 5.29 and 5.30 demonstrate that the factors influencing DFMR over one cycle are ordered as follows: Earth's rotation, the coupling effect of Earth's rotation and SAR's inertial motion, and SAR's inertial motion. Additionally, the DFMR exhibits some change depending on its orbital position, and its components show distinct variation patterns concerning the AOL. It is noteworthy that at most locations, the DFMR exhibits a downward trend with respect to the look angle, and so do the DFMR components generated by SAR's inertial motion and Earth's rotation. In comparison, the DFMR component associated with the coupling effect exhibits the opposite variation trend with an increasing look angle compared to the other two DFMR components.

The next goal is to analyze how the side-looking direction affects DFMR; Figure 5.31 provides a clear depiction of such an influence. It is evident that the variation regularity of DFMR versus AOL is dependent on the look angle and side-looking direction. Interestingly, the left-looking DFMR and

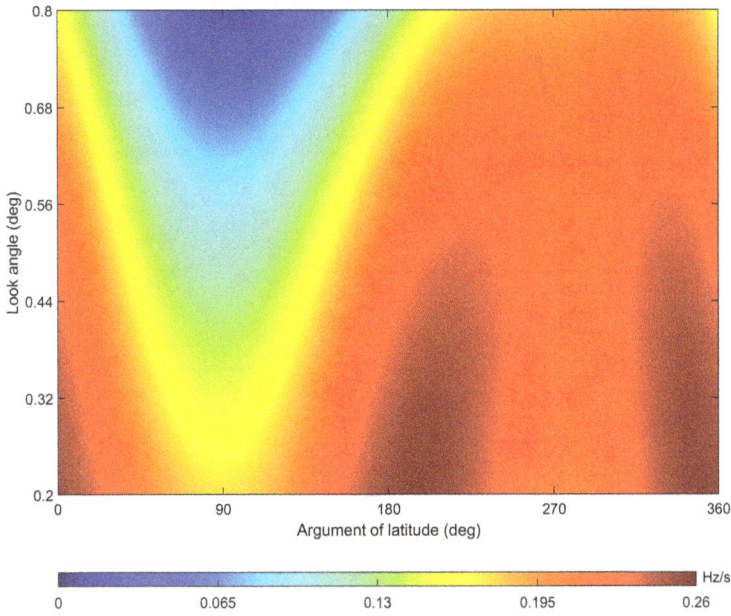

FIGURE 5.29 The DFMR of the MBSAR versus AOL and look angle in cycle T_{lc1}.

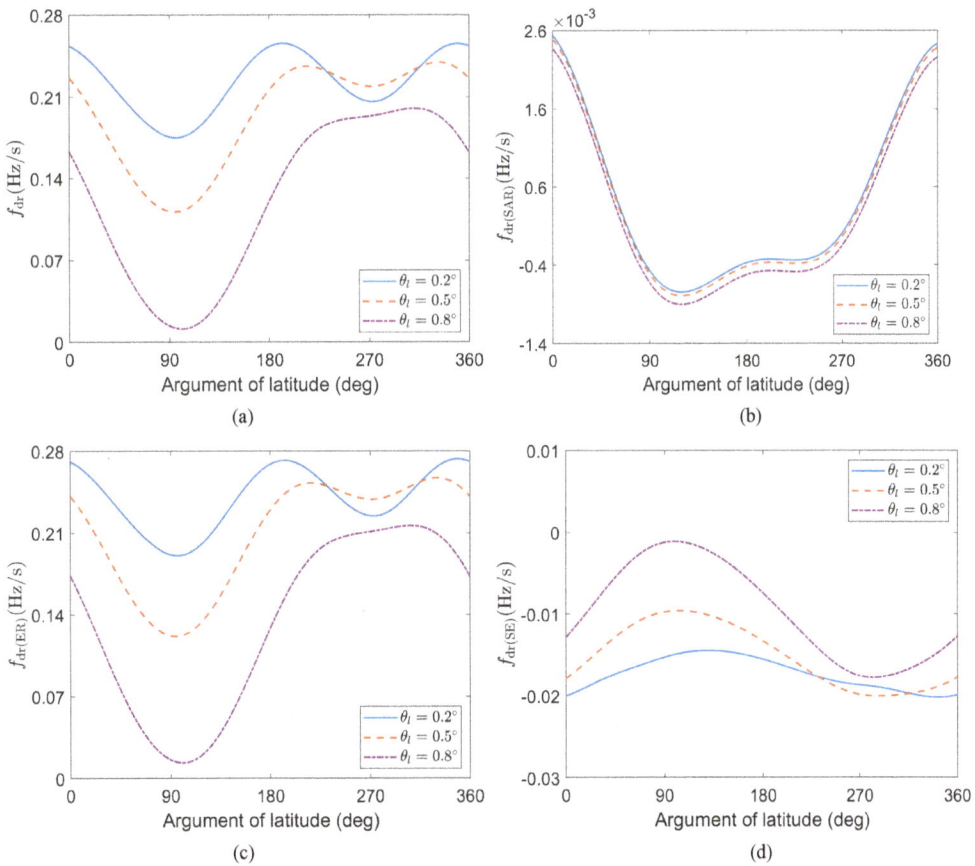

FIGURE 5.30 The DFMR and its components of the MBSAR with various look angles in cycle T_{lc1}, (a) f_{dr} versus AOL, (b) $f_{dr(ER)}$ versus AOL, (c) $f_{dr(SAR)}$ versus AOL, (d) $f_{dr(SE)}$ versus AOL.

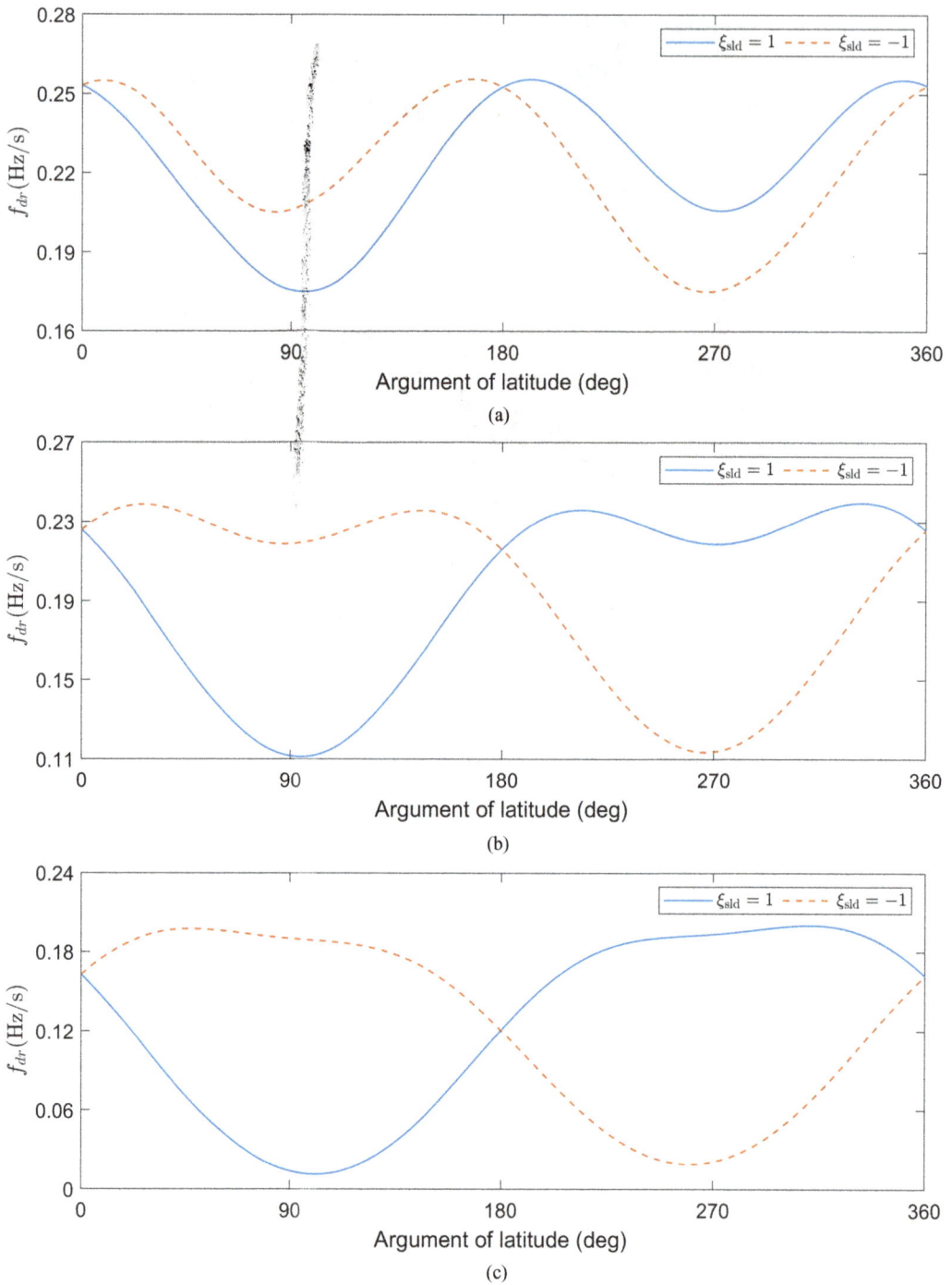

FIGURE 5.31 The left-looking and right-looking DFMRs versus AOL with look angles of (a) 0.2°, (b) 0.5°, (c) 0.8°.

right-looking one exhibit axial symmetry about the AOL of 180°. Therefore, even when the beam center points to the zero-Doppler plane, the left-looking Doppler shift differs from the right-looking one at most locations. Once again, it is asserted that the MBSAR seldom encounters left–right ambiguity.

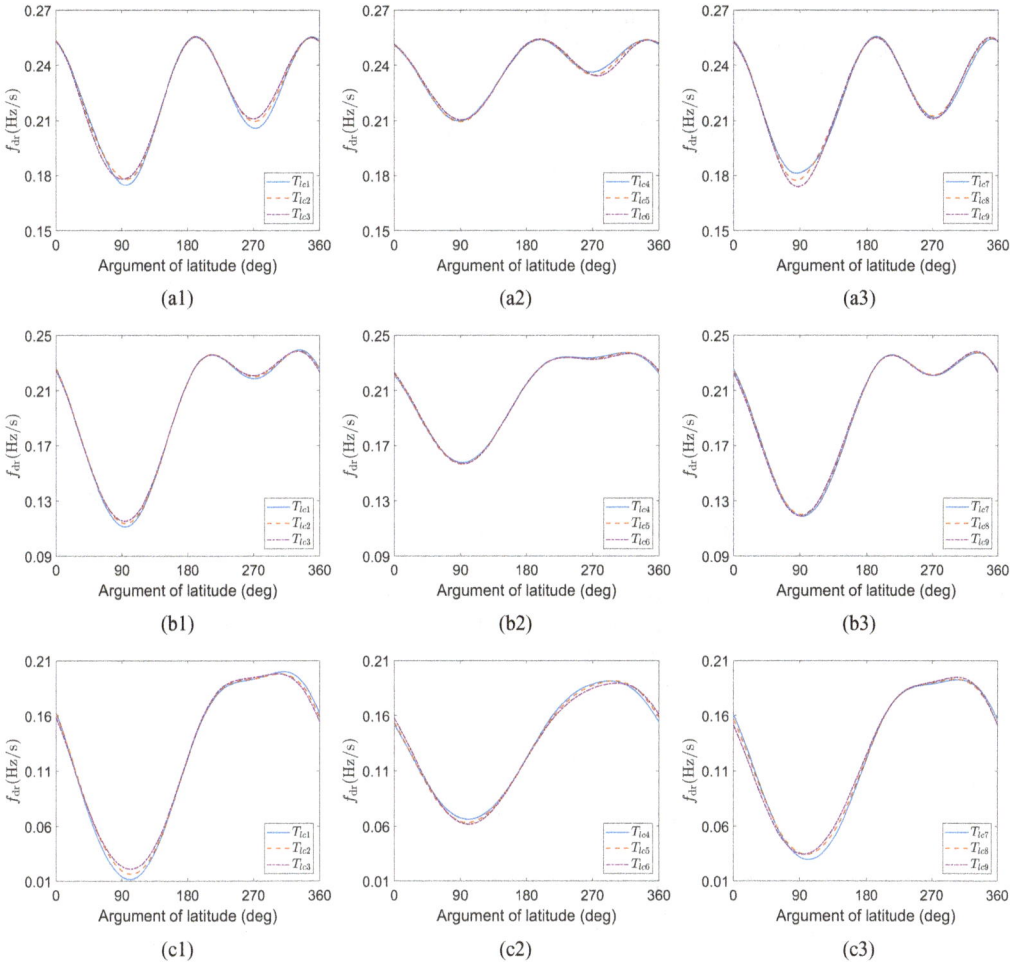

FIGURE 5.32 At various look angles, the DFMR versus AOL across various cycles, where the 1st (a1–a3), 2nd (b1–b3), and 3rd (c1–c3) rows stand for look angles of 0.2°, 0.5°, and 0.8°, while the 1st (a1–c1), 2nd (a2–c2), and 3rd (a3–c3) columns represent epochs of $T_{lc1} \sim T_{lc3}$, $T_{lc4} \sim T_{lc6}$, and $T_{lc7} \sim T_{lc9}$, respectively.

To investigate the impact of orbital perturbations on the MBSAR's DFMR, we conduct further analysis of DFMR over cycles from T_{lc1} to T_{lc9}. Plots demonstrating the DFMR versus the AOL at various look angles across these cycles are illustrated in Figure 5.32.

As seen, orbital perturbations could impact the DFMR even in consecutive cycles, albeit to a small degree. Nevertheless, the MBSAR's image focusing may be potentially affected by these perturbations, as will be discussed in Chapter 7. Once the long-term Earth observation is considered, considerable fluctuations can be observed in the DFMR, and further, a larger look angle could lead to more significant fluctuations. Therefore, it is essential to consider the orbital perturbations in the MBSAR system design and signal processing.

5.3 ZERO-DOPPLER STEERING FOR THE MBSAR

The previous analysis showed that the non-zero-Doppler centroid appearing in broadside looking MBSAR is due to the perturbed elliptical orbit and Earth's rotation. This non-zero-Doppler centroid results in significant range cell migration, compounded by the long SAT and high coupling between azimuth and range, deteriorating the imaging quality [25]. Consequently, it is crucial to eliminate the non-zero-Doppler centroid to achieve optimal imaging quality, as detailed below.

5.3.1 THEORETICAL DERIVATION OF ZERO-DOPPLER CENTROID

Quite a few Doppler steering methods have been put forward to minimize the Doppler centroid of spaceborne SAR systems, which is typically achieved through the attitude steering of the satellite [23, 26–29]. However, these methods are not necessarily applicable in the case of the MBSAR due to the following two factors:

1) The platform of MBSAR is a celestial body, making it impossible to steer the platform's attitude to compensate for the Doppler centroid;
2) The conventional methods rely on the condition that the platform's rotational angular velocity is much greater than, or at least close to, Earth's rotational angular velocity.

The MBSAR has a far smaller rotational angular velocity during its inertial motion than that of the Earth's rotation, rendering the conventional zero-Doppler steering methods no longer applicable to the MBSAR [7]. Even if conventional methods are equivalently applied by steering the beam pointing direction, the processed MBSAR's beam fails to point toward the zero-Doppler plane. In some positions, it might be impossible to illuminate Earth. One possible solution to tackle the aforementioned challenges inherent in the MBSAR is to implement the phase scan based on its orbital characteristics, which allows the antenna beam to be steered to the desired pointing direction (the zero-Doppler plane) without steering the platform's attitude [30–32]. In this section, we propose a general zero-Doppler steering approach that employs the phase scan to reduce the Doppler centroid to zero Hz for the MBSAR.

Recall the definition of Doppler frequency: in the zero-Doppler plane, we have:

$$\xi_{\text{sld}} V_{\text{SX,ACS}} \sin \theta_l \cos \phi + V_{\text{SZ,ACS}} \cos \theta_l + \omega_E \xi_{\text{sld}} R_{\text{SAR}} \sin \theta_l \left(\sin i \cos u \sin \phi - \cos \phi \cos i \right) = 0 \qquad (5.43)$$

According to Eq. (5.43), we can reduce the Doppler frequency to zero Hz by ensuring that the antenna azimuth and look angles satisfy the following relation:

$$T_{\text{zd1}} \cdot \cos \phi + T_{\text{zd2}} \cdot \sin \phi + T_{\text{zd3}} \cot \theta_l = 0 \qquad (5.44)$$

with:

$$\begin{cases} T_{\text{zd1}} = \omega_E^{-1} R_{\text{SAR}}^{-1} V_{\text{SX,ACS}} - \cos i \\ T_{\text{zd2}} = \sin i \cos u \\ T_{\text{zd3}} = \xi_{\text{sld}} \omega_E^{-1} R_{\text{SAR}}^{-1} V_{\text{SZ,ACS}} \end{cases} \qquad (5.45)$$

The solution of Eq. (5.44) provides the necessary azimuth angle for the zero-Doppler centroid beam pointing, such that:

$$\phi = \cos^{-1} \left\{ \frac{-T_{\text{zd1}} T_{\text{zd3}} \cot \theta_l \pm \Re_{\text{sign}} \cdot T_{\text{zd2}} \left(T_{\text{zd1}}^2 + T_{\text{zd2}}^2 - T_{\text{zd3}}^2 \cot^2 \theta_l \right)^{0.5}}{T_{\text{zd1}}^2 + T_{\text{zd2}}^2} \right\} \qquad (5.46)$$

where:

$$\Re_{\text{sign}} = \begin{cases} 1, & \cos u \geq 0 \\ -1, & \cos u < 0 \end{cases} \tag{5.47}$$

The sign "$\pm$" in front of R_{sign} is dependent on the SAR platform's altitude, with the "$-$" sign applying for the spaceborne SAR, while the "$+$" sign is for the MBSAR. Once the orbital elements are determined, Eq. (5.46) can be used to set the desired antenna azimuth angle for a specified look angle. This stage can be implemented through the phase scan involving amplitude and phase weighting by the phase shifter in the phased array antenna. Consequently, the Doppler centroid frequency can be reduced to zero Hz.

Note that the antenna azimuth angle indicates the direction where the antenna beam is pointing, which does not affect the antenna performance. Therefore, there are no restrictions on the antenna azimuth angle. The SAR look angle, however, must be limited to a specific range comprising the upper and lower bounds. Specifically, the valid echoes with a zero-Doppler centroid require the look angle to adhere to the following relation:

$$\theta_l > \cot^{-1}\left[\frac{\left(T_{\text{zd1}}^2 + T_{\text{zd2}}^2\right)^{0.5}}{|T_{\text{zd3}}|}\right] \tag{5.48}$$

The antenna beam should cover only Earth, and to ensure this, the following inequality must be held:

$$\theta_l < \sin^{-1}\left(\frac{R_{\text{TOI}}}{R_{\text{SAR}}}\right) \tag{5.49}$$

As the Earth's radius spatially varies according to the TOI's position, the Earth's radius R_{TOI} in Eq. (5.49) could be replaced by R_p, the polar radius of Earth ellipsoid, to ensure the antenna beam illuminates Earth reliably.

Combining Eqs. (5.48) and (5.49), we arrive at the lower and upper bounds of the look angle required for the zero-Doppler steering:

$$\theta_l^{\text{low}} < \theta_l < \theta_l^{\text{up}} \tag{5.50}$$

$$\begin{cases} \theta_l^{\text{low}} = \cot^{-1}\left[\frac{\left(T_{\text{zd1}}^2 + T_{\text{zd2}}^2\right)^{0.5}}{|T_{\text{zd3}}|}\right] \\ \theta_l^{\text{up}} = \sin^{-1}\left(\frac{R_p}{R_{\text{SAR}}}\right) \end{cases} \tag{5.51}$$

Note that the look angle must fall within the bounds specified by Eq. (5.50) to perform zero-Doppler steering for yielding valid echoes with a zero-Doppler centroid.

Figure 5.33 shows both the lower and upper bounds as a function of the AOL across various cycles. We observe that they exhibit different patterns of variation over various cycles: The upper bound depends solely on the ratio of Earth's polar radius to the Earth–MBSAR distance. In contrast, the lower bound is sensitive to temporally varying orbital elements. As a result, the fluctuation of the lower bound is more noticeable than that of the upper bound. Quantitative evaluation shows that the

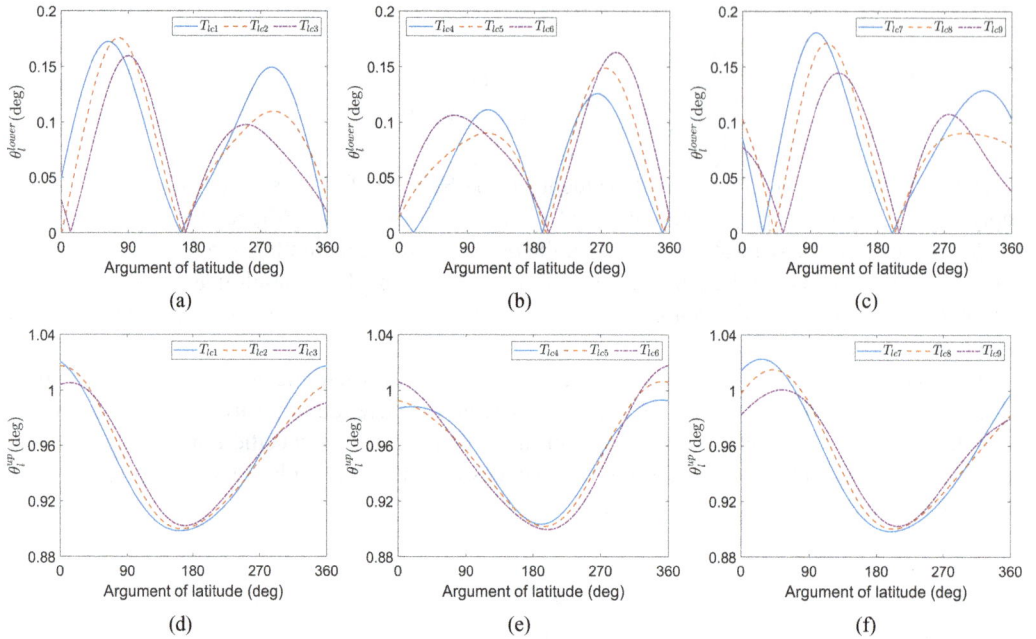

FIGURE 5.33 The bounds of look angle versus AOL in the MBSAR, where the upper (a, b, c) and lower (d, e, f) rows represent the lower and upper bounds of look angle, while the 1st (a, d), 2nd (b, e), and 3rd (c, f) rows stand for the epochs of $T_{lc1} \sim T_{lc3}$, $T_{lc4} \sim T_{lc6}$, $T_{lc7} \sim T_{lc9}$, respectively.

lower bound remains below 0.2°, while the upper bound exceeds 0.8°. Based on these findings, the range of look angle can be set from 0.2° to 0.8° in the MBSAR.

Next, using TerraSAR-X method as an example, we explain why the conventional attitude steering is inadequate for MBSAR. The TerraSAR-X method is equivalently achieved through the phase scan, as the Moon's attitude cannot be steered. Figure 5.34 shows the equivalent look angle and Doppler frequency residual after the MBSAR is processed by the TerraSAR-X method; both are presented as a function of AOL across different cycles. In this case, we set the look angle of the MBSAR to 0.5°.

Figure 5.34 demonstrates that after steering by the TerraSAR-X method, the steered look angle exceeds the upper bound of look angle at most positions, indicating that the Earth cannot be illuminated by the MBSAR at those locations. In some positions where the MBSAR illuminates Earth, the Doppler centroid is far from zero Hz. Such findings hold for other look angles when applying the TerraSAR-X method to the MBSAR. As a result, the foundation of the TerraSAR-X method is entirely ineffective in the case of MBSAR.

We will now evaluate the effectiveness of the proposed method for zero-Doppler steering. Figure 5.35 displays the magnitudes of Doppler frequency residuals versus the AOL and look angle across various cycles when the steered MBSAR is left-looking. In addition, we examine the magnitudes of steered Doppler residuals under the scenario where MBSAR looks from the right-hand side, as shown in Figure 5.36.

According to Figures 5.35 and 5.36, the magnitude of Doppler frequency residual changes depending on the position and look angle of MBSAR. The maximum magnitude of the Doppler frequency residual occurs at the AOL approaches 90°, regardless of the orbital cycle or side-looking direction. Further, the maximal Doppler frequency residual consistently remains below 10^{-8} Hz. Consequently, the proposed method can effectively reduce the Doppler centroid to zero Hz in the MBSAR.

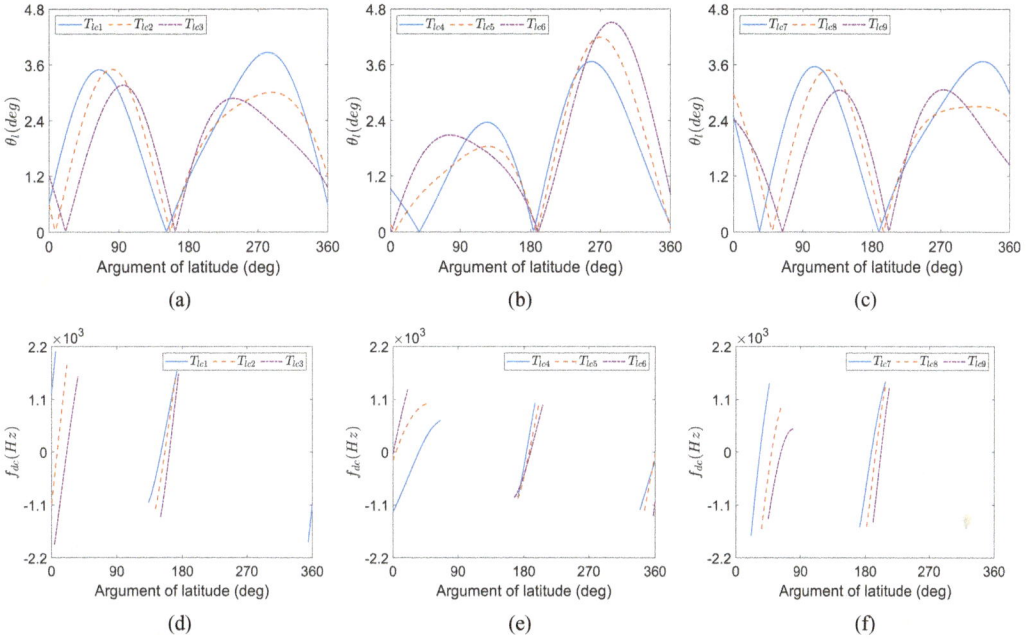

FIGURE 5.34 After processed by TerraSAR-X method, the equivalent look angle and Doppler frequency versus AOL, where the upper (a, b, c) and lower (d, e, f) rows represent the equivalent look angle and Doppler frequency, while the 1st (a, d), 2nd (b, e), and 3rd (c, f) rows stand for the epochs of $T_{lc1} \sim T_{lc3}$, $T_{lc4} \sim T_{lc6}$, $T_{lc7} \sim T_{lc9}$, respectively.

5.3.2 ACCURACY REQUIREMENT FOR THE BEAM POINTING DIRECTION

In general, zero-Doppler steering leads to a decrease in coverage performance. Even though, the MBSAR is still able to cover expansive regions even after Doppler steering. This, in turn, can result in a significant variation in terrain altitude within the image scene, which might pose a challenge to the zero-Doppler steering. Fortunately, the proposed method is independent of the target's altitude, making it well-suited for any terrain [7].

Although the proposed method is not affected by the terrain altitude, it is highly dependent on the beam pointing, making it sensitive to variations in antenna azimuth and look angles. As such, any errors in the look and azimuth angles can lead to a mispointing, thereby resulting in Doppler frequency errors. Such error should be limited below 7.5% of PRF to achieve high image quality in SAR, as recommended in [3]. To prevent any azimuth ambiguity, the PRF should be higher than the Doppler bandwidth [33]. In this regard, we can adopt an upper bound of 7.5% of Doppler bandwidth for the magnitude of Doppler frequency error.

In the following, we check the accuracy requirements of antenna azimuth and look angles in MBSAR, using a look angle of 0.5° and an SAT of 200 s. Figures 5.37 and 5.38 plot the Doppler frequency errors across various cycles under various azimuth angle errors and those under various look angle errors, respectively.

As demonstrated in Figures 5.37 and 5.38, the Doppler frequency error varies versus the AOL and across different cycles, depending on the type of error. The results show that an antenna azimuth angle error below 0.05° enables the resulting Doppler frequency errors to fall within the acceptable range. By contrast, only when the look angle error is limited to 0.001°, the Doppler frequency errors remain within acceptable limits. For larger antenna azimuth angle or look angle errors, the Doppler residual may exceed the upper bound and affect the image quality of the MBSAR accordingly.

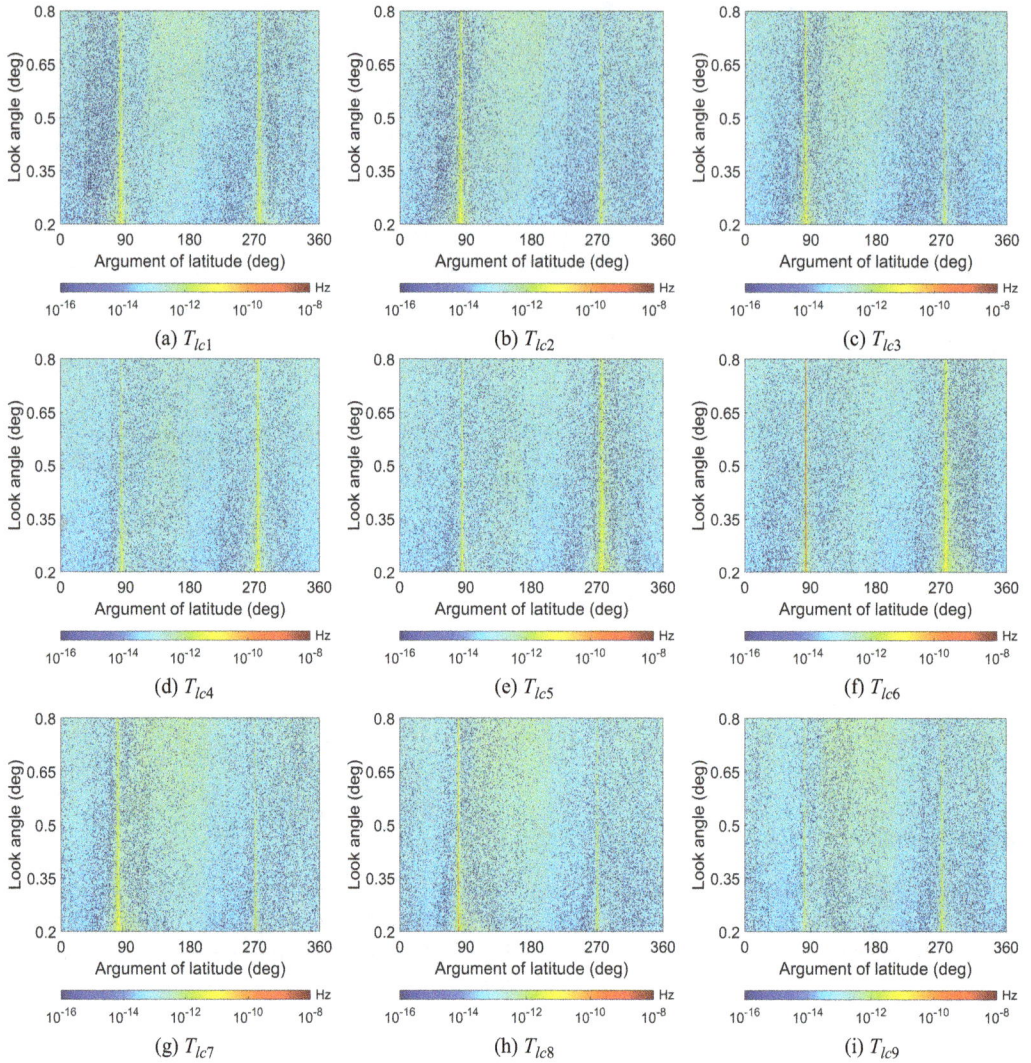

FIGURE 5.35 After processing by the proposed method, the left-looking Doppler frequency residual versus AOL and look angle across cycles from T_{lc1} to T_{lc9}.

In consequence, it can be concluded that the antenna azimuth angle error exerts a relatively small impact on the Doppler frequency. Thus, it provides a more tolerant accuracy requirement: A pointing accuracy of 0.05° for the azimuth angle can be acceptable in the MBSAR. In contrast, the look angle error requires a far higher positioning accuracy level; its error must be constrained within 0.001° to ensure the MBSAR's image quality.

5.4 NON-STOP-AND-GO SIGNAL MODEL IN THE MBSAR

The above analysis employs the stop-and-go assumption, which assumes that both the TOI and SAR remain stationary during signal transmission and reception, moving to new locations when the next pulse is transmitted [34–36]. In MBSAR, however, the round-trip time delay can be several seconds [37], thus, neither the TOI nor the MBSAR can be considered stationary. As a result, adopting the stop-and-go assumption in MBSAR could result in the stop-and-go error in signal. Therefore, before establishing the MBSAR's signal model, it is necessary to examine the stop-and-go assumption and its effects on the MBSAR's imaging performance.

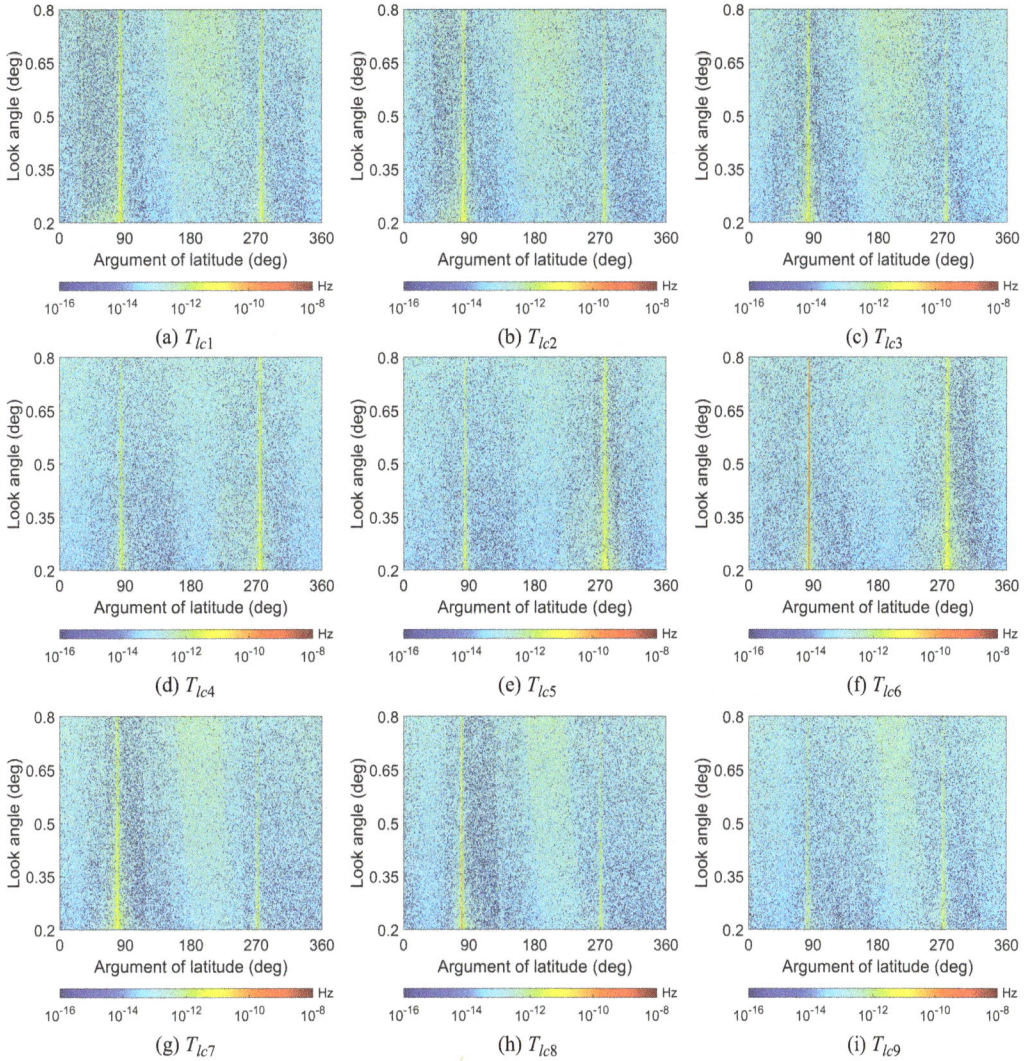

FIGURE 5.36 After processing by the proposed method, the right-looking Doppler frequency residual versus AOL and look angle across cycles from T_{lc1} to T_{lc9}.

5.4.1 IMPACT OF STOP-AND-GO ASSUMPTION ON THE MBSAR'S SIGNAL

The stop-and-go assumption might affect MBSAR signals in two ways [38]: Firstly, the antenna displacement occurs during pulse round-trip time, leading to the carrier frequency shift and FM rate variation in the fast time domain. The second effect is the variation in range history, resulting in temporally varying range error (or Doppler history error) in the slow time domain. For ease of reference, Figure 5.39 presents the transmission and reception of signal in the non-stop-and-go echo model.

According to Figure 5.39, the MBSAR signal with a pulse duration of T_p is transmitted to Earth at the time η_s, when the position vector of SAR (PVS) and the position vector of TOI (PVT) are denoted as $\mathbf{R}_{\mathbf{SAR}, \eta s}$ and $\mathbf{R}_{\mathbf{TOI}, \eta s}$, respectively. After a propagation time delay t_1, the transmitted signal arrives on Earth, at this moment, the PVS and PVT are respectively signified by $\mathbf{R}_{\mathbf{SAR}, \eta m}$ and $\mathbf{R}_{\mathbf{TOI}, \eta m}$ ($\eta_m = \eta_s + t_1$). Note that the radio signal is scattered once it interacts with the TOI, and the scattered signal is received by the MBSAR after a time delay of t_2 from the time η_m (namely $\eta_e = \eta_m + t_2$). Here, the PVS and PVT are represented by $\mathbf{R}_{\mathbf{SAR}, \eta e}$ and $\mathbf{R}_{\mathbf{TOI}, \eta e}$, respectively.

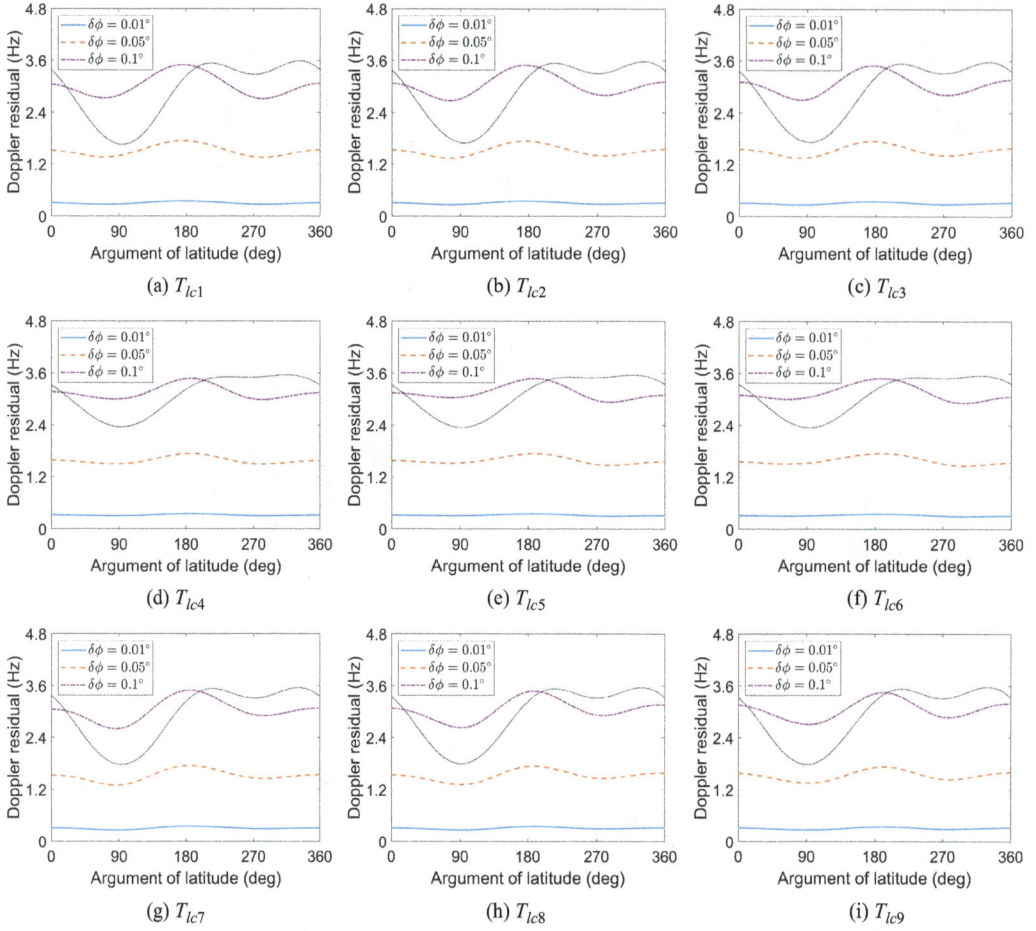

FIGURE 5.37 After zero-Doppler steering, the magnitudes of Doppler residuals with different azimuth angle errors versus the AOL across cycles from T_{lc1} to T_{lc9}.

The range history under the stop-and-go assumption can be expressed as follows:

$$R(\eta_s) = \left\| \mathbf{R}_{\mathrm{SAR},\eta_s} - \mathbf{R}_{\mathrm{TOI},\eta_s} \right\|_2 \qquad (5.52)$$

The actual range history in the non-stop-and-go echo model takes the following form:

$$R_S(\eta_s) = \frac{\left\| \mathbf{R}_{\mathrm{SAR},\eta_s} - \mathbf{R}_{\mathrm{TOI},\eta_m} \right\|_2 + \left\| \mathbf{R}_{\mathrm{SAR},\eta_e} - \mathbf{R}_{\mathrm{TOI},\eta_m} \right\|_2}{2} \qquad (5.53)$$

The time delay in Eq. (5.53) has no close-form solution. Thus, an iterative scheme for computing the round-trip delay is outlined in Table 5.2.

For more general cases, the aforementioned time η_s is replaced by the slow time (azimuth time) η. After obtaining the round-trip propagation time delays, it is possible to calculate the range history error:

$$\Delta R_S(\eta) = R_S(\eta) - R(\eta) \qquad (5.54)$$

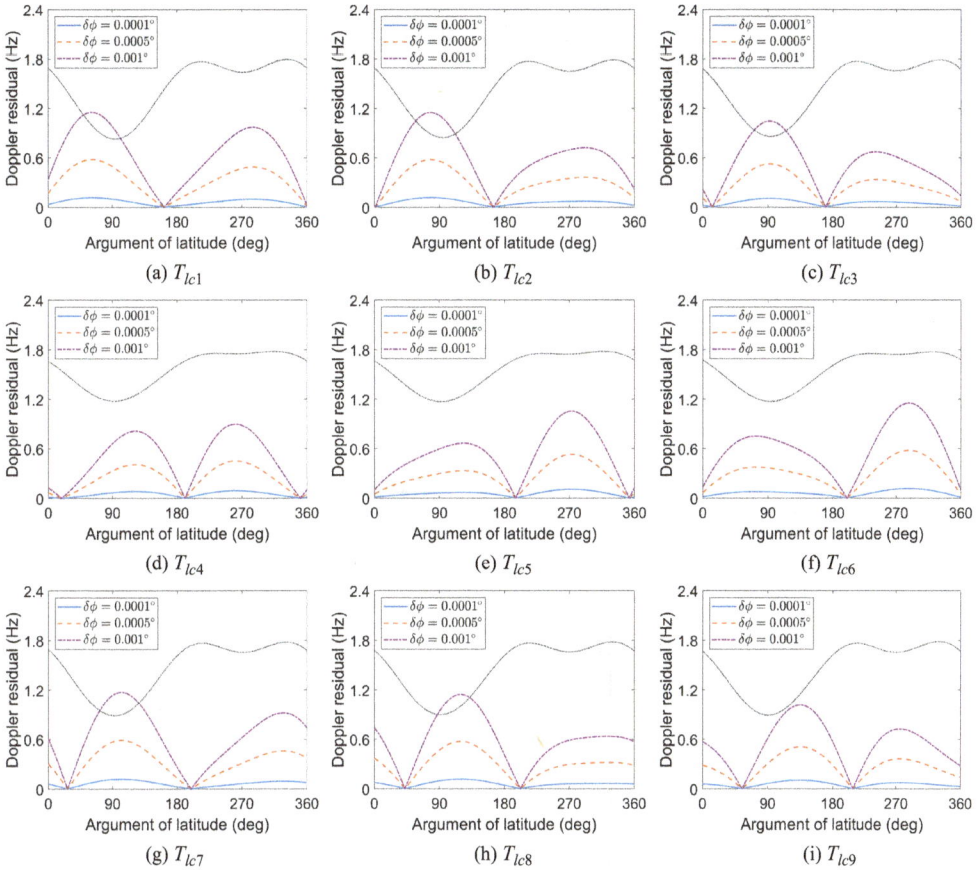

FIGURE 5.38 After zero-Doppler steering, the magnitudes of Doppler residuals with different look angle errors versus the AOL across cycles from T_{lc1} to T_{lc9}.

The range history error might bring about the Doppler frequency and DFMR errors, as:

$$\Delta f_d^{\text{nsg}} = f_d^{\text{nsg}} - f_d^{\text{sg}}; \quad f_d^{\text{nsg}} = -\frac{2}{\lambda} \frac{dR_S(\eta)}{d\eta}\bigg|_{\eta=0}, \quad f_d^{\text{sg}} = -\frac{2}{\lambda} \frac{dR(\eta)}{d\eta}\bigg|_{\eta=0} \quad (5.55)$$

$$\Delta f_{\text{dr}}^{\text{nsg}} = f_{\text{dr}}^{\text{nsg}} - f_{\text{dr}}^{\text{sg}}; \quad f_{\text{dr}}^{\text{nsg}} = \frac{2}{\lambda} \frac{d^2 R_S(\eta)}{d\eta^2}\bigg|_{\eta=0}, \quad f_{\text{dr}}^{\text{sg}} = \frac{2}{\lambda} \frac{d^2 R(\eta)}{d\eta^2}\bigg|_{\eta=0} \quad (5.56)$$

Besides both Doppler errors, the antenna displacement during the round-trip time could cause carrier frequency shift and FM rate variation; both can be estimated by:

$$\Delta f_c = -\frac{2 f_c \cdot \dot{R}_S(\eta)}{c} \quad (5.57)$$

and:

$$\Delta K_r = -2\frac{2 K_r \cdot \dot{R}_S(\eta) + f_c \cdot \ddot{R}_S(\eta)}{c} \quad (5.58)$$

where K_r is the FM rate of the chirp signal.

(a)

(b)

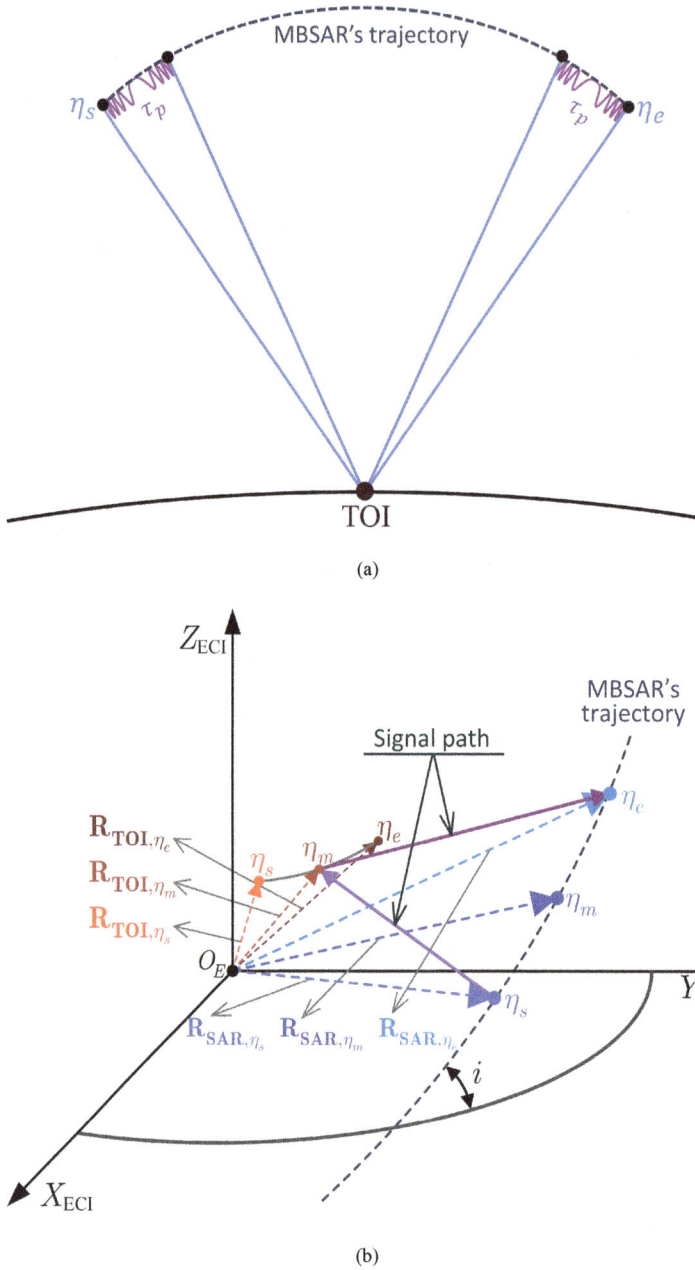

FIGURE 5.39 Schematic diagram of signal propagation in the non-stop-and-go echo model in the (a) TOI-MBSAR trajectory plane, (b) ECI coordinate.

The carrier frequency shift leads to an Extra Range Cell Migration (ERCM), which can be identified by:

$$\Delta R_{\text{ERCM}} = \frac{c}{2} \cdot \frac{\Delta f_c}{K_r} = -\frac{f_c \cdot \dot{R}_S(\eta)}{K_r} \tag{5.59}$$

where f_c is the carrier frequency.

TABLE 5.2

Iterative Process for Calculating the Propagation Time Delays

The iterative process for computing propagation time delay t_1

1) The initial propagation time delay of the transmitted signal:

$$t_{1\_0} = c^{-1} \left\| \mathbf{R}_{\text{SAR},\eta_s} - \mathbf{R}_{\text{TOI},\eta_s} \right\|_2$$

2) The propagation time delay after the ith ($i = 1, 2...$) iteration:

$$t_{1\_i} = c^{-1} \left\| \mathbf{R}_{\text{SAR},\eta_s} - \mathbf{R}_{\text{TOI},\eta_s + t_{1\_(i-1)}} \right\|_2$$

3) The residual time delay and corresponding range error after the ith iteration are:

$$\Delta t_{1\_i} = t_{1\_i} - t_{1\_(i-1)} \text{ and } \Delta r_{1\_i} = c \cdot \Delta t_{1\_i} = \left\| \mathbf{R}_{\text{SAR},\eta_s} - \mathbf{R}_{\text{TOI},\eta_s + t_{1\_(i-1)}} \right\|_2 - c \cdot t_{1\_(i-1)}$$

4) Determining the time delay t_1 by the preset threshold T_{r1}; once the residual range error is below T_{r1}:

$$\Delta r_{1\_i} < T_{r1}$$

    The solution of time delay t_1 can be obtained.

The iterative process for computing propagation time delay t_2

5) The initial time delay of the received signal can be set to attained time delay t_1, namely:

$$t_{2\_0} = t_1$$

6) After the jth ($j = 1, 2...$) iteration, the corresponding time delay for signal propagation becomes:

$$t_{2\_j} = c^{-1} \left\| \mathbf{R}_{\text{SAR},\eta_s + t_1 + t_{2\_(j-1)}} - \mathbf{R}_{\text{TOI},\eta_s + t_1} \right\|_2$$

7) The residual time delay error and corresponding range error after the jth iteration are:

$$\Delta t_{1\_j} = t_{1\_j} - t_{1\_(j-1)} \text{ and } \Delta r_{2\_j} = c \cdot \Delta t_{1\_j} = \left\| \mathbf{R}_{\text{SAR},\eta_s + t_1 + t_{2\_(j-1)}} - \mathbf{R}_{\text{TOI},\eta_s + t_1} \right\|_2 - c \cdot t_{2\_(j-1)}$$

8) Determining the time delay t_2 by the preset threshold T_{r2}; once the residual range error is below T_{r2}:

$$\Delta r_{2\_j} < T_{r2}$$

    The solution of time delay t_2 can be yielded.

    In both cases, the thresholds for residual range errors are set to 10^{-8} m.

For high-quality imaging, the following criterion must be met [9]:

$$\frac{2 \left| \Delta R_{\text{ERCM}} \right|_{\max}}{\rho_r} < 1 \tag{5.60}$$

Eq. (5.60) can be expressed in terms of Eq. (5.59), namely:

$$2 \frac{f_c \cdot \left| \dot{R}_s(\eta) \right|_{\max}}{\left| K_r \right| \rho_r} < 1 \tag{5.61}$$

Recall Eq. (3.1), the range resolution can be expressed in terms of FM rate and pulse duration, then, we obtain an upper bound for pulse duration:

$$T_p < \frac{\lambda}{4 \left| \dot{R}_s(\eta) \right|_{\max}} \tag{5.62}$$

As high-order terms contribute little to the change rate of range history, the following approximation can be applied:

$$\left| \dot{R}_s(\eta) \right|_{\max} = \frac{\lambda}{2} \left| f_d^{\text{nsg}} \right| + \frac{\lambda}{2} \left| f_{\text{dr}}^{\text{nsg}} \right| \cdot \frac{T_{\text{SAR}}}{2} + \cdots \approx \frac{\lambda}{4} \left[2 \left| f_d^{\text{nsg}} \right| + \left| f_{\text{dr}}^{\text{nsg}} \right| \cdot T_{\text{SAR}} \right] \tag{5.63}$$

By inserting Eq. (5.63) into Eq. (5.62), one arrives at the upper bound for pulse duration. This is expressed as:

$$T_p < T_{\text{pb1}}, \quad \text{with} \quad T_{\text{pb1}} = \frac{1}{2\left|f_d^{\text{nsg}}\right| + \left|f_{\text{dr}}^{\text{nsg}}\right| \cdot T_{\text{SAR}}} \tag{5.64}$$

The carrier frequency shift has no bearing on MBSAR imaging, unless the pulse duration exceeds the upper bound T_{pb1}.

We shall examine how the FM rate variation affects the MBSAR's signal. The maximum value allowed for this variation is specified as:

$$\left|\Delta K_r\right|_{\text{max}} = \frac{2\left|2K_r \cdot \dot{R}_S(\eta) + f_c \cdot \ddot{R}_S(\eta)\right|_{\text{max}}}{c} \tag{5.65}$$

The second term in the numerator of Eq. (5.65) is discarded due to its tiny magnitude compared to the first term. Then, (5.56) becomes:

$$\left|\Delta K_r\right|_{\text{max}} \approx \frac{4K_r \cdot \left|\dot{R}_S(\eta)\right|_{\text{max}}}{c} \tag{5.66}$$

Afterward, the maximum phase error resulting from the FM rate variation can be estimated by:

$$\Delta\psi_{\text{range}} = \pi \cdot \left|\Delta K_r\right|_{\text{max}} \cdot \left(0.5T_p\right)^2 \approx \pi c^{-1} B_r T_p \left|\dot{R}_S(\eta)\right|_{\text{max}} \tag{5.67}$$

To prevent range defocusing in the MBSAR, the phase error is required to comply with the relationship:

$$\Delta\psi_{\text{range}} < \pi/4 \tag{5.68}$$

Substituting Eqs. (5.63) and (5.67) into (5.68) yields another upper bound for pulse duration:

$$T_p < T_{\text{pb2}}, \quad \text{with} \quad T_{\text{pb2}} = \frac{f_c}{B_r\left[2\left|f_d^{\text{nsg}}\right| + \left|f_{\text{dr}}^{\text{nsg}}\right| \cdot T_{\text{SAR}}\right]} \tag{5.69}$$

Once the pulse duration surpasses the upper bound T_{pb2}, the FM rate variation could affect the MBSAR imaging.

Now, we can evaluate the impacts of the stop-and-go assumption. We employ an SAT of 200 s, a signal bandwidth of 20 MHz, and zero-Doppler steering. Figure 5.40 presents the Doppler error and bounds of pulse duration as a function of the look angle at different MBSAR positions in the cycle T_{lc1}. For comparison, Figure 5.41 shows the Doppler error and bounds of pulse duration as a function of AOL under various look angles.

According to Figures 5.40 and 5.41, the magnitudes of Doppler frequency error remain below 0.4 Hz. Notwithstanding, small changes in the Doppler frequency can result in azimuth offsets and thus dislocated SAR images. The DFMR errors are persistently below 10^{-5} Hz/s, regardless of the MBSAR's location or look angle. In this regard, neither the azimuth resolution nor the azimuth focusing is affected by the stop-and-go assumption. Regarding the effect of antenna displacement, both bounds of pulse duration, i.e., T_{pb1} and T_{pb2}, exceed 20 ms at all locations regardless of look angle. On the other hand, SAR systems rarely employ a pulse duration exceeding 20 ms. Correspondingly, the range imaging remains mostly unaffected by carrier frequency shift or FM rate variation. Given this, it is appropriate to concentrate solely on the Doppler frequency error induced by the stop-and-go assumption when modeling and processing MBSAR's signals.

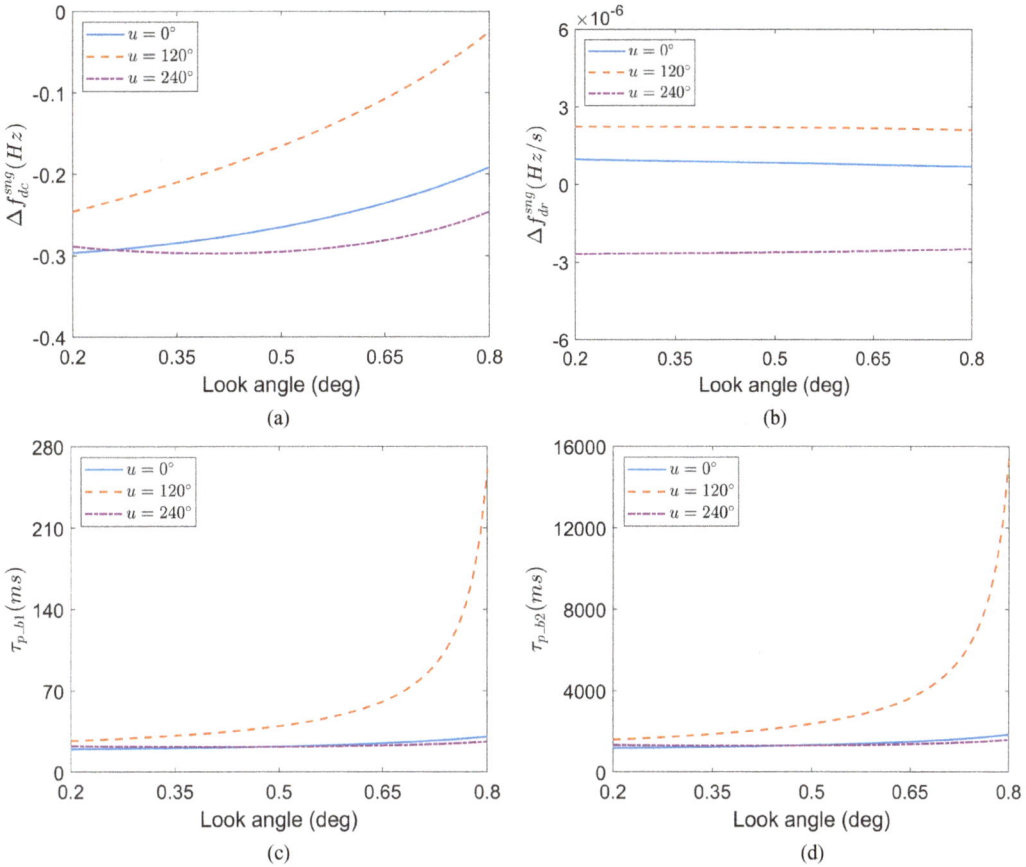

FIGURE 5.40 Due to impacts of stop-and-go assumption, (a) the Doppler frequency error versus look angle; (b) the DFMR error versus look angle; (c) the bound T_{pb1} versus look angle; (d) the bound T_{pb2} versus look angle.

5.4.2 THE NON-STOP-AND-GO SIGNAL MODEL IN MBSAR

A signal model could be developed after identifying the impact of the stop-and-go assumption in the MBSAR. Despite the availability of multiple waveforms, such as Continuous Wave (CW), Frequency-Modulated Continuous Wave (FMCW), and phase-coded waveforms, in contemporary SAR systems [39], our focus is narrowed to the linear frequency modulation (LFM) signal (also known as chirp signal) that is commonly employed in spaceborne SARs.

The LFM signal $s_t(\tau)$ with a pulse duration T_p can be represented by [32]:

$$s_t(\tau) = w_r(\tau) \exp\left[j\left(2\pi f_c \tau + \pi K_r \tau^2 \right) \right] \tag{5.70}$$

with the pulse envelope expressed as:

$$w_r(\tau) = \Pi\left(\frac{\tau}{T_p} \right) = \begin{cases} 1, & \tau \leq 0.5 T_p \\ 0, & \tau > 0.5 T_p \end{cases} \tag{5.71}$$

where τ is the fast time (range time).

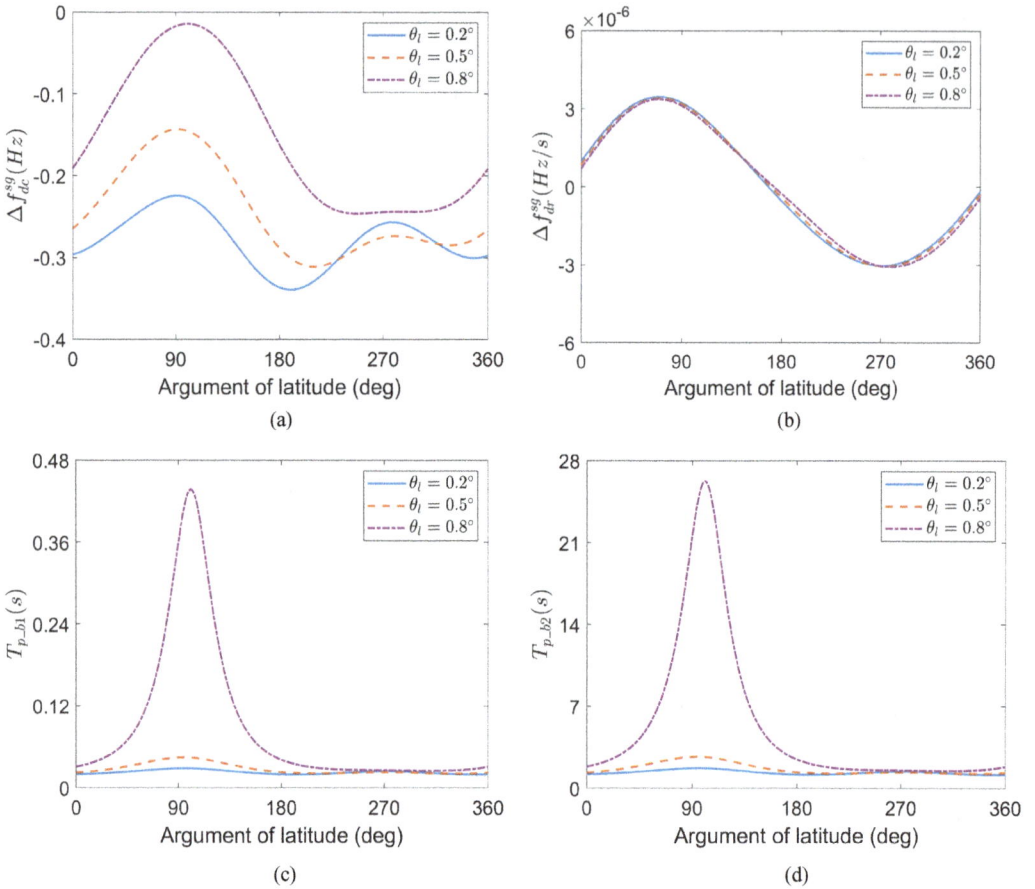

FIGURE 5.41 Due to the impacts of the stop-and-go assumption, (a) the Doppler frequency error versus AOL; (b) the DFMR error versus AOL; (c) the bound T_{pb1} versus AOL; (d) the bound T_{pb2} versus AOL.

Although there are carrier frequency shift and FM rate variation in the MBSAR signal, neither impacts the signal properties and imaging performance. Hence, both the carrier frequency shift and FM rate variation may be excluded from the signal modeling. By contrast, the range errors potentially affect the MBSAR imaging. Thus, the scatter signal received by the MBSAR can be expressed as:

$$
\begin{aligned}
s_r(\tau,\eta) &= A_0 \otimes s_t(\tau) \\
&= A_0 \cdot w_a(\eta - \eta_c) \cdot w_r\left[\tau - \frac{2R_S(\eta)}{c}\right] \\
&\quad \cdot \exp\left\{j\pi K_r\left[\tau - \frac{2R_S(\eta)}{c}\right]^2\right\} \cdot \exp\left[j2\pi f_c\left(\tau - \frac{2R_S(\eta)}{c}\right)\right]
\end{aligned}
\tag{5.72}
$$

where:
  A_0 is the amplitude related to scattering, it is generally regarded as a constant,
  η_c is the beam center crossing time, and
  $w_a(\cdot)$ is the azimuth envelope, which relates to the antenna gain pattern in the slow time domain.

A demodulation will filter out the carrier frequency, which lacks information pertinent to the TOI [40, 41]. The resultant demodulated two-dimensional signal is given by:

$$s_r(\tau,\eta) = A_0 \cdot w_a(\eta - \eta_c) \cdot w_r\left[\tau - \frac{2R_S(\eta)}{c}\right]$$

$$\cdot \exp\left[-j\frac{4\pi f_c R_S(\eta)}{c}\right] \exp\left\{j\pi K_r\left[\tau - \frac{2R_S(\eta)}{c}\right]^2\right\} \tag{5.73}$$

Figure 5.42 displays the profile of MBSAR's signals (real part) from a point target in the slow time domain, depicting both the *stop-and-go* and non-stop-and-go echoes. For numerical purposes,

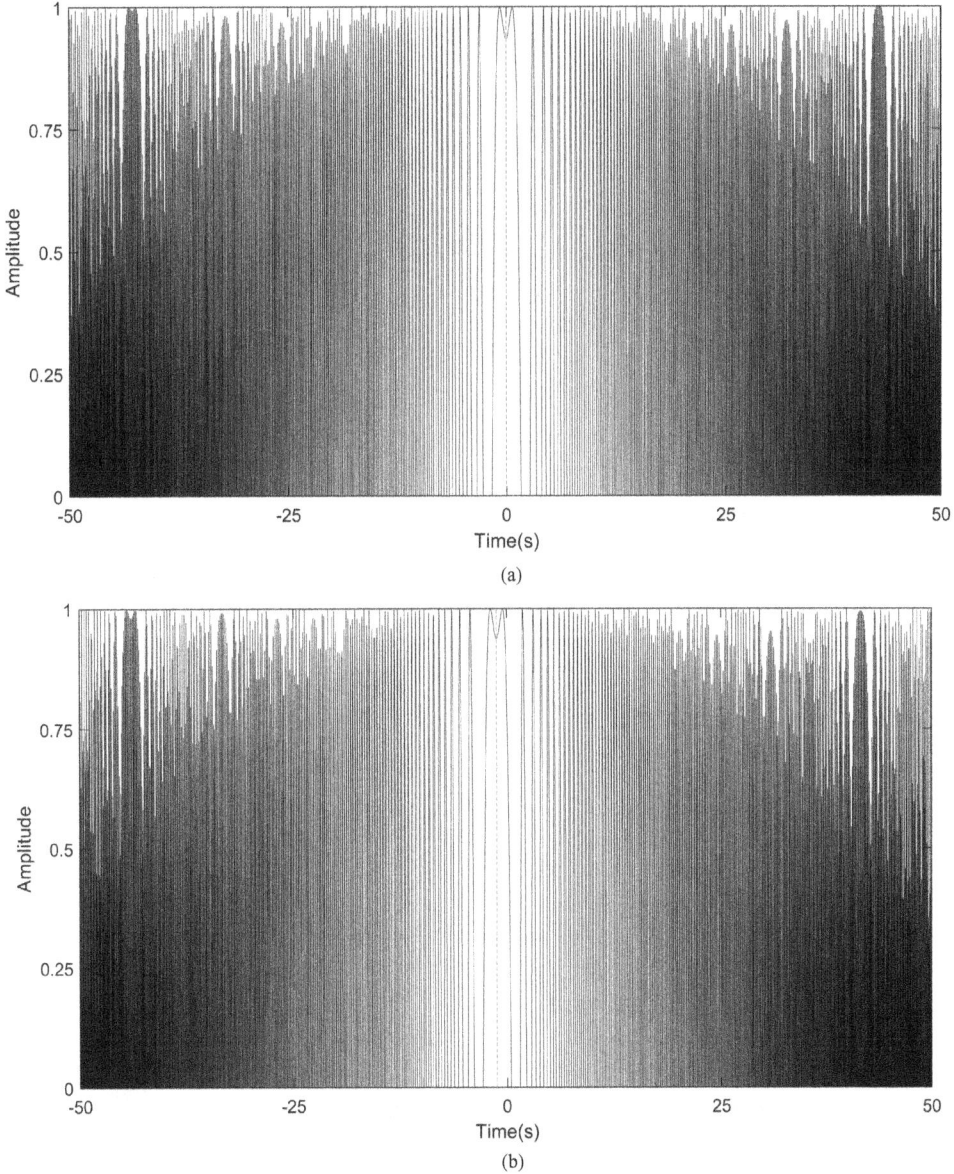

(a)

(b)

FIGURE 5.42 The demodulated signal from a point target in the slow time domain, where (a) is the signal of MBSAR under the stop-and-go assumption, (b) is the non-stop-and-go echo signal of MBSAR.

the parameters are set as: $f_{PRF} = 29$ Hz, $T_{SAR} = 200$ s, $\theta_l = 0.5°$, with the MBSAR located at the ascending node. It is apparent that the stop-and-go error results in a shift of the Doppler centroid, a phenomenon similar to the squint effect found in conventional SAR systems, albeit to a lesser degree.

At this point, we have developed a non-stop-and-go echo model for the MBSAR. The signal processing and image formation for the MBSAR under this model will be detailed in the next chapter.

APPENDIX: DERIVATIONS OF THE DFMR AND DOPPLER FREQUENCY

To simplify the explanation without losing its general applicability, we shall focus on the derivation processes of DFMR and Doppler frequency for left-looking SAR. Regarding the case that the MBSAR looks from the right-hand side, we can readily obtain the required Doppler parameters by replacing θ_l with $-\theta_l$. Besides, given that the DFMR consists of four distinct components, with the fourth component $\mathfrak{R}_4$ demonstrating a strong correlation to the Doppler frequency, we shall first focus on the representation of the DFMR.

Each component within the SAR's DFMR can be mathematically defined as follows:

$$\mathfrak{R}_1 = R_{ST}^{-1} \cdot \left(\mathbf{V}_{SAR} \cdot \mathbf{V}_{SAR} \right) + \left(R_{ST}^{-1} \cdot \mathbf{R}_{ST} \right) \cdot \mathbf{A}_{SAR} \tag{A.1}$$

$$\mathfrak{R}_2 = -2R_{ST}^{-1} \cdot \left(\mathbf{V}_{SAR} \cdot \mathbf{V}_{TOI} \right) \tag{A.2}$$

$$\mathfrak{R}_3 = R_{ST}^{-1} \cdot \mathbf{V}_{TOI} \cdot \mathbf{V}_{TOI} - R_{ST}^{-1} \cdot \mathbf{A}_{TOI} \cdot \mathbf{R}_{ST} \tag{A.3}$$

$$\mathfrak{R}_4 = -R_{ST}^{-3} \cdot \left[\left(\mathbf{V}_{SAR} - \mathbf{V}_{TOI} \right) \cdot \mathbf{R}_{ST} \right]^2 \tag{A.4}$$

Upon a close examination of the first component in Eq. (A.1) for a Keplerian orbit, one can represent the velocity vector of the SAR system in the ACS in the following manner:

$$\mathbf{V}_{SAR(ACS)} = \mathbf{p}_{ACS} \cdot V_{kep} \tag{A.5}$$

with:

$$\mathbf{p}_{ACS} = \left[\left(e \cos v + 1 \right), 0, -e \sin v \right]^T \tag{A.6}$$

$$V_{kep} = \sqrt{\frac{\mu}{a_{kep} \left(1 - e^2 \right)}} \tag{A.7}$$

Upon considering the squared norm of Eq. (A.5), one can derive the ensuing representation:

$$\frac{\mathbf{V}_{SAR} \cdot \mathbf{V}_{SAR}}{R_{ST}} = \frac{V_{kep}^2 \left(1 + e^2 + 2e \cos v \right)}{R_{ST}} \tag{A.8}$$

In the context of a Keplerian orbit, the SAR acceleration vector is attained through the following relation:

$$\mathbf{A}_{SAR} = -\mu \frac{\mathbf{R}_{SAR}}{R_{SAR}^3} \tag{A.9}$$

The scalar product of vectors $\mathbf{A}_{SAR}$ and $R_{ST}^{-1} \cdot \mathbf{R}_{ST}$ is subsequently formulated as:

$$\mathbf{A}_{SAR} \cdot \frac{\mathbf{R}_{ST}}{R_{ST}} = -\mu \frac{\mathbf{R}_{SAR}}{R_{SAR}^3} \cdot \frac{\mathbf{R}_{ST}}{R_{ST}} = -\frac{\mu}{R_{SAR}^2} \cdot \frac{\mathbf{R}_{SAR}\mathbf{R}_{ST}}{R_{SAR}R_{ST}} = -\frac{\mu}{R_{SAR}^2} \cos\theta_l \tag{A.10}$$

By incorporating Eqs. (A.8) and (A.10), it is possible to re-express Eq. (A.1) in the following manner:

$$\mathfrak{R}_1^{kep} = -\frac{\mu}{R_{SAR}^2} \cos\theta_l + \frac{V_{kep}^2 \cdot \left(1 + e^2 + 2e\cos v\right)}{R_{ST}} \tag{A.11}$$

When contemplating the perturbed orbit, the determination of MBSAR's velocity vector in the ECI can be achieved through DE440. Thus, one can derive the following expression:

$$\frac{\mathbf{V}_{SAR} \cdot \mathbf{V}_{SAR}}{R_{ST}} = \frac{V_{SAR}^2}{R_{ST}} \tag{A.12}$$

Eq. (A.1) includes a coefficient that is dependent on the vector $\mathbf{R}_{ST}$; In the ACS, this vector takes the form of:

$$\mathbf{R}_{ST(ACS)} = -\left[R_{ST} \sin\theta_l \cos\phi, -R_{ST} \sin\theta_l \sin\phi, R_{ST} \cos\theta_l \right]^T \tag{A.13}$$

Therefore, it is possible to generate:

$$R_{ST}^{-1} \cdot \mathbf{R}_{ST} = \left[-\sin\theta_l \cos\phi, \sin\theta_l \sin\phi, -\cos\theta_l \right]^T \tag{A.14}$$

Additionally, when the perturbed orbit is considered, expressing the MBSAR's acceleration vector in the ACS results in:

$$\mathbf{A}_{SAR(ACS)} = \mathbf{M}_2^{-1}\mathbf{M}_1^{-1}\mathbf{A}_{SAR(ECI)} \tag{A.15}$$

with:

$$\mathbf{M}_1 = \mathbf{R}_z\left(-\Omega\right)\mathbf{R}_x\left(-i\right)\mathbf{R}_z\left(-u\right) \tag{A.16}$$

$$\mathbf{M}_2 = \begin{bmatrix} 0 & 0 & -1 \\ 1 & 0 & 0 \\ 0 & -1 & 0 \end{bmatrix} \tag{A.17}$$

where $\mathbf{R}_x$ and $\mathbf{R}_z$ represent the right-handed rotation matrices about the x- and z-axes, respectively. The expressions for both matrices have already been provided in Eq. (2.16) in Chapter 2. Besides, $\mathbf{A}_{SAR(ECI)} = [A_{SX}, A_{SY}, A_{SZ}]^T$ represents the SAR acceleration in the ECI coordinate system; it is conceivable to calculate this acceleration vector using numerical methods by utilizing DE440. Subsequently, one can obtain the SAR acceleration in the ACS by means of the coordinate transformation as follows:

$$\begin{cases} A_{SX,ACS} = -A_{SX} \cdot (\sin u \cos\Omega + \cos i \cos u \sin\Omega) \\ \qquad\quad - A_{SY} \cdot (\sin u \sin\Omega - \cos i \cos u \cos\Omega) + A_{SZ} \cdot \sin i \cos u \\ A_{SY,ACS} = -A_{SX} \cdot \sin i \sin\Omega + A_{SY} \cdot \sin i \cos\Omega - A_{SZ} \cdot \cos i \\ A_{SZ,ACS} = -A_{SX} \cdot (\cos u \cos\Omega - \cos i \sin u \sin\Omega) \\ \qquad\quad - A_{SY} \cdot (\cos u \sin\Omega + \cos i \sin u \cos\Omega) - A_{SZ} \cdot \sin i \sin u \end{cases} \tag{A.18}$$

Afterward, one could express the 2nd term in Eq. (A.1) by utilizing Eqs. (A.14) and (A.18), thereby obtaining:

$$\frac{\mathbf{R}_{ST} \cdot \mathbf{A}_{SAR}}{R_{ST}} = -A_{SX,ACS} \sin \theta_l \cos \phi + A_{SY,ACS} \sin \theta_l \sin \phi - A_{SZ,ACS} \cos \theta_l \tag{A.19}$$

Through utilizing Eqs. (A.12) and (A.19), one can deduce an expression for the component $\mathfrak{R}_1$ under the influence of orbital perturbations, as follows:

$$\begin{aligned}\mathfrak{R}_1^{per} = R_{ST}^{-1} \cdot V_{SAR}^2 \\ - A_{SX,ACS} \sin \theta_l \cos \phi + A_{SY,ACS} \sin \theta_l \sin \phi - A_{SZ,ACS} \cos \theta_l\end{aligned} \tag{A.20}$$

Our forthcoming aim is to derive the expression for the coefficient $\mathfrak{R}_2$. To accomplish this, we shall introduce the linear velocity vector of TOI, which is attributable to the rotational motion of Earth, in the following manner.

$$\mathbf{V}_{TOI} = \mathbf{w}_E \times \mathbf{R}_{TOI} = \omega_E \cdot \mathbf{u}_N \times \mathbf{R}_{TOI} \tag{A.21}$$

where $\mathbf{u}_N = [0, 0, 1]^T$, ω_E is the Earth's rotational angular velocity.

According to Eq. (A.21), it is possible to obtain:

$$\mathbf{V}_{SAR} \cdot \mathbf{V}_{TOI} = \omega_E \cdot \left[\mathbf{V}_{SAR} \cdot \left(\mathbf{u}_N \times \mathbf{R}_{TOI} \right) \right] \tag{A.22}$$

Within the ACS framework, the vector $\mathbf{u}_N$ is formulated as follows:

$$\mathbf{u}_{N(ACS)} = \mathbf{M}_2^{-1} \mathbf{M}_1^{-1} \cdot \mathbf{u}_N \tag{A.23}$$

One can represent express Eq. (A.23) in an equivalent form as follows:

$$\mathbf{u}_{N(ACS)} = \left[\sin i \cos u, \, -\cos i, \, -\sin i \sin u \right]^T \tag{A.24}$$

Regarding the TOI's position vector, it is formulated within the ACS framework as follows:

$$\mathbf{R}_{TOI(ACS)} = \left[R_{ST} \sin \theta_l \cos \phi, \, -R_{ST} \sin \theta_l \sin \phi, \, R_{ST} \cos \theta_l - R_{SAR} \right]^T \tag{A.25}$$

Afterward, the cross product of $\mathbf{u}_N$ and $\mathbf{R}_{TOI}$ adopts the following form:

$$\begin{aligned}\mathbf{u}_N \times \mathbf{R}_{TOI} = \mathbf{u}_{N(ACS)} \times \mathbf{R}_{TOI(ACS)} \\ = \begin{bmatrix} \cos i \left(R_{SAR} - R_{ST} \cos \theta_l \right) - R_{ST} \sin i \sin u \sin \theta_l \sin \phi \\ \sin i \cos u \left(R_{SAR} - R_{ST} \cos \theta_l \right) - R_{ST} \sin i \sin u \sin \theta_l \cos \phi \\ -\sin i \cos u \left(R_{ST} \sin \theta_l \sin \phi \right) + R_{ST} \sin \theta_l \cos \phi \cos i \end{bmatrix}\end{aligned} \tag{A.26}$$

It is worth noting that Eq. (A.26) retains its form irrespective of the shape of MBSAR's orbit. Notwithstanding, the orbital perturbations exert certain impacts on the SAR's inertial motion. In the case of a Keplerian orbit, Eq. (A.5) has already been presented as the expression for the SAR velocity vector. As for the SAR velocity that has been perturbed, it can be expressed as follows:

$$\mathbf{V}_{SAR(ACS)} = \mathbf{M}_2^{-1} \mathbf{M}_1^{-1} \mathbf{V}_{SAR(ECI)} \tag{A.27}$$

Employing the same principle as outlined in Eq. (A.15), we can rephrase Eq. (A.27) in the following manner:

$$\begin{cases} V_{SX,ACS} = -V_{SX} \cdot (\sin u \cos\Omega + \cos i \cos u \sin\Omega) \\ \qquad\qquad - V_{SY} \cdot (\sin u \sin\Omega - \cos i \cos u \cos\Omega) + V_{SZ} \cdot \sin i \cos u \\ V_{SY,ACS} = 0 \\ V_{SZ,ACS} = -V_{SX} \cdot (\cos u \cos\Omega - \cos i \sin u \sin\Omega) \\ \qquad\qquad - V_{SY} \cdot (\cos u \sin\Omega + \cos i \sin u \cos\Omega) - V_{SZ} \cdot \sin i \sin u \end{cases} \qquad (A.28)$$

where $\mathbf{V}_{SAR(ECI)} = [V_{SX}, V_{SY}, V_{SY}]^T$ stands for the velocity vector of the SAR system in the ECI coordinate system.

The mathematical derivation of the coefficient $\mathfrak{R}_2$ in the Keplerian orbit can be obtained by utilizing Eqs. (A.2), (A.5), (A.22), and (A.26). The resulting expression is as follows:

$$\begin{aligned} \mathfrak{R}_2^{kep} = & -2R_{ST}^{-1} \cdot \omega_E \cdot V_{kep} \cdot \{ (e\cos\nu + 1) \\ & \cdot [(R_{SAR} - R_{ST}\cos\theta_l)\cos i - R_{ST}\sin\theta_l \sin\phi\sin i \sin u] \\ & + e \cdot R_{ST}\sin\theta_l \sin\nu \cdot (-\cos\phi\cos i + \sin\phi\sin i \cos u) \} \end{aligned} \qquad (A.29)$$

In contrast, the perturbed coefficient $\mathfrak{R}_2$ can be derived through Eqs. (A.2), (A.22), (A.26), and (A.28), which now is articulated as:

$$\begin{aligned} \mathfrak{R}_2^{per} = & -2\omega_E R_{ST}^{-1} \cdot \{ V_{SX,ACS} \\ & \cdot [(R_{SAR} - R_{ST}\cos\theta_l)\cos i - R_{ST}\sin\theta_l \sin\phi\sin i \sin u] \\ & - V_{SZ,ACS} \cdot (R_{ST}\sin\theta_l \cos\phi\cos i - R_{ST}\sin\theta_l \sin\phi\sin i \cos u) \} \end{aligned} \qquad (A.30)$$

We shall now proceed to delve into the representation of the coefficient $\mathfrak{R}_3$. Referring to Eq. (A.21), we can effortlessly obtain the relationship provided below:

$$V_{TOI}^2 = \mathbf{V}_{TOI} \cdot \mathbf{V}_{TOI} = \omega_E^2 \cdot (\mathbf{u}_N \times \mathbf{R}_{TOI}) \cdot (\mathbf{u}_N \times \mathbf{R}_{TOI}) \qquad (A.31)$$

By implanting Eq. (A.26) into Eq. (A.31), one could concisely articulate:

$$\begin{aligned} V_{TOI}^2 = \omega_E^2 \cdot \big(& -2R_{ST}R_{SAR}\cos\theta_l \sin^2 i \cos^2 u \\ & -2R_{ST}R_{SAR}\sin\theta_l \cos\phi\sin^2 i \cos u \sin u \\ & -2R_{ST}R_{SAR}\sin\theta_l \sin\phi\cos i \sin i \sin u \\ & -2R_{ST}R_{SAR}\cos\theta_l \cos^2 i + R_{ST}^2 \sin^2\theta_l \cos^2 i \cos^2\phi \\ & +2R_{ST}^2 \cos\theta_l \sin\theta_l \sin\phi\cos i \sin i \sin u + R_{SAR}^2 \cos^2 i \\ & -2R_{ST}^2 \sin^2\theta_l \cos\phi\sin\phi\cos i \sin i \cos u + R_{ST}^2 \cos^2\theta_l \cos^2 i \\ & +2R_{ST}^2 \cos\theta_l \sin\theta_l \cos\phi\sin^2 i \cos u \sin u + R_{SAR}^2 \sin^2 i \cos^2 u \\ & + R_{ST}^2 \cos^2\theta_l \sin^2 i \cos^2 u + R_{ST}^2 \sin^2\theta_l \cos^2\phi\sin^2 i \sin^2 u \\ & + R_{ST}^2 \sin^2\theta_l \sin^2\phi\sin^2 i \cos^2 u + R_{ST}^2 \sin^2\theta_l \sin^2\phi\sin^2 i \sin^2 u \big) \end{aligned} \qquad (A.31)$$

Uniting and amalgamating analogous terms can result in the streamlining of Eq. (A.32) to:

$$\begin{aligned} V_{TOI}^2 = \omega_E^2 \cdot \Big\{ & (R_{SAR} - R_{ST}\cos\theta_l)^2 \cdot (1 - \sin^2 i \sin^2 u) \\ & + R_{ST}^2 \sin^2\theta_l \cdot (1 - \sin^2\phi\cos^2 i - \cos^2\phi\sin^2 i \cos^2 u) \\ & - 2R_{ST}\sin\theta_l \sin i \cdot [R_{ST}\sin\theta_l \cos\phi\sin\phi\cos i \cos u \\ & + \sin u \cdot (R_{SAR} - R_{ST}\cos\theta_l)(\cos\phi\cos u \sin i + \sin\phi\cos i)] \Big\} \end{aligned} \qquad (A.33)$$

To arrive at an explicit expression for the second term in $\mathfrak{R}_3$, we shall identify the TOI acceleration vector resulting from Earth's rotational motion; this quantity is represented by:

$$
\begin{aligned}
\mathbf{A}_{TOI} &= \mathbf{w}_E \times \mathbf{V}_{TOI} \\
&= \mathbf{w}_E \times \left(\mathbf{w}_E \times \mathbf{R}_{TOI} \right) \\
&= \left(\mathbf{w}_E \cdot \mathbf{R}_{TOI} \right) \cdot \mathbf{w}_E - \mathbf{R}_{TOI} \cdot \left(\mathbf{w}_E \cdot \mathbf{w}_E \right)
\end{aligned}
\tag{A.34}
$$

Taking into account the form of $\mathbf{A}_{TOI}$ and $\mathbf{R}_{ST}$, the corresponding inner product can be articulated by:

$$
\begin{aligned}
\mathbf{A}_{TOI} \cdot \mathbf{R}_{ST} &= \left[(\mathbf{w}_E \cdot \mathbf{R}_{TOI}) \cdot \mathbf{w}_E - \mathbf{R}_{TOI} \cdot (\mathbf{w}_E \cdot \mathbf{w}_E) \right] \cdot \mathbf{R}_{ST} \\
&= (\omega_E \mathbf{u}_N \cdot \mathbf{R}_{TOI}) \cdot \omega_E \mathbf{u}_N \cdot \mathbf{R}_{ST} - \omega_E^2 \cdot \mathbf{R}_{TOI} \cdot \mathbf{R}_{ST} \\
&= \omega_E^2 \cdot \left[(\mathbf{u}_N \cdot \mathbf{R}_{TOI}) \cdot (\mathbf{u}_N \cdot \mathbf{R}_{ST}) - \mathbf{R}_{TOI} \cdot \mathbf{R}_{ST} \right]
\end{aligned}
\tag{A.35}
$$

To facilitate ease of understanding, we hereby introduce the notation presented below:

$$
\mathbf{A}_{TOI} \cdot \mathbf{R}_{ST} = \omega_E^2 \cdot \left(R_{A1} - R_{A2} \right)
\tag{A.36}
$$

with:

$$
R_{A1} = \left(\mathbf{u}_N \cdot \mathbf{R}_{TOI} \right) \cdot \left(\mathbf{u}_N \cdot \mathbf{R}_{ST} \right)
\tag{A.37}
$$

$$
R_{A2} = \mathbf{R}_{TOI} \cdot \mathbf{R}_{ST}
\tag{A.38}
$$

As per Eqs. (A.13), (A.24), and (A.25), it is feasible to re-express the coefficient R_{A1} in the subsequent form:

$$
\begin{aligned}
R_{A1} = &\ 2R_{ST}^2 \cos i \cos \theta_l \sin i \sin \phi \sin \theta_l \sin u \\
&- 2R_{ST}^2 \cos i \cos \phi \cos u \sin i \sin \phi \sin^2 \theta_l \\
&+ 2R_{ST}^2 \cos \phi \cos \theta_l \cos u \sin^2 i \sin \theta_l \sin u \\
&- R_{ST} R_{SAR} \cos i \sin i \sin \phi \sin \theta_l \sin u \\
&- R_{ST} R_{SAR} \cos \phi \cos u \sin^2 i \sin \theta_l \sin u \\
&+ R_{ST} R_{SAR} \cos \theta_l \sin^2 i \sin^2 u - R_{ST}^2 \cos^2 \theta_l \sin^2 i \sin^2 u \\
&- R_{ST}^2 \cos^2 i \sin^2 \phi \sin^2 \theta_l - R_{ST}^2 \cos^2 \phi \cos^2 u \sin^2 i \sin^2 \theta_l
\end{aligned}
\tag{A.39}
$$

Upon consolidating similar terms, Eq. (A.39) can be distilled into a more simplified form as follows:

$$
\begin{aligned}
R_{A1} = &\ R_{ST} \cos \theta_l \sin^2 i \sin^2 u \cdot \left(R_{SAR} - R_{ST} \cos \theta_l \right) \\
&- R_{ST}^2 \sin^2 \theta_l \cdot \left(\cos^2 i \sin^2 \phi + \cos^2 \phi \cos^2 u \sin^2 i \right) \\
&+ 2R_{ST}^2 \sin \theta_l \cos \phi \sin i \cos u \cdot \left(\cos \theta_l \sin i \sin u - \sin \theta_l \sin \phi \cos i \right) \\
&- R_{ST} \sin \theta_l \sin i \sin u \cdot \left(R_{SAR} \cos i \sin \phi + R_{SAR} \cos \phi \cos u \sin i - 2R_{ST} \cos \theta_l \sin \phi \cos i \right)
\end{aligned}
\tag{A.40}
$$

As for Eq. (A.38), it can be re-expressed by:

$$
R_{A2} = \mathbf{R}_{TOI} \cdot \mathbf{R}_{SAR} - \mathbf{R}_{TOI} \cdot \mathbf{R}_{TOI}
\tag{A.41}
$$

Note that the former term in Eq. (A.41) can be articulated in terms of geocentric angle:

$$\mathbf{R_{TOI}R_{SAR}} = R_{TOI}R_{SAR}\left(\frac{\mathbf{R_{TOI} \cdot R_{SAR}}}{\|\mathbf{R_{TOI}}\|_2 \cdot \|\mathbf{R_{SAR}}\|_2}\right) = R_{TOI}R_{SAR}\cos\theta_e \tag{A.42}$$

Moreover, applying the law of cosines, one yields:

$$R_{ST}^2 = R_{SAR}^2 + R_{TOI}^2 - 2R_{SAR}R_{TOI}\cos\theta_e \tag{A.43}$$

$$R_{TOI}^2 = R_{SAR}^2 + R_{ST}^2 - 2R_{SAR}R_{ST}\cos\theta_l \tag{A.44}$$

One can render Eq. (A.43) in an equivalent fashion as follows:

$$\cos\theta_e = 0.5R_{SAR}^{-1}R_{TOI}^{-1} \cdot \left(R_{SAR}^2 + R_{TOI}^2 - R_{ST}^2\right) \tag{A.45}$$

In keeping with the contents of Eqs. (A.42) and (A.45), one can represent Eq. (A.41) as

$$R_{A2} = 0.5 \cdot \left(R_{SAR}^2 - R_{ST}^2 - R_{TOI}^2\right) \tag{A.46}$$

After comparing Eq. (A.46) with Eq. (A.44), the term R_{A2} is rewritten as follows:

$$R_{A2} = R_{SAR}R_{ST}\cos\theta_l - R_{ST}^2 \tag{A.47}$$

As per the information conveyed by Eqs. (A.3), (A.31), and (A.36), it can be surmised that:

$$\mathfrak{R}_3 = R_{ST}^{-1} \cdot \left(V_{TOI}^2 - \omega_E^2 \cdot R_{A1} + \omega_E^2 \cdot R_{A2}\right) \tag{A.48}$$

Now, it is feasible to explicitly articulate Eq. (A.48) by employing Eqs. (A.33), (A.35), (A.40), and (A.47), as demonstrated below:

$$\begin{aligned}
\mathfrak{R}_3 = {}& \omega_E^2 \cdot (R_{SAR}\cos\theta_l - R_{ST}) + \omega_E^2 R_{ST}^{-1} \cdot \\
& \left\{ (R_{SAR} - R_{ST}\cos\theta_l)^2 \cdot (1 - \sin^2 i\sin^2 u) \right. \\
& + R_{ST}^2\sin^2\theta_l \cdot (1 - \sin^2\phi\cos^2 i - \cos^2\phi\sin^2 i\cos^2 u) \\
& - 2R_{ST}\sin\theta_l\sin i \cdot \left[R_{ST}\sin\theta_l\cos\phi\sin\phi \cdot \cos i\cos u \right. \\
& \left. + \sin u \cdot (R_{SAR} - R_{ST}\cos\theta_l) \cdot (\cos\phi\cos u\sin i + \sin\phi\cos i) \right] \right\} \\
& - \omega_E^2 R_{ST}^{-1} \cdot \left[R_{ST}\cos\theta_l\sin^2 i\sin^2 u \cdot (R_{SAR} - R_{ST}\cos\theta_l) \right. \\
& - R_{ST}^2\sin^2\theta_l \cdot (\cos^2 i\sin^2\phi + \cos^2\phi\cos^2 u\sin^2 i) \\
& - R_{ST}\sin\theta_l\sin i\sin u \\
& \cdot (R_{SAR}\cos i\sin\phi + R_{SAR}\cos\phi\cos u\sin i - 2R_{ST}\cos\theta_l\sin\phi\cos i) \\
& \left. + 2R_{ST}^2\sin\theta_l\cos\phi\sin i\cos u \cdot (\cos\theta_l\sin i\sin u - \sin\theta_l\sin\phi\cos i) \right]
\end{aligned} \tag{A.49}$$

Eq. (A.49) exhibits a number of terms that can be consolidated to simplify the expression. As a result, we can eventually arrive at the ensuing formulation:

$$\begin{aligned}
\mathfrak{R}_3 = {}& R_{ST}^{-1}\omega_E^2 \cdot \left\{ R_{SAR}^2 - R_{SAR}R_{ST}\cos\theta_l \right. \\
& + R_{SAR}\sin^2 i\sin^2 u \cdot \left(R_{ST}\cos\theta_l - R_{SAR}\right) \\
& \left. - R_{SAR}R_{ST}\sin\theta_l\sin i\sin u \cdot \left(\cos\phi\sin i\cos u + \sin\phi\cos i\right) \right\}
\end{aligned} \tag{A.50}$$

The coefficient $\mathfrak{R}_3$ solely pertains to the rotational motion of Earth. As such, its formulation remains unaltered irrespective of perturbation effects.

It is now appropriate to derive the expression of the coefficient $\mathfrak{R}_4$. Such a coefficient can be deconstructed into three individual components that are attributable to their respective contributions, namely:

$$\mathfrak{R}_4 = \mathfrak{R}_{4S} + \mathfrak{R}_{4E} + \mathfrak{R}_{4SE} \tag{A.51}$$

$$\mathfrak{R}_{4E} = -\frac{\left(\mathbf{V}_{TOI} \cdot \mathbf{R}_{ST}\right)^2}{R_{ST}^3} \tag{A.52}$$

$$\mathfrak{R}_{4S} = -\frac{\left(\mathbf{V}_{SAR} \cdot \mathbf{R}_{ST}\right)^2}{R_{ST}^3} \tag{A.53}$$

$$\mathfrak{R}_{4SE} = 2 \cdot \frac{\left(\mathbf{V}_{SAR} \cdot \mathbf{R}_{ST}\right) \cdot \left(\mathbf{V}_{TOI} \cdot \mathbf{R}_{ST}\right)}{R_{ST}^3} \tag{A.54}$$

As seen, the coefficient $\mathfrak{R}_4$ associates with two distinct scalar products: $\mathbf{V}_{TOI}\mathbf{R}_{ST}$ and $\mathbf{V}_{SAR}\mathbf{R}_{ST}$, wherein the former is independent of orbital perturbations, while the latter is influenced by such perturbations.

By utilizing Eqs. (A.13) and (A.21), one can represent the scalar product $\mathbf{V}_{TOI}\mathbf{R}_{ST}$ as follows:

$$\mathbf{V}_{TOI}\mathbf{R}_{ST} = \omega_E R_{SAR} R_{ST} \cdot \sin\theta_l \cdot \left(\sin\phi\sin i\cos u - \cos\phi\cos i\right) \tag{A.55}$$

In the context of a Keplerian orbit, to derive the expression for the scalar product $\mathbf{V}_{SAR}\mathbf{R}_{ST}$, one can consult Eqs. (A.5) and (A.13) to express it in the following manner:

$$\mathbf{V}_{SAR}^{kep}\mathbf{R}_{ST} = -R_{ST}V_{kep} \cdot \left[\left(e\cos\nu + 1\right) \cdot \sin\theta_l\cos\phi - e \cdot \sin\nu\cos\theta_l\right] \tag{A.56}$$

To determine the aforementioned scalar product in the perturbed orbit scenario, one can utilize Eqs. (A.13) and (A.28), and express $\mathbf{V}_{SAR} \cdot \mathbf{R}_{ST}$ as follows:

$$\mathbf{V}_{SAR}^{per} \cdot \mathbf{R}_{ST} = -R_{ST} \cdot \left(V_{SX,ACS} \cdot \sin\theta_l\cos\phi + V_{SZ,ACS} \cdot \cos\theta_l\right) \tag{A.57}$$

We shall now turn our attention to the coefficient $\mathfrak{R}_{4E}$. As stated in Eq. (A.55), it possesses a constant representation despite its orbital shape and is expressed as follows:

$$\mathfrak{R}_{4E} = -\omega_E^2 R_{SAR}^2 R_{ST}^{-1} \sin^2\theta_l$$
$$\cdot \left(\sin^2\phi\sin^2 i\cos^2 u + \cos^2\phi\cos^2 i - 2\sin\phi\cos\phi\sin i\cos i\cos u\right) \tag{A.58}$$

Regarding the coefficients $\mathfrak{R}_{4S}$ and $\mathfrak{R}_{4SE}$, both are intricately linked with the SAR inertial motion. In the case of a Keplerian orbit, it is imperative that both coefficients consider Eq. (A.56) and can be mathematically expressed as follows:

$$\mathfrak{R}_{4S}^{kep} = -R_{ST}^{-1} \cdot V_{kep}^2 \cdot \left[\left(1 + e\cos\nu\right)^2 \cos^2\theta_l + e^2 \sin^2\nu\right.$$
$$\left. \cdot \sin^2\theta_l\cos^2\phi + 2e \cdot \sin\nu \cdot \left(e\cos\nu + 1\right) \cdot \sin\theta_l\cos\theta_l\cos\phi\right] \tag{A.59}$$

$$\Re_{4SE}^{kep} = -2\omega_E R_{SAR} R_{ST}^{-1} V_{kep} \sin\theta_l \cdot (\sin\phi \sin i \cos u$$
$$-\cos\phi\cos i)\left[\left(e\cos v+1\right)\sin\theta_l \cos\phi + e\sin v \cos\theta_l\right] \tag{A.60}$$

In the scenario where the orbit is subjected to orbital perturbations, Eq. (A.57) dictates that the coefficients $\Re_{4S}$ and $\Re_{4SE}$ are expressed in the following manner:

$$\Re_{4S}^{per} = -\frac{V_{SX,ACS}^2 \sin^2\theta_l \cos^2\phi + V_{SZ,ACS}^2 \cos^2\theta_l + 2V_{SX,ACS}V_{SZ,ACS}\sin\theta_l \cos\theta_l \cos\phi}{R_{ST}} \tag{A.61}$$

$$\Re_{4SE}^{per} = 2\omega_E R_{SAR} R_{ST}^{-1} \sin\theta_l$$
$$\cdot\left(V_{SX,ACS}\sin\theta_l \cos\phi + V_{SZ,ACS}\cos\theta_l\right)\cdot\left(\cos\phi\cos i - \sin\phi\sin i \cos u\right) \tag{A.62}$$

Hitherto, the DFMR's explicit expression has been thoroughly delved into, wherein both perturbed and Keplerian orbits are considered. Further, an additional aspect that warrants consideration in relation to the Doppler shift is the Doppler frequency. By definition, the Doppler frequency can be formulated as follows:

$$f_d = f_{d(SAR)} + f_{d(ER)} = -\frac{2}{\lambda}\frac{\mathbf{V_{SAR}}\cdot\mathbf{R_{ST}}}{R_{ST}} + \frac{2}{\lambda}\frac{\mathbf{V_{TOI}}\cdot\mathbf{R_{ST}}}{R_{ST}} \tag{A.63}$$

According to Eq. (A.55), the component of the Doppler frequency produced by the rotational motion of the Earth is:

$$f_{d(ER)} = \frac{2}{\lambda}\omega_E R_{SAR}\sin\theta_l \cdot\left(\sin\phi\sin i \cos u - \cos\phi\cos i\right) \tag{A.64}$$

Based on Eqs. (A.56) and (A.57), one can determine the Doppler frequency component caused by SAR's inertial motion for both Keplerian and perturbed orbits. Specifically, the corresponding equations are as follows:

$$f_{d(SAR)}^{kep} = \frac{2}{\lambda}V_{kep}\cdot\left[\left(e\cos v+1\right)\cdot\sin\theta_l \cos\phi - e\sin v\cos\theta_l\right] \tag{A.65}$$

$$f_{d(SAR)}^{per} = \frac{2}{\lambda}\left(V_{SX,ACS}\cdot\sin\theta_l \cos\phi + V_{SZ,ACS}\cdot\cos\theta_l\right) \tag{A.66}$$

In a Keplerian orbit, the SAR velocity can be stated as follows:

$$V_{SAR} = V_{kep}\sqrt{\left(1+2e\cos v+e^2\right)} \tag{A.67}$$

The Doppler frequency resulting from SAR's inertial motion in a Keplerian orbit is ultimately presented as:

$$f_{d(SAR)}^{kep} = \frac{2}{\lambda}\cdot V_{SAR}\cdot\left(\cos\gamma\cdot\sin\theta_l \cos\phi - \sin\gamma\cos\theta_l\right) \tag{A.68}$$

with:

$$\cos\gamma = \frac{e\cos v+1}{\sqrt{1+2e\cos v+e^2}}, \quad \sin\gamma = \frac{e\sin v}{\sqrt{1+2e\cos v+e^2}}, \quad \tan\gamma = \frac{e\sin v}{e\cos v+1} \tag{A.69}$$

REFERENCES

[1] K. S. Chen, *Principles of Synthetic Aperture Radar: A System Simulation Approach*. Boca Raton: CRC Press, 2015.

[2] A. Moreira, et al., "A Tutorial on Synthetic Aperture Radar," *IEEE Geoscience and Remote Sensing Magazine*, vol. 1, no. 1, pp. 6–43, Mar. 2013.

[3] I. G. Cumming and F. H. Wong, *Digital Signal Processing of Synthetic Aperture Radar Data: Algorithms and Implementation*. Boston: Artech House, 2005.

[4] Z. Xu and K. S. Chen, "On Signal Modeling of Moon-Based Synthetic Aperture Radar (SAR) Imaging of Earth," *Remote Sensing*, vol. 10, no. 3, p. 486, Mar. 2018.

[5] Z. Xu and K. S. Chen, "Effects of the Earth's Curvature and Lunar Revolution on the Imaging Performance of the Moon-Based Synthetic Aperture Radar," *IEEE Transactions on Geoscience and Remote Sensing*, vol. 57, no. 8, pp. 5868–5882, Mar. 2019.

[6] A. Moccia and A. Renga, "Synthetic Aperture Radar for Earth Observation from a Lunar Base: Performance and Potential Applications," *IEEE Transactions on Aerospace and Electronic Systems*, vol. 46, no. 3, pp. 1034–1051, Jul. 2010.

[7] Z. Xu, K. S. Chen, and G. Q. Zhou, "Zero-Doppler Centroid Steering for the Moon-Based Synthetic Aperture Radar: A Theoretical Analysis," *IEEE Geoscience and Remote Sensing Letters*, vol. 17, no. 7, Jul. 2020.

[8] S. V. Tsynkov, "On the Use of Start-Stop Approximation for Spaceborne SAR Imaging," *SIAM Journal on Imaging Sciences*, vol. 2, no. 2, pp. 646–669, 2009.

[9] J. C. Curlander and R. N. McDonough, *Synthetic Aperture Radar: Systems and Signal Processing*. New York: Wiley, 1991.

[10] C. V. Jakowatz, et al., *Spotlight-Mode Synthetic Aperture Radar: A Signal Processing Approach*. New York, NY: Springer, 2012.

[11] Z. Xu, K. S. Chen, and H. Guo, "On Azimuthal Resolution of the Lunar-Based SAR under the Orbital Perturbation Effects," *IEEE Transactions on Geoscience and Remote Sensing*, vol. 61, Apr. 2023, doi: 10.1109/TGRS.2023.3266548

[12] W. M. Folkner, et al., "The Planetary and Lunar Ephemerides DE430 and DE431," *Interplanetary Network Progress Report*, vol. 196, no. 1, pp. 1–81, Feb. 2014.

[13] R. G. Cionco and D. A. Pavlov, "Solar Barycentric Dynamics from a New Solar-Planetary Ephemeris," *Astronomy & Astrophysics*, vol. 615, p. A153, Jul. 2018.

[14] R. S. Park, W. M. Folkner, J. G. Williams, and D. H. Boggs, "The JPL Planetary and Lunar Ephemerides DE440 and DE441," *The Astronomical Journal*, vol. 161, no. 3, p. 105, Feb. 2021, doi: 10.3847/1538-3881/abd414

[15] O. Montenbruck, E. Gill, and F. Lutze, *Satellite Orbits: Models, Methods, and Applications*. Berlin: Springer, 2000.

[16] Z. Xu, K. S. Chen, G. Liu, and H. Guo, "Spatiotemporal Coverage of a Moon-Based Synthetic Aperture Radar: Theoretical Analyses and Numerical Simulations," *IEEE Transactions on Geoscience and Remote Sensing*, vol. 58, no. 12, pp. 8735–8750, 2020.

[17] P. Bello, "Joint Estimation of Delay, Doppler, and Doppler Rate," *IRE Transactions on Information Theory*, vol. 6, no. 3, pp. 330–341, Jun. 1960, doi: 10.1109/TIT.1960.1057562

[18] R. K. Raney, "A Comment on Doppler FM Rate," *International Journal of Remote Sensing*, vol. 8, no. 7, pp. 1091–1092, Jul. 1987, doi: 10.1080/01431168708954755

[19] A. Renga, "Configurations and Performance of Moon-Based SAR Systems for Very High Resolution Earth Remote Sensing," *Proc. AIAA Pegasus Aerospace Conference*, Naples, Italy, pp. 12–13, Apr. 2007.

[20] A. Renga and A. Moccia, "Preliminary Analysis of a Moon-based Interferometric SAR System for Very High Resolution Earth Remote Sensing," *Proc. 9th ILEWG International Conference on Exploration and Utilisation of the Moon*, Sorrento, Italy, pp. 22–26, Oct. 2007.

[21] G. Fornaro, et al., "Potentials and Limitations of Moon-Borne SAR Imaging," *IEEE Transactions on Geoscience and Remote Sensing*, vol. 48, no. 7, pp. 3009–3019, Apr. 2010.

[22] A. Renga and A. Moccia, "Moon-Based Synthetic Aperture Radar: Review and Challenges," *Proc. IEEE International Geoscience and Remote Sensing Symposium*, pp. 3708–3711, Jul. 2016.

[23] E. Boerner, H. Fiedler, G. Krieger, and J. Mittermayer, "A New Method for Total Zero Doppler Steering," *Proc. IEEE International Geoscience and Remote Sensing Symposium*, Anchorage, AK, USA, vol. 2, pp. 1526–1529, Sep. 2004.

[24] J. Dong, et al., "Spatio-Temporal Distribution of the Zero-Doppler Line of Lunar-Based SAR," *Remote Sensing Letters*, vol. 12, no. 2, pp. 113–121, Dec. 2021.
[25] M. A. Richards, *Fundamentals of Radar Signal Processing*. New York: McGraw-Hill Education, 2014.
[26] R. K. Raney, "Doppler Properties of Radars in Circular Orbits," *International Journal of Remote Sensing*, vol. 7, no. 9, pp. 1153–1162, Sep. 1986, doi: 10.1080/01431168608948916
[27] H. Fiedler, E. Boerner, J. Mittermayer, and G. Krieger, "Total Zero Doppler Steering-A New Method for Minimizing the Doppler Centroid," *IEEE Geoscience and Remote Sensing Letters*, vol. 2, no. 2, pp. 141–145, Apr. 2005.
[28] H. Fiedler, T. Fritz, and R. Kahle, "Verification of the Total Zero Doppler Steering," *Proc. International Conference on Radar*, Adelaide, SA, Australia, pp. 340–342, Sep. 2008, doi: 10.1109/RADAR.2008.4653943
[29] J. Mittermayer, G. Krieger, and S. Wollstadt, "Numerical Calculation of Doppler Steering Laws in Bi- and Multistatic SAR," *IEEE Transactions on Geoscience and Remote Sensing*, vol. 60, 2022, doi: 10.1109/TGRS.2021.3058554
[30] C. Roemer, "Introduction to a New Wide Area SAR Mode Using the F-SCAN Principle," *Proc. IEEE International Geoscience and Remote Sensing Symposium*, pp. 3844–3847, 23–28 July 2017 2017.
[31] C. Roemer, R. Gierlich, J. Marquez-Martinez, and M. Notter, "Frequency Scanning applied to Wide Area SAR Imaging," *Proc. 12th European Conference on Synthetic Aperture Radar*, Aachen, Germany, pp. 1–5, Jun. 2018.
[32] F. T. Ulaby and D. G. Long, *Microwave Radar and Radiometric Remote Sensing*. Ann Arbor: Univ. Michigan Press, 2014.
[33] G. Franceschetti and R. Lanari, *Synthetic Aperture Radar Processing*. Boca Raton: CRC press, 1999.
[34] A. Meta, P. Hoogeboom, and L. P. Ligthart, "Signal Processing for FMCW SAR," *IEEE Transactions on Geoscience and Remote Sensing*, vol. 45, no. 11, pp. 3519–3532, Nov. 2007, doi: 10.1109/TGRS.2007.906140
[35] P. Prats-Iraola, et al., "High Precision SAR Focusing of TerraSAR-X Experimental Staring Spotlight Data," *Proc. IEEE International Geoscience and Remote Sensing Symposium*, Munich, Germany, pp. 3576–3579, Jul. 2012, doi: 10.1109/IGARSS.2012.6350644
[36] R. Lorusso and G. Milillo, "Stop-and-Go Approximation Effects on COSMO-SkyMed Spotlight SAR Data," *Proc. IEEE International Geoscience and Remote Sensing Symposium*, Milan, Italy, pp. 1797–1800, Jul. 2015.
[37] Z. Xu, K. S. Chen, and H. Guo, "Doppler Estimation with "Non-Stop-and-Go" Assumption in Moon-Based SAR Imaging," *Proc. 2018 IEEE International Geoscience and Remote Sensing Symposium*, Valencia, Spain, pp. 7809–7812, Jul. 2018.
[38] D. Liang, et al., "Processing of Very High Resolution GF-3 SAR Spotlight Data with Non-Start–Stop Model and Correction of Curved Orbit," *IEEE Journal of Selected Topics in Applied Earth Observations and Remote Sensing*, vol. 13, pp. 2112–2122, May. 2020, doi: 10.1109/JSTARS.2020.2986862
[39] M. I. Skolnik, *Radar Handbook*, 3rd ed. New York: McGraw-Hill Education, 2008.
[40] U. Pillai, K. Y. Li, and B. Himed, *Space Based Radar: Theory and Applications*. New York: McGraw-Hill, 2008.
[41] L. J. Cantafio, *Space-Based Radar Handbook*. Norwood, MA: Artech House, 1989.

6 Signal Processing and Image Focusing

6.1 INTRODUCTION

When considering the image formation or focusing of target responses, there exist critical issues that are interwoven in the Moon-Based SAR (MBSAR) [1]. To be more specific, the intricate geometry, affected by the Earth's rotation and MBSAR's inertial motion, exerts a notable influence on the phase history [2, 3]. Particularly, the MBSAR's extremely high orbit results in an extraordinarily long synthetic aperture time (SAT) and a significant slant range [4]. The SAT leads to a curved trajectory rather than a linear flight path, while the slant range causes an extra-long round-trip time delay that can extend up to several seconds; the combination of both factors renders the stop-and-go assumption invalid [5]. Consequently, the MBSAR signal exhibits a stop-and-go error, which, as elucidated in Chapter 5, gives rise to Doppler errors (mainly the Doppler frequency error), thereby rendering it stiff to characterize the MBSAR's phase history using the range model currently employed in spaceborne SAR.

The MBSAR produces a distinct 2-dimensional (2-D) spectrum that differs from spaceborne SAR due to the non-stop-and-go range history under a curved trajectory, making conventional frequency domain-focusing algorithms no longer applicable [6, 7]. Notably, the time-domain focusing algorithms, namely the back projection (BP) algorithm, can be utilized for imaging formation without considering the range model [8]. However, the extensive coverage provided by MBSAR undoubtedly generates quite a large amount of echo data [9], making the BP algorithm inefficient and unsuitable for signal processing of MBSAR. Thus, it is preferred to employ frequency-domain algorithms in MBSAR signal processing to attain the desired imaging quality efficiently. This requires incorporating a proper range model and developing a suitable focusing algorithm tailored specifically for MBSAR.

In this chapter, we offer a systematic viewpoint on MBSAR imaging, accompanied by numerical simulations of target responses under various circumstances. The commencement of this chapter entails a discourse on the compensation method to rectify the stop-and-go error. It puts forth two methods to demonstrate the factual non-stop-and-go range history of MBSAR. Subsequently, this chapter delves into the formulation of MBSAR's range model, duly accounting for its curved trajectory that may arise. In this context, two range models are developed, one derived through Taylor series expansion and the other developed from the hyperbolic range equation (HRE). Built upon the derived range model, this chapter derives the 2-D signal spectrum, which forms the framework for constructing a frequency domain-focusing algorithm tailored explicitly for MBSAR. Finally, the target responses under various conditions are simulated to demonstrate the MBSAR imaging performance.

6.2 MODELING THE NON-STOP-AND-GO RANGE HISTORY

This section discusses the non-stop-and-go slant range history of the MBSAR, and the range model for characterizing the range history under various conditions. The following sections will present the compensation method for addressing the stop-and-go error in the signal of MBSAR.

DOI: 10.1201/9781003308430-6

6.2.1   COMPENSATION METHOD FOR THE STOP-AND-GO ERROR

In Chapter 5 we have discussed the MBSAR signal compromised by the Doppler frequency shift caused by the *stop-and-go* error, which must be compensated in line with image focusing. However, it is laborious and computationally expensive to estimate the phase history for the entire image scene of MBSAR by means of iterative approaches. While some methods have been proposed to compensate for the stop-and-go error in airborne and spaceborne SAR systems [10, 11], they might not be suitable for MBSAR due to their disparate imaging geometry. Hence, we present two methods to compensate for the stop-and-go error for estimating the range history of MBSAR, which will be presented below.

6.2.1.1   The Compensation Based on Signal Propagation

Figure 6.1 demonstrates how the iteration error changes as the number of iterations increases when calculating the round-trip time delay. The analysis involves three look angles, with varying MBSAR locations at the AOLs of $0°$, $120°$, and $180°$ in cycle T_{lc1}. Clearly, after one iteration, both signal transmission and reception delay experience a significant reduction in their iteration errors. The range error can be less than 10^{-6} m after two iterations. As a result, a single-step approximation might yield good accuracy in estimating the non-stop-and-go range history of the MBSAR.

The non-stop-and-go range history can be approximated to:

$$R_S(\eta) \approx R_{SA1}(\eta) \tag{6.1}$$

with:

$$\begin{cases} R_{SA1}(\eta) = 0.5\left\|\mathbf{R}_{SAR,\eta} - \mathbf{R}_{TOI,\eta+t_{1a}}\right\|_2 + 0.5\left\|\mathbf{R}_{SAR,\eta+t_{1a}+t_{2a}} - \mathbf{R}_{TOI,\eta+t_{1a}}\right\|_2 \\ t_{1a} = c^{-1}\cdot\left\|\mathbf{R}_{SAR,\eta} - \mathbf{R}_{TOI,\eta}\right\|_2, \qquad t_{2a} = c^{-1}\cdot\left\|\mathbf{R}_{SAR,\eta+2t_{1a}} - \mathbf{R}_{TOI,\eta+t_{1a}}\right\|_2 \end{cases} \tag{6.2}$$

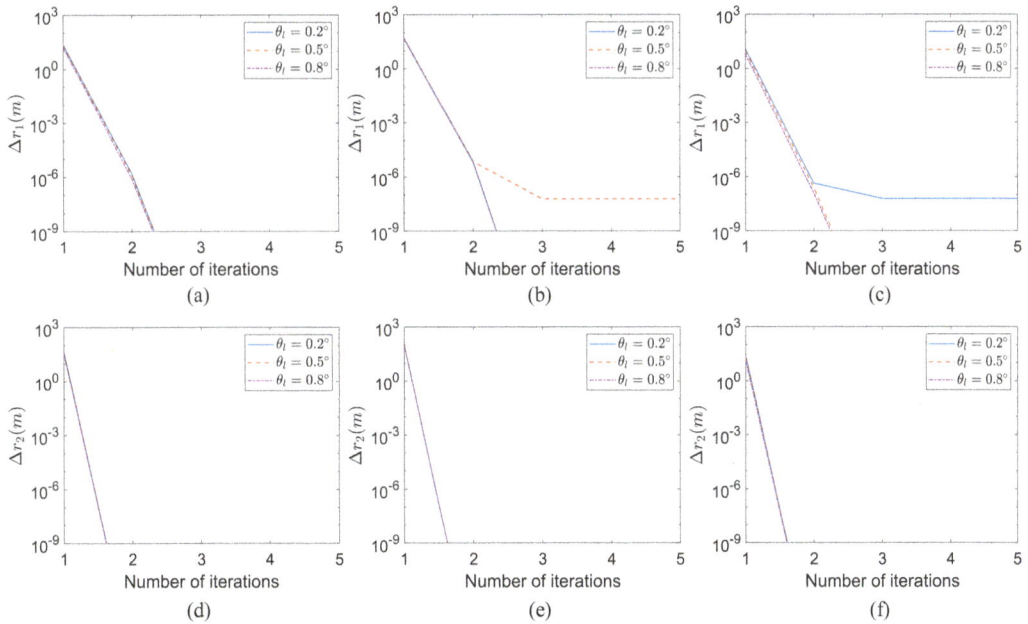

FIGURE 6.1   At various look angles, the magnitudes of residual range errors with respect to the iterative number, where the 1st (a, b, c) and 2nd (d, e, f) rows represent the residual range errors for numerically calculating the signal transmission and reception, respectively; the 1st (a, d), 2nd (b, e) and 3rd (c, f) columns stand for the residual range errors at the AOL of $0°$, $120°$, $180°$ in cycle T_{lc1}.

A residual phase error (RPE) for this approximation is given by:

$$\Delta\varphi_{\mathrm{RPE1}}\left(\eta\right) = -4\pi\lambda^{-1}\left[R_S\left(\eta\right) - R_{\mathrm{SA1}}\left(\eta\right)\right] \tag{6.3}$$

The above residual phase error is constrained to a Rayleigh criterion of $\pi/4$ [12, 13].

We shall examine $\Delta\varphi_{\mathrm{RPE1}}$ in the MBSAR at diverse look angles (0.2°, 0.5°, and 0.8°) in the zero-Doppler steering mode, detailed procedure can be found in Chapter 5. To ensure a complete validation of the proposed method, we shall adopt a synthetic aperture time (SAT) of 1800 s that can achieve an azimuthal resolution ranging from the submeter to the meter level. Figure 6.2 presents the RPE $\Delta\varphi_{\mathrm{RPE1}}$ versus azimuth time at various orbital positions. Figure 6.3 plots the maximum magnitude of $\Delta\varphi_{\mathrm{RPE1}}$ against AOL. In this and the following cases, the MBSAR's position specifically refers to the AOL at zero-azimuth time.

Figure 6.2 shows that $\Delta\varphi_{\mathrm{RPE1}}$ has a strong correlation with the azimuth time, while the MBSAR's position also exerts an impact on it. Even though, the magnitude of $\Delta\varphi_{\mathrm{RPE1}}$ and its temporal variation are negligibly small, regardless of the MBSAR's position. Figure 6.3 indicates the maximum magnitude of $\Delta\varphi_{\mathrm{RPE1}}$ remains below 2×10^{-3} rad within an SAT of 1800 s, and it is expected to be smaller for a shorter SAT. Such a small phase error has negligible impact on MBSAR imaging, indicating that the proposed method accurately describes the actual range history of MBSAR that employs zero-Doppler steering.

To provide insights into the performance of compensating for stop-and-go error when the MBSAR operates in the squint-looking mode, Figure 6.4 depicts the change in $\Delta\varphi_{\mathrm{RPE1}}$ with respect to azimuth time under various antenna azimuth angles (45°, 60°, 120°, and 135°) when the MBSAR locates at the ascending node. Figures 6.5 and 6.6 present the maximum magnitudes of the $\Delta\varphi_{\mathrm{RPE1}}$ (within an SAT of 1800 s) as a function of antenna azimuth angle at different locations and those as a function of AOL under various antenna azimuth angles, respectively.

Results in Figures 6.4–6.6 indicate that the magnitude of $\Delta\varphi_{\mathrm{RPE1}}$ is correlated to the look angle, SAT, and position of MBSAR, respectively. Notwithstanding, the magnitude of $\Delta\varphi_{\mathrm{RPE1}}$ is persistently smaller than 0.04 rad within an SAT of 1800 s, regardless of the beam pointing direction and

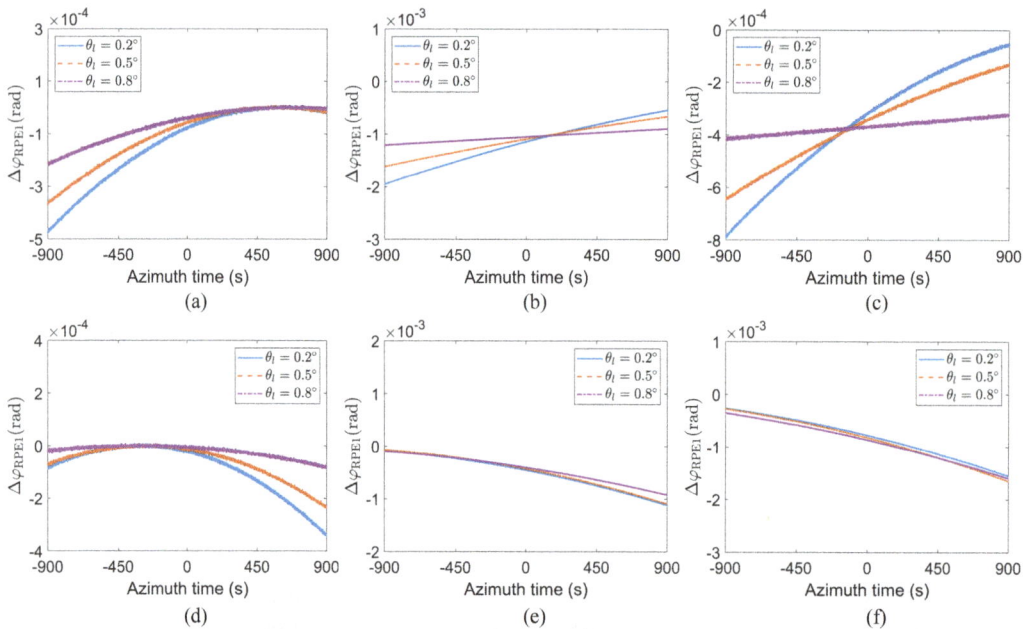

FIGURE 6.2 In the zero-Doppler steering mode, $\Delta\varphi_{\mathrm{RPE1}}$ in the signal of the MBSAR with various look angles against azimuth time at the AOL of (a) 0°, (b) 60°, (c) 120°, (d) 180°, (e) 240°, (f) 300°.

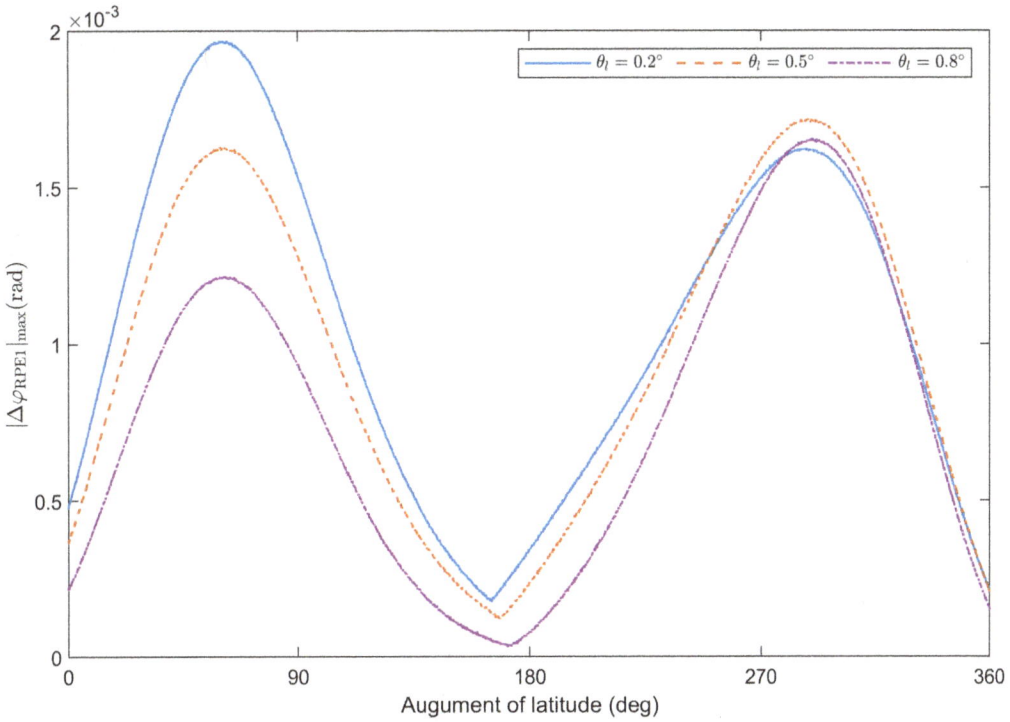

FIGURE 6.3 In the zero-Doppler steering mode, the maximum amplitudes of $\Delta\varphi_{RPE1}$ within an SAT of 1800 s versus the AOL at various look angles in the MBSAR.

position of the MBSAR. Generally, the real SAT adopted by the MBSAR is far shorter than 1800 s, resulting in a smaller magnitude of $\Delta\varphi_{RPE1}$. The phase error of such a magnitude exerts negligible impact on the focusing quality of the MBSAR even when it operates in a squint-looking mode. Therefore, Eq. (6.1) can effectively compensate for the stop-and-go error in the MBSAR.

6.2.1.2 The Compensation Based on Doppler Shift

As shown in Chapter 5, the stop-and-go error induces a Doppler frequency shift but exerts negligible impact on the higher-order Doppler parameters like DFMR. Hence, the stop-and-go error can be compensated for by clarifying the central slant range and Doppler frequency errors. To reduce computational complexity and enhance efficiency, we adopt an alternative approach to compensate for the stop-and-go error in the signal. A strategy for mitigating the stop-and-go error derived from [14] is presented below.

The non-stop-and-go range history can be approximated as:

$$R_{SA2}(\eta) \approx R(\eta) + \delta r_{nsg0} + \delta r_{nsg1}\eta \tag{6.4}$$

with:

$$\delta r_{nsg0} = \frac{R_c}{2}\frac{cV_r^{-1}-\sin\theta_r}{c^2V_r^{-2}-1}\left(\frac{cV_r^{-1}\cos^2\theta_r-\cos^2\theta_r\sin\theta_r}{c^2V_r^{-2}-1}-2\sin\theta_r\right) \tag{6.5}$$

$$\delta r_{nsg1} = \frac{2cR_c\cos^2\theta_r-2R_cV_r\cos^2\theta_r\sin\theta_r}{c^2V_r^{-2}-1} \tag{6.6}$$

where V_r and θ_r denote the equivalent velocity and squint angle, as given in Appendix B.

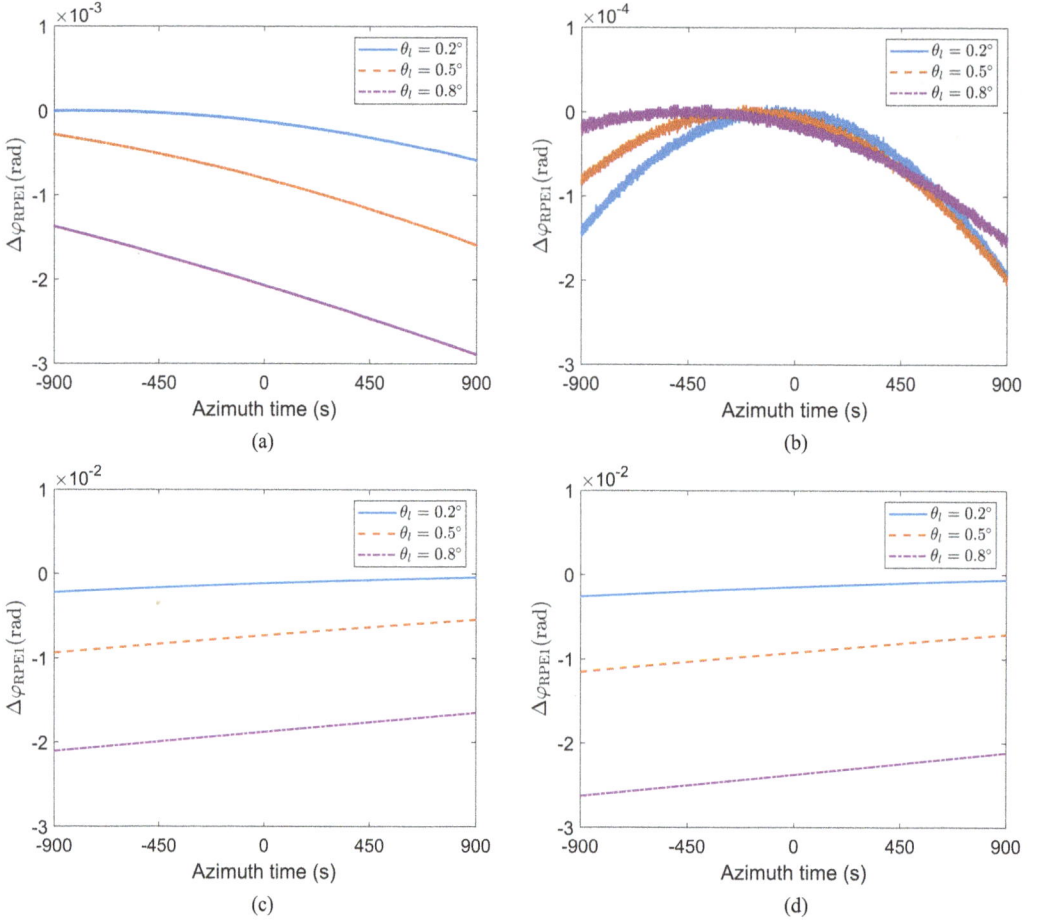

FIGURE 6.4 At the ascending node, the RPE $\Delta\varphi_{\mathrm{RPE1}}$ with various look angles against the azimuth time with antenna azimuth angles of (a) 45°, (b) 60°, (c) 120°, (d) 135°.

Since the SAR focusing is susceptible to the temporal variability of range error [15], the following RPE is defined to assess the approximate non-stop-and-go model in Eq. (6.4):

$$\Delta\varphi_{\mathrm{RPE2}}\left(\eta\right) = -\frac{4\pi}{\lambda}\left[R_S\left(\eta\right) - R_c - \delta r_{\mathrm{nsg1}} \cdot \eta\right]\right] \tag{6.7}$$

where R_c is the slant range of the beam center.

Figure 6.7 illustrates the spatial variability of $\Delta\varphi_{\mathrm{RPE2}}$ at various AOLs as a function of azimuth time. In this case, the MBSAR operates at zero-Doppler steering mode, and three look angles of 0.2°, 0.5°, and 0.8° are employed. As seen, $\Delta\varphi_{\mathrm{RPE2}}$ shows an upward trend with respect to azimuth time, and further, the pattern of ascending depends on the look angle and position of the MBSAR. A comparison between Figures 6.7 and 6.2 highlights that the $\Delta\varphi_{\mathrm{RPE2}}$ is close to $\pi/4$ bound when the SAT approaches or exceeds 600 s. Accordingly, the approximated non-stop-and-go model in Eq. (6.4) comes with a limited application range.

Figure 6.8 illustrates the maximum magnitude of $\Delta\varphi_{\mathrm{RPE2}}$ with respect to AOL under different SATs of the MBSAR in zero-Doppler centroid beam pointing. It is evident that the magnitude of $\Delta\varphi_{\mathrm{RPE2}}$ exhibits substantial variability with respect to the AOL, but it is affected, to a much lesser extent, by the look angle. Besides, this RPE may exceed $\pi/4$ in the case of an SAT of 600 s, implying

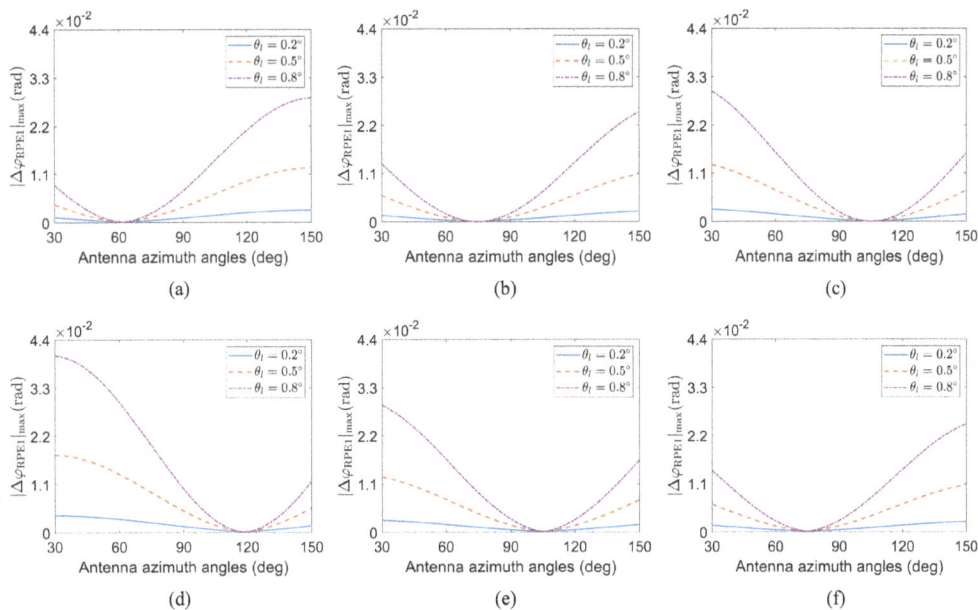

FIGURE 6.5 Within an SAT of 1800 s, the maximal magnitude of $\Delta\varphi_{\mathrm{RPE1}}$ with various look angles versus antenna azimuth angles at the AOL of (a) 0°, (b) 60°, (c) 120°, (d) 180°, (e) 240°, (f) 300°.

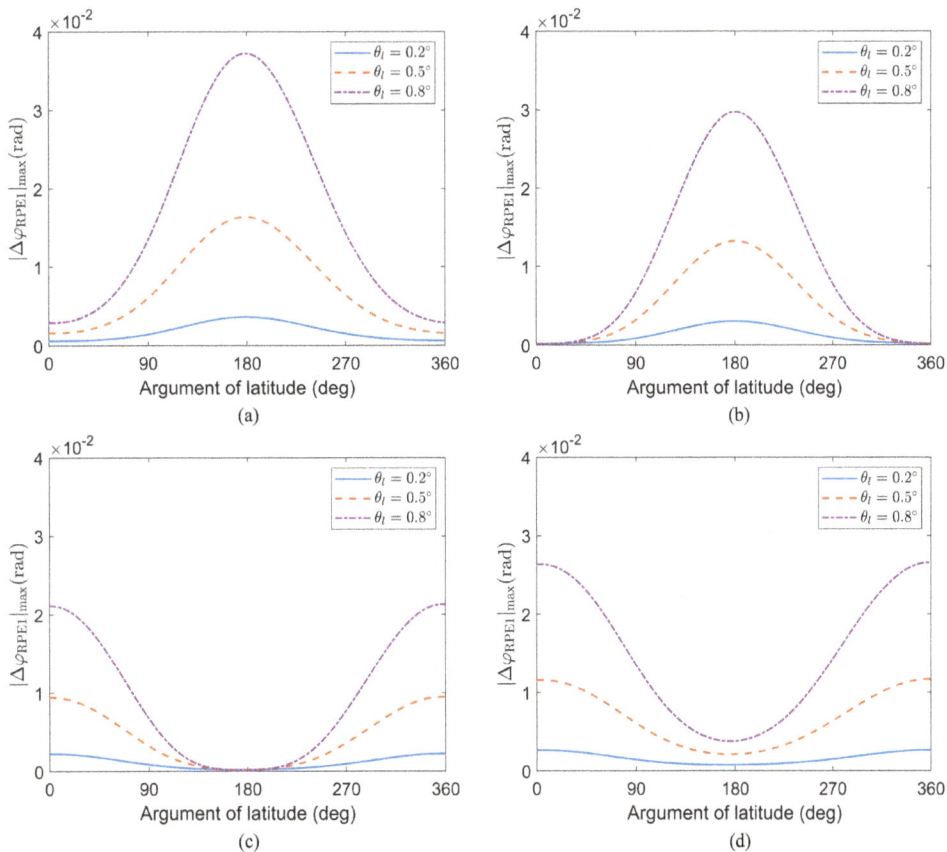

FIGURE 6.6 Within an SAT of 1800 s, the maximal magnitude of $\Delta\varphi_{\mathrm{RPE1}}$ with various look angles versus the AOL under antenna azimuth angles of (a) 45°, (b) 60°, (c) 120°, (d) 135°.

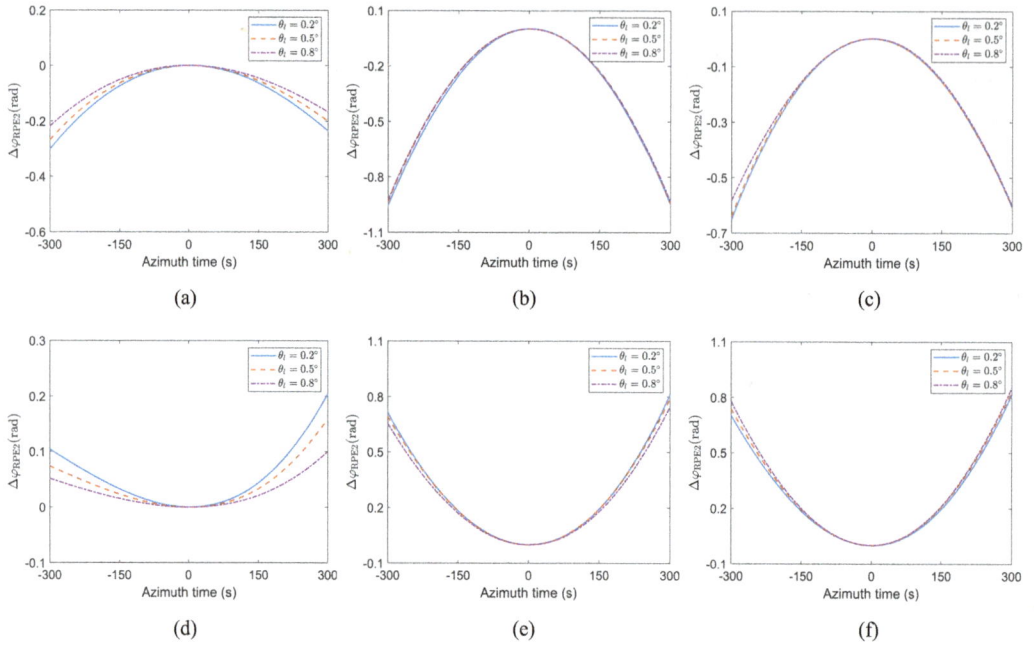

FIGURE 6.7 In the zero-Doppler steering mode, the RPE $\Delta\varphi_{RPE2}$ under various look angles against azimuth time at the AOL of (a) 0°, (b) 60°, (c) 120°, (d) 180°, (e) 240°, (f) 300°.

that the method presented in Eq. (6.4) is not applicable for compensating the stop-and-go error of the MBSAR with a fine azimuthal resolution. If the SAT is around or shorter than 400 s, the magnitude of RPE $\Delta\varphi_{RPE2}$ remains below $\pi/4$. Hence, we can model the non-stop-and-go range history within an SAT of 400 s using Eq. (6.4) in zero-Doppler steering mode.

The fluctuations in the maximum magnitude of $\Delta\varphi_{RPE2}$ versus AOL in squint-looking mode are depicted in Figure 6.9, where a look angle of 0.2° and antenna azimuth angles of 45°, 60°, 120°, and 135°, and various SATs are shown.

We can observe that the variation of $\Delta\varphi_{RPE2}$ is associated with the beam pointing direction and position of the MBSAR. Besides, for an SAT of 400 s or 600 s, the magnitude of $\Delta\varphi_{RPE2}$ can be over $\pi/4$ for certain locations. By contrast, the maximum RPE stays within the $\pi/4$ threshold when the SAT is below 200 s, regardless of the MBSAR's location. Therefore, the approach proposed in Eq. (6.4) should be adopted with a shorter SAT, say 200 s or below, in the squint-looking mode.

Thus far, this section has introduced two approaches to compensate for the stop-and-go error in the MBSAR. The compensation based on signal propagation, a time-domain method, exhibits commendable proficiency in precisely depicting the range history of MBSAR within a long SAT in diverse scenarios. The compensation based on Doppler frequency shift shows a constraint in its applicability range, despite less complex computation and costs. Moreover, the squint effect narrows down the scope, rendering it applicable only for an SAT of 400 s or even shorter. The next section will look into the range model a non-stop-and-go range history.

6.2.2 THE ESTABLISHMENT OF MBSAR'S RANGE MODEL

Recall the MBSAR's azimuthal resolution, a function of SAT and rotating angular velocity of slant range [16], is:

$$\rho_a = \frac{\lambda}{2\omega_{syn}T_{SAR}} \tag{6.8}$$

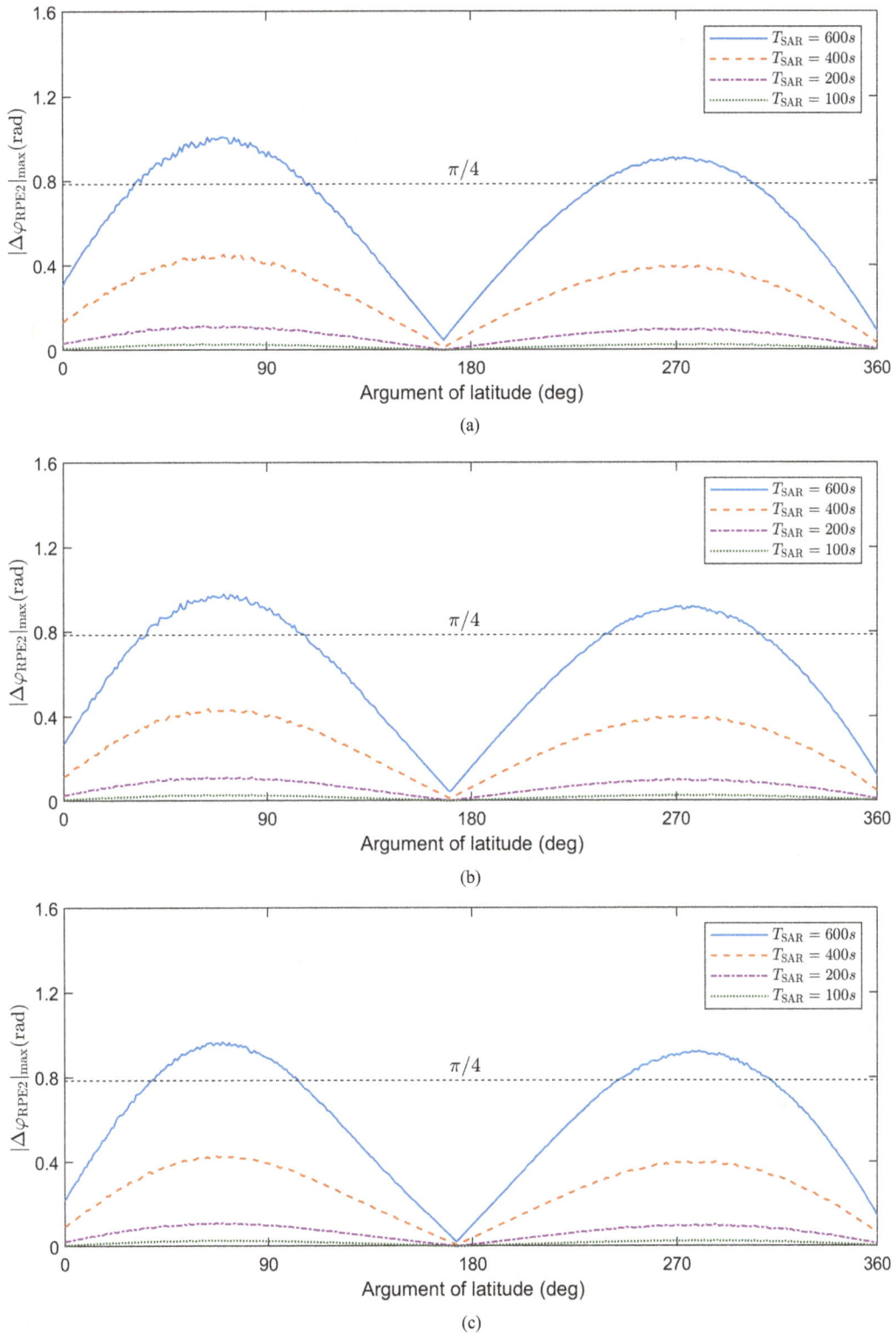

FIGURE 6.8 In the zero-Doppler steering mode, the maximal magnitude of $\Delta\varphi_{RPE2}$ within various time spans versus the AOL under the look angles of (a) 0.2°, (b) 0.5°, (c) 0.8°.

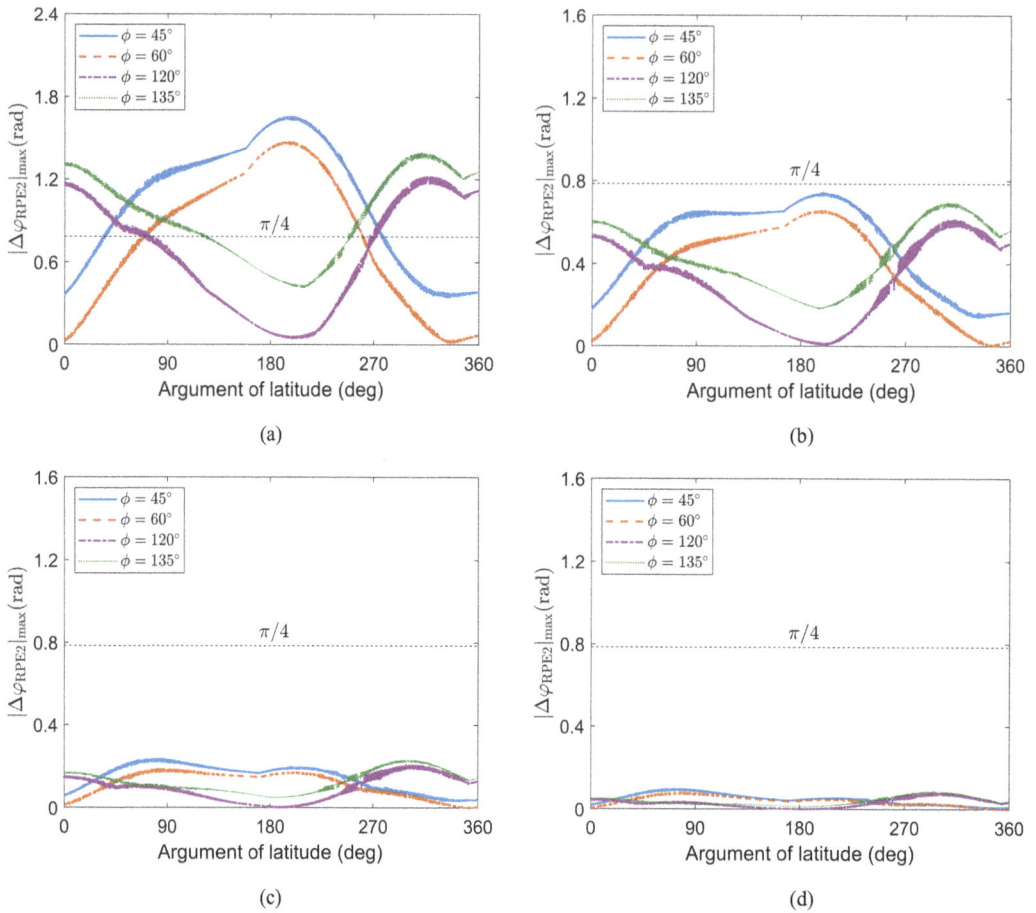

FIGURE 6.9 With a look angle of 0.2°, the maximum magnitude of $\Delta\varphi_{RPE2}$ with various antenna azimuth angles versus the AOL under the SAT of (a) 600 s, (b) 400 s, (c) 200 s, (d) 100 s.

Figure 6.10 shows the interdependence of azimuthal resolution on the SAT and AOL in the MBSAR, using a carrier frequency of 1.2 GHz and a reference look angle of 0.5°. It is worth noting that in this scenario, the MBSAR system utilizes the zero-Doppler steering method. As seen, the azimuthal resolution of MBSAR varies according to its location along the orbit for a specified SAT. Besides, the MBSAR offers a uniform azimuthal resolution that depends on the SAT for the imaging region when the Doppler centroid approaches zero. The SAT of MBSAR is much longer than that of spaceborne SAR, even for a coarser azimuthal resolution. For instance, an MBSAR may take an SAT of hundreds of seconds to achieve a tens-meter azimuthal resolution, and requires an SAT of thousands of seconds for a meter-level azimuth resolution. Further, the MBSAR will suffer from strong fluctuations in SAT with respect to AOL, and, thus, pronounced variation exists in the resolution cell.

We now examine the range models in line with SAT: the polynomial range model (PRM) and the extended hyperbolic range equation (EHRE). The PRM is simply a Taylor series expansion to the range history. Based on the order of expansion, the PRM can be categorized as: PRM2 (up to 2nd order), PRM3 (up to 3rd order), and PRM4 (up to 4th order). The EHRE is derived from the HRE by adjusting the equivalent velocity and squint angle, along with two curved factors compensating for higher-order range errors.

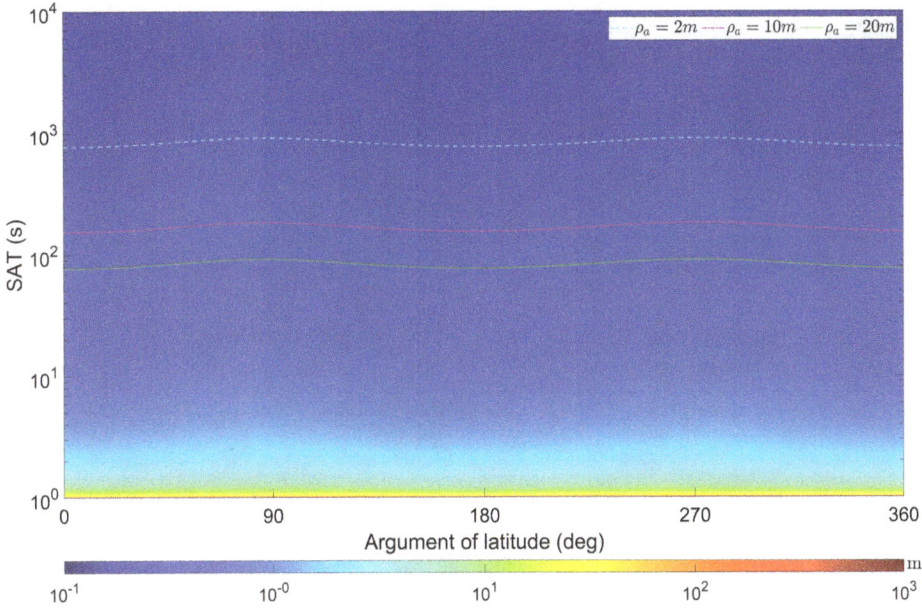

FIGURE 6.10 The azimuthal resolution versus the AOL and SAT in the MBSAR with a carrier frequency of 1.2 GHz and a look angle of 0.5°.

6.2.2.1 The Polynomial Range Model of 4th-Order

The slant range history of MBSAR can be represented as a polynomial equation of azimuth time. The PRMs from 2nd to 4th orders can be obtained by Taylor series expansion about $\eta = 0$:

$$\begin{cases} R_{\text{PRM2}} = R_c + K_1 \cdot \eta + K_2 \cdot \eta^2 \\ R_{\text{PRM3}} = R_c + K_1 \cdot \eta + K_2 \cdot \eta^2 + K_3 \cdot \eta^3 \\ R_{\text{PRM4}} = R_c + K_1 \cdot \eta + K_2 \cdot \eta^2 + K_3 \cdot \eta^3 + K_4 \cdot \eta^4 \end{cases} \tag{6.9}$$

where K_i are the ith, $i = 1, \cdots, 4$ order polynomial coefficients that the Doppler parameters related to, each of which is specified by the look angle θ_l and antenna azimuth angle ϕ.

To determine what orders are sufficient for a specific situation, we define the phase error (using the absolute value for simplicity) for PRMs from the 2nd to 4th orders:

$$\begin{cases} \Delta\psi_{\text{PRM2}} = 4\pi\lambda^{-1} \cdot \left| R_S(\eta) - R_{\text{PRM2}}(\eta) \right|_{\max}, \\ \Delta\psi_{\text{PRM3}} = 4\pi\lambda^{-1} \cdot \left| R_S(\eta) - R_{\text{PRM3}}(\eta) \right|_{\max}, \quad -\frac{\eta}{2} < T_{\text{SAR}} < \frac{\eta}{2} \\ \Delta\psi_{\text{PRM4}} = 4\pi\lambda^{-1} \cdot \left| R_S(\eta) - R_{\text{PRM4}}(\eta) \right|_{\max}, \end{cases} \tag{6.10}$$

The phase error should be smaller than the threshold of $\pi/4$ [17] for the SAR image focusing. By this threshold, one can thus determine the required order of expansion under a particular beam pointing direction. Figures 6.11–6.13 compare the PRM4 to the PRM2 and PRM3 of the MBSAR at different look angles under zero-Doppler steering mode, from which we may draw the following points:

1) The phase errors exhibit distinct variation patterns in response to changes in the position and look angle of MBSAR with different orders of polynomial expansion;

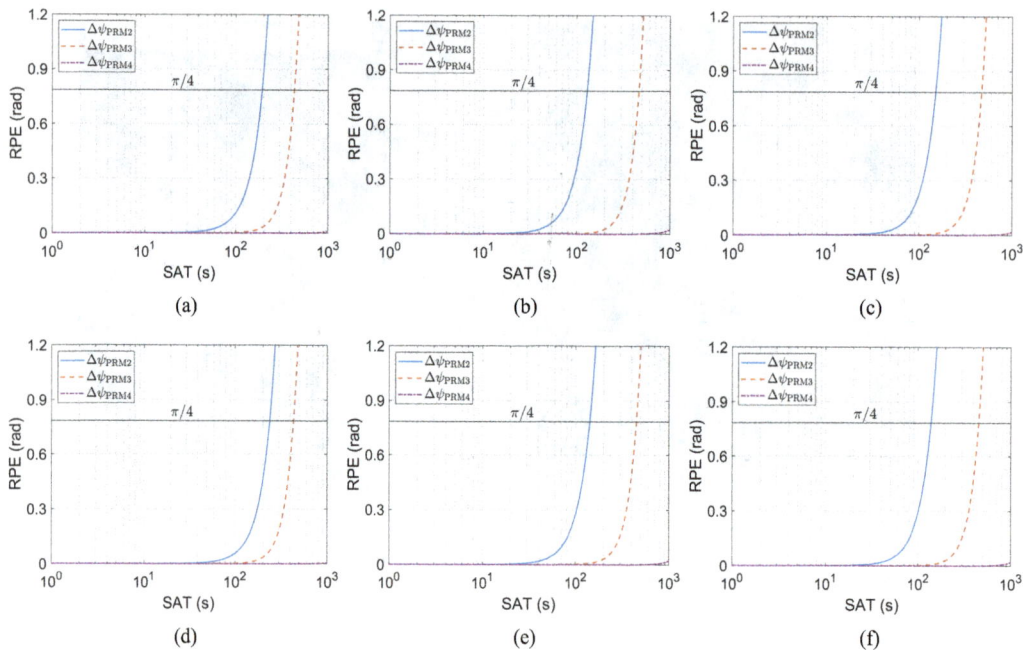

FIGURE 6.11 In the MBSAR with a look angle of 0.2°, the phase errors of various PRMs versus the SAT at the AOL of (a) 0°, (b) 60°, (c) 120°, (d) 180°, (e) 240°, (f) 300°.

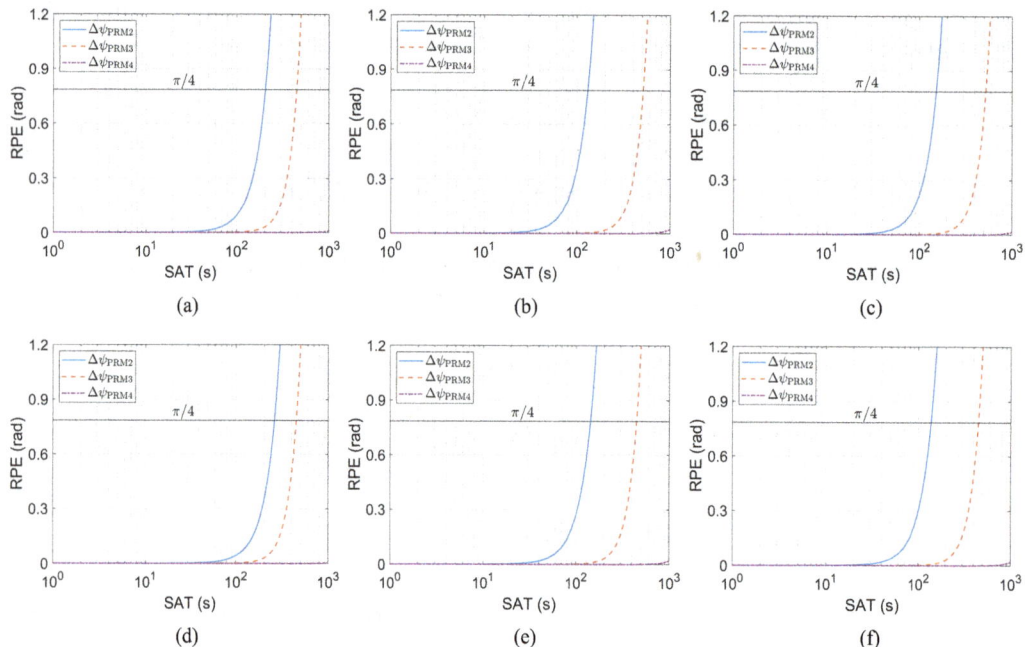

FIGURE 6.12 In the MBSAR with a look angle of 0.5°, the phase errors of various PRMs versus the SAT at the AOL of (a) 0°, (b) 60°, (c) 120°, (d) 180°, (e) 240°, (f) 300°.

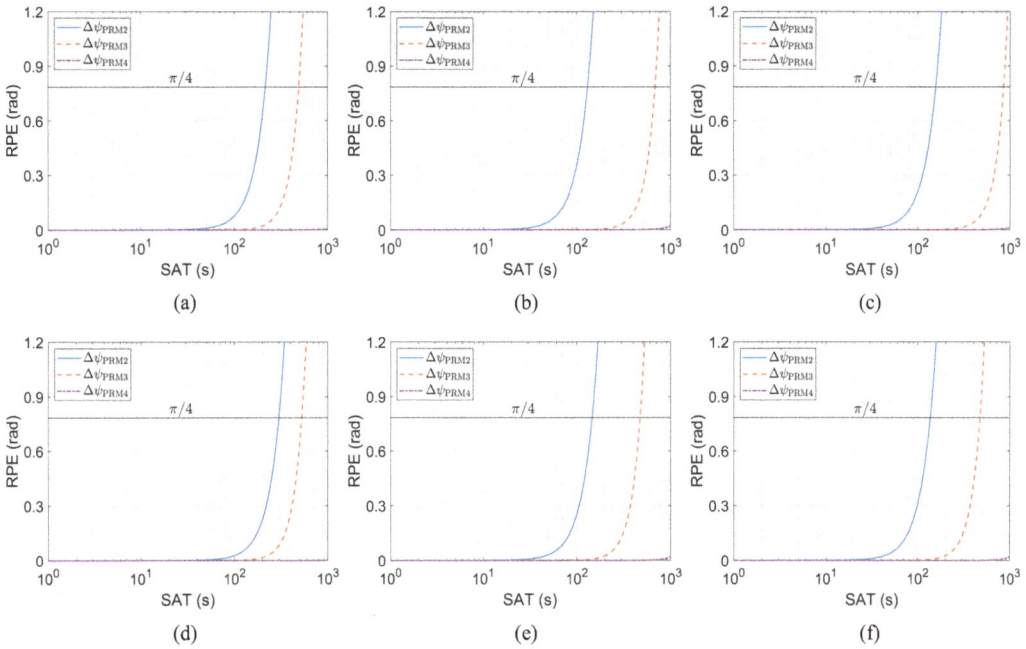

FIGURE 6.13   In the MBSAR with a look angle of 0.8°, the phase errors of various PRMs versus the SAT at the AOL of (a) 0°, (b) 60°, (c) 120°, (d) 180°, (e) 240°, (f) 300°.

2) A PRM2 is sufficient, with the SAT below 100 s for the phase error being relatively small. If the SAT exceeds 100 s, the phase error significantly increases and thus needs higher-order polynomial expansion;

3) A PRM3 is sufficient, with SAT below 400 s, where the phase error maintains below $\pi/4$. If the SAT is beyond 400 s, the phase error is over $\pi/4$;

4) A PRM4 can satisfy the focusing requirements of MBSAR even if the SAT stays around 1000 s with an acceptable phase error level.

To further validate the accuracy of the PRM4 for modeling the range history of MBSAR, we analyze the phase errors versus the AOL and look angles under zero-Doppler steering mode with an SAT of 1000 s, as presented in Figure 6.14. From Figure 6.14, it is evident that the PRM4 has a phase error below 3×10^{-2} rad, far smaller than the threshold of $\pi/4$, regardless of MBSAR location. Therefore, this range model is accurate enough to satisfy the requirements for MBSAR imaging.

It should be noted that if MBSAR operates in the squint-looking mode, the valid range of the PRM in terms of expansion orders is impaired. Hence, we need to know how the phase error changes in response to SAT and squint angle at different MBSAR locations. Figures 6.15–6.17, respectively, present the phase errors of PRMs from 2nd to 4th orders under different antenna azimuth angles at the look angles of 0.2°, 0.5°, and 0.8°.

Figures 6.15–6.17 show that the phase errors in PRMs of different orders change with respect to SAT at various antenna azimuth angles. The PRM2 has the most significant changes in phase error because of the lower order of expansion. This suggests that the PRM2 is most affected by the squint effects, and perhaps only acceptable when the SAT is below 60 s. As the expansion order increases, the phase error decreases under a specified SAT. Still, the PRM3 has limited applicability and fails

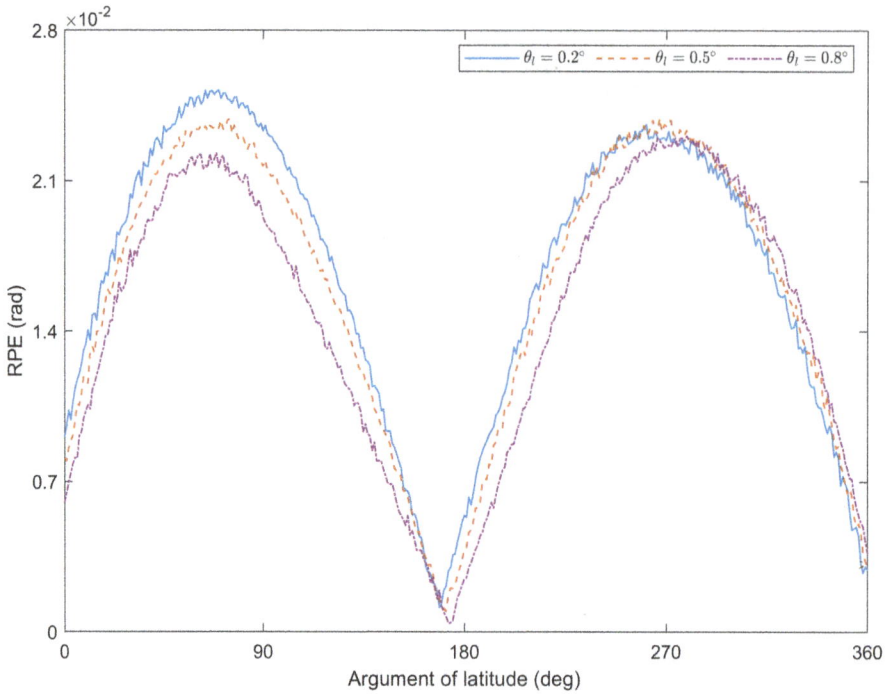

FIGURE 6.14 With an SAT of 1000 s, the phase errors of the PRM4 versus AOL in the MBSAR under various look angles.

to achieve high-azimuthal resolution in MBSAR imaging. By contrast, the PRM4 only induces minor phase error even at an SAT of 1000 s, making it well suited for describing the range history of MBSAR when it operates in squint-looking mode.

Next, we select three look angles: 0.2°, 0.5°, and 0.8° to simulate the maximum phase error of the PRM4 as a function of AOL for an SAT of 1000 s, as presented in Figure 6.18. In this case, the MBSAR works in the squint-looking mode, and various antenna azimuth angles are considered.

Figure 6.18 shows that the phase error increases with the increasing look angle, regardless of the antenna azimuth angle or location of MBSAR. Besides, the phase error of PRM4 in the squint-looking mode is larger than that in the zero-Doppler steering mode. Though the phase error is below 0.15 rad, it is expected to be even smaller when the SAT is shorter than 1000 s. Hence, the PRM4 is appropriate for the signal processing of MBSAR, whether with zero-Doppler steering or squint-looking mode.

6.2.2.2 The Extended Hyperbolic Range Equation Model

The HRE is a range model commonly used in SAR signal processing [18–21]; it originated from the geometry of airborne SAR and provides a general formula relating the slant range history of a SAR system to its velocity, squint angle, and slow time by [22]

$$R_{\mathrm{HRE}}(\eta) = \sqrt{R_c^2 - 2R_c V_{\mathrm{SAR}} \sin\theta_{\mathrm{sq}}(\eta - \eta_c) + V_{\mathrm{SAR}}^2 (\eta - \eta_c)^2} \qquad (6.11)$$

where V_{SAR} and θ_{sq} are the velocity and squint angle of the SAR. The HRE can be further simplified by setting the beam center crossing time to zero, namely $\eta_c = 0$.

In spaceborne SAR that operates in Low Earth Orbit (LEO), the HRE is still applicable, but with modifications according to its orbit [23]. Specifically, the velocity V_{SAR} and squint angle θ_{sq} in the airborne

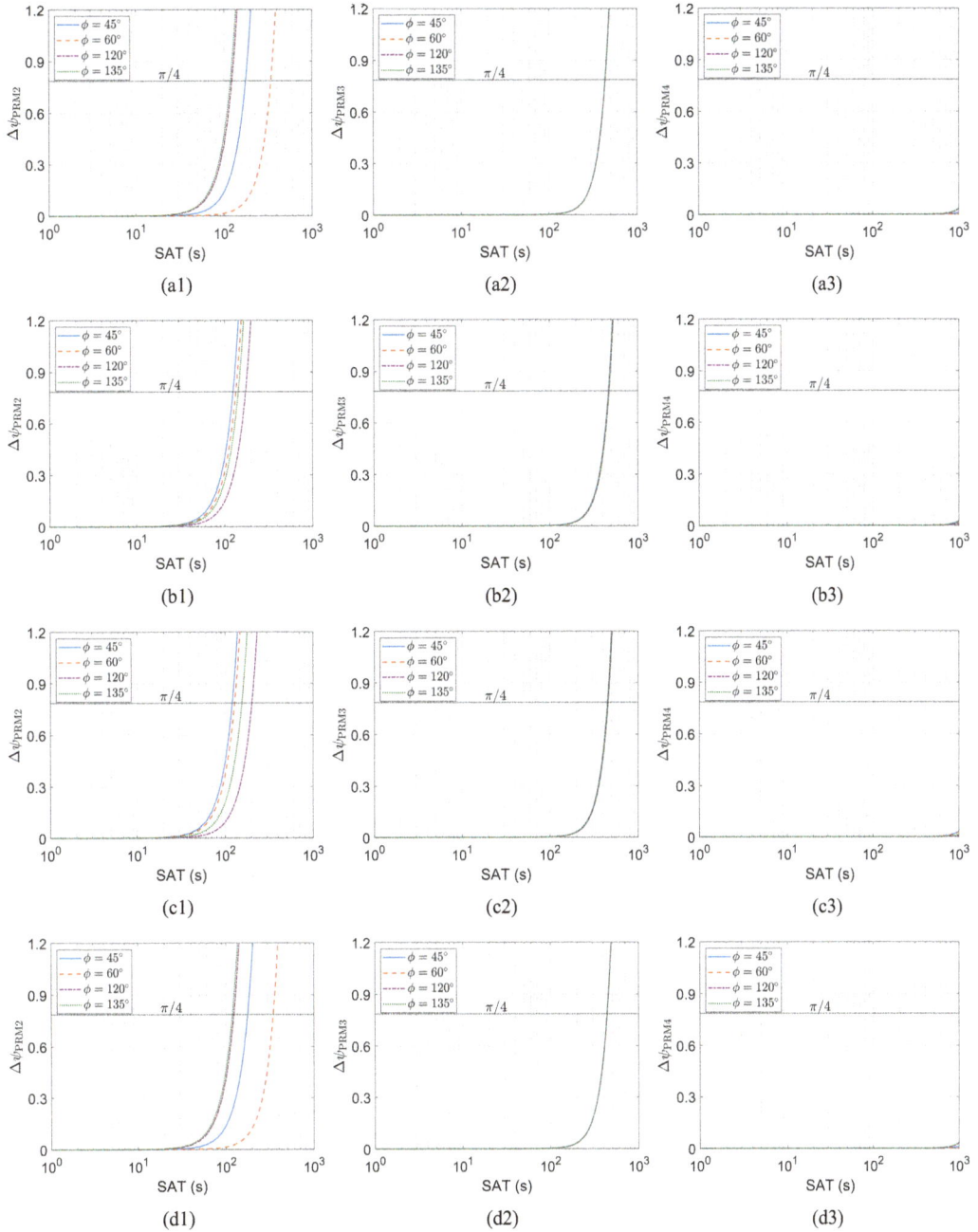

FIGURE 6.15 In the MBSAR with a look angle of 0.2°, the phase errors with various antenna azimuth angles versus the SAT at various locations, where the 1st (a1–d1), 2nd (a2–d2) and 3rd (a3–d3) columns represent the phase errors of PRM2, PRM3, and PRM4, while 1st (a1–a3), 2nd (b1–b3), 3rd (c1–c3) and 4th (d1–d3) rows stand for the AOL of 0°, 120°, 240°, 360°.

SAR case must be replaced with the equivalent velocity V_r and equivalent squint angle θ_r. The new HRE employed in the LEO spaceborne SAR becomes [24]:

$$R_{\text{HRE}}(\eta) = \sqrt{R_c^2 - 2R_c V_r \sin\theta_r\eta + V_r^2\eta^2} \tag{6.12}$$

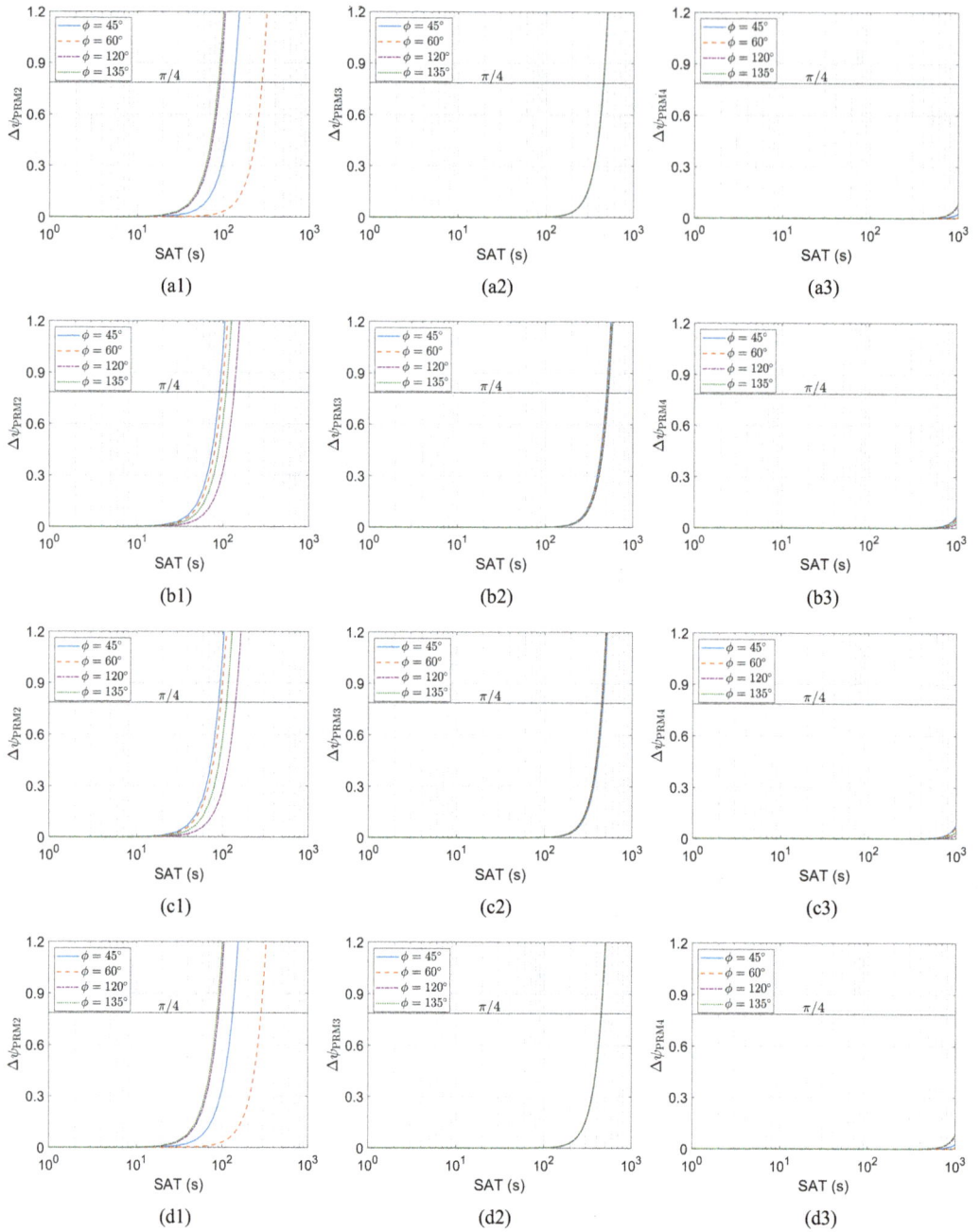

FIGURE 6.16 In the MBSAR with a look angle of 0.5°, the phase errors with various antenna azimuth angles versus the SAT at various locations, where the 1st (a1–d1), 2nd (a2–d2) and 3rd (a3–d3) columns represent the phase errors of PRM2, PRM3, and PRM4, while 1st (a1–a3), 2nd (b1–b3), 3rd (c1–c3) and 4th (d1–d3) rows stand for the AOL of 0°, 120°, 240°, 360°.

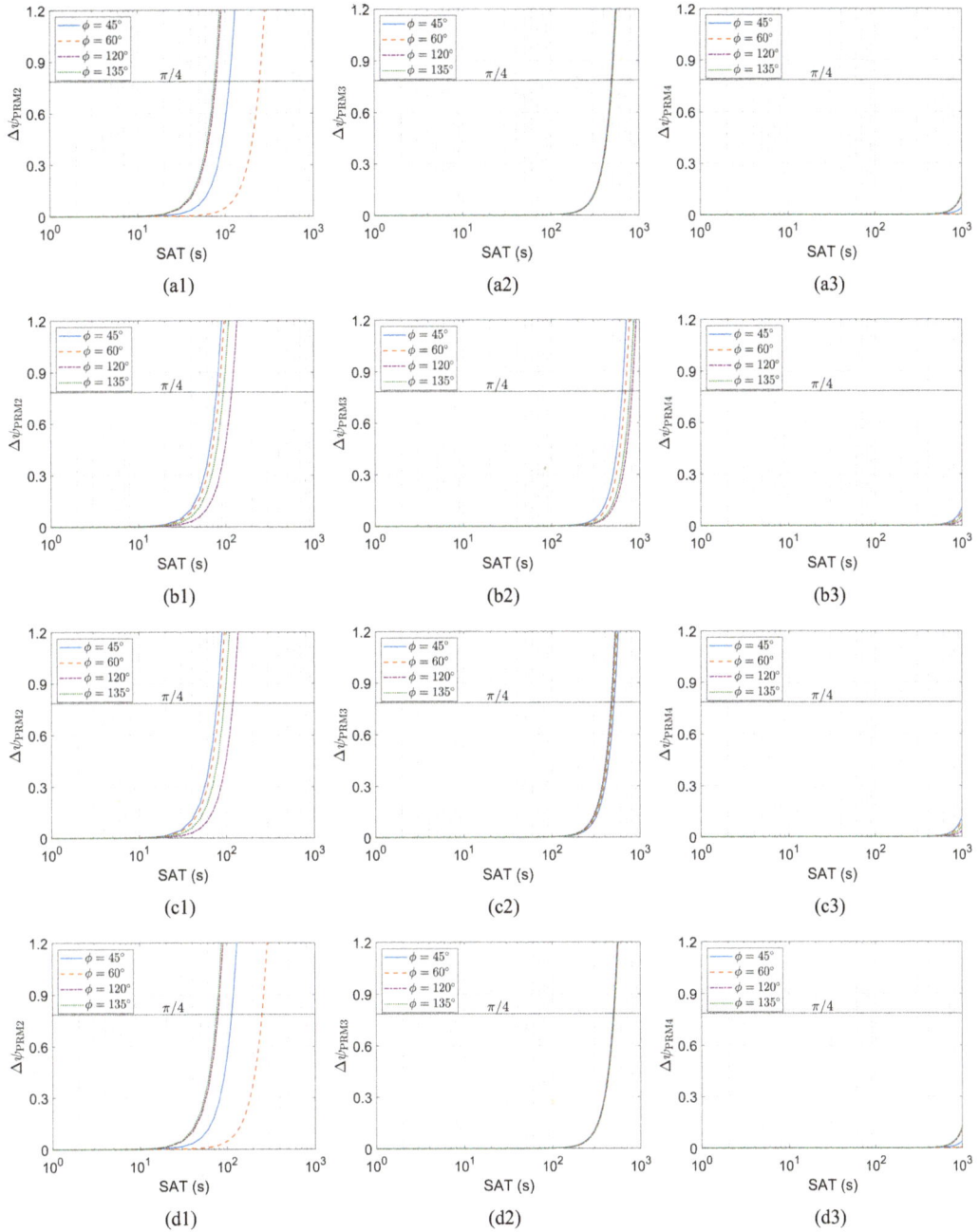

FIGURE 6.17 In the MBSAR with a look angle of 0.8°, the phase errors with various antenna azimuth angles versus the SAT at various locations, where the 1st (a1–d1), 2nd (a2–d2) and 3rd (a3–d3) columns represent the phase errors of PRM2, PRM3, and PRM4, while 1st (a1–a3), 2nd (b1–b3), 3rd (c1–c3) and 4th (d1–d3) rows stand for the AOL of 0°, 120°, 240°, 360°.

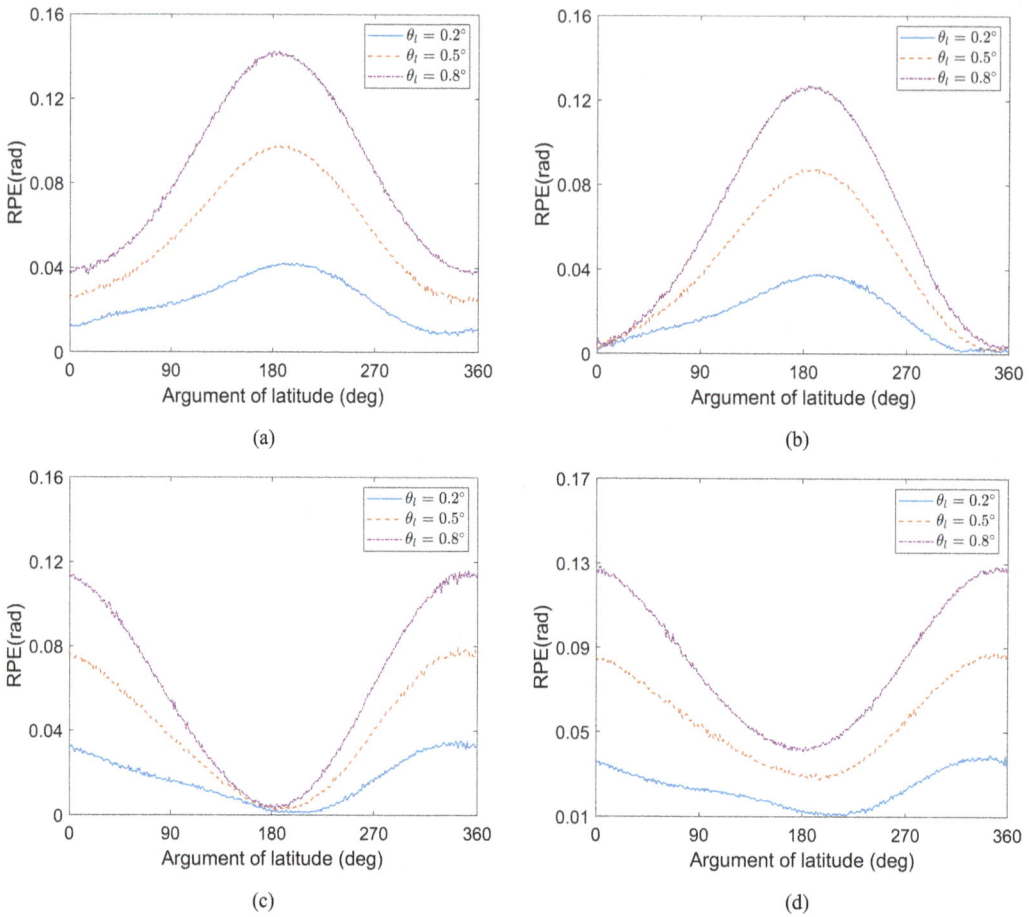

FIGURE 6.18 In the PRM4 with an SAT of 1000 s, the phase errors with various look angles versus AOL under antenna azimuth angles of (a) 45°, (b) 60°, (c) 120°, (d) 135°.

with:

$$V_r = 0.5\sqrt{\lambda^2 f_{\text{dc}}^2 + 2\lambda R_c f_{\text{dr}}} \qquad (6.13)$$

and:

$$\theta_r = \sin^{-1}\left(\frac{f_{dc}}{\sqrt{f_{\text{dc}}^2 + 2\lambda^{-1}R_c f_{\text{dr}}}}\right) \qquad (6.14)$$

where f_{dc} and f_{dr} are the Doppler centroid and DFMR, respectively.

When a spaceborne SAR operates in a circular or quasi-circular orbit with zero-Doppler centroid beam pointing, the equivalent velocity and squint angle are approximated to [22, 25]

$$V_r \approx \sqrt{V_{\text{SAR}}V_g} \qquad (6.15)$$

$$\theta_r \approx \frac{V_r}{V_g}\theta_{\text{sq}} = \frac{V_{\text{SAR}}}{V_r}\theta_{\text{sq}} \qquad (6.16)$$

In the HRE, only the quadratic term is compensated, leaving out the cubic and higher-order phase errors of the range history. This, in effect, leads to noticeable range error as synthetic aperture time increases, particularly at higher altitude orbits. To mitigate this issue, the advanced HRE (AHRE) is proposed to facilitate the signal processing of spaceborne SAR in medium and high Earth orbits, by adding a factor to compensate for the residual cubic phase error [26–28]:

$$R_{\mathrm{AHRE}}(\eta) = \sqrt{R_c^2 - 2R_c \sin\theta_{ar} V_{ar}\eta + V_{ar}^2\eta^2} + \Delta_l\eta \tag{6.17}$$

with:

$$V_{ar} = \sqrt{\left(\frac{2R_c f_{drr}}{3f_{dr}}\right)^2 + \frac{\lambda R_c f_{dr}}{2}} \tag{6.18}$$

$$\theta_{ar} = \sin^{-1}\left(\frac{2R_c f_{drr}}{3V_{ar} f_{dr}}\right) \tag{6.19}$$

$$\Delta_l = -\frac{\lambda f_{dc}}{2} + \frac{2R_c f_{drr}}{3f_{dr}} \tag{6.20}$$

where f_{drr} is the rate of DFMR.

We note that the AHRE is only applicable in middle-to-high Earth orbit, specifically those with an altitude below 10,000 km [29]. For MBSAR, the cubic range error in HRE is more prominent, resulting in a significant deviation from the actual trajectory of the MBSAR. Hence, we introduce an EHRE below to tackle the phase error from the cubic and quartic terms [3]:

$$R_{\mathrm{EHRE}}(\eta) = \sqrt{R_c^2 - 2R_c \sin\theta_r V_r\eta + V_r^2\eta^2} + \Delta_c\eta^3 + \Delta_q\eta^4 \tag{6.21}$$

with:

$$\Delta_c = \frac{\lambda f_{drr}}{12} - \frac{\lambda^2 f_{dr} f_{dc}}{8R_c} \tag{6.22}$$

$$\Delta_q = \frac{\lambda}{48} f_{drrr} + \frac{\lambda^2 f_{dr}^2}{32R_c} - \frac{\lambda^3 f_{dc}^2 f_{dr}}{16R_c^2} \tag{6.23}$$

where f_{drrr} denotes the 2nd-order rate of the DFMR. The equivalent velocity and equivalent squint angle in Eq. (6.21) are the same as those in Eqs. (6.13) and (6.14). For a thorough derivation of EHRE, please refer to Appendix B.

In the following, we shall evaluate and compare the application ranges of the HRE, AHRE, and EHRE in the MBSAR. To illustrate this, Figure 6.19 presents phase errors for these range models ($\Delta\psi_{\mathrm{HRE}}$, $\Delta\psi_{\mathrm{AHRE}}$, $\Delta\psi_{\mathrm{EHRE}}$ for HRE, AHRE, and EHRE, respectively), plotted against SAT with various look angles. Further, Figure 6.20 presents the phase errors of the PRM4 and EHRE as a function of AOL in the cycle T_{lc1} under various look angles. In both cases, the MBSAR's beam center points at the zero-Doppler plane.

Intuitively, all phase errors vary based on the position and look angle of the MBSAR, as shown in Figure 6.19. Accordingly, the validity of those range models is disturbed by the beam pointing direction and orbit parameters of the MBSAR: the phase error for the AHRE has little difference from, or is slightly higher than that of the HRE. In this regard, the AHRE exhibits similar or even poorer performance in modeling the range history of MBSAR compared to the HRE. By contrast,

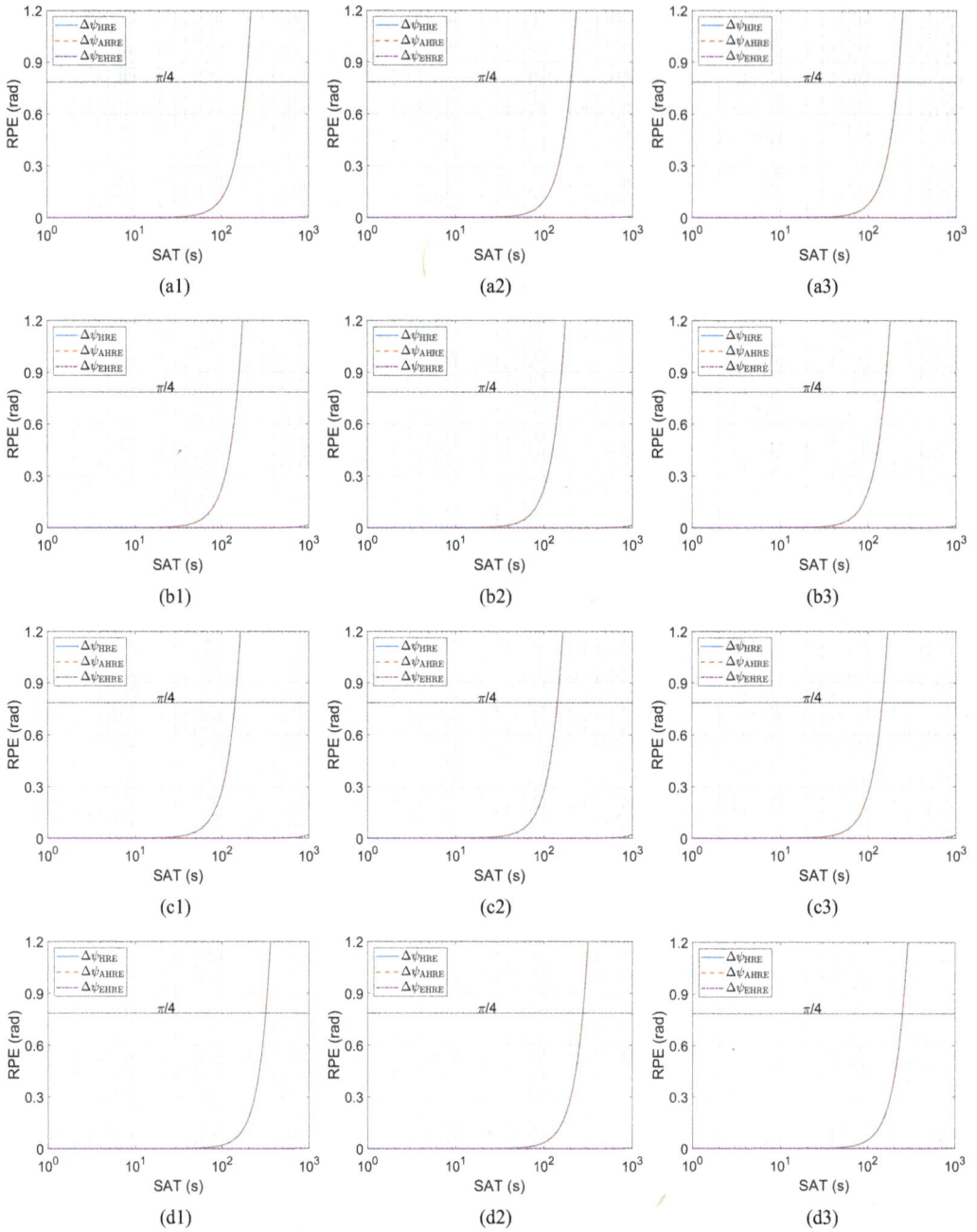

FIGURE 6.19 In the zero-Doppler steering mode, the phase errors of the HRE, AHRE, and EHRE versus the SAT under various look angles at different locations, where the 1st (a1–d1), 2nd (a2–d2) and 3rd (a3–d3) columns represent the look angles of 0.2°, 0.5°, and 0.8°, while 1st (a1–a3), 2nd (b1–b3), 3rd (c1–c3) and 4th (d1–d3) rows stand for the AOL of 0°, 120°, 240°, 360°.

the EHRE performs well with a pretty small phase error within an SAT of 1000 s. In this aspect, among those three models, only the EHRE fulfills the focusing requirements for a high-azimuthal resolution MBSAR. Moreover, the results in Figure 6.20 suggest that the EHRE and PRM4 exhibit similar application ranges in the zero-Doppler steering mode.

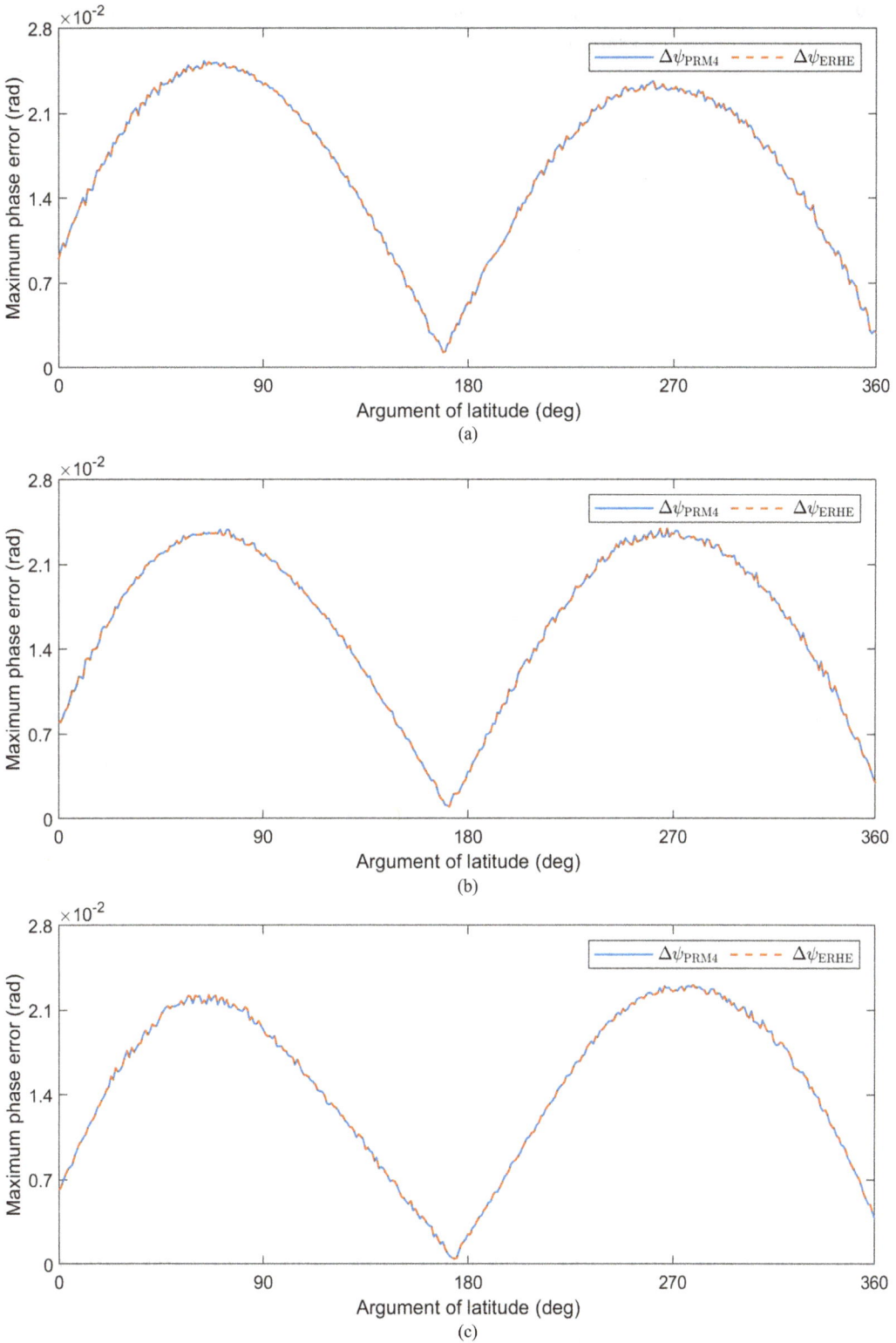

FIGURE 6.20 With an SAT of 1000 s, the phase errors of EHRE and PRM4 versus AOL under the look angle of (a) 0.2°, (b) 0.5°, (c) 0.8°.

We select an SAT of 1000 s, the antenna azimuth angles of $45°$, $60°$, $120°$, and $135°$, and the AOL spanning from $0°$ to $360°$ to examine the impact of various beam pointing directions on EHRE and PRM4 at different MBSAR locations. Figure 6.21 through Figure 6.23 presents the phase errors of both range models over various cycles (T_{lc1}, T_{lc4}, and T_{lc7}).

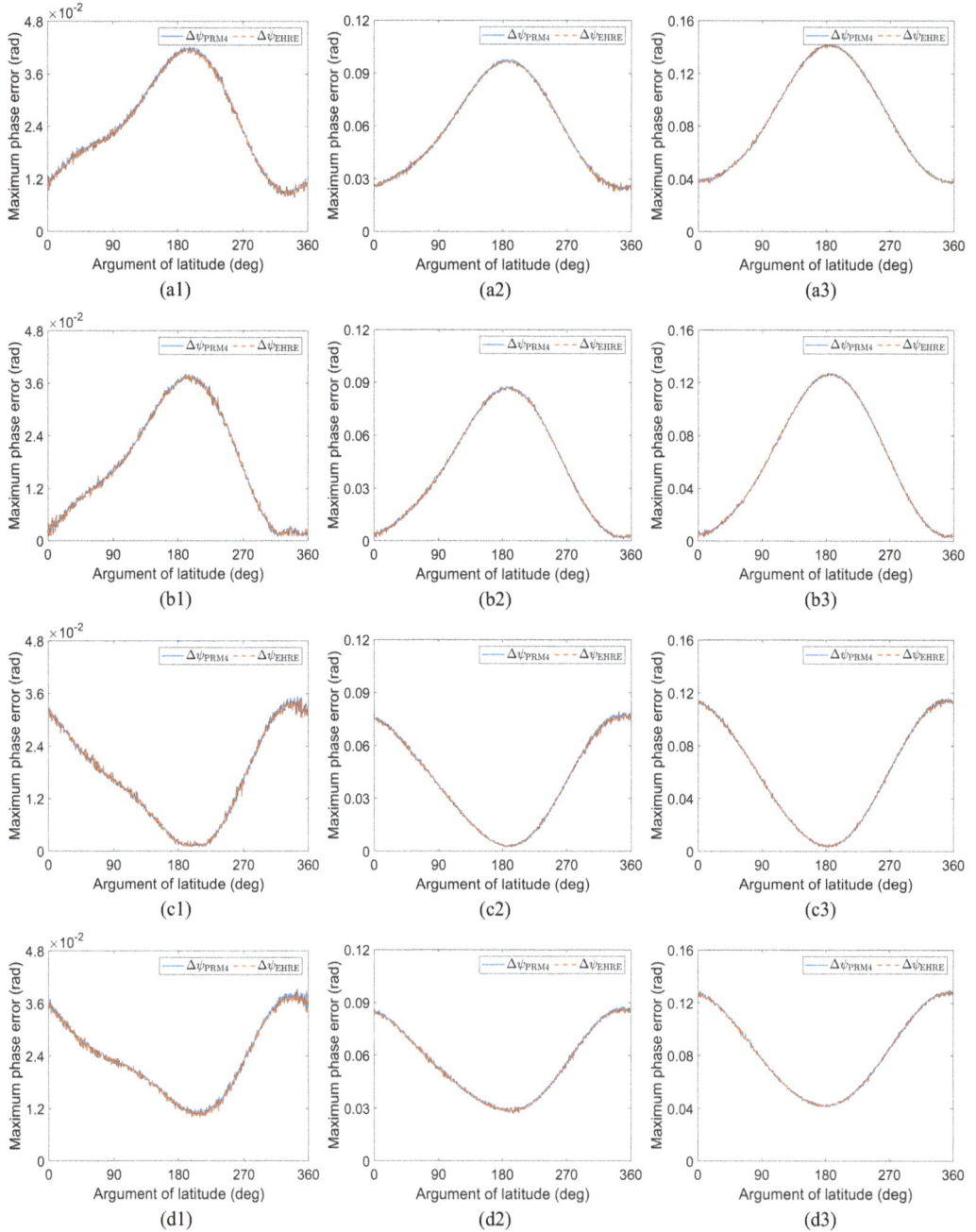

FIGURE 6.21 Within the cycle T_{lc1}, the phase errors of the EHRE and PRM4 versus AOL under various antenna azimuth and look angles, where the 1st (a1–d1), 2nd (a2–d2) and 3rd (a3–d3) columns represent the look angles of $0.2°$, $0.5°$, and $0.8°$, while the 1st (a1–a3), 2nd (b1–b3), 3rd (c1–c3), and 4th (d1–d3) rows stand for the antenna azimuth angles of (a) $45°$, (b) $60°$, (c) $120°$, (d) $135°$.

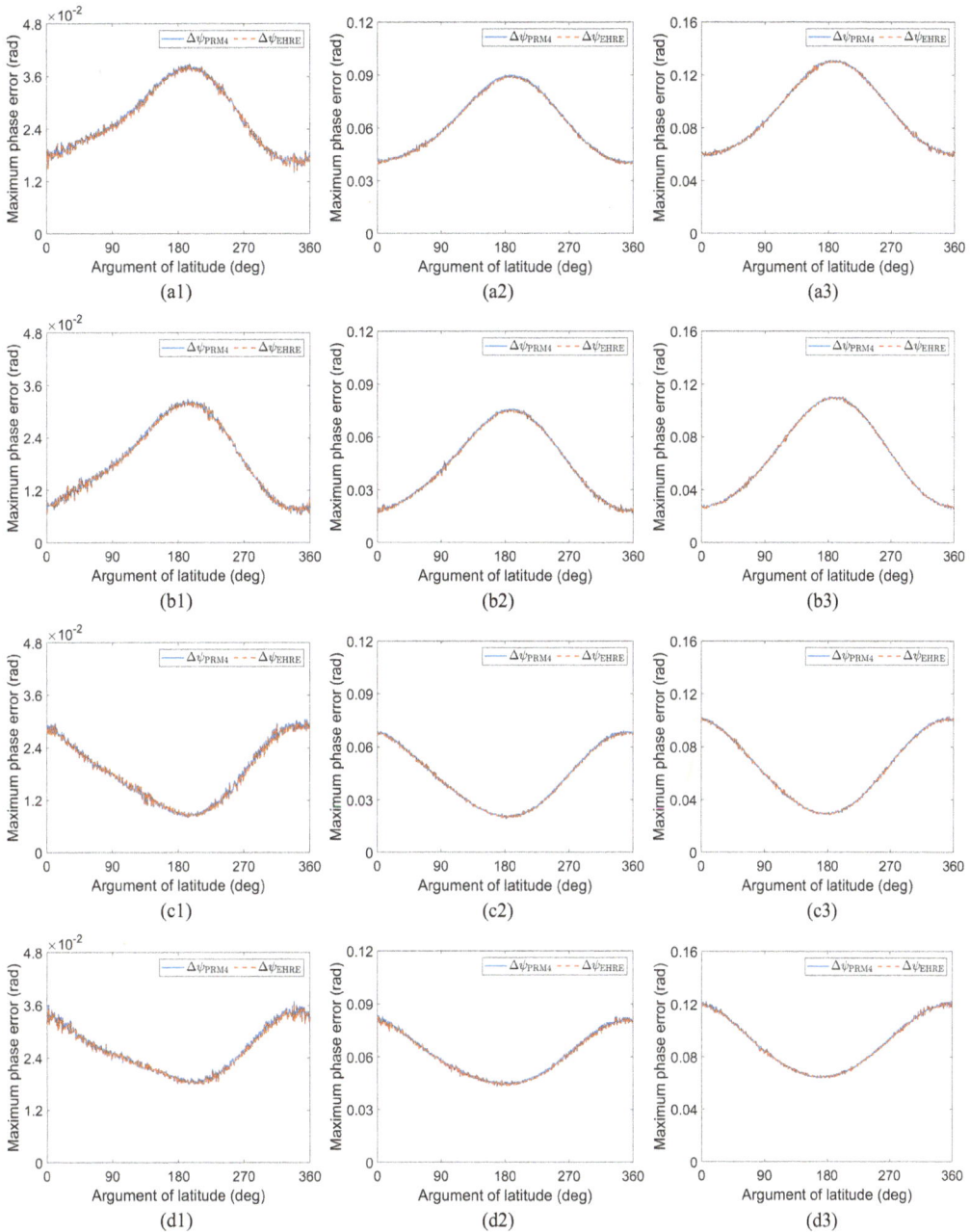

FIGURE 6.22 Within the cycle T_{lc4}, the phase errors of the EHRE and PRM4 versus AOL under various antenna azimuth and look angles, where the 1st (a1–d1), 2nd (a2–d2) and 3rd (a3–d3) columns represent the look angles of 0.2°, 0.5°, and 0.8°, while the 1st (a1–a3), 2nd (b1–b3), 3rd (c1–c3), and 4th (d1–d3) rows stand for the antenna azimuth angles of (a) 45°, (b) 60°, (c) 120°, (d) 135°.

Figures 6.21–6.23 demonstrate that phase errors for both PRM4 and EHRE have almost identical variations with respect to the AOL despite the beam pointing direction of MBSAR. Besides, the EHRE and PRM4 have been found effective in accurately fitting the squint-looking MBSAR's trajectory with an SAT of 1000 s, as evidenced by the largest phase error being consistently below 0.15 rad at various positions across distinct cycles. This suggests that both the PRM4 and EHRE are appropriate to model MBSAR's range history.

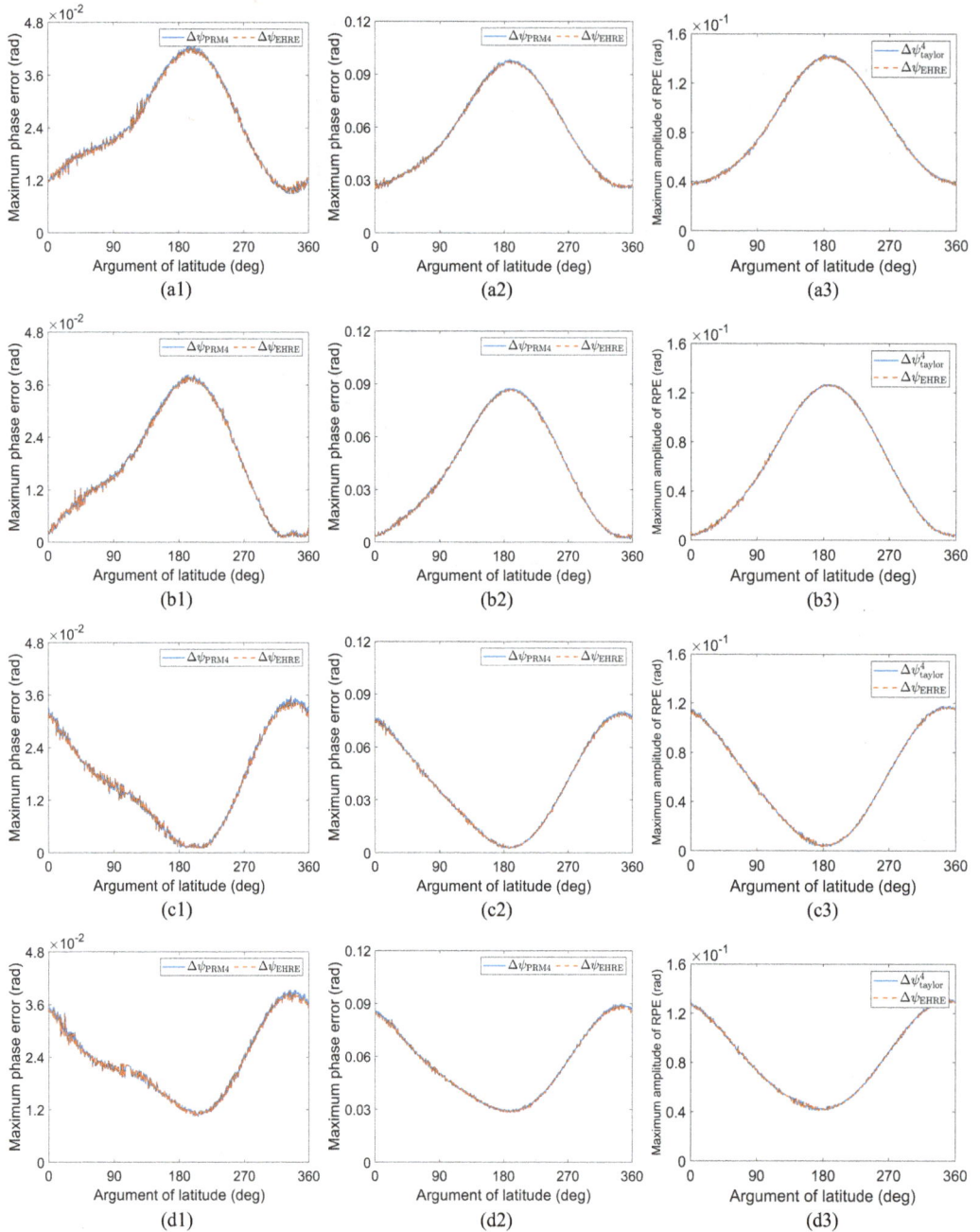

FIGURE 6.23 Within the cycle T_{lc7}, the phase errors of the EHRE and PRM4 versus AOL under various antenna azimuth and look angles, where the 1st (a1–d1), 2nd (a2–d2) and 3rd (a3–d3) columns represent the look angles of 0.2°, 0.5°, and 0.8°, while the 1st (a1–a3), 2nd (b1–b3), 3rd (c1–c3), and 4th (d1–d3) rows stand for the antenna azimuth angles of (a) 45°, (b) 60°, (c) 120°, (d) 135°.

In order to deal with the MBSAR's signal using the EHRE, we derive a 2-D spectrum by the principle of stationary phase (POSP) [30] and Method of Series Reversion (MSR) [31, 32]. We shall expand the EHRE into a polynomial due to the presence of higher-order factors:

$$R_{\mathrm{EHRE}}\left(\eta\right) = R_c + H_1\eta + H_2\eta^2 + H_3\eta^3 + H_4\eta^4 \tag{6.24}$$

where:

$$\begin{cases} H_1 = -V_r \sin \theta_r \\[2mm] H_2 = V_r^2 \dfrac{\cos^2 \theta_r}{2R_c} \\[4mm] H_3 = V_r^3 \dfrac{\cos^2 \theta_r \sin \theta_r}{2R_c^2} + \Delta_c \\[4mm] H_4 = V_r^4 \dfrac{6\sin^2 \theta_r - 5\sin^4 \theta_r - 1}{8R_c^3} + \Delta_q \end{cases} \tag{6.25}$$

Note that the EHRE model differs from the PRM4 only in cases where it is expanded below the 4th-order. In the following section, we derive the explicit 2-D signal spectrum of the MBSAR.

6.3   SIGNAL SPECTRUM AND FOCUSING ALGORITHM

The MBSAR encounters a distinct 2-D signal spectrum that differs from that of spaceborne SAR due to its non-stop-and-go range history with a curved trajectory. The non-stop-and-go range history can be expanded with reference to $\eta = 0$, namely:

$$R_S(\eta) = R_c + R_1\eta + R_2\eta^2 + R_3\eta^3 + R_4\eta^4 \tag{6.26}$$

$$R_i = \begin{cases} K_i \ (\text{PRM4}) \\ H_i \ (\text{EHRE}) \end{cases}, \ i = 1, \cdots, 4 \tag{6.27}$$

The echo signal of the SAR system takes on the following form [33]:

$$s_r(\tau,\eta) = A_0 \cdot w_a(\eta - \eta_c) \cdot w_r \left[\tau - \frac{2R_S(\eta)}{c} \right]$$

$$\cdot \exp\left[-j\frac{4\pi f_c R_S(\eta)}{c} \right] \cdot \exp\left\{ j\pi K_r \left[\tau - \frac{2R_S(\eta)}{c} \right]^2 \right\} \tag{6.28}$$

To process the MBSAR's signal, we apply the Fourier Transform (FT) in the range direction to Eq. (6.28) yielding:

$$S_r(f_\tau,\eta) = \int_{-\infty}^{\infty} s_r(\tau,\eta) \cdot \exp\left[-j2\pi f_\tau \tau \right] d\tau$$

$$= \int_{-\infty}^{\infty} A_0 \cdot w_a(\eta - \eta_c) \cdot w_r \left[\tau - \frac{2R_S(\eta)}{c} \right] \exp\left[j\Phi(\tau,\eta) \right] d\tau \tag{6.29}$$

Rearrange the phase term in Eq. (6.29) we have:

$$\Phi(\tau,\eta) = -\frac{4\pi f_c R_S(\eta)}{c} + \pi K_r \left[\tau - \frac{2R_S(\eta)}{c} \right]^2 - 2\pi f_\tau \tau \tag{6.30}$$

We then apply the POSP to Eq. (6.30) for an approximate time-frequency relation in range direction, as:

$$\frac{d\Phi(\tau,\eta)}{d\tau} = 0 \rightarrow 2\pi K_r \left[\tau - \frac{2R_S(\eta)}{c} \right] - 2\pi f_\tau \tau = 0 \rightarrow \tau = \frac{f_\tau}{K_r} + \frac{2R_S(\eta)}{c} \tag{6.31}$$

By substituting the fast time in Eq. (6.31) into Eq. (6.29) and applying the shift property of FT [34], we get:

$$S_r(f_\tau,\eta) = A_0 W_r(f_\tau) w_a(\eta) \exp\left(-j\pi \frac{f_\tau^2}{K_r} \right) \exp\left[-j4\pi \frac{(f_c + f_\tau)R_S(\eta)}{c} \right] \tag{6.32}$$

where $W_r(\cdot)$ is the range envelope in the frequency domain.

We now apply the Fourier Transform to Eq. (6.32) in the azimuth direction so that:

$$S_r(f_\tau,f_\eta) = \int_{-\infty}^{\infty} S_r(f_\tau,\eta) \cdot \exp\left[-j2\pi f_\eta \eta \right] d\eta \tag{6.33}$$

The phase term of Eq. (6.33) is then expressed as:

$$\begin{aligned}
\Phi(f_\tau,\eta) &= -\pi \frac{f_\tau^2}{K_r} - 4\pi \frac{(f_c + f_\tau)}{c} R_S(\eta) - 2\pi f_\eta \eta \\
&= -\pi \frac{f_\tau^2}{K_r} - 4\pi \frac{(f_c + f_\tau)}{c} \left(R_c + R_1\eta + R_2\eta^2 + R_3\eta^3 + R_4\eta^4 \right) - 2\pi f_\eta \eta
\end{aligned} \tag{6.34}$$

By virtue of the same procedure as to Eq. (6.30), we take the partial derivative of Eq. (6.34) with respect to η and apply the POSP, arriving at:

$$\frac{d\Phi(\eta)}{d\eta} = 0 \rightarrow 2\pi f_\eta = -4\pi \frac{(f_c + f_\tau)}{c} \left(R_1 + 2R_2\eta + 3R_3\eta^2 + 4R_4\eta^3 \right) \tag{6.35}$$

To apply MSR, we need to eliminate the term R_1 in Eq. (6.35) so that:

$$2R_2\eta + 3R_3\eta^2 + 4R_4\eta^3 = -\frac{c}{2(f_c + f_\tau)} f_\eta \tag{6.36}$$

By MSR, we arrive at an azimuth time-frequency relation as follows:

$$\begin{aligned}
\eta = &-\frac{1}{2R_2}\left[\frac{c}{2(f_c + f_\tau)} f_\eta \right] - \frac{3R_3}{8R_2^3}\left[\frac{c}{2(f_c + f_\tau)} f_\eta \right]^2 \\
&- \frac{9R_3^2 - 4R_2R_4}{16R_2^5}\left[\frac{c}{2(f_c + f_\tau)} f_\eta \right]^3
\end{aligned} \tag{6.37}$$

We give the MSR in Appendix C for ease of reference.

As stated in Eq. (6.34), there is a linear phase term. By utilizing FT's skew and shift properties [35], replacing f_η with $f_\eta + 2R_1 \dfrac{f_c + f_\tau}{c}$, the relation between the slow time and azimuth frequency in MBSAR's signal is given by:

$$
\eta\left(f_\eta\right) = -\frac{1}{2R_2}\left[\frac{c}{2\left(f_c + f_\tau\right)}\left(f_\eta + 2R_1\frac{f_c + f_\tau}{c}\right)\right]
$$

$$
-\frac{3R_3}{8R_2^3}\left[\frac{c}{2\left(f_c + f_\tau\right)}\left(f_\eta + 2R_1\frac{f_c + f_\tau}{c}\right)\right]^2 \tag{6.38}
$$

$$
-\frac{9R_3^2 - 4R_2R_4}{16R_2^5}\left[\frac{c}{2\left(f_c + f_\tau\right)}f_\eta\left(f_\eta + 2R_1\frac{f_c + f_\tau}{c}\right)\right]^3
$$

By substituting the azimuth time-frequency relation from Eq.(6.38) into Eq. (6.34), resulting in a 2-D signal spectrum:

$$
S_r\left(f_\tau, f_\eta\right) = A_0 \cdot W_r\left(f_\tau\right)W_a\left(f_\eta\right)\exp\left\{j\Phi\left(f_\tau, f_\eta\right)\right\} \tag{6.39}
$$

where $W_a(\cdot)$ is the azimuth envelope in the frequency domain, and the phase term is expressed as:

$$
\Phi\left(f_\tau, f_\eta\right) = -\pi\frac{f_\tau^2}{K_r} - \frac{4\pi R_c\left(f_c + f_\tau\right)}{c}
$$

$$
-4\pi\frac{\left(f_c + f_\tau\right)}{c}\left[R_1\eta\left(f_\eta\right) + R_2\eta^2\left(f_\eta\right) + R_3\eta^3\left(f_\eta\right) + R_4\eta^4\left(f_\eta\right)\right] \tag{6.40}
$$

or more explicitly:

$$
\Phi\left(f_\tau, f_\eta\right) = -\pi\frac{f_\tau^2}{K_r} - \frac{4\pi R_c\left(f_c + f_\tau\right)}{c} + 2\pi\frac{1}{8R_2}\frac{c}{2\left(f_c + f_\tau\right)}\left[f_\eta + 2R_1\frac{\left(f_c + f_\tau\right)}{c}\right]^2
$$

$$
+2\pi\frac{R_3}{32R_2^3}\cdot\left[\frac{c}{2\left(f_c + f_\tau\right)}\right]^2\cdot\left[f_\eta + 2R_1\frac{\left(f_c + f_\tau\right)}{c}\right]^3 \tag{6.41}
$$

$$
+2\pi\frac{9R_3^2 - 4R_2R_4}{512R_2^5}\cdot\left[\frac{c}{2\left(f_c + f_\tau\right)}\right]^3\cdot\left[f_\eta + 2R_1\frac{\left(f_c + f_\tau\right)}{c}\right]^4
$$

The phase term of the 2-D signal spectrum has a coupling term of range and azimuth frequencies. To simplify the signal processing in MBSAR, we expand Eq. (6.41) into the Taylor series with respect to $f_\tau = 0$, and after rearranging, we get:

$$
\Phi(f_\tau, f_\eta) = \Phi_{rc}(f_\eta) + \Phi_{ac}(f_\tau) + \Phi_{rcm}(f_\tau, f_\eta) + \Phi_{src}(f_\tau, f_\eta) + \Phi_{res} \tag{6.42}
$$

where $\Phi_{rc}(f_\tau)$ corresponds to the range compression term:

$$
\Phi_{rc}\left(f_\tau\right) = 2\pi\left\{-R_c + \frac{R_1^2}{2R_2} + \frac{R_1^3R_3}{4R_2^3} + \frac{9R_3^2 - 4R_2R_4}{32R_2^5}R_1^4\right\}\frac{f_\tau}{c} - \pi\frac{f_\tau^2}{K_r} \tag{6.43}
$$

The azimuth compression term $\Phi_{ac}(f_\eta)$ in Eq. (6.42) is:

$$\Phi_{ac}(f_\eta) = 2\pi\left\{\left(\frac{R_1}{2R_2} + \frac{3R_1^2 R_3}{8R_2^3} + R_1^3\frac{9R_3^2 - 4R_2 R_4}{16R_2^5}\right)f_\eta\right.$$

$$+ \frac{c}{f_c}\left(\frac{1}{8R_2} + \frac{3R_1 R_3}{16R_2^3} + 3R_1^2\frac{9R_3^2 - 4R_2 R_4}{64R_2^5}\right)f_\eta^2$$

$$\left.+ \frac{c^2}{f_c^2}\left(\frac{R_3}{32R_2^3} + R_1\frac{9R_3^2 - 4R_2 R_4}{64R_2^5}\right)f_\eta^3 + \frac{c^3}{f_c^3}\cdot\frac{9R_3^2 - 4R_2 R_4}{512R_2^5}f_\eta^4\right\}$$

$$(6.44)$$

$\Phi_{rcm}(f_\tau, f_\eta)$ in Eq. (6.42) represents the range cell migration term that is explicitly expressed by:

$$\Phi_{rcm}(f_\tau, f_\eta) = 2\pi\left\{-\frac{c}{f_c^2}\left(\frac{1}{8R_2} + \frac{3R_1 R_3}{16R_2^3} + 3R_1^2\frac{9R_3^2 - 4R_2 R_4}{64R_2^5}\right)f_\eta^2\right.$$

$$\left.-\frac{c^2}{f_c^3}\left(R_1\frac{9R_3^2 - 4R_2 R_4}{32R_2^5} + \frac{R_3}{16R_2^3}\right)f_\eta^3 - 3\frac{c^3}{f_c^4}\frac{9R_3^2 - 4R_2 R_4}{512R_2^5}f_\eta^4\right\}f_\tau$$

$$(6.45)$$

$\Phi_{src}(f_\tau, f_\eta)$ in Eq. (6.42) is related to the secondary range compression:

$$\Phi_{src}(f_\tau, f_\eta) = 2\pi\left\{\left[\frac{c}{f_c^3}\left(\frac{1}{8R_2} + \frac{3R_1 R_3}{16R_2^3} + 3R_1^2\frac{9R_3^2 - 4R_2 R_4}{64R_2^5}\right)f_\eta^2\right.\right.$$

$$\left.+ \frac{c^2}{f_c^4}\left(\frac{3R_3}{32R_2^3} + 3R_1\frac{9R_3^2 - 4R_2 R_4}{64R_2^5}\right)f_\eta^3 + 3\frac{c^3}{f_c^4}\frac{9R_3^2 - 4R_2 R_4}{256R_2^5}f_\eta^4\right]f_\tau^2$$

$$- \left[\frac{c}{f_c^4}\left(\frac{1}{8R_2} + \frac{3R_1 R_3}{16R_2^3} + 3R_1^2\frac{9R_3^2 - 4R_2 R_4}{64R_2^5}\right)f_\eta^2 + \right.$$

$$\left.\left.\frac{c^2}{f_c^5}\left(\frac{R_3}{8R_2^3} + R_1\frac{9R_3^2 - 4R_2 R_4}{16R_2^5}\right)f_\eta^3 + 5\frac{c^3}{f_c^6}\frac{9R_3^2 - 4R_2 R_4}{256R_2^5}f_\eta^4\right]f_\tau^3\right\}$$

$$(6.46)$$

Finally, Φ_{res} represents the residual phase error that bears little on either azimuth or range compression:

$$\Phi_{res} = 2\pi\frac{f_c}{c}\left(-2R_c + \frac{R_1^2}{2R_2} + \frac{R_1^3 R_3}{4R_2^3} + R_1^4\frac{9R_3^2 - 4R_2 R_4}{32R_2^5}\right)$$

$$(6.47)$$

Based on the 4th-order range model, we have now presented a 2-D signal spectrum for the MBSAR. Following this, a frequency domain-focusing algorithm can be applied for the image focusing of MBSAR, to be described in detail in what follows.

Figure 6.24 displays a flowchart of the MBSAR image-focusing algorithm. The algorithm is designed by utilizing the 2-D signal spectrum with the scene center serving as the reference point. The main steps of the focusing algorithm are outlined below.

1) Transform the raw echo data into the range frequency domain by employing the range Fast Fourier Transform (FFT), and range compression is then applied by the matched filter:

$$S_{rc}(f_\tau, \eta) = \text{FFT}_{range}\{s_r(\tau, \eta)\}\cdot H_{rc}(f_\tau)$$

$$(6.48)$$

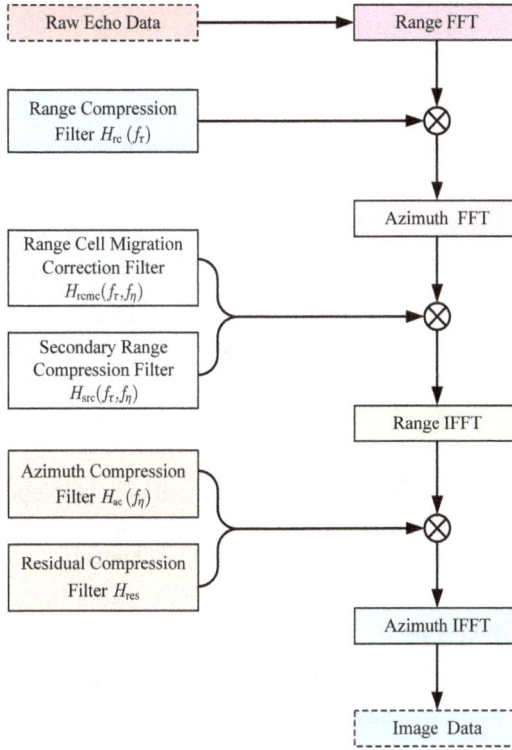

FIGURE 6.24 The flowchart of MBSAR imaging focusing.

with:

$$H_{rc}\left(f_{\tau}\right) = \exp\left\{-j\Phi_{rc}\left(f_{\tau}\right)\right\} \tag{6.49}$$

2) Transform the raw echo data into the azimuth (Doppler) frequency domain by azimuth FFT, followed by range cell migration correction (RCMC) and secondary range compression (SRC) using corresponding matched filters:

$$S_{rcmc}\left(f_{\tau},f_{\eta}\right) = \text{FFT}_{azimuth}\left\{S_{rc}\left(f_{\tau},\eta\right)\right\}H_{rcmc}\left(f_{\tau},f_{\eta}\right) \tag{6.50}$$

$$S_{src}\left(f_{\tau},f_{\eta}\right) = S_{rcmc}\left(f_{\tau},f_{\eta}\right)H_{src}\left(f_{\tau},f_{\eta}\right) \tag{6.51}$$

with:

$$H_{rcmc}\left(f_{\tau},f_{\eta}\right) = \exp\left\{-j\Phi_{rcm}\left(f_{\tau},f_{\eta}\right)\right\} \tag{6.52}$$

$$H_{src}\left(f_{\tau},f_{\eta}\right) = \exp\left\{-j\Phi_{src}\left(f_{\tau},f_{\eta}\right)\right\} \tag{6.53}$$

3) Transform the above signal into the range-Doppler (RD) domain by performing the Inverse Fast Fourier Transform (IFFT) in range direction, followed by an azimuth compression with the matched filter $H_a(f_\eta)$ and compensation for the residual phase error. These processes are mathematically represented by:

$$S_{ac}\left(\tau,f_{\eta}\right) = \text{IFFT}_{range}\left\{S_{src}\left(f_{\tau},f_{\eta}\right)\right\}H_{ac}\left(f_{\eta}\right) \tag{6.54}$$

$$S_{\mathrm{res}}\left(\tau,f_{\eta}\right)=S_{\mathrm{ac}}\left(\tau,f_{\eta}\right)H_{\mathrm{res}} \tag{6.55}$$

with:

$$H_{\mathrm{ac}}\left(f_{\eta}\right)=\exp\left\{-j\Phi_{\mathrm{ac}}\left(f_{\eta}\right)\right\} \tag{6.56}$$

$$H_{\mathrm{res}}=\exp\left\{-j\Phi_{\mathrm{res}}\right\} \tag{6.57}$$

4) Finally, transform the signal in Eq. (6.55) into the time domain by azimuth IFFT:

$$s_{\mathrm{SAR}}\left(\tau,\eta\right)=\mathrm{IFFT}_{\mathrm{azimuth}}\left\{S_{\mathrm{res}}\left(\tau,f_{\eta}\right)\right\} \tag{6.58}$$

When imaging a large scene with the MBSAR, one crucial step is to update the spatially varying Doppler parameters along the range and azimuth directions. Such 2-D spatially varying signals are subject to beam pointing direction, swath width, and orbital elements.

6.4 IMAGING SIMULATIONS

To demonstrate the above focusing algorithm, this section presents numerical illustrations using a point target response with a signal bandwidth of 20 MHz at a carrier frequency of 1.2 GHz.

6.4.1 SIMULATIONS OF TARGET RESPONSES

6.4.1.1 Single-point Target Response

The single-point target responses of the MBSAR with a look angle of 0.5° are simulated using the PRM2 with various SATs under the non-stop-and-go signal model.

Figure 6.25 illustrates the simulation results for the MBSAR (located at the AOL of 0°) operating in the zero-Doppler steering mode, in which the PRM2 is employed. The figures of merit for image quality check include the peak-to-sidelobe ratio (PSLR), integrated sidelobe ratio (ISLR), and impulse response width (IRW) along azimuth and range directions. We can see that the range imaging performance remains little affected regardless of the SAT, but the azimuth focusing depends on the SAT: The PRM2 can yield good focusing quality in the MBSAR with an SAT of 100 s or shorter. By contrast, there is a discernible increase in sidelobes and a slight rise in ISLR when the SAT exceeds 200 s. This phenomenon can be attributed to the influence of the third-order phase error presented in the PRM2, which induces asymmetric distortion in the phase of the MBSAR signal. The point target response is expected to deteriorate as the SAT further increases. At an SAT of 400 s, an offset and complete defocusing manifest along the azimuth direction, rendering MBSAR imaging unviable. Consequently, it can be deduced that the PRM2 is only suitable for MBSAR imaging with relatively coarse azimuthal resolutions.

Figure 6.26 shows the simulated point target response in MBSAR using a PRM4 to assess and compare the performance of different range models in the MBSAR. From Figure 6.26, we find that at SATs of 100 and 200 s, the ISLR and PSLR of point target response are more desirable than those in Figure 6.25. For an SAT of 400 s, the asymmetric sidelobes, as those appeared in Figure 6.25, are greatly suppressed and the ISLR values of the point target response are much lower. As the SAT is over 400 s, corresponding to an azimuthal resolution of 4 m, the ISLR and PSLR values are very satisfactory. Hence, applying the PRM4 to the image focusing can achieve a superior quality of focused image with meter scale azimuthal resolution.

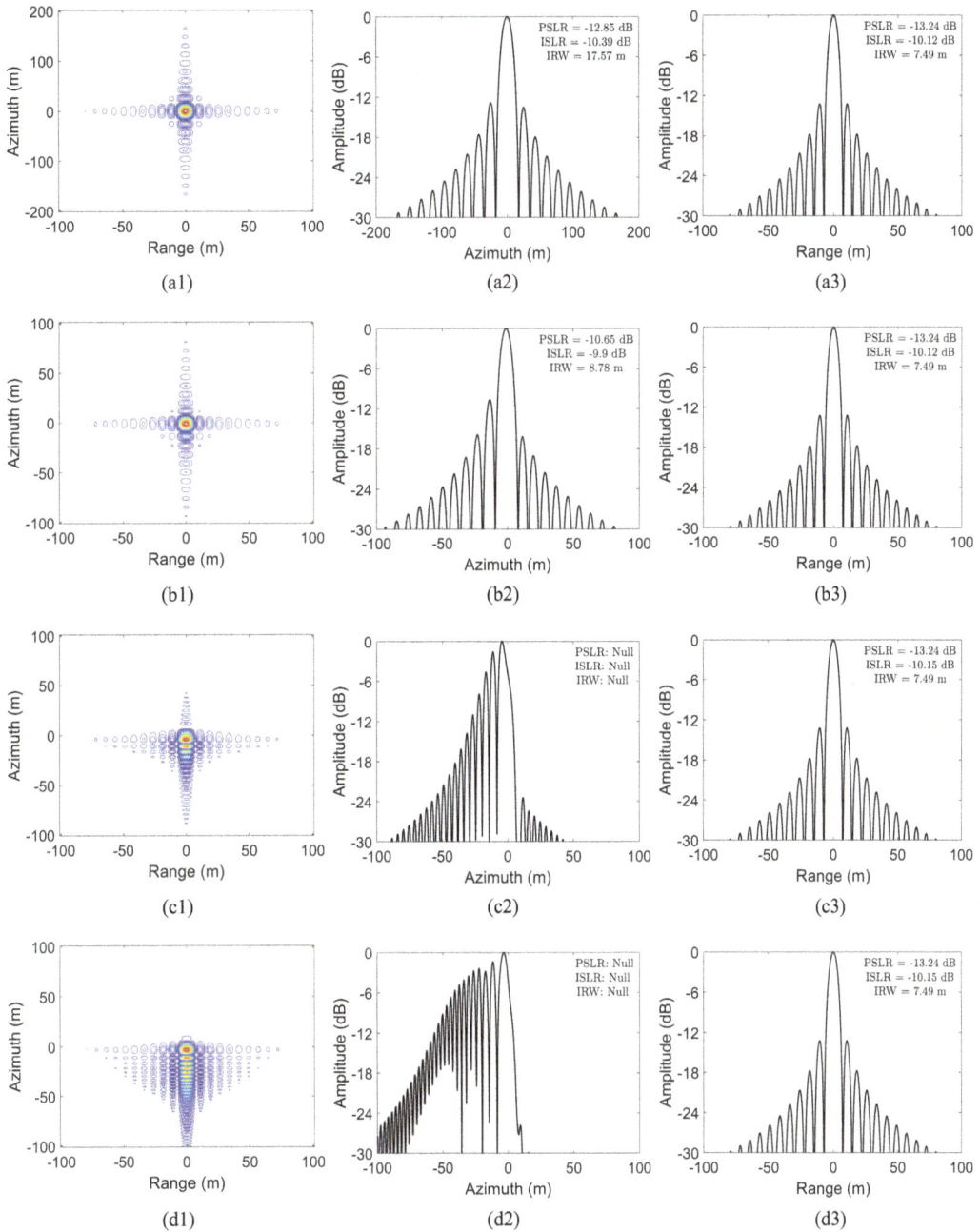

FIGURE 6.25 The single-point target response of the MBSAR by employing the PRM2 where the 1st (a1–d1), 2nd (a2–d2), and 3rd (a3–d3) columns respectively represent the response contour, azimuth, and range profiles, while 1st (a1–a3), 2nd (b1–b3), 3rd (c1–c3) and 4th (d1–d3) rows stand for the SATs of 100 s, 200 s, 400 s, and 600 s. In this context, "Null" refers to the unavailability of MBSAR imaging.

Figure 6.27 presents the simulations of point target responses using HRE and EHRE at varying locations of MBSAR. Various SATs are employed to evaluate the HRE and EHRE in handling MBSAR's signal and to demonstrate the imaging characteristics of MBSAR along its orbit. In this case, the beam pointing of MBSAR is regulated using the zero-Doppler steering method [36]. As the

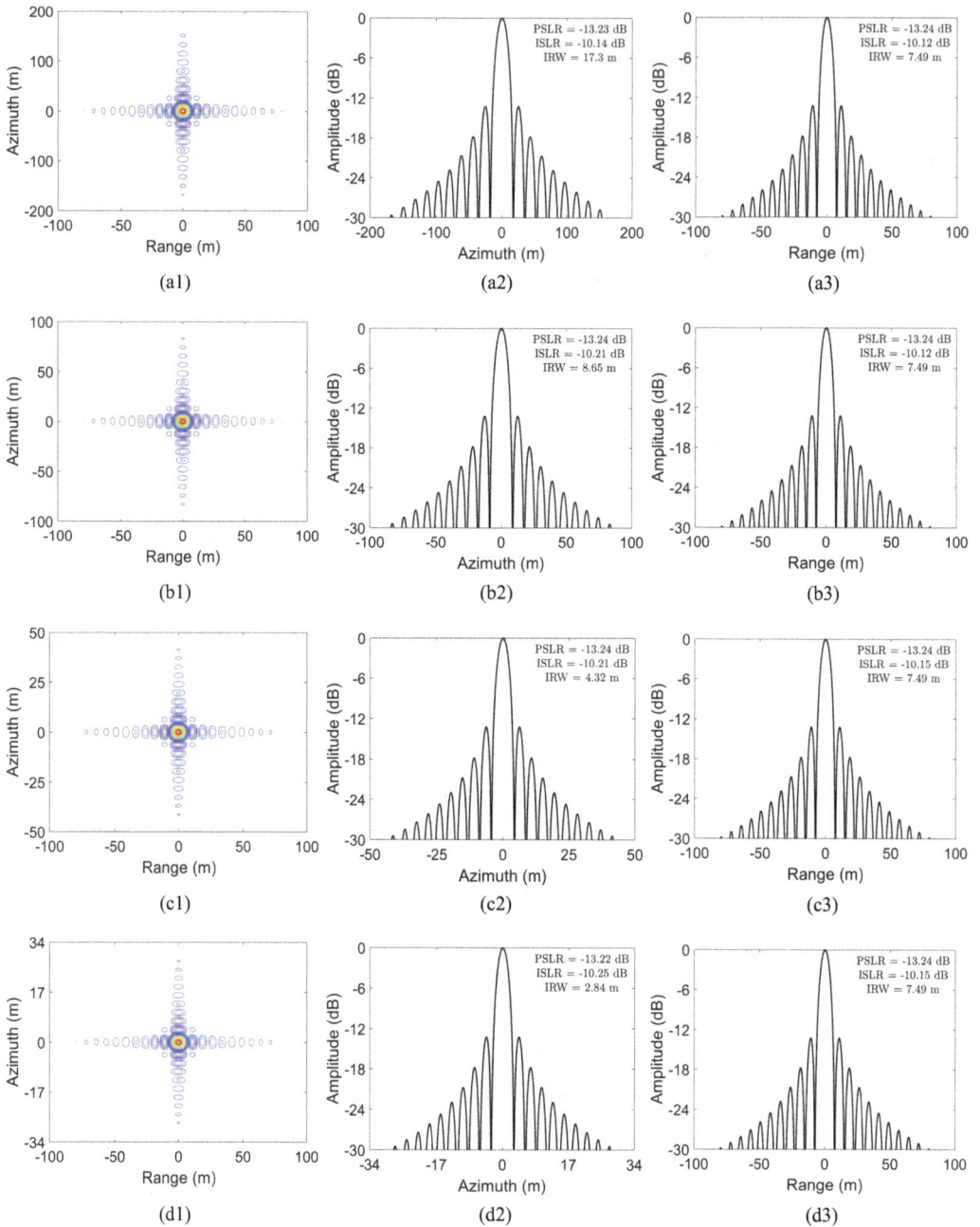

FIGURE 6.26 The single-point target response of the MBSAR by employing the PRM4 where the 1st (a1–d1), 2nd (a2–d2), and 3rd (a3–d3) columns respectively represent the response contour, azimuth, and range profiles, while 1st (a1–a3), 2nd (b1–b3), 3rd (c1–c3) and 4th (d1–d3) rows stand for the SATs of 100 s, 200 s, 400 s, and 600 s.

range resolution is primarily determined by the signal bandwidth and is little affected by the range model, we only present the azimuth profile here.

As seen, under a coarse azimuthal resolution, e.g., 100 s of SAT, HRE can achieve the MBSAR imaging, although its focusing quality is inferior to that obtained by the EHRE. Furthermore, a fully defocusing phenomenon is observed when applying HRE under a finer azimuthal resolution corresponding to an SAT of 600 s. By contrast, the EHRE can achieve good focusing quality with satisfactory

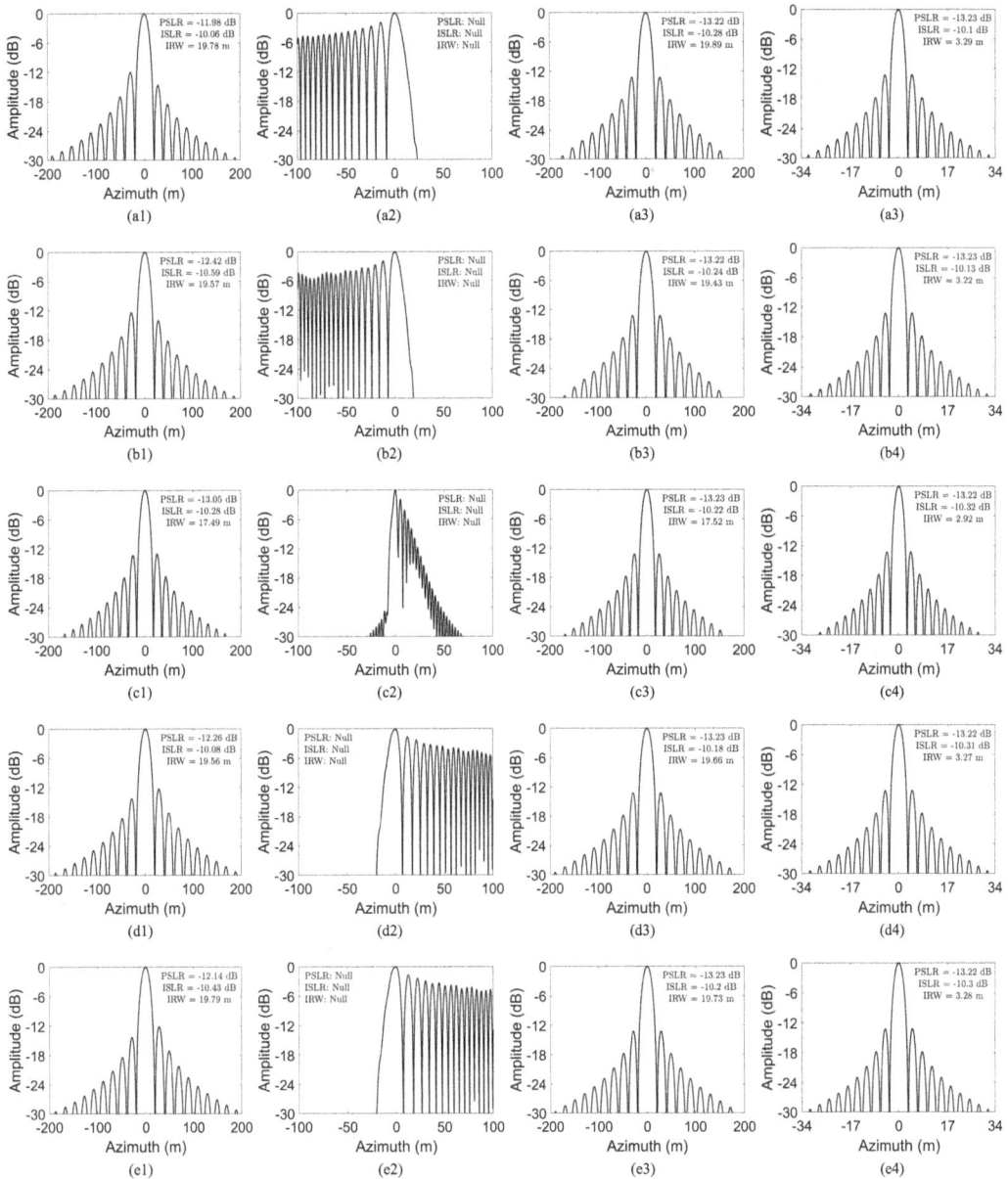

FIGURE 6.27 The single-point target response of the MBSAR, where the 1st (a1–e1), 2nd (a2–e2) represent azimuth profiles simulated by HRE under SATs of 100 and 600 s, 3rd (a3–e3) and 4th (a4–e4) columns stand for azimuth profiles simulated by EHRE under SATs of 100 and 600 s; while 1st (a1–a4), 2nd (b1–b4), 3rd (c1–c4), 4th (d1–d4), and 5th (e1–e4) rows indicates the AOLs of 60°, 120°, 180°, 240°, and 300°. "Null" refers to the unavailability of MBSAR imaging.

PSLR and ISLR values, though the azimuthal resolution suffers from small fluctuations with respect to the AOL. We see that imaging with EHRE can effectively yield well-focused images in the MBSAR.

Next, the responses of point targets under the stop-and-go and non-stop-and-go modes are simulated and compared to show the need to compensate for the stop-and-go error. In both cases, the MBSAR is located at the AOL of 90° at the zero-azimuth time. To clarify the impact of stop-and-go error, the SAT is set to 200 s and each look angle corresponds to a scene coordinate of (0, 0). Figures 6.28–6.30 displays the point target responses at look angles of 0.2°, 0.5°, and 0.8°, respectively.

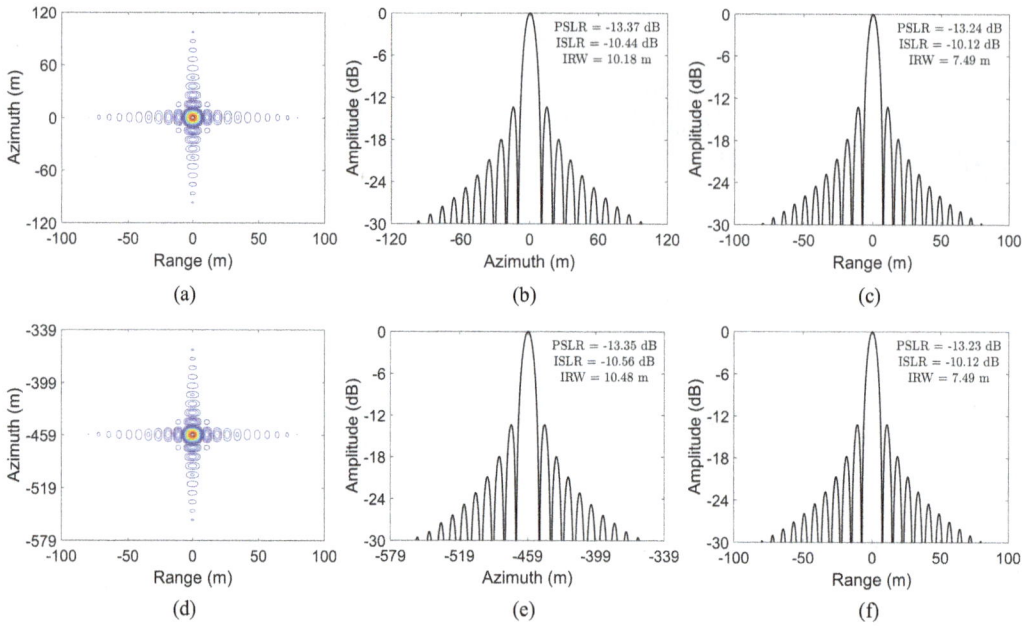

FIGURE 6.28 The single-point target response of the MBSAR with a look angle of 0.2°, where the 1st (a, d), 2nd (b, e), and 3rd (c, f) columns represent the response contour, azimuth, and range profiles, while 1st (a-c) and 2nd (d-f) rows stand for the target response under the stop-and-go assumption and that in non-stop-and-go mode.

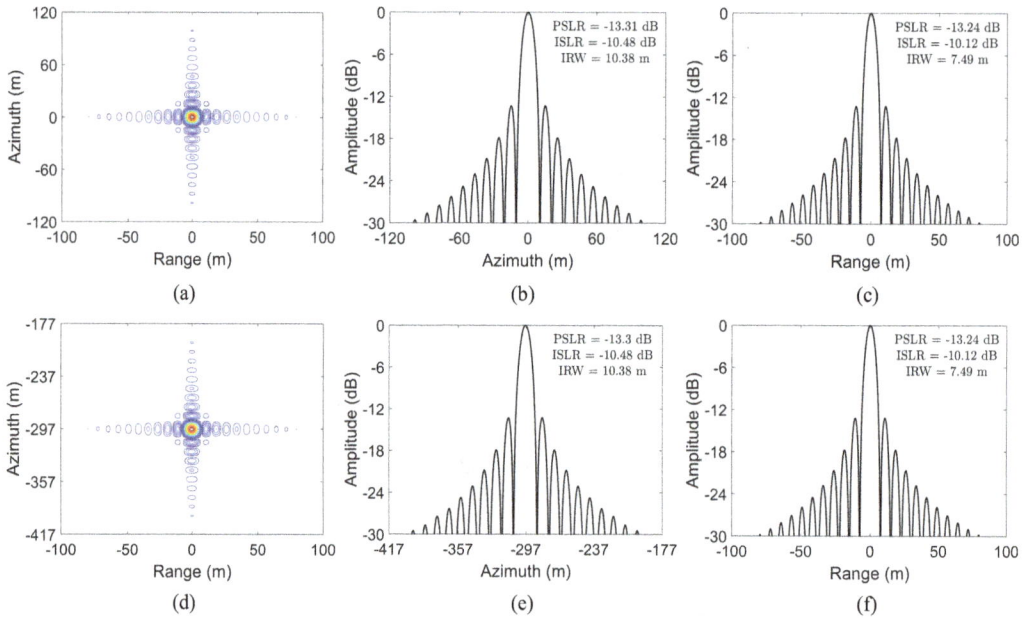

FIGURE 6.29 The single-point target response of the MBSAR with a look angle of 0.5°, where the 1st (a, d), 2nd (b, e) and 3rd (c, f) columns represent the response contour, azimuth, and range profiles, while 1st (a–c) and 2nd (d–f) rows stand for the target response under the stop-and-go assumption and that in non-stop-and-go mode.

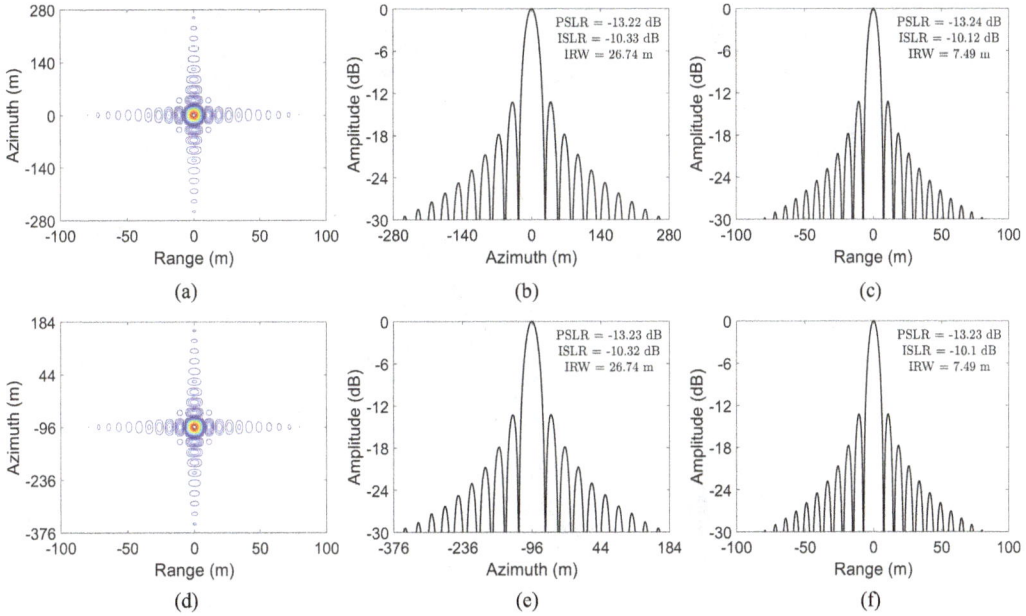

FIGURE 6.30 The single-point target response of the MBSAR with a look angle of 0.8°, where the 1st (a, d), 2nd (b, e) and 3rd (c, f) columns represent the response contour, azimuth, and range profiles, while 1st (a–c) and 2nd (d–f) rows stand for the target response under the stop-and-go assumption and that in non-stop-and-go mode.

It is clear that the range and azimuth resolutions, under the stop-and-go assumption, closely approximate their ideal values with desirable PSLR and ISRL. However, there is an azimuth shift that varies with the look angle, indicating that the stop-and-go error can result in spatially varying geometric deviation in the azimuth direction. In contrast, the geometric dislocation along the range direction is negligibly small, regardless of the look angle. It can be concluded that image pixels will be dislocated to an unacceptable level in the azimuth direction, if the stop-and-go error is not properly compensated. Therefore, it is imperative to prioritize addressing the stop-and-go error in the MBSAR image focusing.

6.4.1.2 Responses of Array Point Targets

We conduct simulations of array point targets using the PRM4 to further illustrate the focusing performance of MBSAR. The results are presented in Figure 6.31, with an SAT of 200 s or an azimuthal resolution of 9.74 m for zero-Doppler centroid beam pointing and a bandwidth of 15 MHz corresponding to a range resolution of 10 m. Besides, the zero-azimuth time of the MBSAR is set to 00:00:00 on Mar. 20, 2024 TDB, where the Moon is located at around 119.2°. Figure 6.31 demonstrates that the focusing algorithm is capable of effectively focusing both center point and edge point targets in the image scene with PRM4.

6.4.2 EFFECTS OF MBSAR'S INERTIAL MOTION ON TARGET RESPONSES

Under the stationary moon assumption, the MBSAR remains stationary within the SAT, with the Doppler shift induced by Earth's rotational motion [37–39]. In reality, the stationary moon assumption considerably distorts imaging formation and quality, or even invalidate the imaging capability of the MBSAR [3]. Specifically, the DFMR errors resulting from MBSAR's inertial motion could lead to image defocusing if not adequately compensated. Furthermore, the errors in both DFMR and Doppler frequency can alter the azimuthal resolution and introduce additional squint effects. As a result,

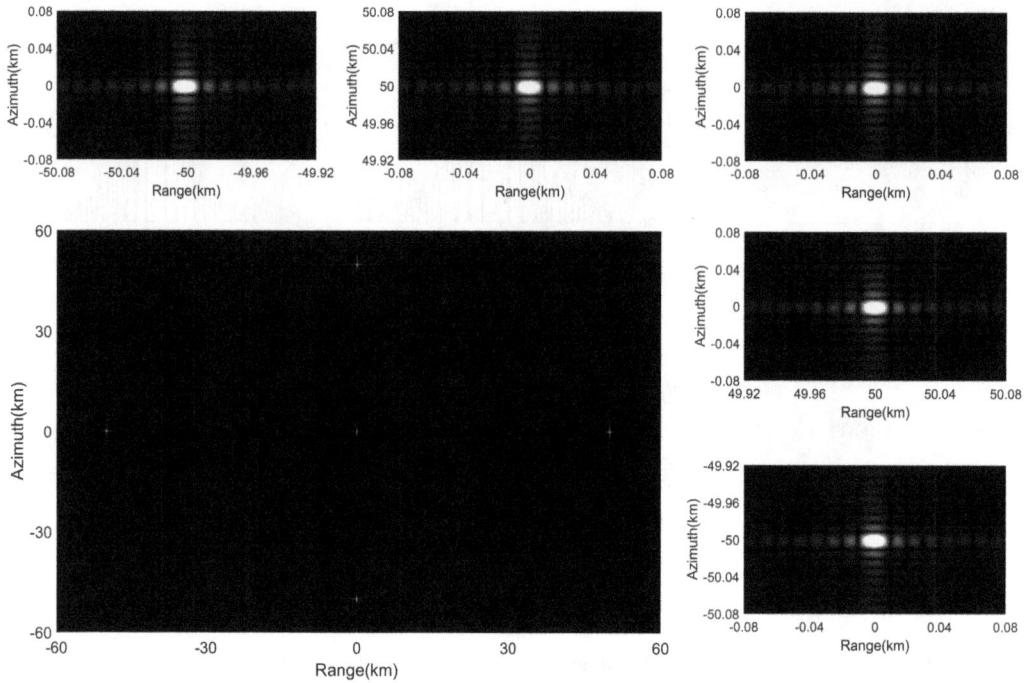

FIGURE 6.31 The responses of point array targets in the MBSAR with an image scene of (60 km × 60 km).

there can be noticeable differences between the actual azimuthal resolution and skewed angles of the target response compared to those assumed under the stationary moon assumption. This section will delve into the impacts of MBSAR's inertial motion on imaging performance.

6.4.2.1   The Effects of MBSAR's Inertial Motion on Focusing Quality

We assess the impact of MBSAR's inertial motion on the focusing quality by simulating point target responses, considering the DFMR error only while ignoring the Doppler frequency error. Using the same parameters as Figure 6.31, Figure 6.32 presents the point target responses at various look angles with the zero-Doppler centroid beam pointing. Figure 6.33 presents the point target response of MBSAR in the squint-looking mode with various beam pointing directions, both with and without MBSAR's inertial motion. In both cases, only the contours of target responses are provided for simplicity. Besides, the zero-azimuth time corresponds to 00:00:00 on Mar. 20, 2024, TDB.

As illustrated in Figures 6.32 and 6.33, the imaging performance of MBSAR is notably affected by its inertial motion, with the influence degree highly depending on the beam pointing direction. It is worth noting that in cases where the MBSAR's inertial motion is not factored in, all target responses (with azimuthal resolutions of around 10 m) are entirely defocused, whether the MBSAR is regulated by zero-Doppler steering or works in the squint-looking mode. It can be deduced that the DFMR error makes it unfeasible to generate an image in the MBSAR if the MBSAR's inertial motion is not accounted for. In fact, the imaging quality of the MBSAR deteriorates if the MBSAR's orbital determination is not sufficiently accurate to estimate the induced Doppler shift, as will be elaborated on in Chapter 7.

6.4.2.2   The Variations in Skewed Angle due to MBSAR's Inertial Motion

The stationary Moon assumption can induce errors in the equivalent squint angle and velocity due to the Doppler frequency and DFMR extras. This, in turn, affects the image quality because of additional skewed angles and varying resolution sizes. We simulate the target response with and without the MBSAR's inertial motion by applying the corresponding Doppler parameters in each

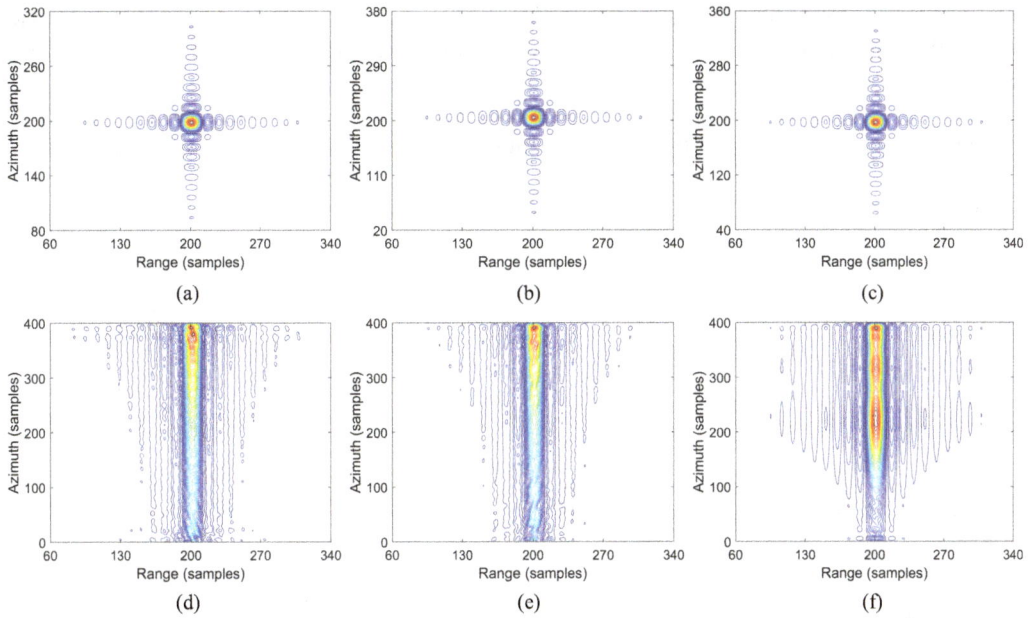

FIGURE 6.32 In the zero-Doppler steering mode, the single-point target responses of the MBSAR, where the upper (a, b, c) and lower (d, e, f) rows represent target responses with and without MBSAR's inertial motion. while 1st (a, d), 2nd (b, e), and 3rd (c, f) columns stand for the look angles of 0.2°, 0.5°, and 0.8°, respectively. The Doppler frequency error is excluded from this case.

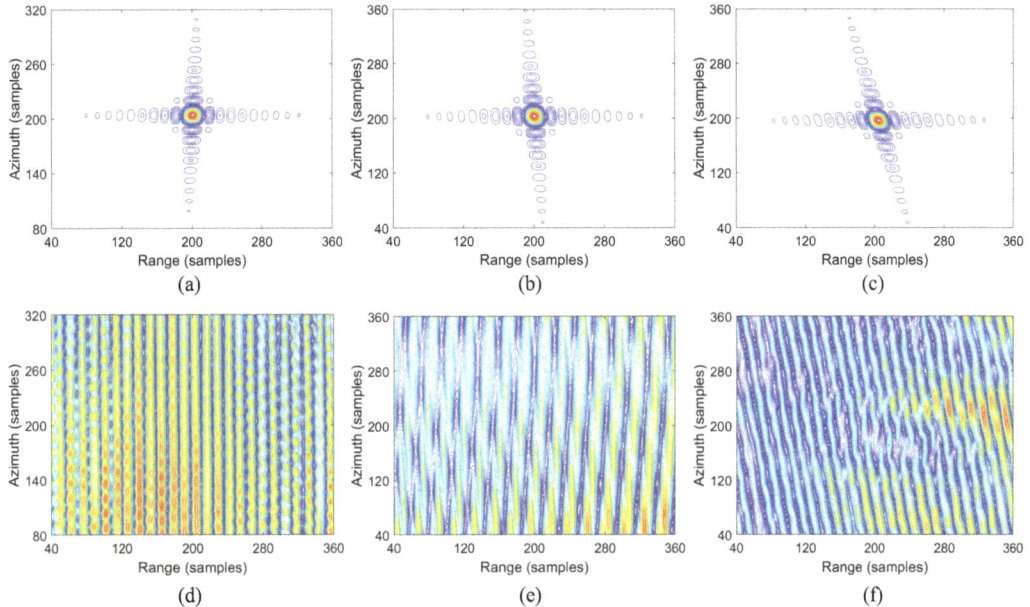

FIGURE 6.33 In the squint-looking mode, the single-point target responses of the MBSAR, where the upper (a, b, c) and lower (d, e, f) rows represent target responses with and without MBSAR's inertial motion. while 1st (a, d), 2nd (b, e), and 3rd (c, f) columns stand for beam pointing direction of $\theta_l = 0.2°$ & $\phi = 120°$, $\theta_l = 0.5°$ & $\phi = 120°$, and $\theta_l = 0.5°$ & $\phi = 135°$, respectively. The Doppler frequency error is excluded from this case.

case. Given that the squint effect depends on the position of the ground target relative to MBSAR, we introduce the factor Λ_{DTS}, the angular difference between the longitude of the TOI and MBSAR's nadir point.

Figure 6.34 shows the single-point target responses in both the presence and absence of MBSAR's inertial motion, and Table 6.1 lists numerical values of azimuthal IRW and skewed angle. Without

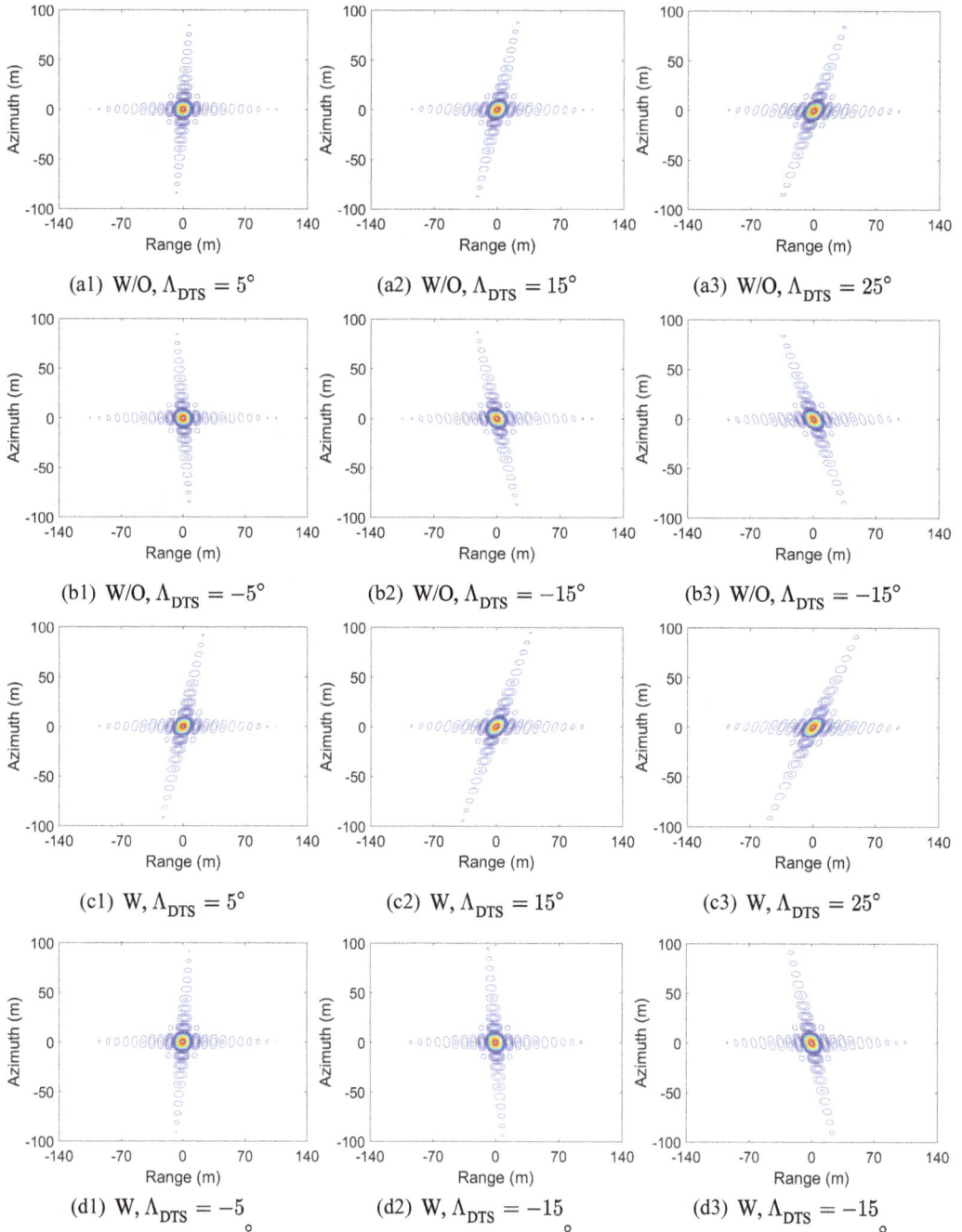

FIGURE 6.34 At a look angle of 0.5°, the point target response of the MBSAR under different conditions, where "W/O" means the target response in the absence of MBSAR's inertial motion, while "W" represents the target response in the context of MBSAR's inertial motion.

TABLE 6.1

Numeric Measures of Azimuthal IRW and Skewed Angle for the Point Target Response under Various Λ_{DTS}

	Parameters	$\Lambda_{DTS} = 5°$	$\Lambda_{DTS} = -5°$	$\Lambda_{DTS} = 15°$	$\Lambda_{DTS} = -15°$	$\Lambda_{DTS} = 25°$	$\Lambda_{DTS} = -25°$
W/O	Azimuthal IRW	8.85 m	8.84 m	9.02 m	8.97 m	9.35 m	9.27 m
	Skewed angle	4.15°	−4.18°	14.76°	−14.72°	25.12°	25.23°
W	Azimuthal IRW	9.41 m	9.58 m	9.39 m	9.91 m	9.53 m	10.38 m
	Skewed angle	12.34°	3.84°	23.12°	−5.78°	30.28°	−13.73°

considering the MBSAR's inertial motion, it is found that a symmetrical distribution in the skewed angle, with the azimuthal resolution remaining identical for both positive and negative values of Λ_{DTS}. This phenomenon is attributed to the fact that the Earth's rotation generates two opposite Doppler centroids, while its contribution to DFMR remains unaltered in either positive or negative Λ_{DTS}. The MBSAR's inertial motion causes image distortions in terms of varying skewed angles and azimuthal resolutions, and the extent of such distortion strongly depends on the value of Λ_{DTS}.

Notably, the MBSAR's inertial motion, in effect, produces a clockwise rotation for the skewed angle. Interestingly, the rotation degree of the skewed angle depends on the magnitude of Λ_{DTS}. Consequently, a higher degree of skewing is observed in the target response for the positive Λ_{DTS} case. In the case where Λ_{DTS} is negative, its magnitude exerts a more pronounced impact. For smaller magnitudes, the skewed angle tends to diminish; while for larger values of Λ_{DTS}, the orientation of the skewed angle undergoes a complete change.

Next, we show how the azimuthal resolution and skewed angle vary with the look angle under the effects of MBSAR's inertial motion by simulating the responses for a single-point target in the absence and presence of MBSAR's inertial motion with a fixed Λ_{DTS}, e.g., $-10°$. In this instance, the look angles of $0.2°$, $0.5°$, and $0.8°$, which correspond to low, middle, and high latitudes, respectively, are chosen. The outcomes of simulations are depicted in Figure 6.35; the measures of azimuthal IRW and skewed angle are given in Table 6.2.

Figure 6.35 and Table 6.2 show that the azimuthal resolution gets finer as the look angle increases when considering only Earth's rotation. The underlying reason is that the degree of relative decrease in DFMR resulting from Earth's rotation is more remarkable than that in beam crossing velocity. In reality, the MBSAR's inertial motion also affects the DFMR and beam crossing velocity, leading to a lesser variation in the azimuthal resolution of MBSAR with respect to the look angle. In addition, the skewed angle experiences change with varying look angles due to the influence of MBSAR's inertial motion. The variation in the skewed angle at look angles of $0.2°$ and $0.5°$, which pertain to mid-to-low latitudes, is comparatively less noteworthy when contrasted with that at a look angle of $0.8°$ that corresponds to higher latitude. This indicates that the effect of MBSAR's inertial motion is relatively trivial, though it cannot be neglected at mid-to-low latitudes. In such circumstances, the imaging performance of MBSAR is influenced mainly by Earth's rotation. As the antenna beam points to high latitudes, the skewed angle increases, signifying that the influence of MBSAR's inertial motion is more apparent at higher latitudes. Hence, considering the effects of MBSAR's inertial motion is essential for observing the Polar Regions.

Finally, the image distortions caused by the stationary Moon assumption are examined for various locations of MBSAR. By so doing, three locations, namely the ascending and descending nodes and the AOL of $90°$ are selected to carry out imaging simulation while keeping the rest parameters the same as those used in Figure 6.35. Figure 6.36 shows the point target responses of the MBSAR with a fixed look angle of $0.5°$ and Λ_{DTS} of $10°$. Table 6.3 summarizes the numeric measures of azimuthal IRW and skewed angles.

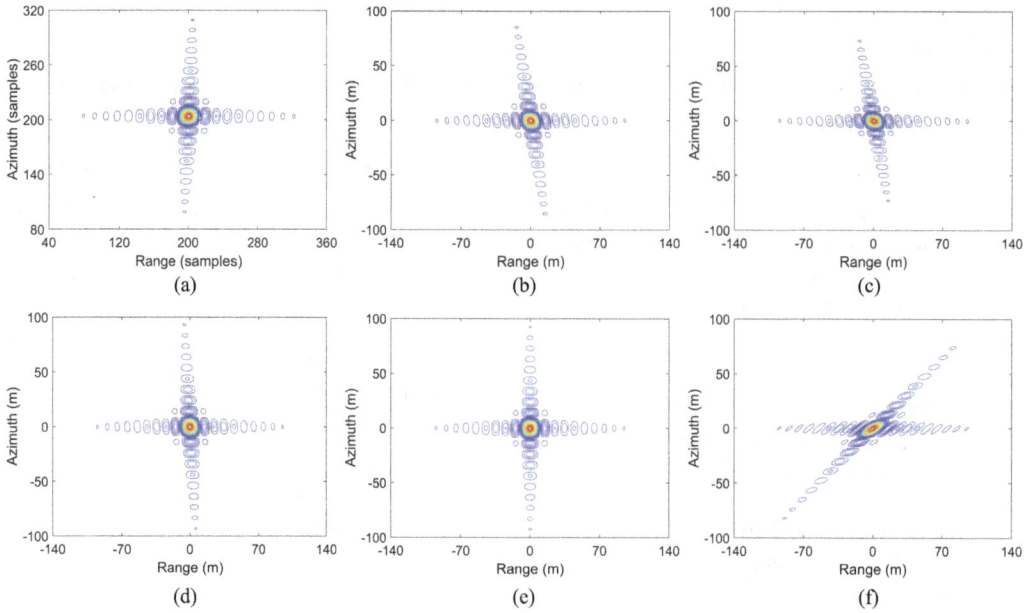

FIGURE 6.35 Under a fixed Λ_{DTS} of $-10°$, the point target response of the MBSAR under various look angles, where the upper (a, b, c) and lower (d, e, f) rows respectively depict the target responses with and without considering the MBSAR's inertial motion, and 1st (a, d), 2nd (b, e) and 3rd (c, f) columns stand for the look angles of 0.2°, 0.5° and 0.8°.

TABLE 6.2

Numeric Measures of Azimuthal IRW and Skewed Angle for the Point Target Response under a Fixed Λ_{DTS} of $-10°$

Look Angle	In the Absence of MBSAR's Inertial Motion		In the Context of MBSAR's Inertial Motion	
	Azimuthal IRW	Skewed Angle	Azimuthal IRW	Skewed Angle
0.2°	9.08 m	−10.86°	9.83 m	−4.19°
0.5°	8.98 m	−7.20°	9.74 m	−0.41°
0.8°	7.69 m	−8.89°	8.67 m	35.58°

It is evident from the point target response that the effects of MBSAR's inertial motion modify both the azimuthal resolution and skewed angle at different MBSAR locations. The MBSAR's inertial motion deteriorates the azimuthal resolution and gives rise to a rotation in the skewed angle; the extent of both variations depends on the MBSAR's location. It is important to note that when the MBSAR is located at the AOL of 90°, where its nadir point has a higher latitude, the variations in both azimuthal resolution and skewed angle are more significant than the rest two AOLs. This finding suggests that the effects of MBSAR's inertial motion become more pronounced as the latitude of the MBSAR's nadir point increases, posing an even greater challenge for producing high-quality SAR images.

MBSAR's inertial motion is subject to the orbital perturbation effects; thus, the Doppler shift and associated imaging performance of MBSAR experience degradation, owing to these effects, as well as the uncertainties in orbital determination. A more detailed discussion in such matters will be presented in Chapter 7.

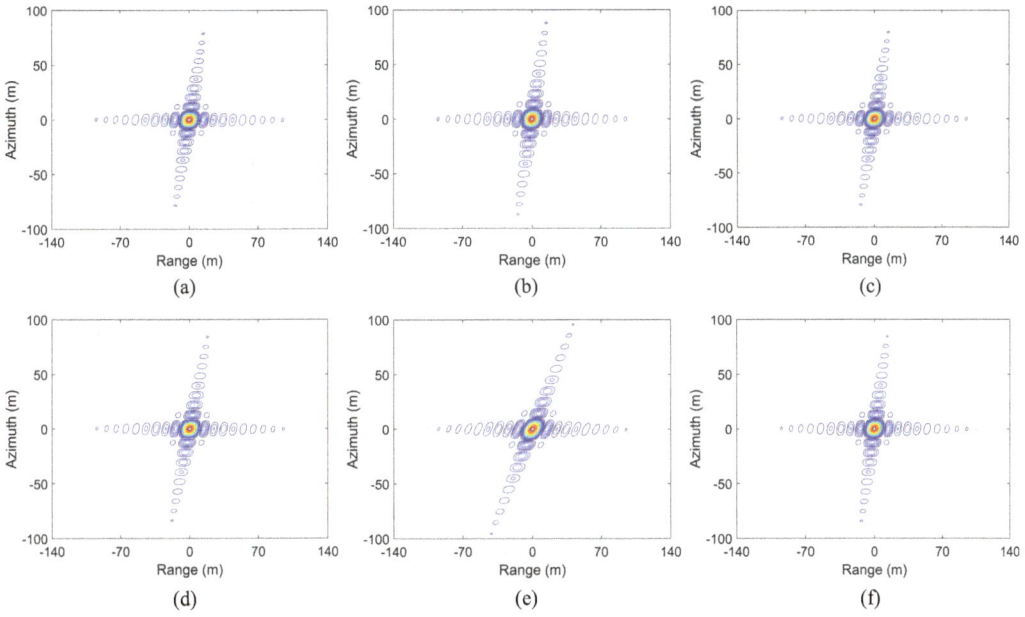

FIGURE 6.36 Under a fixed Λ_{DTS} of $10°$, the point target response of the MBSAR under various look angles, where the left (a, c, e) and right (b, d, f) columns respectively depict the target responses without and with considering the MBSAR's inertial motion, and 1st (a, b), 2nd (c, d) and 3rd (e, f) rows stand for the look angles of $0.2°$, $0.5°$ and $0.8°$.

TABLE 6.3

Numeric Measures of Azimuthal IRW and Skewed Angle for the Point Target Response under a Fixed Λ_{DTS} of $10°$ and a Look Angle of $0.5°$

AOL	In the Absence of MBSAR's Inertial Motion		In the Context of MBSAR's Inertial Motion	
	Azimuthal IRW	Skewed Angle	Azimuthal IRW	Skewed Angle
0°	8.33 m	6.40°	8.89 m	10.83°
90°	9.28 m	7.23°	11.25 m	26.61°
180°	8.39 m	6.94°	8.99 m	8.74°

APPENDIX A: A NEW STOP-AND-GO ERROR COMPENSATION METHOD

The HRE is proficient in describing the MBSAR's range history within a short timespan like round-trip time delay. Therefore, we posit that, during the signal propagation time delay, the range history under the stop-and-go assumption is approximated as follows:

$$R(\eta) \approx \sqrt{R_c^2 - 2R_c \sin\theta_r V_r \eta + V_r^2 \eta^2} \tag{A.1}$$

Once the stop-and-go error is taken into account, where the signal is transmitted at η and received after a round-trip time delay of t_d, the non-stop-and-go range history can be approximated by:

$$R_S(\eta) \approx \frac{R(\eta) + R(\eta + t_d)}{2} \tag{A.2}$$

with round-trip time delay expressed by:

$$t_d = \frac{2R_S(\eta)}{c} \approx \frac{R(\eta) + R(\eta + t_d)}{c} \tag{A.3}$$

With the aforementioned approximation, an explicit solution for the signal propagation time delay can be obtained. According to Eq. (A.3), we have:

$$\left[ct_d - R(\eta)\right]^2 = R^2(\eta + t_d) \tag{A.4}$$

The left-side of Eq. (A.4) can be written as:

$$\left[ct_d - R(\eta)\right]^2 = c^2 t_d^2 - 2ct_d R(\eta) + R^2(\eta) \tag{A.5}$$

The right-side of Eq. (A.4) requires:

$$\begin{aligned} R^2(\eta + t_d) &= R_c^2 - 2R_c V_r \sin\theta_r (\eta + t_d) + V_r^2 (\eta + t_d)^2 \\ &= \left[R_c^2 - 2R_c V_r \sin\theta_r \eta + V_r^2 \eta^2\right] - 2R_c V_r \sin\theta_r t_d + 2V_r^2 \eta t_d + V_r^2 t_d^2 \\ &= R^2(\eta) - 2R_c V_r \sin\theta_r t_d + 2V_r^2 \eta t_d + V_r^2 t_d^2 \end{aligned} \tag{A.6}$$

Taking account of Eqs. (A.5) and (A.6), one yields:

$$R^2(\eta) - 2ct_d R(\eta) + c^2 t_d^2 = R^2(\eta) - 2R_c V_r \sin\theta_r t_d + 2V_r^2 \eta t_d + V_r^2 t_d^2 \tag{A.7}$$

Eq. (A.7) can be rephrased equivalently as:

$$c^2 t_d^2 - V_r^2 t_d^2 = 2ct_d R(\eta) - 2R_c V_r \sin\theta_r t_d + 2V_r^2 \eta t_d \tag{A.8}$$

Solving for (A.8) yields the round-trip time delay, namely:

$$t_d = \frac{2cR(\eta) - 2R_c V_r \sin\theta_r + 2V_r^2 \eta}{c^2 - V_r^2} \tag{A.9}$$

To compensate for the stop-and-go error, it is necessary to focus on the round-trip time delay at $\eta = 0$, yielding the following expression:

$$t_{dc} = \frac{2cR_c - 2R_c V_r \sin\theta_r}{c^2 - V_r^2} \tag{A.10}$$

It follows that:

$$V_r t_{dc} = 2R_c \frac{cV_r^{-1} - \sin\theta_r}{c^2 V_r^{-2} - 1} \tag{A.11}$$

The stop-and-go error (in normal type, not in italic type) induces a Doppler frequency shift but exerts negligible effect on the DFMR, thus the compensation factor for the Doppler frequency error can be approximated by:

$$\begin{aligned} \delta r_{\mathrm{nsg1}} &= \frac{V_r^2 \cos^2\theta_r}{2R_c} \cdot t_{dc} = V_r t_{dc} \cdot \frac{V_r \cos^2\theta_r}{2R_c} \\ &= \frac{2cR_c \cos^2\theta_r - 2R_c V_r \cos^2\theta_r \sin\theta_r}{c^2 V_r^{-2} - 1} \end{aligned} \tag{A.12}$$

Besides, the stop-and-go error gives rise to a central slant range error that is contingent upon the Doppler shift experienced during the propagation time delay; such an error can be approximated by:

$$\delta r_{nsg0} \approx -V_r \sin\theta_r \frac{t_{dc}}{2} + \frac{V_r^2 \cos^2\theta_r}{2R_c}\left(\frac{t_{dc}}{2}\right)^2 \tag{A.13}$$

The factor '$t_{dc}/2$' is attributed to the round-trip time delay of the MBSAR signal. Note (A.13) is an ideal assumption, which may not be the case in practical situations.

Taking account of Eqs. (A.10) and (A.11), yielding:

$$\begin{aligned}\delta r_{sag0} &\approx \frac{V_r t_{dc}}{8R_c}\left(V_r t_{dc}\cos^2\theta_r - 4R_c \sin\theta_r\right)\\ &= \frac{R_c}{2}\cdot\frac{cV_r^{-1}-\sin\theta_r}{c^2 V_r^{-2}-1}\cdot\left(\frac{cV_r^{-1}\cos^2\theta_r - \cos^2\theta_r \sin\theta_r}{c^2 V_r^{-2}-1} - 2\sin\theta_r\right)\end{aligned} \tag{A.14}$$

APPENDIX B: THE DERIVATION OF THE EHRE

In the MBSAR, the EHRE takes the following form:

$$R_{EHRE}(\eta) = \sqrt{R_c^2 - 2R_c \sin\theta_r V_r \eta + V_r^2 \eta^2} + \Delta_c \eta^3 + \Delta_q \eta^4 \tag{A.15}$$

To obtain the EHRE parameters and ensure ease of signal processing, we can perform a Taylor series expansion of the EHRE centered at $\eta = 0$, as:

$$R_{EHRE}(\eta) = R_c + H_1\eta + H_2\eta^2 + H_3\eta^3 + H_4\eta^4 \tag{A.16}$$

where:

$$\begin{cases} H_1 = -V_r \sin\theta_r \\ H_2 = V_r^2 \dfrac{\cos^2\theta_r}{2R_c} \\ H_3 = V_r^3 \dfrac{\cos^2\theta_r \sin\theta_r}{2R_c^2} + \Delta_c \\ H_4 = V_r^4 \dfrac{6\sin^2\theta_r - 5\sin^4\theta_r - 1}{8R_c^3} + \Delta_q \end{cases} \tag{A.17}$$

If the EHRE can accurately describe the range history of MBSAR, then the Doppler parameters need to be adjusted as follows:

$$f_{dc} = -\frac{2}{\lambda}\cdot\Gamma(2)H_1 = \frac{2}{\lambda}V_r \sin\theta_r \tag{A.18}$$

$$f_{dr} = \frac{2}{\lambda}\cdot\Gamma(3)H_2 = \frac{2}{\lambda}\frac{V_r^2 \cos^2\theta_r}{R_c} \tag{A.19}$$

$$f_{ddr} = \frac{2}{\lambda}\cdot\Gamma(4)H_3 = \frac{12}{\lambda}\left(\frac{V_r^3 \cos^2\theta_r \sin\theta_r}{2R_c^2} + \Delta_c\right) \tag{A.20}$$

and:

$$f_{dddr} = \frac{2}{\lambda} \cdot \Gamma(5) H_4 = \frac{48}{\lambda} \left(-\frac{V_r^4 - 6V_r^4 \sin^2 \theta_r + 5V_r^4 \sin^4 \theta_r}{8R_c^3} + \Delta_q \right) \tag{A.21}$$

According to Eqs. (A.19) and (A.20), the square norm of equivalent velocity is:

$$V_r^2 = \left(V_r \sin \theta_r \right)^2 + V_r^2 \cos^2 \theta_r = \left(\frac{\lambda f_{dc}}{2} \right)^2 + \frac{\lambda R_c}{2} f_{dr} \tag{A.22}$$

Then:

$$V_r = 0.5 \sqrt{\lambda^2 f_{dc}^2 + 2\lambda R_c f_{dr}} = 0.5\lambda \sqrt{f_{dc}^2 + 2\lambda^{-1} R_c f_{dr}} \tag{A.23}$$

By substituting Eq. (A.24) into Eq. (A.19), we have:

$$\theta_r = \sin^{-1} \left(\frac{\lambda f_{dc}}{2V_r} \right) = \sin^{-1} \left(\frac{f_{dc}}{\sqrt{f_{dc}^2 + 2\lambda^{-1} R_c f_{dr}}} \right) \tag{A.24}$$

From Eq.(A.21), one can derive the cubical factor in EHRE as:

$$\Delta_c = \frac{\lambda}{12} f_{ddr} - \frac{V_r^3 \cos^2 \theta_r \sin \theta_r}{2R_c^2} = \frac{\lambda}{12} f_{ddr} - \frac{1}{8R_c} \cdot \left(2\frac{V_r^2 \cos^2 \theta_r}{R_c} \right) \cdot \left(2V_r \sin \theta_r \right) \tag{A.25}$$

Taking account of Eqs. (A.19) and (A.20), Eq. (A.25) can be re-expressed as:

$$\Delta_c = \frac{\lambda}{12} f_{ddr} - \frac{\lambda^2}{8R_c} f_{dr} f_{dc} \tag{A.26}$$

From Eq.(A.22), the quartic factor in EHRE has the following expression:

$$\Delta_q = \frac{\lambda f_{dddr}}{48} + \frac{V_r^4 - 6V_r^4 \sin^2 \theta_r + 5V_r^4 \sin^4 \theta_r}{8R_c^3} \tag{A.27}$$

The numerator in the 2nd-term of Eq. (A.27) can be re-written as:

$$\begin{aligned} V_r^4 - 6V_r^4 \sin^2 \theta_r + 5V_r^4 \sin^4 \theta_r &= V_r^4 \left(\left[1 - \sin^2 \theta_r - 5 \cdot \left(\sin^2 \theta_r - \sin^4 \theta_r \right) \right] \right) \\ &= V_r^4 \left(\cos^2 \theta_r - 5\sin^2 \theta_r \cos^2 \theta_r \right) \\ &= V_r^4 \left[\cos^2 \theta_r \left(1 - \sin^2 \theta_r \right) - 4\sin^2 \theta_r \cos^2 \theta_r \right] \\ &= V_r^4 \cos^4 \theta_r - 4V_r^4 \sin^2 \theta_r \cos^2 \theta_r \end{aligned} \tag{A.28}$$

It follows that:

$$\begin{aligned} \Delta_q &= \frac{\lambda f_{dddr}}{48} + \frac{V_r^4 \cos^4 \theta_r - 4V_r^4 \sin^2 \theta_r \cos^2 \theta_r}{8R_c^3} \\ &= \frac{\lambda f_{dddr}}{48} + \frac{1}{32R_c} \frac{4V_r^4 \cos^4 \theta_r}{R_c^2} - \frac{1}{16R_c^2} \frac{2V_r^2 \cos^2 \theta_r}{R_c} \cdot 4V_r^2 \sin^2 \theta_r \end{aligned} \tag{A.29}$$

Taking account of Eqs. (A.19)–(A.20), one arrives at:

$$\Delta_q = \frac{\lambda f_{dddr}}{48} + \frac{\lambda^2 f_{dr}^2}{32 R_c} - \frac{\lambda^3 f_{dr} f_{dr}^2}{16 R_c^2} \tag{A.30}$$

APPENDIX C: THE PRINCIPLE OF MSR

Consider a series function without the constant term that follows the form of:

$$y = a_1 x + a_2 x^2 + a_3 x^3 + \cdots \tag{A.31}$$

The given expression provides the series expansion for the inverse function of Eq. (A.31), as:

$$x = A_1 y + A_2 y^2 + A_3 y^3 + \cdots \tag{A.32}$$

After substituting Eq. (A.31) into Eq. (A.32), we have:

$$
\begin{aligned}
y = a_1 \left(A_1 y + A_2 y^2 + A_3 y^3 \right) \\
+ a_2 \left(A_1 y + A_2 y^2 + A_3 y^3 \right)^2 + a_3 \left(A_1 y + A_2 y^2 + A_3 y^3 \right)^3 + \cdots
\end{aligned}
\tag{A.33}
$$

Expanding Eq. (6.33) and collecting similar terms yields:

$$y = a_1 A_1 y + \left(a_2 A_1^2 + a_1 A_2 \right) y^2 + \left(a_3 A_1^3 + 2 a_1 A_1 A_2 + a_1 A_3 \right) y^3 + \cdots \tag{A.34}$$

Comparing the left and right sides of Eq. (A.34), one arrives at:

$$
\begin{cases}
a_1 A_1 = 1 \\
a_2 A_1^2 + a_1 A_2 = 0 \\
a_3 A_1^3 + 2 a_1 A_1 A_2 + a_1 A_3 = 0
\end{cases}
\tag{A.35}
$$

It is ready to yield the coefficients of Eq. (A.32) from Eq. (A.35), namely:

$$
\begin{cases}
A_1 = a_1^{-1} \\
A_2 = -a_1^{-3} a_2 \\
A_3 = a_1^{-5} \left(2 a_2^2 - a_1 a_3 \right)
\end{cases}
\tag{A.36}
$$

REFERENCES

[1] Z. Xu and K. S. Chen, "On Signal Modeling of Moon-Based Synthetic Aperture Radar (SAR) Imaging of Earth," *Remote Sensing*, vol. 10, no. 3, p. 486, Mar. 2018.
[2] Z. Xu, K. S. Chen, and G. Q. Zhou, "Effects of the Earth's Irregular Rotation on the Moon-Based Synthetic Aperture Radar Imaging," *IEEE ACCESS*, vol. 7, pp. 155014–155027, Oct. 2019.
[3] Z. Xu and K. S. Chen, "Effects of the Earth's Curvature and Lunar Revolution on the Imaging Performance of the Moon-Based Synthetic Aperture Radar," *IEEE Transactions on Geoscience and Remote Sensing*, vol. 57, no. 8, pp. 5868–5882, Mar. 2019.
[4] A. Renga and A. Moccia, "Moon-Based Synthetic Aperture Radar: Review and Challenges," *Proc. IEEE International Geoscience and Remote Sensing Symposium*, pp. 3708–3711, Jul. 2016.

[5] Z. Xu, K. S. Chen, and H. Guo, "Doppler Estimation with "Non-Stop-and-Go" Assumption in Moon-Based SAR Imaging," *Proc. 2018 IEEE International Geoscience and Remote Sensing Symposium*, Valencia, Spain, pp. 7809–7812, Jul. 2018.

[6] Z. Xu and K. S. Chen, "Effects of the Stop-and-Go Approximation on the Lunar-Based SAR Imaging," *IEEE Geoscience and Remote Sensing Letters*, Apr. 2021, doi: 10.1109/LGRS.2021.3070323

[7] Z. Xu and K. S. Chen, "Temporal-Spatial Varying Background Ionospheric Effects on the Moon-Based Synthetic Aperture Radar Imaging: A Theoretical Analysis," *IEEE ACCESS*, vol. 6, pp. 66767–66786, Jul. 2018.

[8] L. M. H. Ulander, H. Hellsten, and G. Stenstrom, "Synthetic Aperture Radar Processing Using Fast Factorized Back Projection," *IEEE Transactions on Aerospace and Electronic Systems*, vol. 39, no. 3, pp. 760–776, Jul. 2003, doi: 10.1109/TAES.2003.1238734

[9] J. Dong, et al., "An Analysis of Spatiotemporal Baseline and Effective Spatial Coverage for Lunar-Based SAR Repeat-Track Interferometry," *IEEE Journal of Selected Topics in Applied Earth Observations and Remote Sensing*, vol. 12, no. 9, pp. 3458–3469, Sep. 2019.

[10] S. V. Tsynkov, "On the Use of Start–Stop Approximation for Spaceborne SAR Imaging," *SIAM Journal on Imaging Sciences*, vol. 2, no. 2, pp. 646–669, 2009.

[11] D. Liang, et al., "Processing of Very High Resolution GF-3 SAR Spotlight Data with Non-Start–Stop Model and Correction of Curved Orbit," *IEEE Journal of Selected Topics in Applied Earth Observations and Remote Sensing*, vol. 13, pp. 2112–2122, May. 2020, doi: 10.1109/JSTARS.2020.2986862

[12] F. T. Ulaby and D. G. Long, *Microwave Radar and Radiometric Remote Sensing*. Ann Arbor: Univ. Michigan Press, 2014.

[13] A. Moreira, et al., "A Tutorial on Synthetic Aperture Radar," *IEEE Geoscience and Remote Sensing Magazine*, vol. 1, no. 1, pp. 6–43, Mar. 2013.

[14] R. Wang, et al., "Focus FMCW SAR Data Using the Wavenumber Domain Algorithm," *IEEE Transactions on Geoscience and Remote Sensing*, vol. 48, no. 4, pp. 2109–2118, Apr. 2010.

[15] J. C. Curlander and R. N. McDonough, *Synthetic Aperture Radar: Systems and Signal Processing*. New York: Wiley, 1991.

[16] Z. Xu, K. S. Chen, and H. Guo, "On Azimuthal Resolution of the Lunar-Based SAR under the Orbital Perturbation Effects," *IEEE Transactions on Geoscience and Remote Sensing*, vol. 61, Apr. 2023, doi: 10.1109/TGRS.2023.3266548

[17] F. M. Henderson and A. J. Lewis, *Manual of Remote Sensing: Principles and Applications of Imaging Radar*. New York: Wiley, 1998.

[18] A. Moreira and H. Yonghong, "Airborne SAR Processing of Highly Squinted Data Using a Chirp Scaling Approach with Integrated Motion Compensation," *IEEE Transactions on Geoscience and Remote Sensing*, vol. 32, no. 5, pp. 1029–1040, Sep. 1994, doi: 10.1109/36.312891

[19] M. I. Skolnik, *Radar Handbook*, 3rd ed. New York: McGraw-Hill Education, 2008.

[20] M. A. Richards, *Fundamentals of Radar Signal Processing*. New York: McGraw-Hill Education, 2014.

[21] G. C. Sun, et al., "Spaceborne Synthetic Aperture Radar Imaging Algorithms: An Overview," *IEEE Geoscience and Remote Sensing Magazine*, vol. 10, no. 1, pp. 161–184, Mar. 2022.

[22] K. S. Chen, *Principles of Synthetic Aperture Radar: A System Simulation Approach*. Boca Raton: CRC Press, 2015.

[23] A. Moreira, J. Mittermayer, and R. Scheiber, "Extended Chirp Scaling Algorithm for Air- and Spaceborne SAR Data Processing in Stripmap and ScanSAR Imaging Modes," *IEEE Transactions on Geoscience and Remote Sensing*, vol. 34, no. 5, pp. 1123–1136, Sep. 1996, doi: 10.1109/36.536528

[24] Y. Luo, et al., "A Novel High-Order Range Model and Imaging Approach for High-Resolution LEO SAR," *IEEE Transactions on Geoscience and Remote Sensing*, vol. 52, no. 6, pp. 3473–3485, Jun. 2014, doi: 10.1109/TGRS.2013.2273086

[25] I. G. Cumming and F. H. Wong, *Digital Signal Processing of Synthetic Aperture Radar Data: Algorithms and Implementation*. Boston: Artech House, 2005.

[26] M. Bao, M. Xing, Y. Wang, and Y. Li, "Two-Dimensional Spectrum for MEO SAR Processing Using a Modified Advanced Hyperbolic Range Equation," *Electronics Letters*, vol. 47, no. 18, pp. 1043–1045, Sep. 2011.

[27] L. Huang, X. Qiu, D. Hu, and C. Ding, "Focusing of Medium-Earth-Orbit SAR with Advanced Nonlinear Chirp Scaling Algorithm," *IEEE Transactions on Geoscience and Remote Sensing*, vol. 49, no. 1, pp. 500–508, Jan. 2011.

[28] L. Huang, X. Qiu, D. Hu, and C. Ding, "An Advanced 2-D Spectrum for High-Resolution and MEO Spaceborne SAR," *Proc. 2009 2nd Asian-Pacific Conference on Synthetic Aperture Radar*, Xi'an, China, pp. 447–450, Oct. 2009, doi: 10.1109/APSAR.2009.5374284

[29] B. Zhao, et al., "An Accurate Range Model Based on the Fourth-Order Doppler Parameters for Geosynchronous SAR," *IEEE Geoscience and Remote Sensing Letters*, vol. 11, no. 1, pp. 205–209, Jan. 2014, doi: 10.1109/LGRS.2013.2252878

[30] E. Key, E. Fowle, and R. Haggarty, "A Method of Designing Signals of Large Time-Bandwidth Product," *IRE Int. Conv. Rec*, vol. 4, pp. 146–155, Mar. 1961.

[31] Y. L. Neo, F. Wong, and I. G. Cumming, "A Two-Dimensional Spectrum for Bistatic SAR Processing Using Series Reversion," *IEEE Geoscience and Remote Sensing Letters*, vol. 4, no. 1, pp. 93–96, Jan. 2007.

[32] Y. L. Neo, F. H. Wong, and I. G. Cumming, "Processing of Azimuth-Invariant Bistatic SAR Data Using the Range Doppler Algorithm," *IEEE Transactions on Geoscience and Remote Sensing*, vol. 46, no. 1, pp. 14–21, Jan. 2008, doi: 10.1109/TGRS.2007.909090

[33] G. Franceschetti and R. Lanari, *Synthetic Aperture Radar Processing*. Boca Raton: CRC Press, 1999.

[34] G. B. Arfken, H. J. Weber, and F. E. Harris, *Mathematical Methods for Physicists: A Comprehensive Guide*. Oxford, UK: Academic press, 2011.

[35] B. Boashash, *Time-Frequency Signal Analysis and Processing: A Comprehensive Reference*. Oxford: Academic Press, 2015.

[36] Z. Xu, K. S. Chen, and G. Q. Zhou, "Zero-Doppler Centroid Steering for the Moon-Based Synthetic Aperture Radar: A Theoretical Analysis," *IEEE Geoscience and Remote Sensing Letters*, vol. 17, no. 7, Jul. 2020.

[37] G. Fornaro, et al., "Potentials and Limitations of Moon-Borne SAR Imaging," *IEEE Transactions on Geoscience and Remote Sensing*, vol. 48, no. 7, pp. 3009–3019, Apr. 2010.

[38] A. Moccia and A. Renga, "Synthetic Aperture Radar for Earth Observation from a Lunar Base: Performance and Potential Applications," *IEEE Transactions on Aerospace and Electronic Systems*, vol. 46, no. 3, pp. 1034–1051, Jul. 2010.

[39] H. Guo, et al., "Conceptual Study of Lunar-Based SAR for Global Change Monitoring," *Science China Earth Sciences*, vol. 57, no. 8, pp. 1771–1779, Aug. 2014.

7 Orbital Perturbation Effects in the MBSAR

7.1 INTRODUCTION

The lunar orbit is subject to various perturbing forces, which, in effect, give rise to temporal fluctuations in the Moon's orbital elements [1]. As a result, the inertial motion of Moon-Based SAR (MBSAR), which is susceptible to the perturbations of the lunar orbit, differs from that in a Keplerian orbit [2]. It should be noted that spaceborne SAR is also subject to orbital perturbations [3]. However, the Doppler shift and related imaging performance are affected, to a much lesser extent, by these perturbations [4]. One instance is the orbital precession, which exerts little influence on the imaging of a spaceborne SAR even when it operates in a geosynchronous orbit [5]. Therefore, it is reasonable to use the osculating orbit (the osculating orbit is a theoretical Keplerian orbit that has the identical orbital elements as the actual one at a given instant) when processing the signal of spaceborne SAR orbiting in low Earth orbit [6, 7].

When it comes to dealing with orbital perturbations in the MBSAR, the highly perturbed orbit, together with the distinctive imaging geometry, exerts a notable influence on the signal propagation and associated imaging performance [8]. Such effects diverge significantly from those encountered by spaceborne SAR systems, and thus, a precise orbital measurement is essential [9]. The ephemerides have been made available to facilitate exploration of the Moon; their precision is confined within the sub-meter accuracy under perturbation effects [10–13]. Note that primary sources of uncertainty in the lunar ephemeris arise from the orbital perturbations, thus we can explore the use of lunar ephemeris in MBSAR by checking the impact of orbital perturbations on the MBSAR's position and velocity determinations [2, 14, 15].

This chapter presents the lunar orbital precessions as case studies of perturbation effects. The precession refers to the modification in the orientation of an object's rotational axis [16]. The orbital precessions that impact MBSAR's inertial motion occur in two major forms within the lunar orbit: the apsidal precession, whose period is 8.85 years; and the nodal precession, with a cycle of 18.6 years [17]. The apsidal precession leads to an eastward shift of the major axis [18], giving rise to variations in the argument of perigee (AOP) over time. The nodal precession is governed by the relative motion of the plane in which the Moon revolves around Earth [19], resulting in temporal fluctuations in both the orbital inclination and RAAN.

As a result of the orbital precession effects, the MBSAR system experiences coordinate drifts compared to the scenario of an osculating orbit. These coordinate drifts cause changes in phase history, leading to potential Doppler errors in the MBSAR signal. Correspondingly, the MBSAR system encounters two inquiries under the effect above: firstly, whether or not the imaging performance is affected by orbital precessions; secondly, if the degradation in imaging performance is confirmed, what are the precision prerequisites for orbital determination to ensure good image quality. The crux here pertains to how the orbital precession effects impact the determinations of position and velocity in the MBSAR.

We present a quantitative analysis to ascertain the phase errors resulting from orbital precessions and evaluate their effects on MBSAR imaging. Subsequently, we establish a set of criteria for the orbital determination that comply with achieving satisfactory image quality under the precession effects. Expanding on this basis, we shall explore the precision demands for the MBSAR's position and velocity determinations. To facilitate the analysis above, we formulate a procedure for evaluating the perturbation effects per MBSAR's signal model.

DOI: 10.1201/9781003308430-7

7.2   THE EVALUATION METHODOLOGY FOR ORBITAL PERTURBATION EFFECTS

The MBSAR's inertial motion is subject to numerous perturbation forces [20]. It is not feasible to analyze effects of all those orbital perturbation factors on the imaging of MBSAR. Orbitally speaking, the perturbation effects can bring about fluctuations in orbital elements over time, eventually resulting in temporally varying range errors in the MBSAR signal when compared to osculating orbits. In this regard, it is possible to analyze orbital perturbation effects by scrutinizing the range history error that occurs with the corresponding temporal variation of the orbital element [2]. Below we develop a model to analyze the orbital perturbation effects on MBSAR imaging.

From Chapter 6, the slant range history of MBSAR is written as:

$$R_S\left(\eta\right) = R_c + R_1\eta + R_2\eta^2 + R_3\eta^3 + \cdots \tag{7.1}$$

Considering the orbital perturbations, Eq. (7.1) becomes:

$$R_{\text{SP}}\left(\eta\right) = \Delta R_p\left(\eta\right) + R_c + R_1\eta + R_2\eta^2 + R_3\eta^3 + \cdots \tag{7.2}$$

where the range history error caused by the orbital perturbations takes the form:

$$\Delta R_p\left(\eta\right) = r_{p0} + r_{p1}\eta + r_{p2}\eta^2 + \cdots \tag{7.3}$$

In general, the effects of orbital perturbation primarily give rise to range errors of 1st and 2nd orders that have a detrimental effect on MBSAR imaging. In contrast, the higher-order range errors exist, but they are negligibly small and have minimal impact on MBSAR imaging, as will be demonstrated in the subsequent sections.

By Eqs. (7.2) and (7.3), the MBSAR's range history can be reformulated as:

$$R_{\text{SP}}\left(\eta\right) = P_0 + P_1\eta + P_2\eta^2 + P_3\eta^3 + \cdots \tag{7.4}$$

where $P_0 = R_c + r_{p0}$, $P_1 = R_1 + r_{p1}$, $P_2 = R_2 + r_{p2}$, $P_3 = R_3$.

Figure 7.1 shows an illustrative comparison of MBSAR trajectories and corresponding slant range history with and without considering the orbital perturbation effects. It is clear that the Doppler history and properties of the MBSAR would be affected by such effects. As a result, severe image distortions would result without considering the orbital perturbation.

Following the procedures in Chapter 6, we can establish the signal model and 2-D spectrum of MBSAR in the context of range errors resulting from orbital perturbations. When accounting for the orbital perturbation effects, the effects of secondary range compression and range cell migration terms on imaging performances are comparatively small. Moreover, the residual phase term in the signal spectrum exerts no discernible impact on the imaging of MBSAR, though it is subjected to orbital perturbation. Therefore, the phase terms above will be excluded from the analysis, and their specific expressions, whose general forms can be found in Chapter 6, are not provided here. By comparison, the MBSAR's imaging performances are primarily linked to the azimuth compression (AC) and RC terms [2, 21]. Specifically, the AC term is dominant in determining the azimuth imagery quality, while the range compression (RC) term is responsible for the range imaging performance.

The RC and AC terms, together with the range history error caused by the orbital perturbation effects, are given by:

$$\Phi_{\text{rc\_per}}\left(f_\tau\right) = 2\pi\left[\left(-2P_0 + \frac{P_1^2}{2P_2} + \frac{P_1^3 P_3}{4P_2^3}\right)\frac{f_\tau}{c} - \frac{f_\tau^2}{2K_r}\right] \tag{7.5}$$

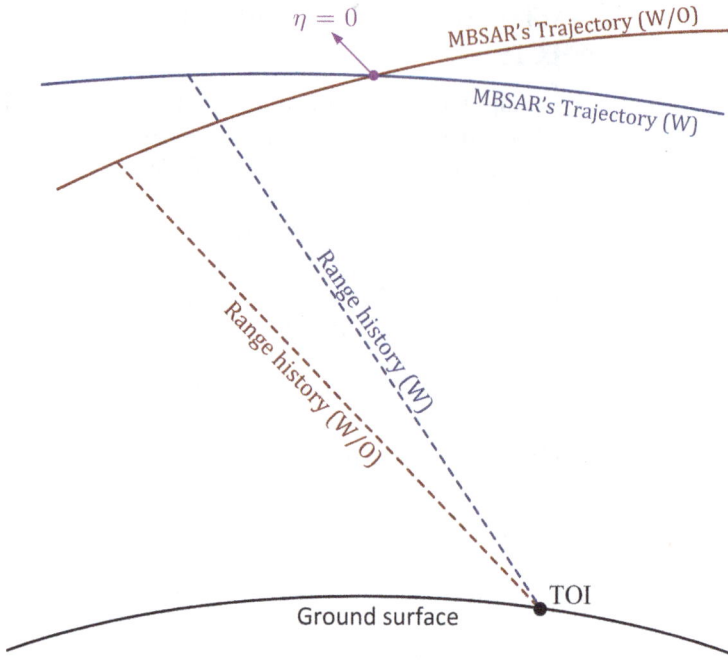

FIGURE 7.1 Schematic diagram of the MBSAR's trajectory and associated range history with (W) and without (W/O) orbital perturbations.

$$\Phi_{ac\_per}\left(f_\eta\right) = 2\pi\left[\left(\frac{P_1}{2P_2} + \frac{3P_1^2 P_3}{8P_2^3}\right)f_\eta + \frac{c}{f_c}\left(\frac{1}{8P_2} + \frac{3P_1 P_3}{16P_2^3}\right)f_\eta^2\right] \tag{7.6}$$

Given the RC and AC phase terms, the orbital perturbations cause geometric dislocation and distortion of the images, which can be assessed in both range and azimuth domains, as will be discussed in what follows.

7.2.1 Range Imaging Distortion under Orbital Perturbations

Let us consider only the range compression. The signal model under the orbital perturbation is:

$$s_{SAR\_per}\left(\tau,\eta\right) = IFFT_{range}\left\{S_{r\_per}\left(f_\tau,\eta\right)\exp\left[-j\Phi_{rc}\left(f_\tau\right)\right]\right\} \tag{7.7}$$

with:

$$\Phi_{rc}\left(f_\tau\right) = 2\pi\left[\left(-R_c + \frac{R_1^2}{2R_2} + \frac{R_1^3 R_3}{4R_2^3}\right)\frac{f_\tau}{c} - \frac{f_\tau^2}{2K_r}\right] \tag{7.8}$$

Eq. (7.7) can be re-expressed as:

$$s_{SAR\_per}\left(\tau,\eta\right) = IFFT_{range}\left\{S_r\left(f_\tau,\eta\right)\exp\left[-j\Phi_{rc}\left(f_\tau\right)\right]\exp\left[-j\left(\Phi_{rc}\left(f_\tau\right) - \Phi_{rc\_per}\left(f_\tau\right)\right)\right]\right\} \tag{7.9}$$

Taking account of Eqs. (7.5) and (7.8), yielding:

$$s_{\text{SAR\_per}}(\tau,\eta) = \text{IFFT}_{\text{range}}\left\{S_r(f_\tau,\eta)\exp\left[-j\Phi_{\text{rc}}(f_\tau)\right]\right.$$
$$\left.\exp\left[-j2\pi\frac{1}{c}\left(-2R_c + \frac{R_1^2}{2R_2} + \frac{R_1^3R_3}{4R_2^3} + 2P_0 - \frac{P_1^2}{2P_2} - \frac{P_1^3P_3}{4P_2^3}\right)f_\tau\right]\right\}$$

(7.10)

To derive the orbital perturbation effects on the range imaging, we shall make use of the shift property of the Fourier transform [22]:

$$g(t) \leftrightarrow G(f) \Rightarrow g(t - t_{\text{offset}}) \leftrightarrow G(f)\cdot\exp\left(-j2\pi f \cdot t_{\text{offset}}\right)$$

(7.11)

where t_{offset} is an offset in time. Eq. (7.11) illustrates that multiplying a signal by a complex exponential function with a negative linear phase term in the frequency domain results in a rightward (positive) shift in the time domain, and vice versa [23].

Note that the SAR data without considering the orbital perturbation effects satisfy the following relation [24]:

$$\begin{cases} s_{\text{SAR}}(\tau,\eta) = \text{IFFT}_{\text{range}}\left\{S_r(f_\tau,\eta)\exp\left[-j\Phi_{\text{rc}}(f_\tau)\right]\right\} \\ S_r(f_\tau,\eta)\exp\left[-j\Phi_{\text{rc}}(f_\tau)\right] = \text{FFT}_{\text{range}}\left\{s_{\text{SAR}}(\tau,\eta)\right\} \end{cases}$$

(7.12)

Using Eqs. (7.10) and (7.12), we have:

$$s_{\text{SAR\_per}}(\tau,\eta) = \text{IFFT}_{\text{range}}\left\{\text{FFT}_{\text{range}}\left\{s_{\text{SAR}}(\tau,\eta)\right\}\right.$$
$$\left.\cdot\exp\left[-j2\pi\frac{1}{c}\left(-2R_c + \frac{R_1^2}{2R_2} + \frac{R_1^3R_3}{4R_2^3} + 2P_0 - \frac{P_1^2}{2P_2} - \frac{P_1^3P_3}{4P_2^3}\right)f_\tau\right]\right\}$$

(7.13)

As seen, in the situation having only a linear phase error, the orbital perturbation effects exhibit no discernible influence on the range focusing. However, an image offset occurring along the range direction, namely a range offset, appears in the target response of the MBSAR. According to Eqs. (7.11) and (7.13), the range offset is given by:

$$L_{rg} = \frac{1}{2}\left(-2R_c + \frac{R_1^2}{2R_2} + \frac{R_1^3R_3}{4R_2^3} + 2P_0 - \frac{P_1^2}{2P_2} - \frac{P_1^3P_3}{4P_2^3}\right)$$

(7.14)

When the zero-Doppler steering mode is devised, the range offset can be simplified in the following form:

$$L_{rg} = r_{p0} - \frac{R_1}{2R_2}r_{p1} - \frac{r_{p1}^2}{4R_2}$$

(7.15)

Eq. (7.15) indicates that the range offset is correlated with the constant term and time-varying components of range history error and the Doppler parameters of the MBSAR.

7.2.2 Azimuth Imaging Distortion under Orbital Perturbations

We now treat the azimuth compression by recognizing the following expression:

$$s_{\text{SAR\_per}}(\tau,\eta) = \text{IFFT}_{\text{azimuth}}\left\{S_{r\_\text{per}}(\tau,f_\eta)\exp\left[-j\Phi_{\text{ac}}(f_\eta)\right]\right\}$$

(7.16)

with

$$\Phi_{ac}\left(f_\eta\right) = 2\pi\left[\left(\frac{R_1}{2R_2} + \frac{3R_1^2 R_3}{8R_2^3}\right)f_\eta + \frac{c}{f_c}\left(\frac{1}{8R_2} + \frac{3R_1 R_3}{16R_2^3}\right)f_\eta^2\right] \quad (7.17)$$

In the azimuth compression term in Eq. (7.17), the third-order and higher-order phase terms are ignored, as they contribute little to the orbital perturbation effects.

Similar to Eq. (7.12), the processed SAR data adhere to the following relationship [25]:

$$\begin{cases} s_{SAR}(\tau,\eta) = \text{IFFT}_{azimuth}\left\{S_r(\tau,f_\eta)\exp\left[-j\Phi_{ac}(f_\eta)\right]\right\} \\ S_r(\tau,f_\eta)\exp\left[-j\Phi_{ac}(f_\eta)\right] = \text{FFT}_{azimuth}\left\{s_{SAR}(\tau,\eta)\right\} \end{cases} \quad (7.18)$$

Subsequently, the signal model in Eq. (7.16) can be reformulated as:

$$s_{SAR\_per}(\tau,\eta) = \text{IFFT}_{azimuth}\left\{\text{FFT}_{azimuth}\left\{s_{SAR}(\tau,\eta)\right\}\exp\left[-j\left(\Phi_{ac}(f_\tau) - \Phi_{ac\_iono}(f_\tau)\right)\right]\right\}$$

$$= \text{IFFT}_{azimuth}\left\{\text{FFT}_{azimuth}\left\{s_{SAR}(\tau,\eta)\right\}\exp\left[-j2\pi\left[\left(\frac{R_1}{2R_2} + \frac{3R_1^2 R_3}{8R_2^3}\right.\right.\right.\right.$$

$$\left.\left.\left.\left. - \frac{P_1}{2P_2} - \frac{3P_1^2 P_3}{8P_2^3}\right)f_\eta + \frac{c}{f_c}\left(\frac{1}{8R_2} + \frac{3R_1 R_3}{16R_2^3} - \frac{1}{8P_2} - \frac{3P_1 P_3}{16P_2^3}\right)f_\eta^2\right]\right]\right\} \quad (7.19)$$

From Eq. (7.19), the time-varying component of range history error would likely cause geometric dislocation and defocusing phenomenon in azimuth imaging. The manifestation of geometric distortion can be perceived through a discernible azimuth offset, the magnitude of which is ascertained by:

$$L_{az} = V_g\left(\frac{R_1}{2R_2} + \frac{3R_1^2 R_3}{8R_2^3} - \frac{P_1}{2P_2} - \frac{3P_1^2 P_3}{8P_2^3}\right)$$

$$= -V_g\left(\frac{(R_1+r_{p1})}{2(R_2+r_{p2})} - \frac{R_1}{2R_2} + \frac{3(R_1+r_{p1})^2 R_3}{8(R_2+r_{p2})^3} - \frac{3R_1^2 R_3}{8R_2^3}\right) \quad (7.20)$$

where V_g is the beam-crossing velocity.

It is worth noting that the higher-order terms in Eq. (7.20) contribute little to the azimuth offset when the antenna beam is directed toward the zero-Doppler plane. To facilitate further analysis, we may approximate the azimuth offset to:

$$L_{az} \approx -V_g\frac{\left(2R_2^2 + 3R_1 R_3\right)}{4R_2^3}r_{p1} \quad (7.21)$$

Upon a closer examination of Eq. (7.21), it is evident that the azimuth offset is connected to the beam-crossing velocity, the 1st-order temporal derivative of the range error, and various orders of Doppler parameters.

The SAR principle suggests that the azimuth Quadratic Phase Error (QPE) in effect broadens the main lobe symmetrically. In addition, if the azimuth Cubic Phase Error (CPE) is sufficiently large, it could broaden the main lobe asymmetrically [21]. In the context of MBSAR, it may be unnecessary to delve into the azimuth CPE as the 3rd-order range history error, on which the azimuth CPE depends, is insignificant under orbital perturbation effects. Therefore, we may exclude the azimuth CPE term in our analysis. When considering the azimuth QPE, the 2nd-order phase error embedded in the AC phase term may lead to azimuth defocusing. Referring to Eq. (7.19), we can determine the azimuth QPE:

$$
\begin{aligned}
\mathrm{QPE}_{az} &= \frac{2\pi c}{f_c} \left| \frac{1}{8P_2} + \frac{3P_1 P_3}{16P_2^3} - \frac{1}{8R_2} - \frac{3R_1 R_3}{16R_2^3} \right| \left(\frac{B_D}{2} \right)^2 \\
&= \frac{2\pi c}{f_c} \left| \frac{1}{8(R_2 + r_{p2})} - \frac{1}{8R_2} + \frac{3(R_1 + r_{p1})R_3}{16(R_2 + r_{p2})^3} - \frac{3R_1 R_3}{16R_2^3} \right| f_{dr}^2 \left(\frac{T_{SAR}}{2} \right)^2
\end{aligned}
\tag{7.22}
$$

Subsequently, it is possible to make a further approximation of azimuth QPE as follows:

$$
\mathrm{QPE}_{az} \approx \frac{\pi f_c}{c} \left| -r_{p2} + \frac{3}{2} \frac{R_3}{R_2} r_{p1} \right| T_{SAR}^2
\tag{7.23}
$$

We note from Eq. (7.23) that the azimuth focusing is influenced by several factors, including the 1st-order and 2nd-order range history errors caused by the orbital perturbations, and the Doppler parameters, carrier frequency, and SAT of the MBSAR.

Thus far, we have established a framework for scrutinizing the orbital perturbation effects on SAR imaging quality. Henceforth, we shall employ this established framework to examine how two prevalent perturbations, nodal precession and apsidal precession, affect the imaging performance of MBSAR.

7.3 THE PRECESSION EFFECTS ON THE MBSAR IMAGING PERFORMANCE

Two predominant precessional movements of the lunar orbit affect the MBSAR's inertial motion, known as the apsidal and nodal precessions [26], wherein the apsidal precession causes temporal fluctuations in the AOP, and the nodal precession induces variations in the orbital inclination and RAAN over time. Therefore, it is ensured that each precession bears distinctive ramifications on MBSAR imaging, even though both phenomena lead to coordinate drifts and consequent Doppler errors in the signal of MBSAR. To delve into the effects of each precession, one may assume that the MBSAR's orbit remains impervious to other perturbations. Further, to conduct this analysis and maintain coherence with the prior ones [14, 15], the MBSAR is configured to be left-looking, with the zero-azimuth time set at 00:00:00 on March 20, 2024, TDB throughout this chapter. The forthcoming section shall first explicate the effects of nodal precession on MBSAR imaging.

7.3.1 THE NODAL PRECESSION EFFECTS ON MBSAR IMAGING

The nodal precession could result in cyclic variations in the orbital inclination and RAAN of the MBSAR over a time span of 18.6 years [15]. Figure 7.2 illustrates the temporal fluctuations in the orbital inclination and RAAN during three 18.6-year cycles. It is worth noting that the results presented here are similar but not entirely identical to those shown in Figure 2.11 since the latter represents the corresponding orbital elements of the Moon's barycenter.

As observed from Figure 7.2, the nodal precession leads to significant variations in the orbital inclination and RAAN; such variations in the MBSAR distinguish it from the spaceborne SAR, where the changes in the orbital inclination and RAAN are minuscule. Therefore, in spaceborne SAR, despite the presence of orbital precession, its effect on the focalization precision is trivial, given the ensuing minimal range error. In the instance of MBSAR, the period of precession is

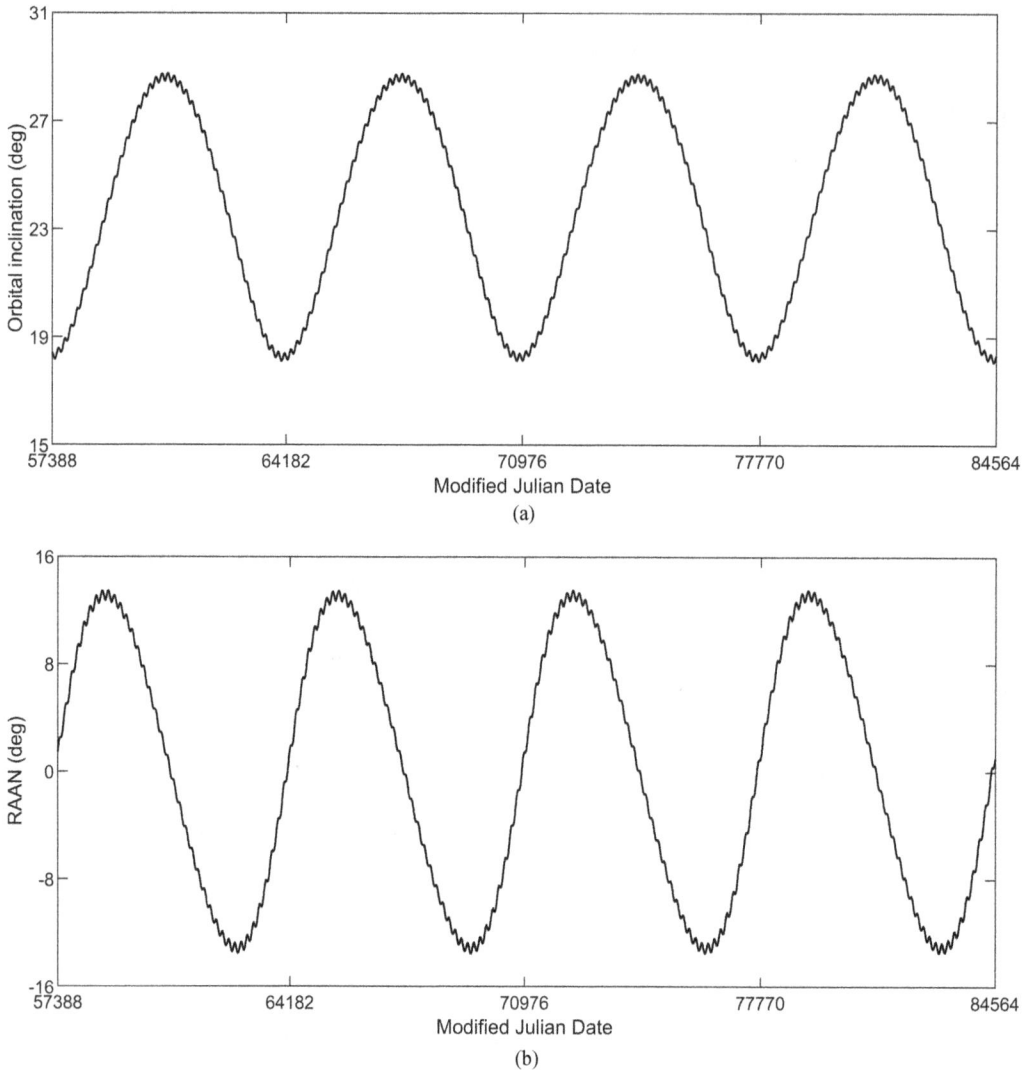

FIGURE 7.2 Within four consecutive cycles of nodal precession, the temporal variations of (a) orbital inclination, (b) RAAN, in the ECI coordinate.

considerably longer compared to spaceborne SAR. However, it is not the duration of the precession cycle that determines the geometric accuracy and focusing quality, but rather the magnitude of range history error. From this perspective, once the resulting range history error surpasses a certain threshold, the imaging capability of MBSAR may be impacted by nodal precession effects. Hence, it is important to examine whether the nodal precession influences the MBSAR's imaging. Such an investigation can be accomplished by examining the MBSAR's coordinate and associated range history errors.

The analysis that follows pertains to the time-varying orbital inclination and RAAN within the SAT of the MBSAR. Figure 7.3 presents the temporal variations of the RAAN and orbital inclination within an SAT of 600 s.

As is demonstrated in Figure 7.3, the most significant changes in the time-varying components of orbital inclination and RAAN are below 8×10^{-6} and 3×10^{-5} degrees, respectively. Besides, it is of significance to mention that the alterations in the orbital inclination and RAAN as a function of

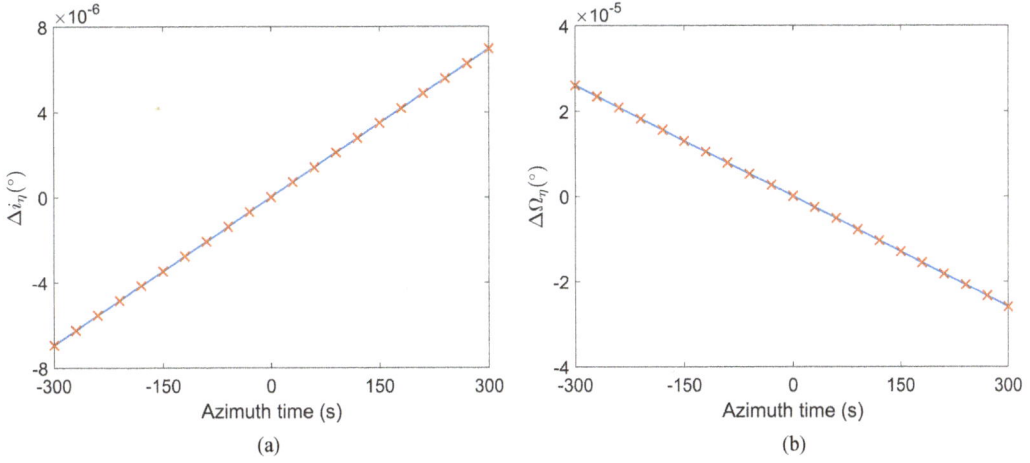

FIGURE 7.3 Within an SAT of 600 s, the time-varying components of (a) orbital inclination; (b) RAAN versus azimuth time, where the symbol "×" signifies the corresponding parameters acquired via DE440, and the line represents that simulated by time-varying components of Eqs. (7.24) and (7.25).

time exhibit a linear trend within an SAT of 600 s; the difference between them is reflected in their change rates over time. Thus, we can estimate the time-varying orbital inclination and RAAN by employing the following relations:

$$i_\eta = i_0 + \Delta i_\eta, \quad \text{with} \quad \Delta i_\eta = k_{\text{inc}} \cdot \eta \tag{7.24}$$

$$\Omega_\eta = \Omega_0 + \Delta \Omega_\eta, \quad \text{with} \quad \Delta \Omega_\eta = k_\Omega \cdot \eta \tag{7.25}$$

where i_0 and Ω_0 represent the constant part of orbit inclination and RAAN, while k_{inc} and k_Ω stand for the change rates of both orbital elements over time, respectively.

Upon accounting for the influence of nodal precession, the MBSAR's position error is given by:

$$\mathbf{R}_{\text{NPERR},\eta} = \mathbf{R}_{\text{SAR},\eta} - \mathbf{R}_{\text{NP},\eta} \tag{7.26}$$

where the additional subscript η means that the corresponding vectors vary with varying azimuth time. The vector $\mathbf{R}_{\text{NPSAR},\eta}$ represents the MBSAR's position vector in the absence of nodal precession referring to geocenter and can be expressed as:

$$\begin{cases} X_{\text{NP},\eta} = R_{\text{SAR},\eta}(\cos\Omega_0 \cos u_\eta - \sin\Omega_0 \cos i_0 \sin u_\eta) \\ Y_{\text{NP},\eta} = R_{\text{SAR},\eta}(\sin\Omega_0 \cos u_\eta + \cos\Omega_0 \cos i_0 \sin u_\eta) \\ Z_{\text{NP},\eta} = R_{\text{SAR},\eta} \sin i_0 \sin u_\eta \end{cases} \tag{7.27}$$

where $R_{\text{SAR},\eta}$ and u_η are the distance and AOL of MBSAR, both a function of azimuth time.

In light of Eqs. (7.24), (7.25), and (7.27), we can rewrite the three components of MBSAR's position errors $\mathbf{R}_{\text{NPERR},\eta}$:

$$\begin{cases} X_{\text{NPERR},\eta} \approx -k_\Omega \cdot Y_{\text{NP},\eta} \cdot \eta + k_{\text{inc}} \cdot \sin\Omega_0 \cdot Z_{\text{NP},\eta} \cdot \eta \\ Y_{\text{NPERR},\eta} \approx k_\Omega \cdot X_{\text{NP},\eta} \cdot \eta - k_{\text{inc}} \cdot \cos\Omega_0 \cdot Z_{\text{NP},\eta} \cdot \eta \\ Z_{\text{NPERR},\eta} \approx k_{\text{inc}} \cdot \cot i_0 \cdot Z_{\text{NP},\eta} \cdot \eta \end{cases} \tag{7.28}$$

Subsequently, we can express the range history error caused by nodal precession effects as:

$$\Delta R_p(\eta) = \frac{\mathbf{R}_{\mathrm{NPERR},\eta} \cdot \mathbf{R}_{\mathrm{TOI},\eta}}{\left\| \mathbf{R}_{\mathrm{SAR},\eta} - \mathbf{R}_{\mathrm{TOI},\eta} \right\|_2} \tag{7.29}$$

Eq. (7.29) establishes a relationship between the MBSAR's coordinate and slant range errors under the nodal precession effects. The detailed derivation of the above equations can be found in Appendix A.

Note that in this and the following scenarios, the TOI's location $\mathbf{R}_{\mathrm{TOI},\eta}$ is ascertained by employing the zero-Doppler steering method to reduce the magnitude of the Doppler centroid. As elucidated in Chapter 6, it is essential to confine the MBSAR's look angle within a particular scope for the successful execution of the zero-Doppler steering method. Thus, the MBSAR is configured to operate at look angles of 0.2° and 0.8° for near and far ranges, respectively. Figure 7.4 demonstrates the coordinate error and corresponding phase error in the MBSAR caused by nodal precession.

Upon inspecting Figure 7.4, it is evident that the varying azimuth time leads to temporal variations in coordinate errors of the MBSAR. Besides, within an SAT of 600 s, the y and z components of MBSAR's position error extend to tens of meters, while the x-component coordinate error exceeds 100 m. Hence, a range history error, corresponding to a phase error in magnitudes of several dozen radians, manifests in the signal of MBSAR. This, in effect, leads to Doppler errors and inevitably affects the MBSAR imaging. Hence, a rational conclusion can be derived that the MBSAR's imaging performance is subject to the influence of nodal precession, notwithstanding its lengthy cycle duration. Further, the phase error exhibits disparate values for various look angles, implying that MBSAR's imaging quality is acutely vulnerable to the phase error that is spatially varying within the image scene.

To achieve a more precise analysis, utilizing the identical parameters displayed in Figure 7.4, Figure 7.5 depicts the 1st-order through 4th-order Doppler errors versus look angle in the MBSAR under the nodal precession effects.

According to Figure 7.5, all those perturbed Doppler errors correlate with the MBSAR's look angles regardless of their order. In addition, the nodal precession primarily gives rise to the Doppler frequency and DFMR errors. Both Doppler errors, in effect, correspond to the 1st-order phase error in the range of dozens of radians and the 2nd-order phase error in the order of several radians for an SAT of 600 s. Hence, the MBSAR's geometric accuracy and focusing quality could deteriorate if both errors are uncompensated. Upon examining the 3rd-order and 4th-order Doppler errors, we

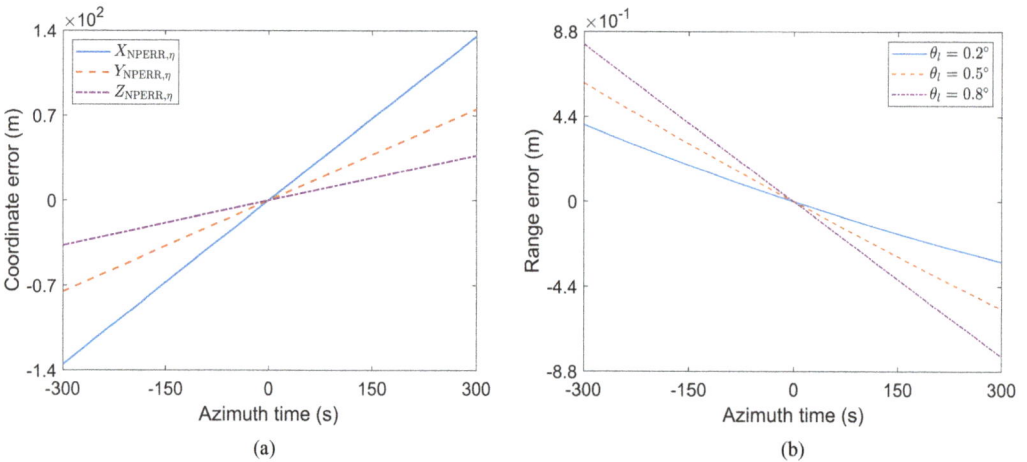

FIGURE 7.4 The nodal precession effects on the MBSAR's signal, (a) the coordinate errors versus azimuth time, (b) the range history error versus azimuth time at various look angles.

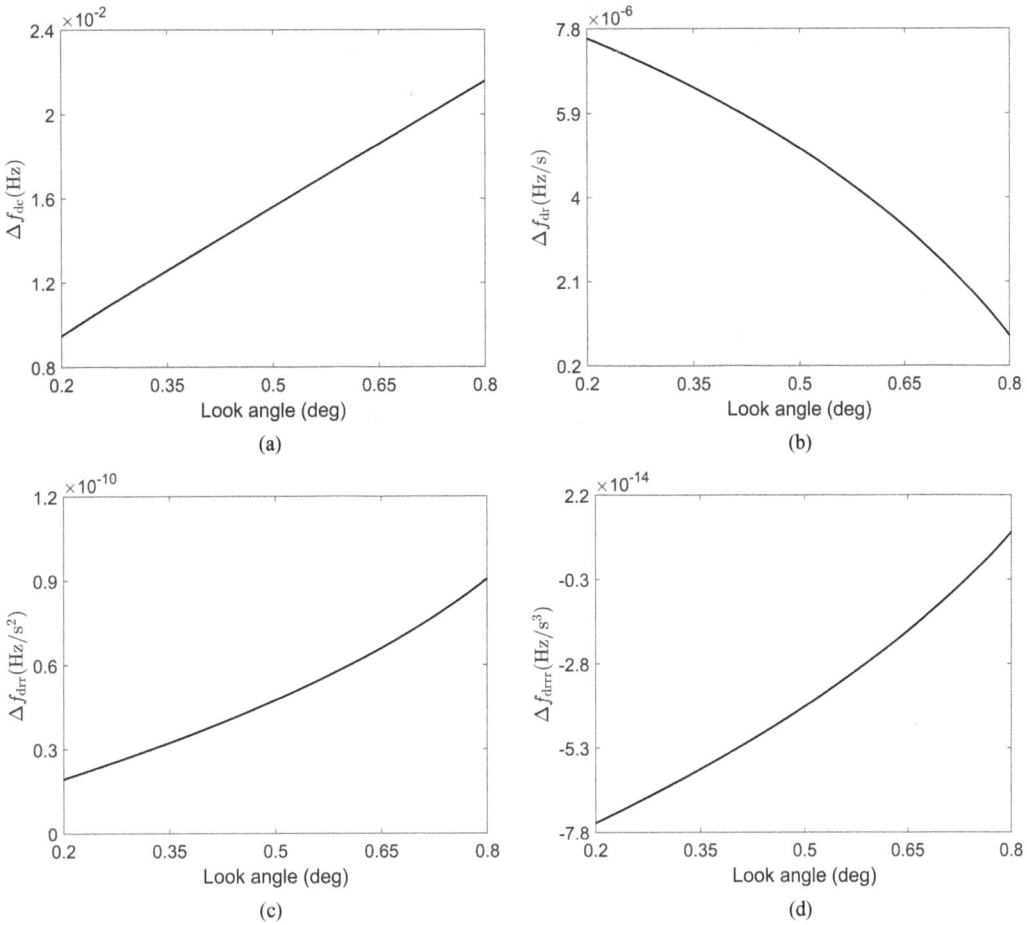

FIGURE 7.5 In the context of nodal precession effects, the Doppler errors of various orders against the look angle in the MBSAR: (a) Doppler frequency error, (b) DFMR error, (c) 3rd-order Doppler error, (d) 4th-order Doppler error.

observe that they are associated with phase errors whose magnitudes are below 10^{-3} and 10^{-5} rad. Correspondingly, the MBSAR's imaging remains unaffected by either of them. As a result, both are excluded from further analysis.

It should be noted that the perturbed range history error discussed above is obtained under the stop-and-go assumption, as it is impossible to express the phase history of MBSAR analytically in the scenario of the non-stop-and-go echo model [27–29]. To ensure the accurate modeling of perturbed error, we introduce a residual range error (RRE) as:

$$\delta R_{\mathrm{RRE}}(\eta) = \Delta R_p^{\mathrm{nsg}}(\eta) - \Delta R_p(\eta) \tag{7.30}$$

where $\Delta R_p^{\mathrm{nsg}}$ denotes the perturbed range history error in the context of the non-stop-and-go signal model.

Figure 7.6 illustrates the RREs at different look angles against azimuth time using the parameters as in Figure 7.5. It is clear that the RRE remains insignificant irrespective of the look angle, with magnitudes below 1×10^{-3} m in general throughout the SAT of 600 s. Considering the limited magnitude of the RRE, it is reasonable to conclude that the impact of the stop-and-go error on MBSAR imaging performance is negligibly small. The comparable results can be observed when

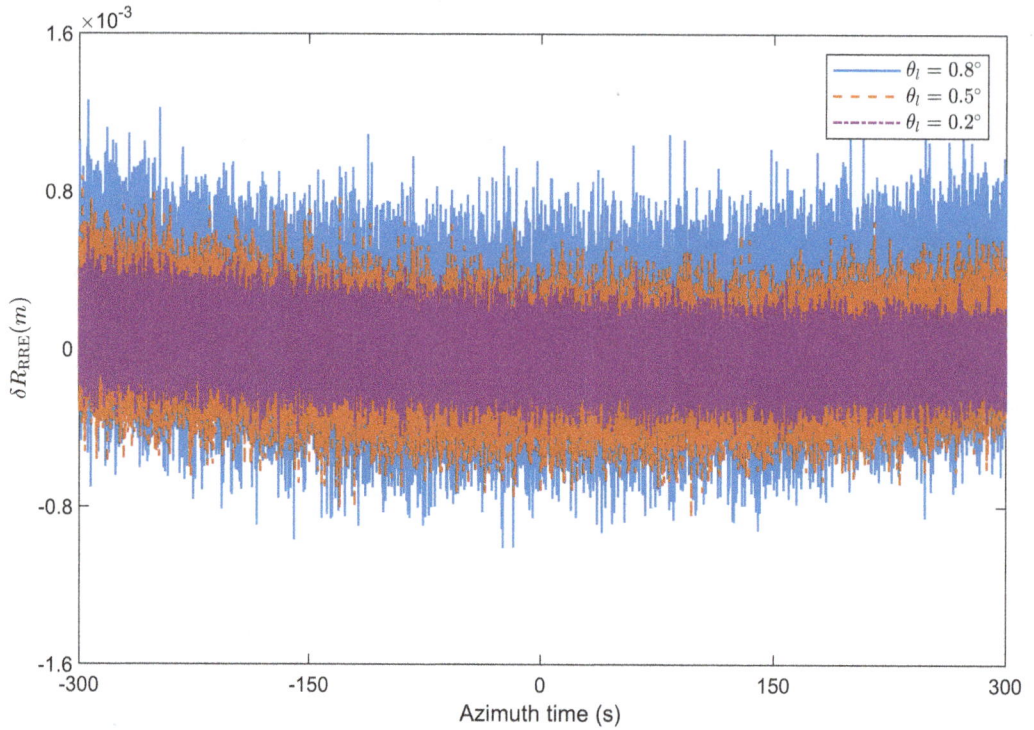

FIGURE 7.6 At various look angles, the RREs due to the stop-and-go error as a function of the azimuth time.

other perturbation effects are analyzed, the stop-and-go assumption can be reasonably adopted in those similar studies.

We have examined the coordinate error and associated range history error caused by nodal precession. Hence, the extent of MBSAR image distortion associated with the nodal precession effects can be assessed. The geometric distortion includes image offsets in range and azimuth directions, as defined in Eqs. (7.15) and (7.21), respectively. Figure 7.7 shows the range and azimuth offsets with respect to the look angle.

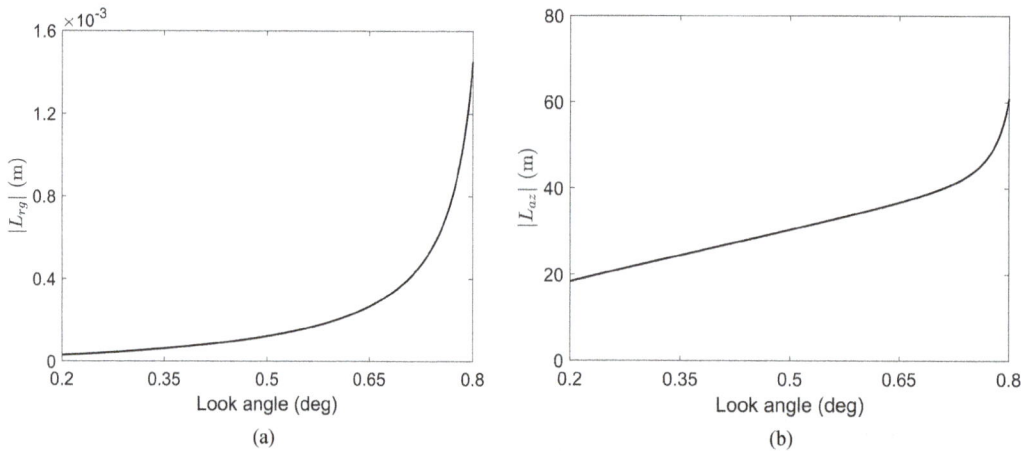

FIGURE 7.7 In the context of nodal precession effects, the magnitude of 2-D image offsets versus the look angle: (a) range offset, (b) azimuth offset.

From Figure 7.7, we find that the range offset remains below 2×10^{-3} m, indicating that the nodal precession exerts a negligible impact on the range geometric precision in the MBSAR. On the other hand, the azimuth offset has magnitudes ranging from tens to dozens of meters. Further, as the look angle increases, the degree of azimuth displacement expands, indicating a spatial fluctuation in its manifestation. The disparity in azimuth offset is minor within a swath width narrower than 100 km, but it must be considered when employing a wider swath width.

Regarding the nodal precession effects on the focusing quality, it is superfluous to scrutinize the azimuth CPE since its influence is minuscule. By contrast, it is essential to thoroughly evaluate the impact of azimuth QPE on the MBSAR focusing. Following Eq. (7.23), Figure 7.8 depicts the azimuth QPE at different look angles as a function of SAT from 0 to 600 s.

There exists a negative correlation between the azimuth QPE and look angle, suggesting that the spatial variance of the nodal precession could influence focusing quality. Specifically, when the look angle is 0.8°, the azimuth QPE is consistently below $\pi/4$. This suggests that the focusing quality remains unaffected in this scenario. In regard to the look angles of 0.2° and 0.5°, the azimuth focusing is contingent upon the SAT. If an SAT shorter than 300 s is utilized, the QPE would never surpass the $\pi/4$ threshold, implying it does not impede the focusing quality in this scenario. However, the situation changes when a longer SAT, e.g., 600 s, is employed: this results in a considerable increase in the QPE, surpassing $\pi/4$ and potentially leading to the deterioration of azimuth imaging. Therefore, although the nodal precession results in nearly linear temporal changes in both orbital inclination and RAAN, the azimuth focusing quality is degraded by it, especially at high azimuthal resolution.

Until now, we have evaluated the MBSAR's imaging performance in the presence of nodal precession effects. Note that other precessional movements of the lunar orbit can affect the MBSAR's inertial motion, including the apsidal precession. In the subsequent section, we shall explore the impact of apsidal precession on the MBSAR imaging.

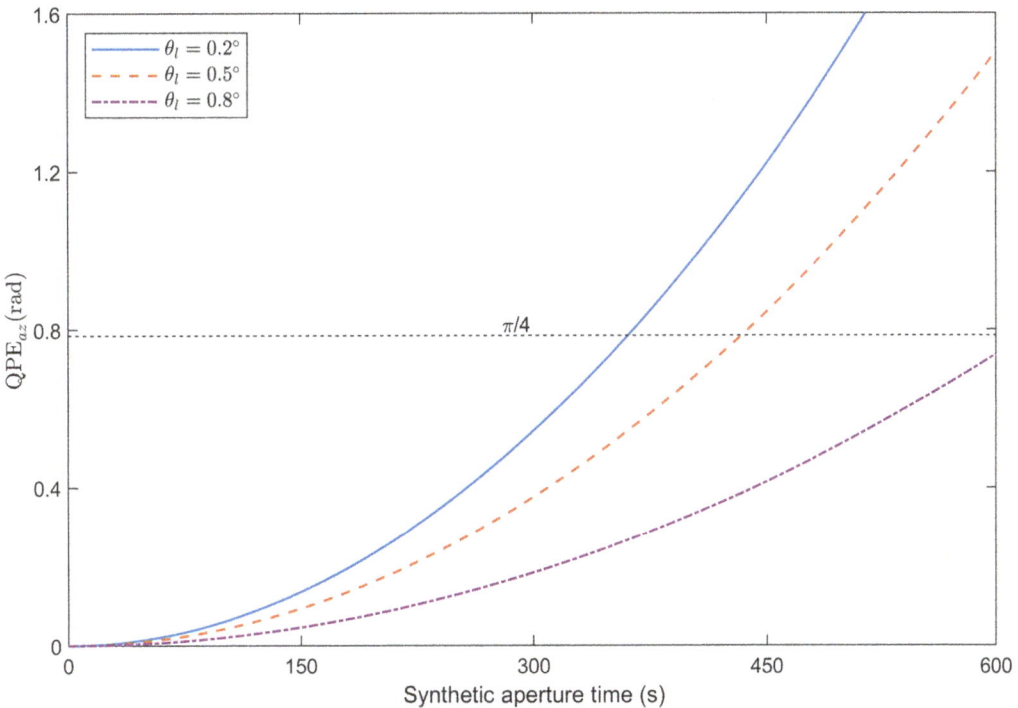

FIGURE 7.8   In the context of nodal precession effects, the magnitude of azimuth QPE versus SAT.

7.3.2 The Apsidal Precession Effects on MBSAR Imaging

In this section, we perform a quantitative analysis regarding the effect of apsidal precession, pertained to the gradual eastward shift of major axis in the lunar orbit with a cycle of 8.85 years, on the MBSAR imaging. Figure 7.9 visually portrays the variation in the MBSAR's AOP over three 8.85-year periods. As observed, the AOP undergoes variations over time due to the apsidal precession, further causing coordinate drift in the MBSAR. Our primary concern lies in determining the effects of apsidal precession on the MBSAR's signal within the SAT. Figure 7.10 illustrates the changes occurring over time in the AOP within an SAT of 600 s.

Upon examining Figures 7.10 and 7.3, it is apparent that the AOP experiences a greater degree of temporal variation compared to the orbital inclination and RAAN under nodal precession effects. As a result, it is plausible that the apsidal precession has a more pronounced impact on MBSAR imaging than the nodal precession. Notwithstanding, the AOP still displays a predominantly linear trend over time. Hence, the time-varying AOP within the SAT can be approximately characterized by:

$$\omega_\eta = \omega_0 + \Delta\omega_\eta, \quad \text{with} \quad \Delta\omega_\eta = k_\omega \cdot \eta \tag{7.31}$$

Hence, the MBSAR's position error induced by the apsidal precession can be obtained and written as follows:

$$\mathbf{R}_{\text{APERR},\eta} = \mathbf{R}_{\text{SAR},\eta} - \mathbf{R}_{\text{AP},\eta} \tag{7.32}$$

with $\mathbf{R}_{\text{AP},\eta} = [X_{\text{APERR},\eta}, Y_{\text{APERR},\eta}, Z_{\text{APERR},\eta}]^T$ given by:

$$\begin{cases} X_{\text{APERR},\eta} = R_{\text{SAR},\eta} \cdot \left[\cos\Omega_\eta \cos(\omega_0 + v_\eta) - \sin\Omega_\eta \cos i_\eta \sin(\omega_0 + v_\eta) \right] \\ Y_{\text{APERR},\eta} = R_{\text{SAR},\eta} \cdot \left[\sin\Omega_\eta \cos(\omega_0 + v_\eta) + \cos\Omega_\eta \cos i_\eta \sin(\omega_0 + v_\eta) \right] \\ Z_{\text{APERR},\eta} = R_{\text{SAR},\eta} \cdot \sin i_\eta \sin(\omega_0 + v_\eta) \end{cases} \tag{7.33}$$

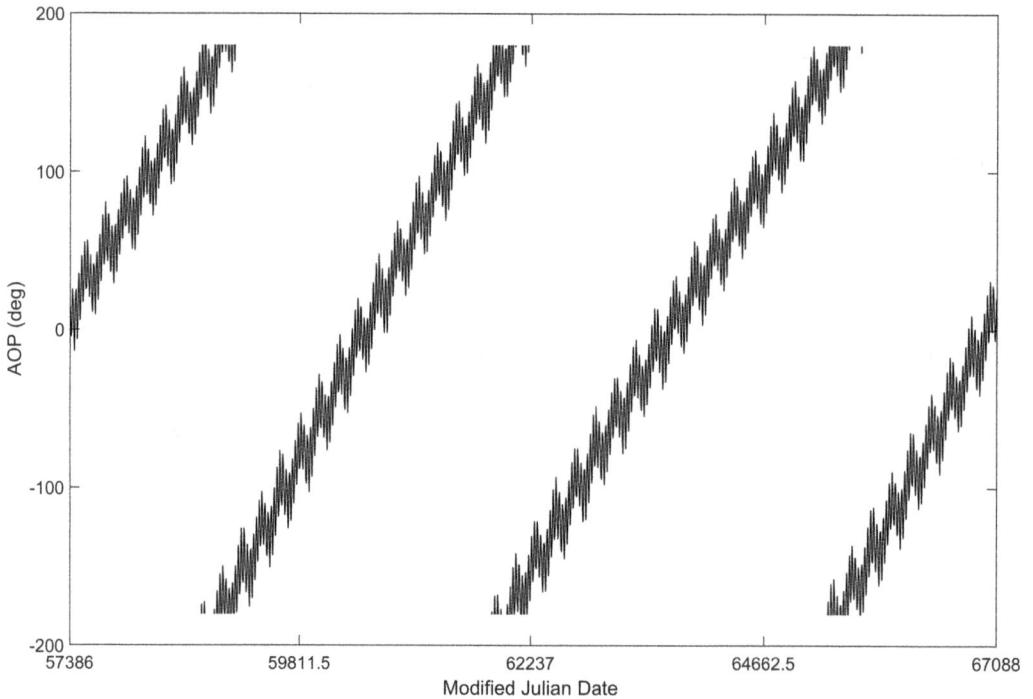

FIGURE 7.9 In the ECI coordinate, the temporal variations in the AOP within the apsidal precession cycle.

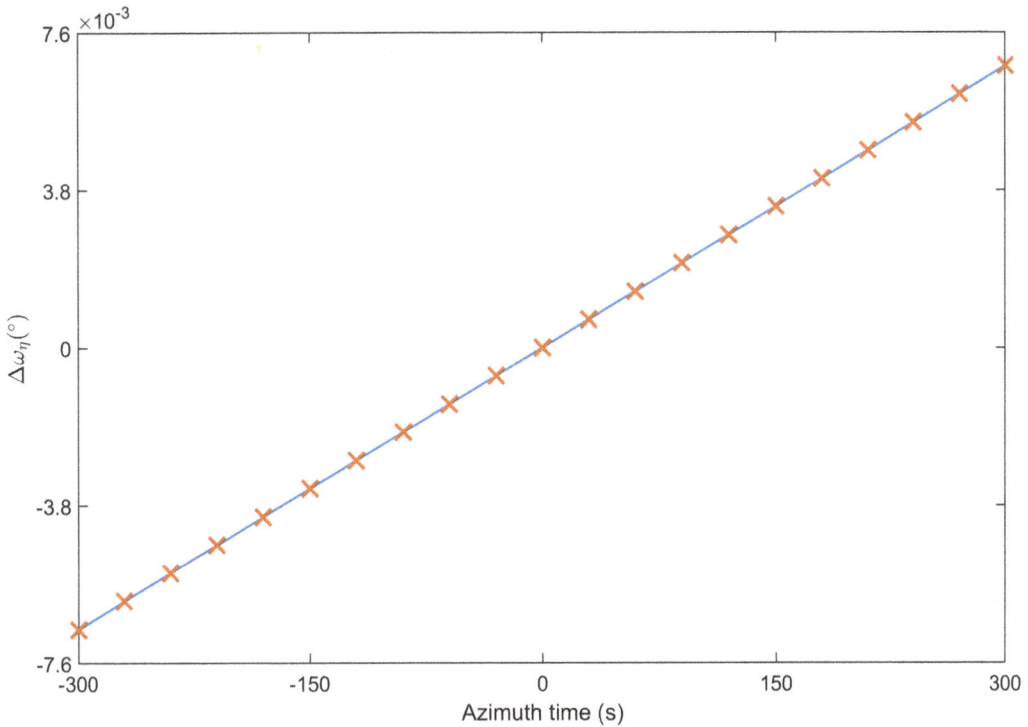

FIGURE 7.10 Within an SAT of 600 s, the time-varying components of AOP versus azimuth time, where the symbol "×" signifies the parameter acquired via DE440, and the line represents that simulated by time-varying components of Eq. (7.31).

While the MBSAR signal is subject to a range history error caused by apsidal precession, we can express it mathematically as follows:

$$\Delta R_p(\eta) = \frac{2\mathbf{R}_{\text{APERR},\eta}(\mathbf{R}_{\text{AP},\eta} - \mathbf{R}_{\text{TOI},\eta}) + \|\mathbf{R}_{\text{APERR},\eta}\|_2^2}{\|\mathbf{R}_{\text{AP},\eta} - \mathbf{R}_{\text{TOI},\eta}\|_2 + \|\mathbf{R}_{\text{SAR},\eta} - \mathbf{R}_{\text{TOI},\eta}\|_2} \tag{7.34}$$

Figure 7.11 illustrates the MBSAR's coordinate errors that arise due to the apsidal precession, followed by the resultant range history errors at various look angles. In both cases, the results are plotted as a function of azimuth time.

From Figure 7.11, it appears that there is a correlation between the position and range history errors and azimuth time in the MBSAR. The range history error is also influenced by the look angle, leading to spatially varying Doppler errors. A comparison of Figures 7.11–7.14 further highlights that the apsidal precession leaves a more pronounced impact on the MBSAR's signal when compared to nodal precession.

The apsidal precession can affect signal coherence through the introduction of Doppler errors. Figure 7.12 illustrates the Doppler errors from 1st-order through 4th-order as a function of look angle.

We observe the apsidal precession predominantly causes errors in Doppler frequency and DFMR. Conversely, the Doppler errors of the 3rd-order and 4th-order are comparatively insignificant. In addition, as the look angle increases, the magnitude of the Doppler frequency error increases while that of the DFMR error diminishes. Henceforth, the association between image offset and look angle is dissimilar to that between defocusing degree and look angle.

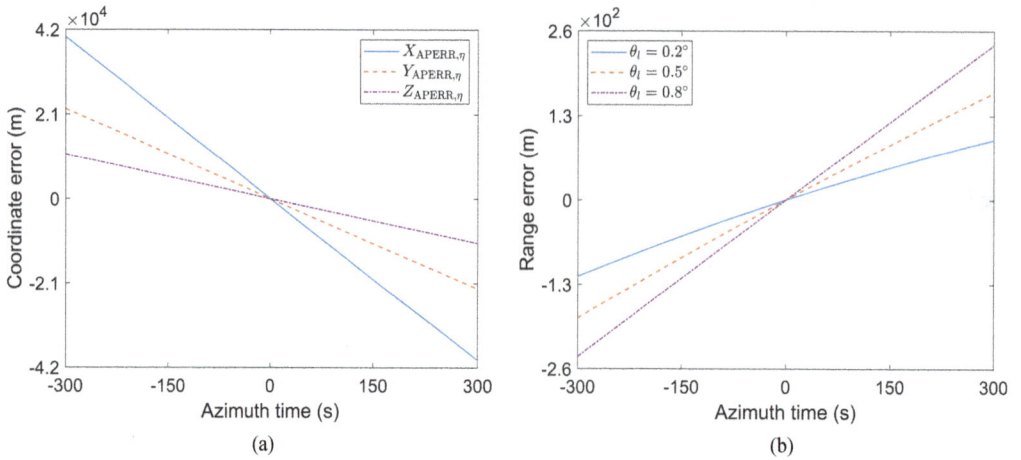

FIGURE 7.11 The apsidal precession effects on the MBSAR's signal, (a) the coordinate errors versus azimuth time, (b) the range history error versus azimuth time at various look angles.

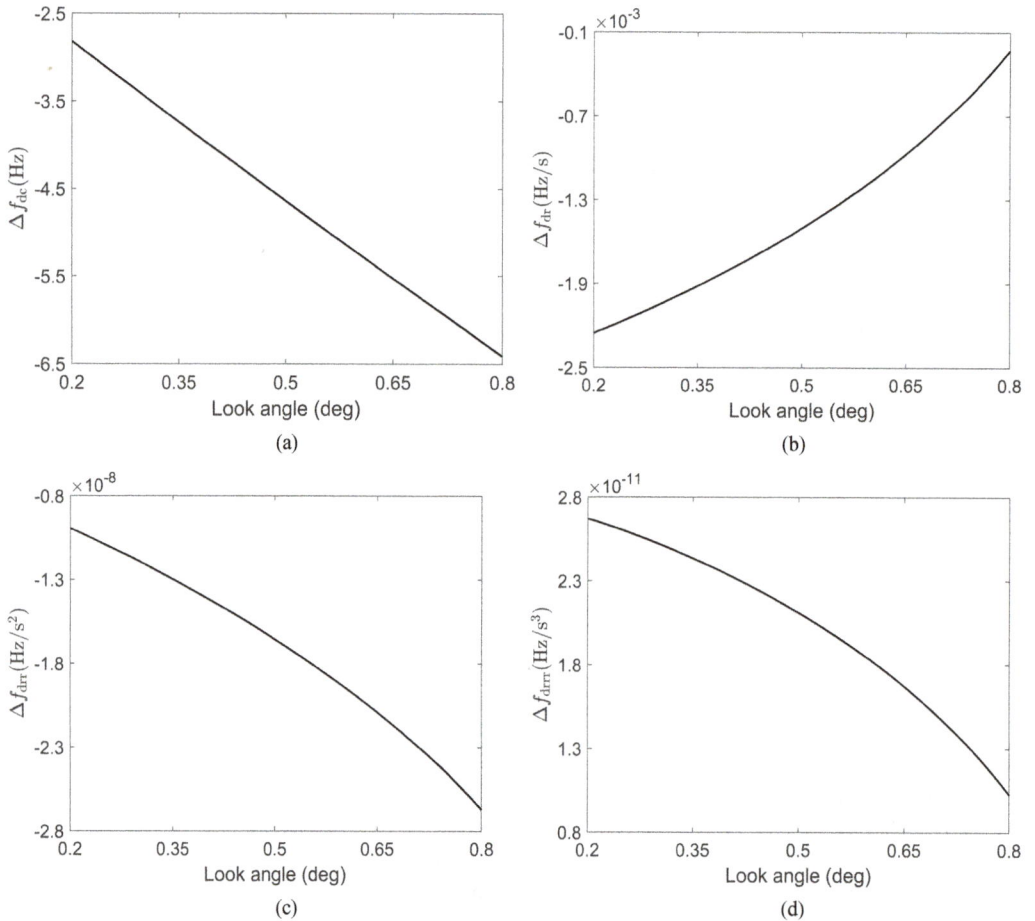

FIGURE 7.12 In the context of apsidal precession effects, the Doppler errors of various orders against the look angle in the MBSAR: (a) Doppler error, (b) DFMR error, (c) 3rd-order Doppler error, (d) 4th-order Doppler error.

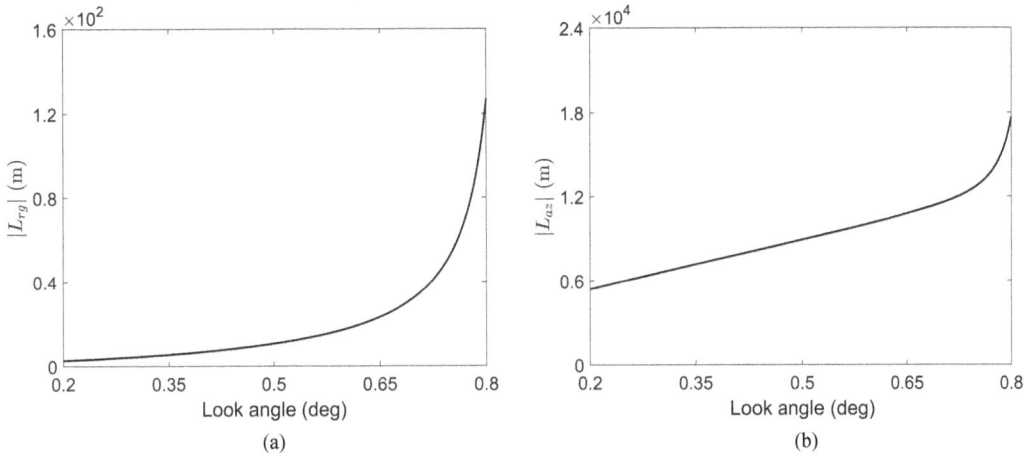

FIGURE 7.13   In the context of apsidal precession effects, the magnitudes of 2-D image offsets versus the look angle: (a) range offset, (b) azimuth offset.

The image offsets in range and azimuth directions that result from apsidal precession as a function of look angle are examined in Figure 7.13. It is shown that the apsidal precession can cause a range offset ranging from several meters to over 100 m for the look angle from $0.2°$ to $0.8°$. Further, this range offset exhibits a non-linear increase with the rising look angle in the MBSAR. Regarding the azimuth offset, it is noteworthy that it can extend up to several, or even tens of, kilometers. The azimuth offset exhibits a linear correlation with the look angle when it is below $0.70°$. A sharp rise in azimuth offset occurs at a look angle exceeding $0.7°$. Although there is a linear increase in the Doppler frequency error as the look angle increases, the azimuth offset does not reveal a similar pattern. It can be inferred that the apsidal precession effects on the geometric accuracy of TOI are more pronounced than nodal precession effects.

When considering the effects of apsidal precession on MBSAR focusing, Figure 7.14 depicts the azimuth QPE as a function of SAT. As nodal precession has minimal impact on MBSAR's focusing when the SAT is below 300 s; for ease of comparison, we set the maximum SAT to 300 s in this case. As illustrated in Figure 7.14, there is a decrease in azimuth QPE as the look angle increases when considering the apsidal precession effects. Besides, the azimuth QPE remains significant even for the SAT shorter than 300 s. Provided the SAT exceeds 35 s, the QPE can surpass its $\pi/4$ threshold regardless of the look angle. This phenomenon emphasizes the considerable influence of apsidal precession on the MBSAR's focusing quality in contrast to the nodal precession effects.

The current work has entailed conducting theoretical analyses regarding the nodal and apsidal precession effects on MBSAR imaging. In the following section, we shall simulate the point target responses to demonstrate the analysis above results visually.

7.3.3   THE TARGET RESPONSES UNDER ORBITAL PRECESSION EFFECTS

This section intends to simulate and assess the point target responses under precession effects based on the earlier theoretical analysis. For simulating target responses of the MBSAR, we set the signal bandwidth to 15 MHz and carrier frequency to 1.2 GHz. In the simulation, diverse SATs are employed to demonstrate how the nodal precession affects the MBSAR's focusing quality. Based on this groundwork, the single-point target responses of the MBSAR are simulated and shown in Figures 7.15–7.17 at look angles of $0.2°$, $0.5°$, and $0.8°$, with reference to an image center located at $(0, 0)$.

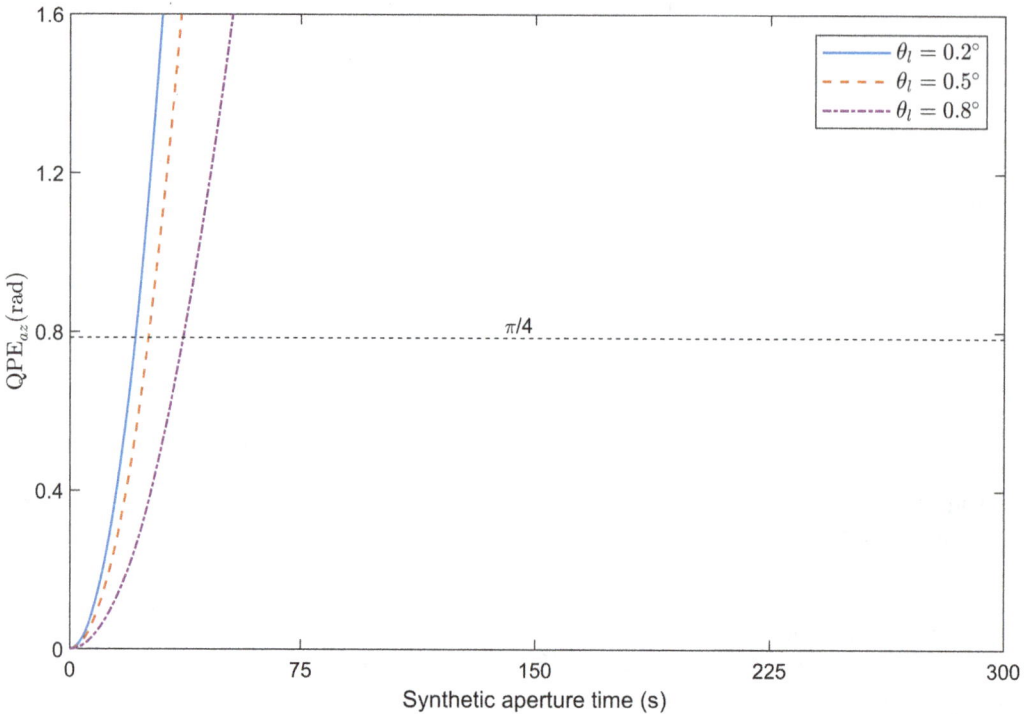

FIGURE 7.14 In the context of apsidal precession effects, the magnitude of azimuth QPE versus SAT.

Figures 7.15–7.17 show no geometric deviation in the range imaging. The quality of range focusing, regarding range PSLR and ISLR, is considered satisfactory. From this perspective, the range imaging of the MBSAR remains unaffected despite the presence of nodal precession. On the other hand, azimuth imaging experiences a certain degree of deterioration due to nodal precession. An occurrence of azimuth offset, which is independent of the SAT, occurs in the target response. Upon comparing Figures 7.15–7.17, it is apparent that the azimuth offset is contingent on the look angle, suggesting that the image offset is spatially varying within the image scene of the MBSAR. When considering azimuth focusing, it is closely contingent on the MBSAR's look angle and SAT.

When the look angle is 0.2°, the azimuth QPE has no detrimental effect on either the azimuth PSLR or ISLR, provided that the SAT equals to 200 s. Upon implementation of an SAT of 400 s, both the azimuth PSLR and ISLR rise, indicating the presence of defocusing phenomena in the point target response. Furthermore, it is unfeasible to achieve satisfactory azimuth PSLR and ISLR for MBSAR when employing an SAT of 600 s in the MBSAR.

In the case of a look angle of 0.5°, the MBSAR's image quality regarding the PSLR and ISLR is acceptable, given that the SAT is set at 200 s. If the SAT of the MBSAR increases to 400 s or 600 s, the azimuth PSLR and ISLR experience deteriorations, the severity of which is less compared to that at a look angle of 0.2°.

As to a look angle of 0.8°, the MBSAR's azimuth imaging undergoes quite a slight degradation. More precisely, the azimuth PSLR and ISLR of the target response maintain their satisfactory values, even when configuring the MBSAR's SAT to 600 s. The shorter SAT and larger look angle could potentially facilitate attaining a high-quality MBSAR image amid the nodal precession effects.

To analyze the spatial variations of phase error caused by nodal precession effects, we shall proceed with the simulation of the point array targets' responses, as depicted in Figure 7.18. In this instance, the center of the MBSAR scene at coordinate (0, 0) corresponds to a look angle of 0.5°.

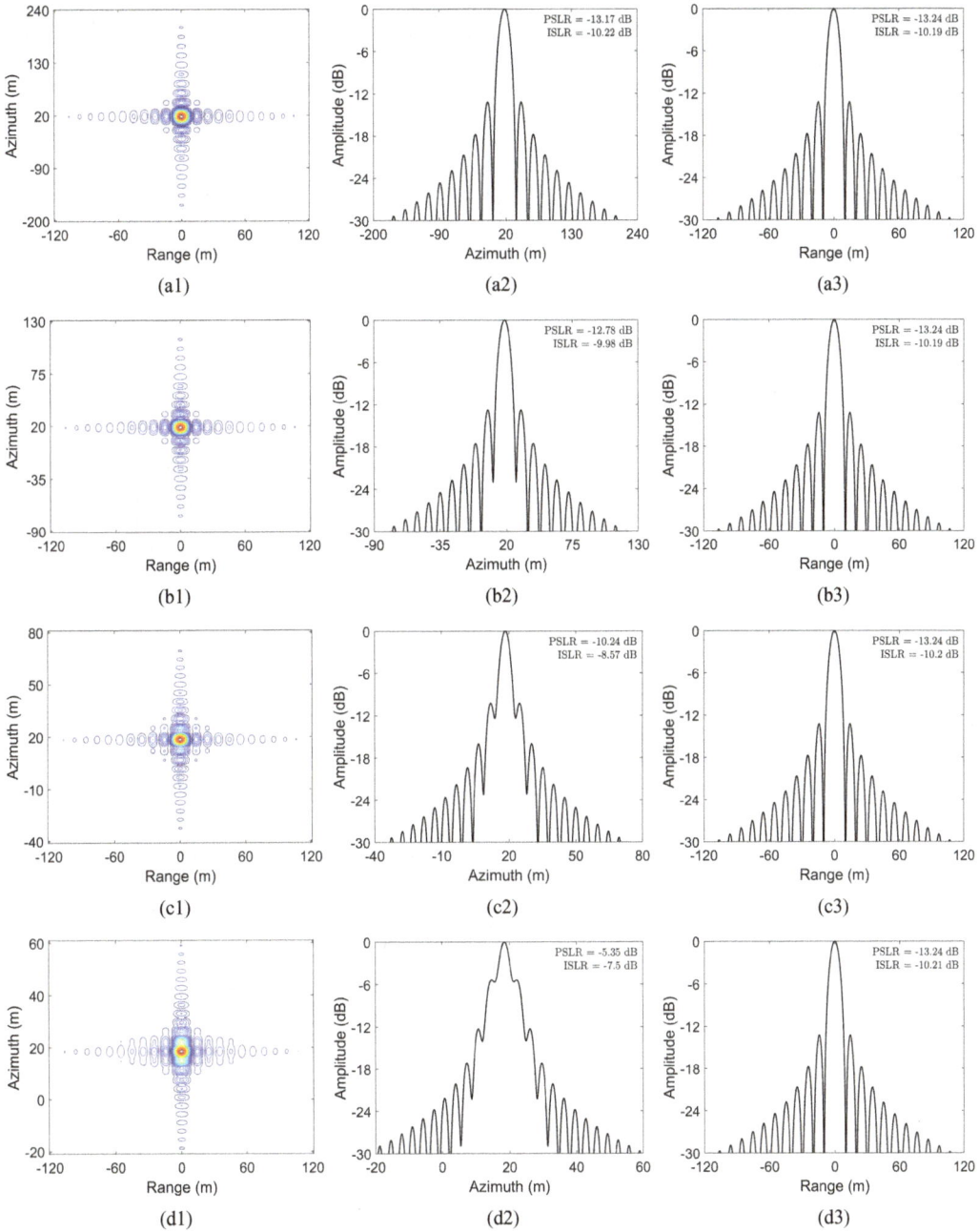

FIGURE 7.15 At a look angle of 0.2°, the point target response under nodal precession, where the 1st (a1–d1), 2nd (a2–d2), and 3rd (a3–d3) columns respectively represent the response contour, azimuth, and range profiles, while 1st (a1–a3), 2nd (b1–b3), 3rd (c1–c3) and 4th (d1–d3) rows stand for the SATs of 100 s, 200 s, 400 s, and 600 s.

The nodal precession has the potential to cause azimuth offset, whose magnitude varies spatially depending on the TOI's location, as illustrated in Figure 7.18. The extent of spatial heterogeneity in the azimuth offset is restricted when the image scene is diminutive. The quantitative metric indicates that the spatial disparity in azimuth offset across a swath width of 100 km (ranging from −50 km to 50 km) is less than 0.1 m. Moreover, the spatial discrepancy in the image offset across the azimuth direction is insignificant. Hence, the image distortion resulting from the inconsistencies of azimuth

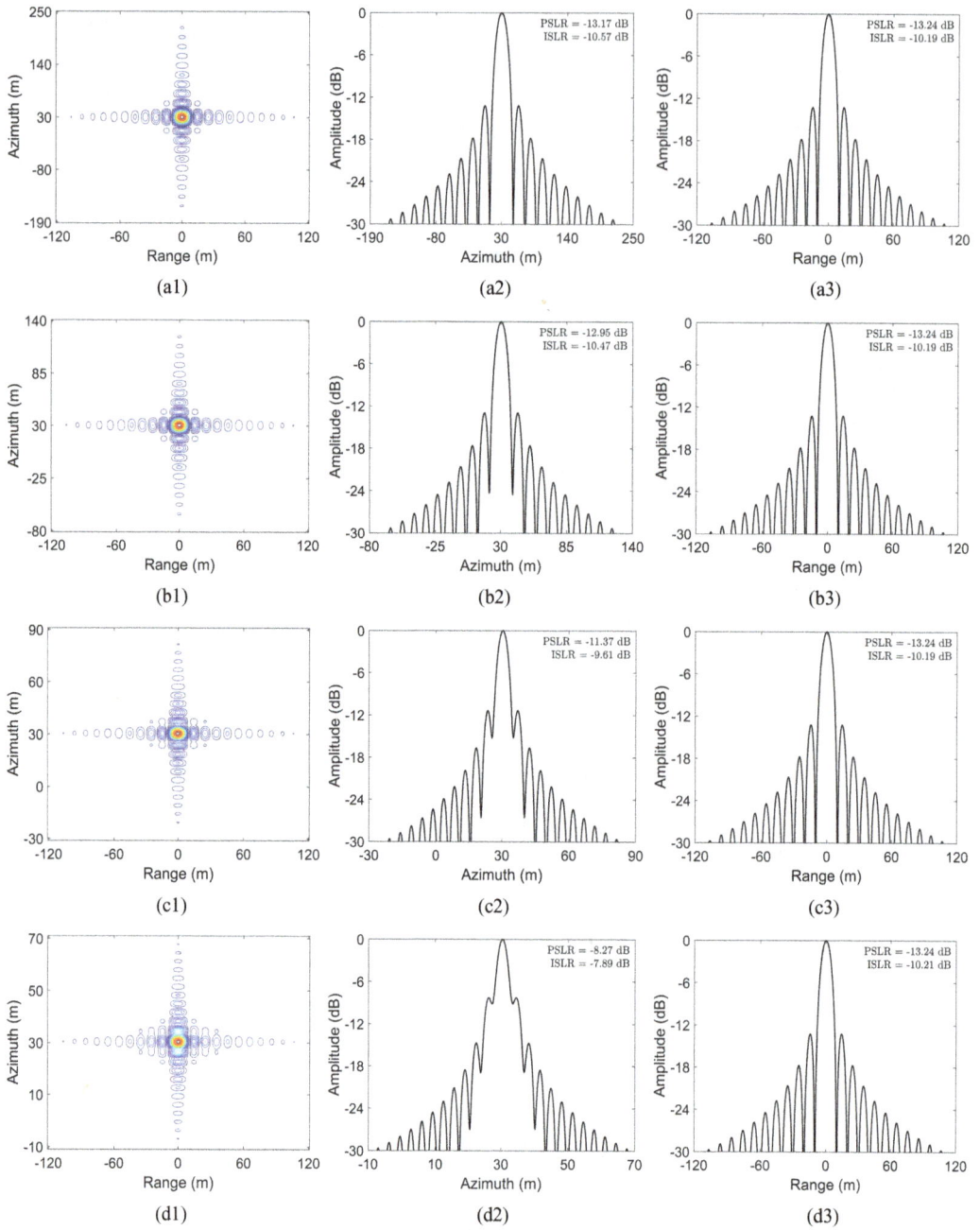

FIGURE 7.16 At a look angle of $0.5°$, the point target response under nodal precession, where the 1st (a1–d1), 2nd (a2–d2), and 3rd (a3–d3) columns respectively represent the response contour, azimuth, and range profiles, while 1st (a1–a3), 2nd (b1–b3), 3rd (c1–c3) and 4th (d1–d3) rows stand for the SATs of 100 s, 200 s, 400 s, and 600 s.

offset is relatively small compared to the overall image scene. As the simulation results presented above align with theoretical analysis, for brevity, we shall refrain from repeating the simulations of target responses under the apsidal precession.

Thus far, we have probed into the phase error that arises from the orbital precessions in terms of the apsidal and nodal precessions. Upon this analysis, it is found that the apsidal precession has a more

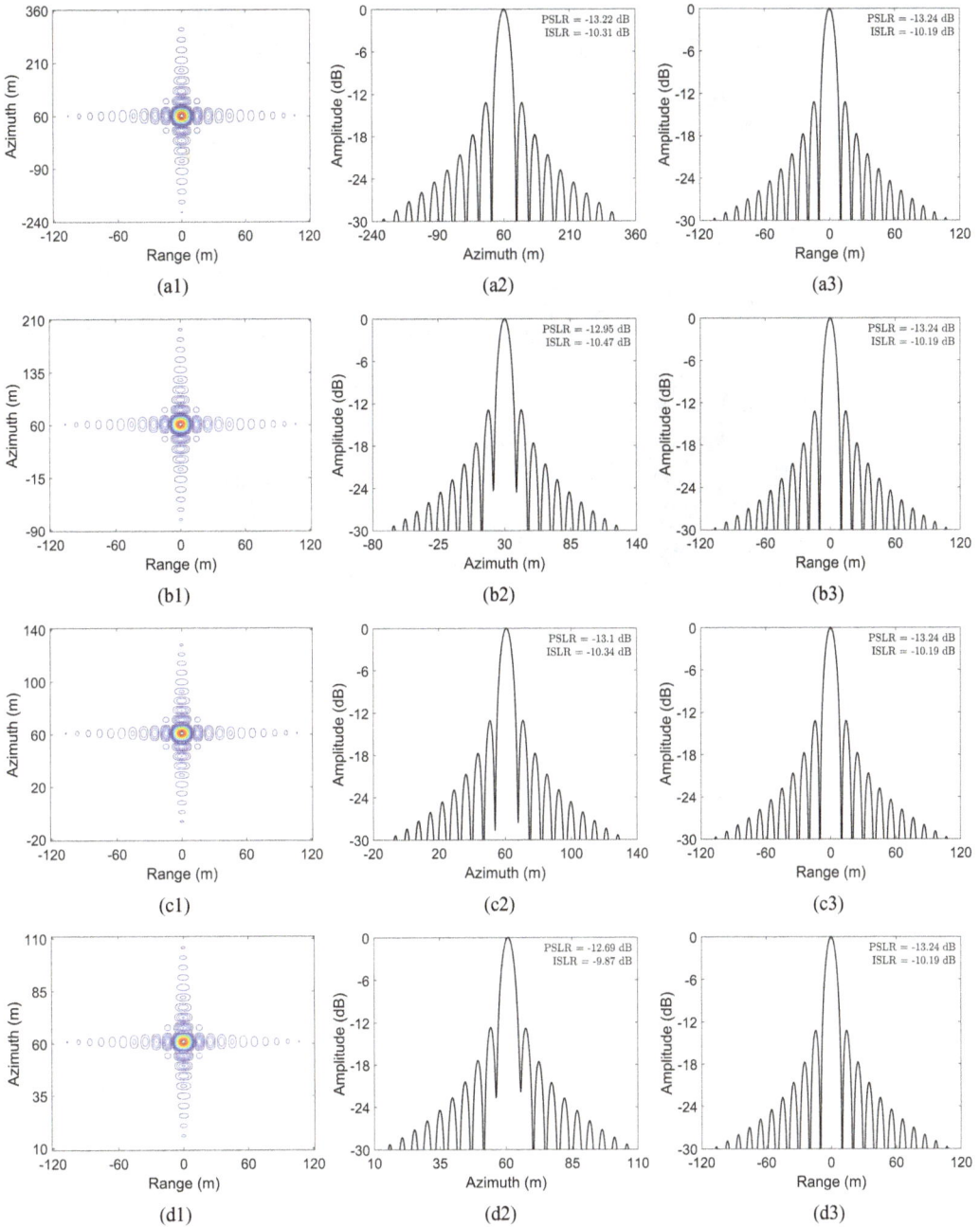

FIGURE 7.17 At a look angle of 0.8°, the point target response under nodal precession, where the 1st (a1–d1), 2nd (a2–d2), and 3rd (a3–d3) columns respectively represent the response contour, azimuth, and range profiles, while 1st (a1–a3), 2nd (b1–b3), 3rd (c1–c3) and 4th (d1–d3) rows stand for the SATs of 100 s, 200 s, 400 s, and 600 s.

significant impact on the MBSAR imaging. To be more precise, the nodal precession does not disrupt range imaging, whereas the range imaging is affected by an image offset given rise by the apsidal precession. Moreover, azimuth imaging is subject to significant levels of image offset and defocusing when the apsidal precession is taken into account. Utilizing a shorter SAT could possibly mitigate the issue of azimuth defocusing; however, the image offset, independent of SAT, remains

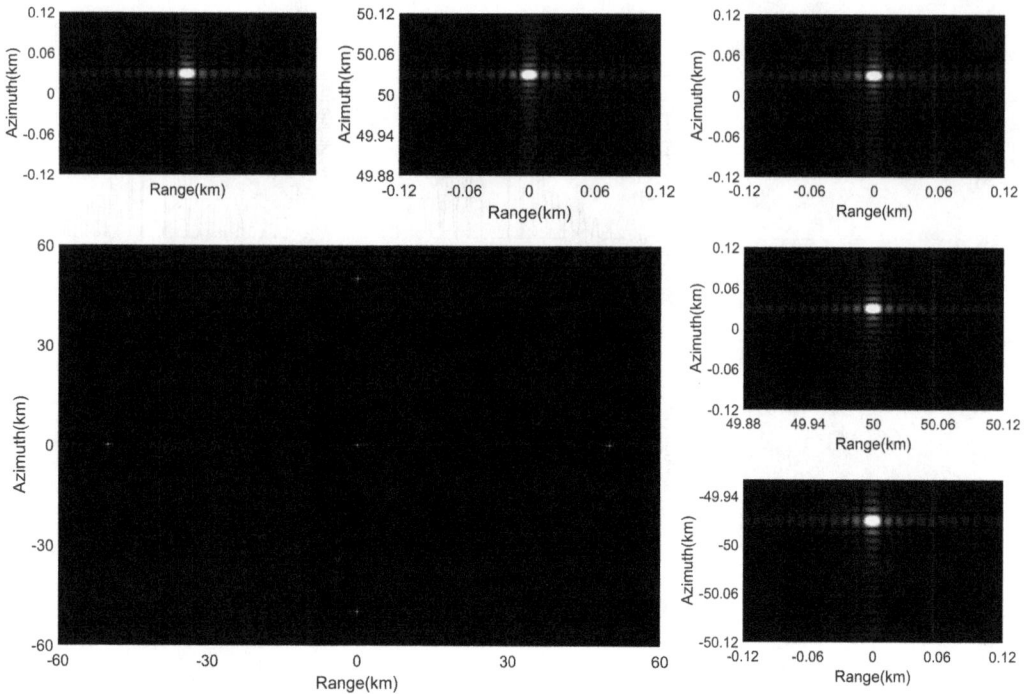

FIGURE 7.18 At a look angle of 0.5°, the responses of point array targets in the MBSAR in the context of the nodal precession effects.

insufferable in the MBSAR. To enhance the geometric positioning accuracy, we are required to employ orbital determination with satisfactory accuracy. In the subsequent section, we shall scrutinize the accuracy requirement for orbital determination in the MBSAR amid orbital precession effects.

7.4 THE ORBITAL DETERMINATION UNDER PRECESSION EFFECTS

The preceding section has demonstrated that the orbital precessions may render the MBSAR unsuitable for Earth observation applications, as the TOI's positioning accuracy and focusing quality are potentially impacted. One possible solution is to carry out meticulous orbital determinations in terms of position and velocity determinations for the MBSAR [15]. As the MBSAR's orbit is susceptible to uncertainties stemming from orbital perturbations, we can ascertain the requisite accuracy level for determining the MBSAR's orbit in the presence of certain orbital perturbation effects. To commence the following discussion, we shall first explore the precision required for the orbital determination considering the nodal precession effects.

7.4.1 Criteria for Orbital Determination under Nodal Precession Effects

As per the previous section, the orbital inclination and RAAN exhibit a linear correlation with time within the SAT. Therefore, when contemplating the orbital determination under the influence of nodal precession effects, it is plausible that potential uncertainties in the orbital inclination and RAAN vary linearly with time. Both uncertainties are expressed in the form of:

$$\delta i_\eta = \delta i_0 + \delta k_{\text{inc}} \cdot \eta \tag{7.35}$$

$$\delta \Omega_\eta = \delta \Omega_0 + \delta k_\Omega \cdot \eta \tag{7.36}$$

where δi_0 and $\delta\Omega_0$ represent the constant errors in the orbital inclination and RAAN, δk_{inc} and δk_{Ω} are the change rate errors in the orbital inclination and RAAN, respectively. It is important to uphold the correlations resulting from the interdependence of the variations in the orbital inclination and RAAN:

$$\begin{cases} \delta\Omega_0 = S_1 \cdot \delta i_0 \\ \delta k_{\Omega} = S_1 \cdot \delta k_{inc} + S_2 \cdot \delta i_0 \end{cases} \tag{7.37}$$

with:

$$\begin{cases} S_1 = \cot i_0 \cot \Omega_0 - \csc \Omega_0 \cot \vartheta_{E0} \\ S_2 = -k_{inc} \cdot \csc^2 i_0 \cot \Omega_0 + k_{\Omega} \cdot (\cot \Omega_0 \csc \Omega_0 \cot \vartheta_{E0} - \cot i_0 \csc^2 \Omega_0) \end{cases} \tag{7.38}$$

where ϑ_{E0} is the instantaneous obliquity at the zero-azimuth time. The comprehensive derivations can be found in Appendix B. Eqs. (7.35) and (7.36) stipulate that the uncertainties associated with the nodal precession can be evaluated by examining the error in orbital inclination.

In what follows, we assume that other orbital elements are determined such that we may determine the nodal precession induced errors for both the MBSAR's position and velocity vectors:

$$\begin{cases} \delta X_{SAR} = -\delta\Omega_0 \cdot Y_{SAR} + \delta i_0 \cdot \sin \Omega_0 \cdot Z_{SAR} \\ \delta Y_{SAR} = \delta\Omega_0 \cdot X_{SAR} - \delta i_0 \cdot \cos \Omega_0 \cdot Z_{SAR} \\ \delta Z_{SAR} = \delta i_0 \cdot \cot i_0 \cdot Z_{SAR} \end{cases} \tag{7.39}$$

and:

$$\begin{cases} \delta V_{SX} = \delta i_0 \cdot \sin \Omega_0 \cdot V_{SZ} - \delta\Omega_0 \cdot V_{SY} + \delta k_{inc} \cdot \sin \Omega_0 \cdot Z_{SAR} - \delta k_{\Omega} \cdot Y_{SAR} \\ \delta V_{SY} = -\delta i_0 \cdot \cos \Omega_0 \cdot V_{SZ} + \delta\Omega_0 \cdot V_{SX} - \delta k_{inc} \cdot \cos \Omega_0 \cdot Z_{SAR} + \delta k_{\Omega} \cdot X_{SAR} \\ \delta V_{SZ} = \delta i_0 \cdot \cot i_0 \cdot V_{SZ} + \delta k_{inc} \cdot \cot i_0 \cdot Z_{SAR} \end{cases} \tag{7.40}$$

where $\mathbf{R_{SAR}} = [X_{SAR}, Y_{SAR}, Z_{SAR}]^T$ and $\mathbf{V_{SAR}} = [V_{SX}, V_{SY}, V_{SZ}]^T$ are the MBSAR's position and velocity vectors at the zero-azimuth time, respectively. Hence, the central slant range error is given by:

$$\delta r_{p0} = -\frac{\delta X_{SAR} \cdot X_{TOI} + \delta Y_{SAR} \cdot Y_{TOI} + \delta Z_{SAR} \cdot Z_{TOI}}{R_c} \tag{7.41}$$

The 1st order error in the range history is:

$$r_{p1} = -R_1 \frac{\delta X_{SAR} X_{TOI} + \delta Y_{SAR} Y_{TOI} + \delta Z_{SAR} Z_{TOI}}{R_c^2}$$
$$- \frac{\delta X_{SAR} V_{TX} + \delta Y_{SAR} V_{TY}}{R_c} - \frac{\delta V_{SX} X_{TOI} + \delta V_{SY} Y_{TOI} + \delta V_{SZ} Z_{TOI}}{R_c} \tag{7.42}$$

where $\mathbf{R_{TOI}} = [X_{TOI}, Y_{TOI}, Z_{TOI}]^T$ and $\mathbf{V_{TOI}} = [V_{TX}, V_{TY}, 0]^T$ are the position and velocity vectors of TOI in ECI coordinate system at the zero-azimuth time, respectively. As the defocusing phenomenon is not a concern for the MBSAR when experiencing nodal precession effects with an SAT of less than 200 s, the higher-order range errors can be ignored [15].

The 2-D image offsets arising from uncertainties in the orbital inclination and RAAN can be expressed as follows:

$$\begin{cases} \delta L_{rg} \approx |\delta r_{p0}| \\ \delta L_{az} \approx \mathfrak{R}_{az} \cdot |\delta r_{p1}| \end{cases} \tag{7.43}$$

with:

$$\mathfrak{R}_{az} \approx V_g \left| \frac{2R_2^2 + 3R_1R_3}{4R_2^3} \right| \tag{7.44}$$

Note that the higher-order terms in the range offset are ignored due to their negligibly small magnitudes.

By Eqs. (7.39)–(7.42), the range offset can be easily seen as

$$\delta L_{rg} = \frac{|\xi_{i0}| \cdot |\delta i_0|}{R_c} \tag{7.45}$$

with:

$$\xi_{i0} = \sin\Omega_0 \cdot Z_{SAR}X_{TOI} - \cos\Omega_0 \cdot Z_{SAR}Y_{TOI} \\ + \cot i_0 \cdot Z_{SAR}Z_{TOI} + S_1 \cdot (X_{SAR}Y_{TOI} - Y_{SAR}X_{TOI}) \tag{7.46}$$

Similarly, the image offset in azimuth direction is given by:

$$\delta L_{az} \approx \frac{\mathfrak{R}_{az} \cdot |\delta i_0 \cdot \zeta_{i0} + \delta k_{inc} \cdot \zeta_{ki}|}{R_c} \tag{7.47}$$

with:

$$\zeta_{i0} = S_1 \cdot (X_{SAR}V_{TY} - Y_{SAR}V_{TX} - V_{SY}X_{TOI} + V_{SX}Y_{TOI}) \\ + \cot i_0 \cdot V_{SZ}Z_{TOI} + S_2 \cdot (X_{SAR}Y_{TOI} - Y_{SAR}X_{TOI}) \\ + \cos\Omega_0 \cdot (Z_{SAR}V_{TY} - V_{SZ}Y_{TOI}) + \sin\Omega_0 \cdot (V_{SZ}X_{TOI} - Z_{SAR}V_{TX}) \tag{7.48}$$

$$\zeta_{ki} = S_1 \cdot (X_{SAR}Y_{TOI} - Y_{SAR}X_{TOI}) + Z_{SAR}X_{TOI}\sin\Omega_0 \\ - Z_{SAR}Y_{TOI}\cos\Omega_0 + Z_{SAR}Z_{TOI}\cot i_0 \tag{7.49}$$

To ensure a satisfactory geometric accuracy in the MBSAR image, we limit the range and azimuth offsets to a specific level, e.g.:

$$\begin{cases} \delta L_{rg} \le T_{rg} \\ \delta L_{az} \le T_{az} \end{cases} \tag{7.50}$$

We can then ascertain the necessary precision for constant error of orbital inclination and its change rate by imposing the thresholds outlined in Eq. (7.50), as:

$$\begin{cases} |\delta i_0| \le T_{i0} \\ |\delta k_{inc}| \le T_{ki} \end{cases} \tag{7.51}$$

with:

$$\begin{cases} T_{i0} = \dfrac{R_c \cdot T_{rg}}{|\xi_{i0}|} \\ T_{ki} = \dfrac{R_c}{|\zeta_{ki}|} \cdot \left(\dfrac{T_{az}}{\mathfrak{R}_{az}} - \dfrac{T_{rg} \cdot |\zeta_{i0}|}{|\xi_{i0}|} \right) \end{cases} \tag{7.52}$$

By the above thresholds, we can ascertain the requisite precision for the MBSAR's orbital determination in terms of position and velocity determinations by Eqs. (7.39) and (7.40).

7.4.2 Criteria for Orbital Determination under Apsidal Precession Effects

To proceed with orbital determination under apsidal precession, we assume that there is a linear relationship between the uncertainty in AOP and azimuth time:

$$\delta\omega_\eta = \delta\omega_0 + \delta k_\omega \cdot \eta \tag{7.53}$$

where $\delta\omega_0$ and δk_ω are the constant error and error rate of AOP.

The temporally varying MBSAR's coordinate drift can be determined by:

$$\begin{cases} \delta X_{\mathrm{SAR},\eta} = \delta\omega_\eta \cdot C_{X,\eta} \\ \delta Y_{\mathrm{SAR},\eta} = \delta\omega_\eta \cdot C_{Y,\eta} \\ \delta Z_{\mathrm{SAR},\eta} = \delta\omega_\eta \cdot C_{Z,\eta} \end{cases} \tag{7.54}$$

with:

$$\begin{cases} C_{X,\eta} = -R_{\mathrm{SAR},\eta} \cdot (\cos\Omega_\eta \sin u_\eta + \sin\Omega_\eta \cos i_\eta \cos u_\eta) \\ C_{Y,\eta} = -R_{\mathrm{SAR},\eta} \cdot (\sin\Omega_\eta \sin u_\eta - \cos\Omega_\eta \cos i_\eta \cos u_\eta) \\ C_{Z,\eta} = R_{\mathrm{SAR},\eta} \cdot \sin i_\eta \cos u_\eta \end{cases} \tag{7.55}$$

Here, the vector $\mathbf{C}_\eta = [C_{X,\eta}, C_{Y,\eta}, C_{Z,\eta}]^T$ is denoted as $\mathbf{C}_0 = [C_{X0}, C_{Y0}, C_{Z0}]^T$ at the zero-azimuth time. Further, the rate of $\mathbf{C}_\eta$ is represented by $\mathbf{V}_C = [V_{CX}, V_{CY}, V_{CZ}]^T$ and is written by:

$$\begin{cases} V_{CX} \approx -w_{\mathrm{SAR}} \cdot X_{\mathrm{SAR}} \\ V_{CY} \approx -w_{\mathrm{SAR}} \cdot Y_{\mathrm{SAR}} \\ V_{CZ} \approx -w_{\mathrm{SAR}} \cdot Z_{\mathrm{SAR}} \end{cases} \tag{7.56}$$

where w_{SAR} is the angular velocity of SAR inertial motion at the zero-azimuth time.

Afterward, the position and velocity errors of the MBSAR are:

$$\begin{cases} \delta X_{\mathrm{SAR}} = \delta\omega_0 \cdot C_{X0} \\ \delta Y_{\mathrm{SAR}} = \delta\omega_0 \cdot C_{Y0} \\ \delta Z_{\mathrm{SAR}} = \delta\omega_0 \cdot C_{Z0} \end{cases} \tag{7.57}$$

and:

$$\begin{cases} \delta V_{SX} = \delta k_\omega \cdot C_{X0} + \delta\omega_0 \cdot V_{CX} \\ \delta V_{SY} = \delta k_\omega \cdot C_{Y0} + \delta\omega_0 \cdot V_{CY} \\ \delta V_{SZ} = \delta k_\omega \cdot C_{Z0} + \delta\omega_0 \cdot V_{CZ} \end{cases} \tag{7.58}$$

Given that the apsidal precession has a more significant impact on MBSAR imaging, we account for the second-order range error and associated azimuth QPE. As a result, the range error that arises from the errors in AOP can be approximated to:

$$\delta R_p(\eta) = \delta r_{p0} + \delta r_{p1} \cdot \eta + \delta r_{p2} \cdot \eta^2, \tag{7.59}$$

where δr_{p0} and δr_{p1} take the same forms of Eqs. (7.41) and (7.42), while the parameter δr_{p2} is:

$$\delta r_{p2} \approx -\frac{\delta V_{SX} \cdot V_{TX} + \delta V_{SY} \cdot V_{TY}}{R_c} \tag{7.60}$$

The criteria for the MBSAR's position and velocity determinations are in order. As in previous analyses, it is imperative to restrict the 2-D image offsets to a certain level to ensure geometric accuracy. Besides, in order to achieve a well-focused image, a threshold of $\pi/4$ is linked to azimuth QPE is employed. We can express the criteria mentioned above as:

$$\begin{cases} \delta L_{rg} \leq T_{rg} \\ \delta L_{ag} \leq T_{az} \\ \mathrm{QPE}_{az} < \pi/4 \end{cases} \tag{7.61}$$

Based on the limit of range offset, it is possible to derive a boundary for the constant error of AOP, as:

$$\left| \delta \omega_0 \right| \leq T_{\omega 0} \tag{7.62}$$

with:

$$T_{\omega 0} = R_c \cdot T_{rg} \cdot \left| \zeta_\omega \right|^{-1} \tag{7.63}$$

$$\zeta_\omega = C_{X0} \cdot X_{\mathrm{TOI}} + C_{Y0} \cdot Y_{\mathrm{TOI}} + C_{Z0} \cdot Z_{\mathrm{TOI}} \tag{7.64}$$

In contrast, the error rate in AOP is constrained by the azimuth offset and QPE. The azimuth offset can be used to establish a limitation as:

$$\left| \delta k_\omega \right| \leq T_{k\omega 1} \tag{7.65}$$

with:

$$T_{k\omega 1} = \frac{\Re_{az}^{-1} R_c \cdot T_{az} + T_{\omega 0} \cdot \left| \xi_{k\omega 1} \right|}{\left| \zeta_\omega \right|} \tag{7.66}$$

$$\begin{aligned} \xi_{k\omega 1} = {} & C_{X0} \cdot V_{\mathrm{TX}} + C_{Y0} \cdot V_{\mathrm{TY}} \\ & + R_1 \cdot R_c^{-1} \left(C_{X0} \cdot X_{\mathrm{TOI}} + C_{Y0} \cdot Y_{\mathrm{TOI}} + C_{Z0} \cdot Z_{\mathrm{TOI}} \right) \\ & - w_{\mathrm{SAR0}} \cdot \left(X_{\mathrm{SAR}} \cdot X_{\mathrm{TOI}} + Y_{\mathrm{SAR}} \cdot Y_{\mathrm{TOI}} + Z_{\mathrm{SAR}} \cdot Z_{\mathrm{TOI}} \right) \end{aligned} \tag{7.67}$$

By determining the threshold for the azimuth QPE, we can deduce another limit for the error rate in AOP that ensures the azimuth focusing quality of the MBSAR.

$$\left| \delta k_\omega \right| < T_{k\omega 2} \tag{7.68}$$

with:

$$T_{k\omega 2} = \frac{\lambda \cdot R_c \cdot T_{\mathrm{SAR}}^{-2} + 4 \cdot T_{\omega 0} \cdot \left| \xi_{k\omega 2} \right|}{4 \cdot \left| \zeta_{k\omega 2} \right|} \tag{7.69}$$

$$\zeta_{k\omega 2} = C_{X0} \cdot V_{\mathrm{TX}} + C_{Y0} \cdot V_{\mathrm{TY}} - 1.5 R_3 R_2^{-1} \cdot \left(C_{X0} \cdot X_{\mathrm{TOI}} + C_{Y0} \cdot Y_{\mathrm{TOI}} + C_{Z0} \cdot Z_{\mathrm{TOI}} \right) \tag{7.70}$$

$$
\begin{aligned}
\xi_{k\omega2} = {} & w_{SAR} \cdot \left(X_{SAR} \cdot V_{TX} + Y_{SAR} \cdot V_{TY} \right) \\
& + 1.5 R_3 R_2^{-1} \cdot [-w_{SAR} \cdot \left(X_{SAR} X_{TOI} + Y_{SAR} Y_{TOI} + Z_{SAR} Z_{TOI} \right) \\
& + C_{X0} V_{TX} + C_{Y0} V_{TY} + R_1 R_c^{-1} \cdot \left(C_{X0} X_{TOI} + C_{Y0} Y_{TOI} + C_{Y0} Z_{TOI} \right)]
\end{aligned}
\tag{7.71}
$$

Given that the threshold for the rate of AOP error is an upper boundary, it follows that the threshold for determining the error rate of AOP takes the following form:

$$
\left| \delta k_\omega \right| < \min \left\{ T_{k\omega1}, T_{k\omega2} \right\}
\tag{7.72}
$$

The derivation for the criteria above is given in Appendix E. Eq. (7.63) stipulates that the precision for a constant AOP error is correlated to range positioning accuracy. The precision required for a specific rate of AOP error is contingent upon the azimuth positioning accuracy and focusing quality.

7.4.3 Accuracy Requirement for Orbital Determination

We evaluate the precision requirements for orbital determination in the MBSAR based on the above criteria, followed by simulating the target responses.

7.4.3.1 Required Orbit Accuracy under Nodal Precession

Figure 7.19 illustrates the range offsets at various look angles arising from uncertainties in orbital inclination versus the constant error of orbital inclination. The range offset exhibits a positive and linear dependence on constant errors in orbital inclination. Moreover, the level of the range offset also displays a dependence on the look angle, with a larger range offset observed at a larger look angle for the same orbital error.

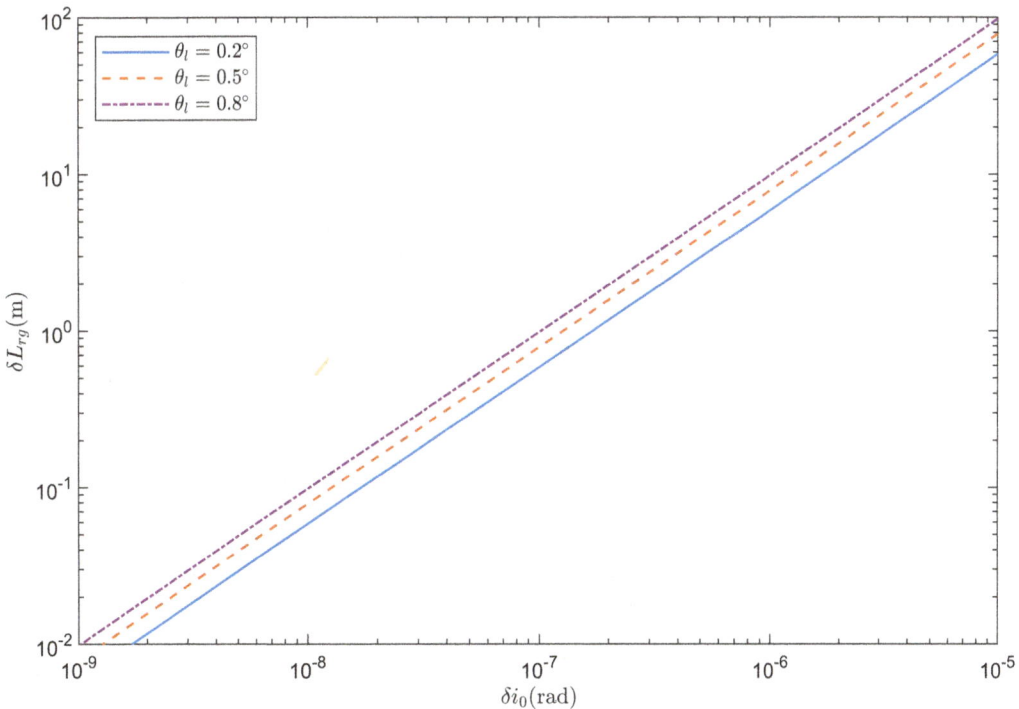

FIGURE 7.19 Owing to the uncertainty in the orbital inclination, the range offset of the MBSAR versus constant error of orbital inclination.

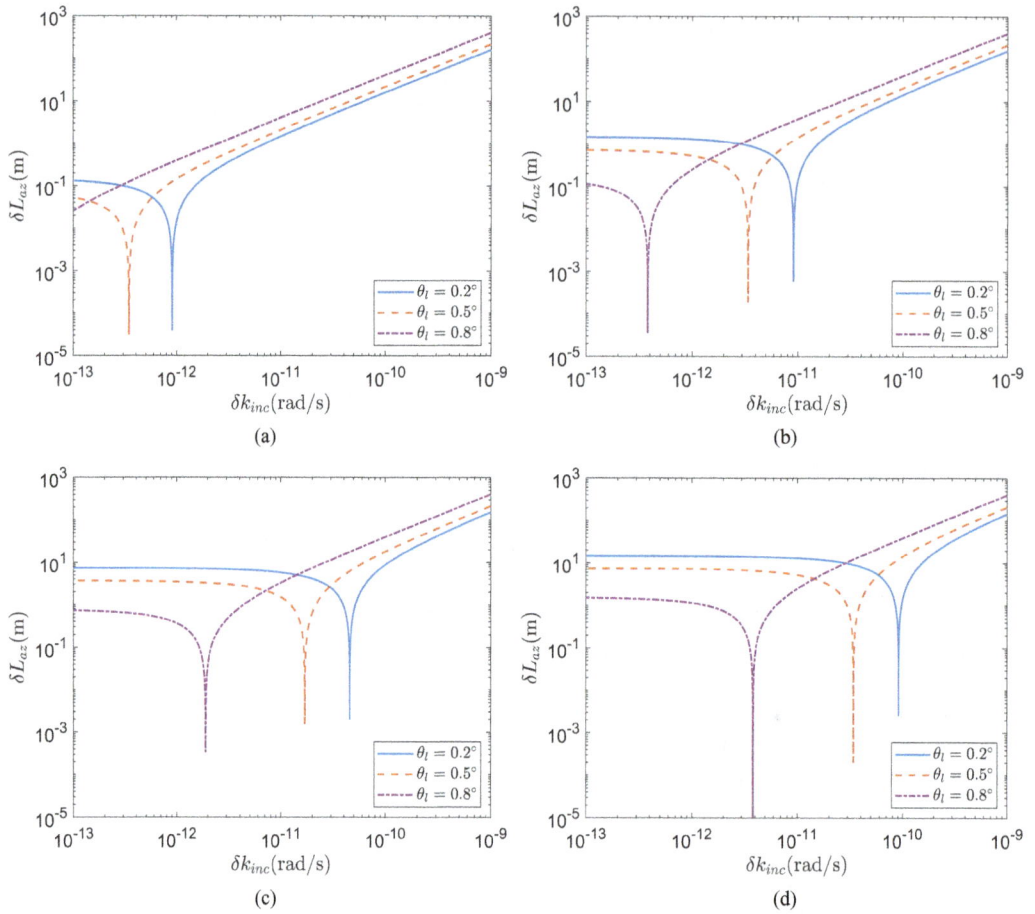

FIGURE 7.20 Owing to the uncertainties in the orbital inclination, the azimuth offset of the MBSAR versus error rate of orbital inclination when the constant error of orbital inclination corresponds to range offsets of (a) 0.01 m, (b) 0.10 m, (c) 0.50 m, (d) 1.00 m.

Figure 7.20 depicts the azimuth offsets against the error rate of orbital inclination, with varying range offsets. As seen, the azimuth offsets correlate with the constant and change rate of orbital inclination error. Also, the azimuth offset increases proportionally with the look angle unless the error rate in orbital inclination is exceptionally low. When considering both the azimuth and range offsets, a larger look angle necessitates a higher accuracy requirement in the MBSAR. Hence, the far look angle could be deemed advantageous for both the position and velocity determinations of the MBSAR. Therefore, we employ a far look angle of $0.8°$ for the orbital determination of the MBSAR.

Figure 7.21 shows the threshold of constant error in orbital inclination as a function of range positioning accuracy. The results indicate that the threshold T_{inc} rises proportionately with the TOI's positioning accuracy along range direction. Therefore, as the precision of range positioning improves, position determination becomes increasingly challenging in the MBSAR. The results also suggest that a notable constant error in orbital inclination leads to a considerable deviation in the azimuth direction, even if the rate of error in orbital inclination is minor. Hence, the orbital inclination's error rate is determined by both the azimuth and range positioning accuracies in a mutually dependent manner.

Next, the threshold for the error rate of orbital inclination is plotted against the azimuth positioning accuracy, as shown in Figure 7.22. The range offset of 0.01 m is acceptable in this scenario. The findings suggest that higher azimuth positioning accuracy for the TOI leads to a larger threshold

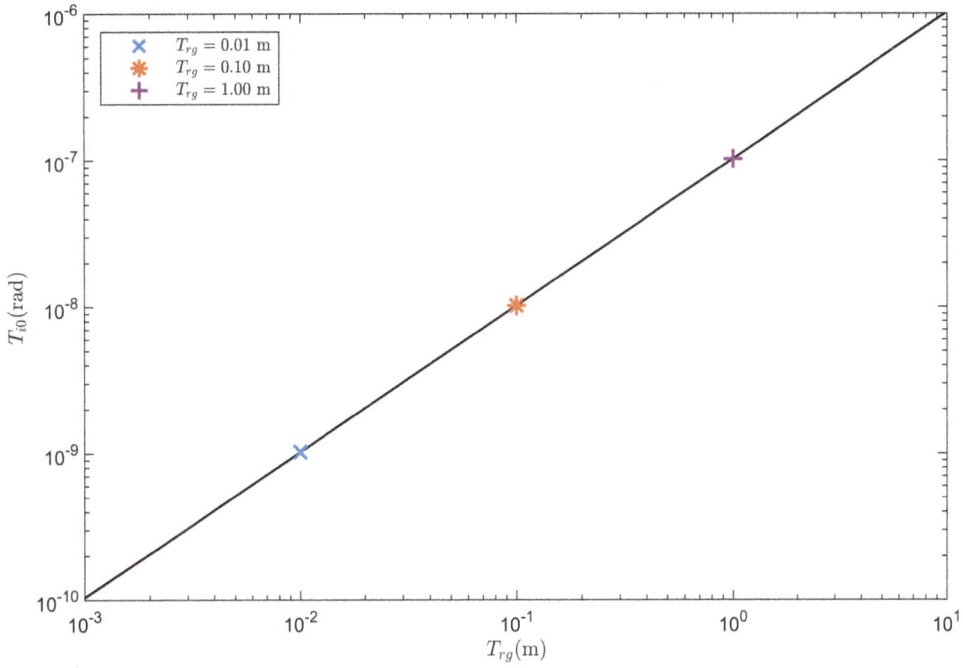

FIGURE 7.21 In the presence of nodal precession, the threshold for constant error in orbital inclination versus range positioning accuracy.

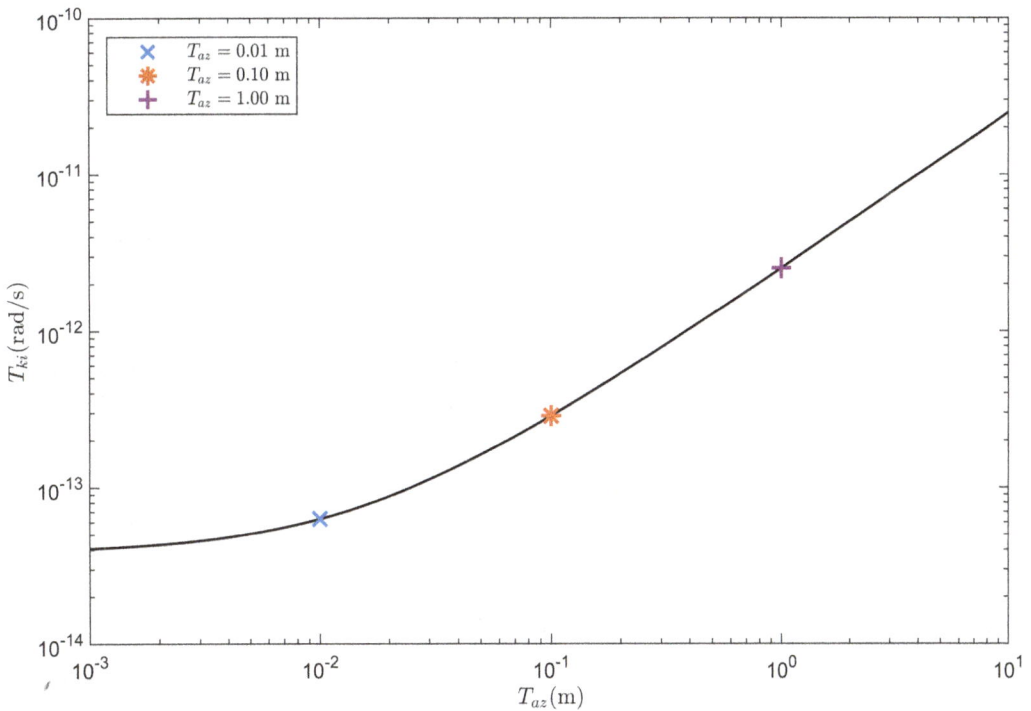

FIGURE 7.22 In the presence of nodal precession, the threshold for the error rate in orbital inclination against azimuth positioning accuracy under a constant orbital inclination error that corresponds to the range positioning accuracy of 0.01 m.

TABLE 7.1

Accuracy Requirement for the Orbital Determination of the MBSAR under Nodal Precession

Orbital Determination	PAR	X-Component	Y-Component	Z-Component
Position Determination	0.01 m	2.79 m	1.81 m	0.31 m
	0.10 m	27.95 m	18.12 m	3.13 m
	1.00 m	279.48 m	181.22 m	31.25 m
Velocity Determination	0.01 m	1.73×10^{-4} m/s	1.12×10^{-4} m/s	1.92×10^{-5} m/s
	0.10 m	7.84×10^{-4} m/s	5.08×10^{-4} m/s	8.76×10^{-5} m/s
	1.00 m	6.89×10^{-3} m/s	4.47×10^{-3} m/s	7.71×10^{-4} m/s

The PAR represents the Positioning Accuracy Requirement (for the TOI), where the position determination corresponds to range PAR while velocity determination is associated with azimuth RAP.

T_r. The property of corresponding regularity, however, differs from that of the threshold T_{inc} versus the range position accuracy. Thus, it follows the MBSAR's velocity determination becomes more formidable as the TOI's azimuth positioning accuracy is augmented.

The criteria depicted in Figures 7.21 and 7.22, together with Eqs. (7.39) and (7.40), are adopted to ascertain the required precision for the position and velocity determinations while considering the effects of nodal precession in the MBSAR, as outlined in Table 7.1.

There exists a correlation between the TOI's positioning accuracy and MBSAR's orbital determination under nodal precession, as revealed in Figures 7.21 and 7.22, and Table 7.1. Specifically, the precision needed for MBSAR's position determination correlates with the TOI's range positioning accuracy. However, the precision required for velocity determination does not exhibit a linear variation pattern concerning the azimuth positioning accuracy. In addition, to ascertain a high level of range positioning accuracy, it is critical to limit the position determination precision to a few meters in the horizontal (x and y) directions and roughly to a decimeter level in the vertical (z) direction. Furthermore, the azimuth positioning accuracy of 0.01 m requires a velocity accuracy on the order of 10^{-4} m/s and 10^{-5} m/s along the x and y directions, respectively. It can be inferred that both the position and velocity determinations in the z-direction pose a greater challenge for the MBSAR.

It should be noted that the accuracy requirement is defined with respect to nodal precession, which can serve as useful indicators to calibrate perturbations of RAAN and orbital inclination. Once other perturbation factors have been accounted for, the required precision for the MBSAR's orbital determination varies depending on specific circumstances. For example, the orbital determination in the presence of apsidal precession and concomitant precision requirements in the MBSAR are detailed in what follows.

7.4.3.2 Required Orbit Accuracy under Apsidal Precession

Regarding the requisite precision for position determination in the context of apsidal precession, the maximum permissible constant error in the AOP is plotted as a function of range positioning accuracy in Figure 7.23.

The results in Figure 7.23 demonstrate a linear relationship between the threshold of constant AOP error and the range positioning accuracy of TOI. This association is analogous to the threshold of constant orbital inclination error under nodal precession effects. It is worth noting that the established threshold for the rate of error in AOP, which is determined based on the azimuth positioning accuracy, is dependent on the constant AOP error. This error's rate, in turn, is associated with the level of positioning accuracy for the TOI along the range direction. To achieve the required precision for the rate of AOP error, we have to maintain a fixed constant AOP error, e.g., 4.93×10^{-9} rad,

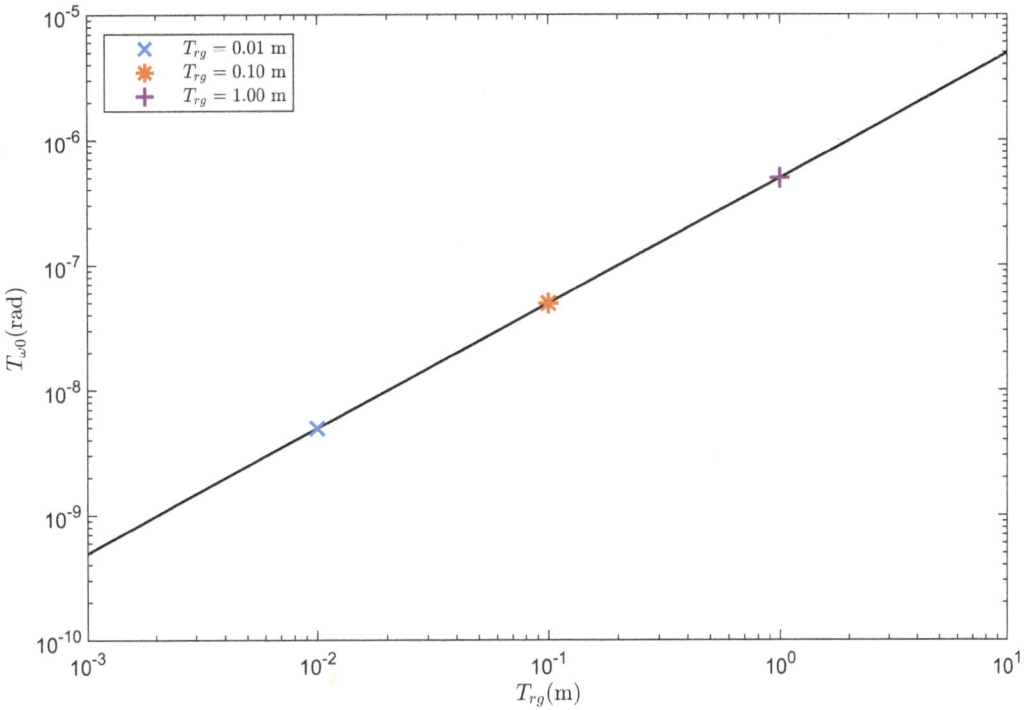

FIGURE 7.23 In the presence of apsidal precession, the threshold for constant AOP error versus the range positioning accuracy.

corresponding to a range positioning accuracy of 0.01 m, as expounded upon in the subsequent discussion.

We examine the requisite precision for the rate of AOP error, its threshold, T_{k1}, versus the azimuth position accuracy is depicted in Figure 7.24. We then take into account another threshold T_{k2}, crucial in ensuring the azimuth focusing quality. It is noteworthy that this threshold is also affected by the constant AOP error associated with range positioning accuracy; hence threshold T_{k2} versus the constant AOP error is shown in Figure 7.25.

Figure 7.24 shows that the azimuth positioning accuracy of TOI is influenced by the rate of AOP error. A discernible non-linear relationship between the two variables can be observed. The corresponding threshold T_{k2} may be affected by the constant AOP error, as exemplified in Figure 7.25. For a constant AOP error spanning from 10^{-10} rad to 10^{-6} rad, the fluctuation of T_{k2} remains below 1.5×10^{-11} rad/s, indicating that the constant AOP error has a minor impact on the threshold for the rate of AOP error. Moreover, by comparing Figures 7.24–7.25, we see that T_{k1} surpasses T_{k2} by a substantial margin, revealing that in the azimuth direction, the positioning accuracy of TOI carries a greater impact on the threshold for the rate of AOP error in contrast to the focusing quality. Therefore, when it comes to dealing with MBSAR's velocity determination, the significance of the azimuth positioning accuracy outweighs that of focusing quality.

By observing Figures 7.24 and 7.25, and using Eqs. (7.57) and (7.58), we can evaluate the precision requirements for position and velocity determinations in the MBSAR under varying degrees of positioning accuracies for TOI. Table 7.2 summarizes the accuracy requirements for the orbital determination of the MBSAR under apsidal precession.

Table 7.2 describes that the orbital determination in the presence of apsidal precession exhibits a strong correlation with the precision of TOI's localization in range and azimuth directions. Besides, the comparison of Table 7.2 with Table 7.1 reveals that the nodal and apsidal precessions have distinct but dissimilar impacts on the position and velocity determinations of the MBSAR. It

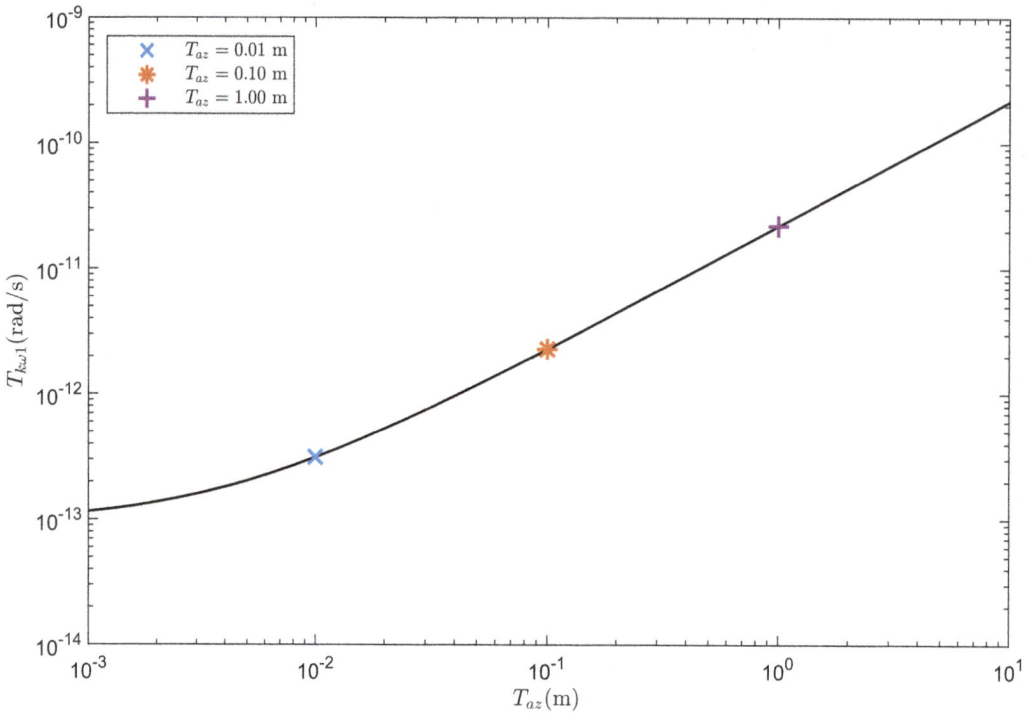

FIGURE 7.24 In the presence of apsidal precession, the threshold for an error rate of AOP T_{k1} versus the azimuth positioning accuracy.

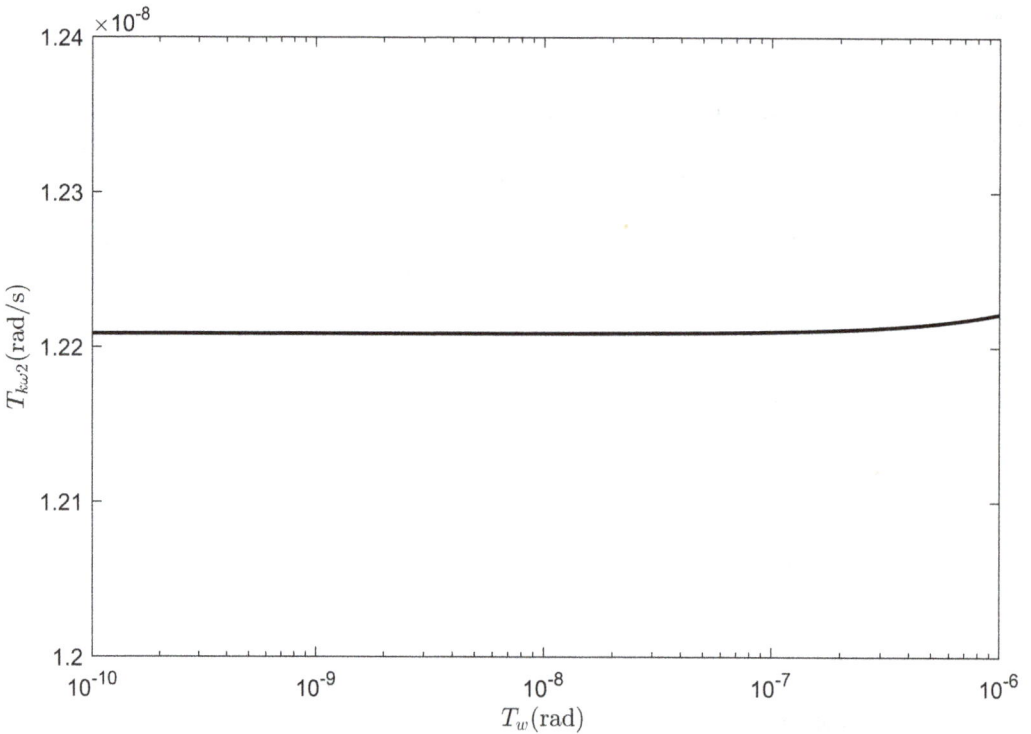

FIGURE 7.25 In the presence of apsidal precession, the threshold for an error rate of AOP T_{k2} versus the constant AOP error.

TABLE 7.2

Accuracy Requirement for the Orbital Determination of the MBSAR under Apsidal Precession

Orbital Determination	PAR	X-Component	Y-Component	Z-Component
Position Determination	0.01 m	1.67 m	0.93 m	0.46 m
	0.10 m	16.72 m	9.30 m	4.59 m
	1.00 m	167.16 m	93.03 m	45.93 m
Velocity Determination	0.01 m	1.03×10^{-4} m/s	6.24×10^{-5} m/s	3.11×10^{-5} m/s
	0.10 m	7.69×10^{-4} m/s	4.33×10^{-4} m/s	2.14×10^{-4} m/s
	1.00 m	7.43×10^{-3} m/s	4.14×10^{-3} m/s	2.04×10^{-3} m/s

is intriguing that the apsidal precession poses a difficulty to the orbital determination in achieving the desired accuracy level in position and velocity of the MBSAR along the z-axis. In contrast, the orbital determination in the x-axis is less strict. The required accuracy in the y-axis lies between those in the x-axis and the z-axis, regardless of the geometric accuracy of TOI. Such a finding is comparable to that encountered in the context of nodal precession.

To attain a range accuracy of 1.00 m, the MBSAR's position determination necessitates accuracies of several hundred meters and a few dozen meters in the x- and z-axes. If a range accuracy of 0.01 m is required, the position accuracy in the x and z directions should be at a level of meters and submeters, respectively. The geometric accuracy of TOI in the azimuth direction sets benchmarks for achieving the requisite precision in velocity determination. For example, consider an azimuth accuracy of 0.01 m. The required velocity accuracy for the MBSAR in the x-axis should be in the order of 10^{-4} m/s, while that in the z-axis would increase to the level of 10^{-5} m/s. For more tolerant azimuth positioning accuracy, such as 1.00 m, the required precision for velocity determination decreases correspondingly, reaching magnitudes of 10^{-3} m/s. In this situation, the velocity determination in the z-axis also proves to be a challenge compared to the other two axes.

At this point, we conclude that the TOI's geometric accuracy dominates MBSAR orbital determination in the context of both nodal and apsidal precessions. It may be advantageous to employ the far look angle for both position and velocity determinations to achieve a high geometric accuracy of the MBSAR image. We also find that both precessions, and other sources of orbital perturbations, are coupling in effect, making the orbital determination a highly challenging task. By the same argument, the above observation holds for other locations of the MBSAR.

APPENDIX A: THE POSITION AND RANGE ERRORS DUE TO NODAL PRECESSION

The coordinates for the MBSAR are presented in the following:

$$X_{SAR,\eta} = R_{SAR,\eta} \cdot (\cos\Omega_\eta \cos u_\eta - \sin\Omega_\eta \cos i_\eta \sin u_\eta) \tag{A.1}$$

$$Y_{SAR,\eta} = R_{SAR,\eta} \cdot (\sin\Omega_\eta \cos u_\eta + \cos\Omega_\eta \cos i_\eta \sin u_\eta) \tag{A.2}$$

$$Z_{SAR,\eta} = R_{SAR,\eta} \cdot \sin i_\eta \sin u_\eta \tag{A.3}$$

By considering the linear variations in both the orbital inclination and RAAN, and employing the Taylor series expansion, one can obtain:

$$\cos\Omega_\eta = \cos(\Omega_0 + k_\Omega \cdot \eta) \approx \cos\Omega_0 - k_\Omega \cdot \sin\Omega_0 \cdot \eta \qquad (A.4)$$

$$\sin\Omega_\eta = \sin\left(\Omega_0 + k_\Omega \cdot \eta\right) \approx \sin\Omega_0 + k_\Omega \cdot \cos\Omega_0 \cdot \eta \qquad (A.5)$$

$$\cos i_\eta = \cos\left(i_0 + k_{\text{inc}} \cdot \eta\right) \approx \cos i_0 - k_{\text{inc}} \cdot \sin i_0 \cdot \eta \qquad (A.6)$$

$$\sin i_\eta = \sin\left(i_0 + k_{\text{inc}} \cdot \eta\right) \approx \sin i_0 + k_{\text{inc}} \cdot \cos i_0 \cdot \eta \qquad (A.7)$$

Given the fact that the minuscule magnitude of the higher-order terms in Eqs. (A.4)–(A.7), it is deemed necessary to retain solely the constant and first-order terms.

Upon substitution of (A.4), (A.5), and (A.6) into (A.1), a revised form of the position vector along the x-axis can be obtained as follows:

$$\begin{aligned} X_{\text{SAR},\eta} \approx R_{\text{SAR},\eta} \cdot (&\cos\Omega_0 \cos u_\eta - k_\Omega \cdot \sin\Omega_0 \cos u_\eta \cdot \eta \\ &- \sin\Omega_0 \cos i_0 \sin u_\eta - k_\Omega \cdot \cos\Omega_0 \cos i_0 \sin u_\eta \cdot \eta \\ &+ k_{\text{inc}} \cdot \sin\Omega_0 \sin i_0 \sin u_\eta \cdot \eta + k_{\text{inc}}k_\Omega \cdot \cos\Omega_0 \sin i_0 \sin u_\eta \cdot \eta^2) \end{aligned} \qquad (A.8)$$

As the second-order terms in Eq. (A.8) are considerably smaller compared to the rest of the terms, they can be neglected, resulting in a simplified form of Eq. (A.8):

$$\begin{aligned} X_{\text{SAR},\eta} \approx\ & R_{\text{SAR},\eta}(\cos\Omega_0 \cos u_\eta - \sin\Omega_0 \cos i_0 \sin u_\eta) \\ & - k_\Omega \cdot R_{\text{SAR},\eta}(\sin\Omega_0 \cos u_\eta + \cos\Omega_0 \cos i_0 \sin u_\eta) \cdot \eta \\ & + k_{\text{inc}} \cdot R_{\text{SAR},\eta} \sin\Omega_0 \sin i_0 \sin u_\eta \cdot \eta \end{aligned} \qquad (A.9)$$

Eq. (A.9) is further reformulated as:

$$X_{\text{SAR},\eta} \approx X_{\text{NP},\eta} - k_\Omega \cdot Y_{\text{NP},\eta} \cdot \eta + k_{\text{inc}} \cdot \sin\Omega_0 \cdot Z_{\text{NP},\eta} \cdot \eta \qquad (A.10)$$

with:

$$X_{\text{NP},\eta} = R_{\text{SAR},\eta} \cdot (\cos\Omega_0 \cos u_\eta - \sin\Omega_0 \cos i_0 \sin u_\eta) \qquad (A.11)$$

The x-coordinate error in the MBSAR can be expressed by:

$$X_{\text{NPERR},\eta} = X_{\text{SAR},\eta} - X_{\text{NP},\eta} \approx -k_\Omega \cdot Y_{\text{NP},\eta} \cdot \eta + k_{\text{inc}} \cdot \sin\Omega_0 \cdot Z_{\text{NP},\eta} \cdot \eta \qquad (A.12)$$

By implementing the identical principle, we can readily derive the formulations for the y and z components of the MBSAR's orbital drift, which are expressed as:

$$\begin{cases} Y_{\text{NPERR},\eta} \approx k_\Omega \cdot X_{\text{NP},\eta} \cdot \eta - k_{\text{inc}} \cdot \cos\Omega_0 \cdot Z_{\text{NP},\eta} \cdot \eta \\ Z_{\text{NPERR},\eta} \approx k_{\text{inc}} \cdot Z_{\text{NP},\eta} \cdot \cot i_0 \cdot \eta \end{cases} \qquad (A.13)$$

with:

$$\begin{cases} Y_{\text{NP},\eta} = R_{\text{SAR},\eta} \cdot (\sin\Omega_0 \cos u_\eta + \cos\Omega_0 \cos i_0 \sin u_\eta) \\ Z_{\text{NP},\eta} = R_{\text{SAR},\eta} \cdot \sin i_0 \sin u_\eta \end{cases} \qquad (A.14)$$

We shall now proceed with the deduction of the range history error. The MBSAR's range histories, with and without the nodal precession, are considered and conveyed as follows:

$$R_S(\eta) = \left[R_{SAR,\eta}^2 + R_{TOI}^2 - 2 \cdot (X_{SAR,\eta} X_{TOI,\eta} + X_{SAR,\eta} Y_{TOI,\eta} + X_{SAR,\eta} Z_{TOI,\eta}) \right]^{0.5} \quad (A.15)$$

$$R_S(\eta) + \Delta R_p(\eta) = \left[R_{SAR,\eta}^2 + R_{TOI}^2 - 2 \cdot (X_{NP,\eta} X_{TOI,\eta} + Y_{NP,\eta} Y_{TOI,\eta} + Z_{NP,\eta} Z_{TOI,\eta}) \right]^{0.5} \quad (A.16)$$

Through the utilization of Eqs. (A.12)–(A.13), it is possible to present the range history in the following manner:

$$R_S(\eta) + \Delta R_p(\eta) = \left\{ R_{SAR,\eta}^2 + R_{TOI}^2 - 2 \cdot \left[(X_{SAR,\eta} - X_{NPERR,\eta}) \cdot X_{TOI,\eta} \right. \right.$$
$$\left. \left. + (Y_{SAR,\eta} - Y_{NPERR,\eta}) \cdot Y_{TOI,\eta} + (Z_{SAR,\eta} - Z_{NPERR,\eta}) \cdot Z_{TOI,\eta} \right] \right\}^{0.5} \quad (A.17)$$

One could employ the square norm of range history in Eqs. (A.15) and (A.17) to achieve the following results:

$$R_S^2(\eta) = R_{SAR,\eta}^2 + R_{TOI}^2 - 2 \cdot \left(X_{SAR,\eta} X_{TOI,\eta} + X_{SAR,\eta} Y_{TOI,\eta} + X_{SAR,\eta} Z_{TOI,\eta} \right) \quad (A.18)$$

$$\left[R_S(\eta) + \Delta R_p(\eta) \right]^2 = R_{SAR,\eta}^2 + R_{TOI}^2$$
$$- 2 \cdot (X_{SAR,\eta} X_{TOI,\eta} + Y_{NP,\eta} Y_{TOI,\eta} + Z_{NP,\eta} Z_{TOI,\eta})$$
$$+ 2 \cdot (X_{NPERR,\eta} X_{TOI,\eta} + Y_{NPERR,\eta} Y_{TOI,\eta} + Z_{NPERR,\eta} Z_{TOI,\eta}) \quad (A.19)$$

By subtracting Eq. (A.18) from Eq. (A.19), one arrives at the following relationship:

$$\Delta R_p(\eta) \cdot \left[R_S(\eta) + 2 \cdot \Delta R_p(\eta) \right] = 2 \cdot (X_{NPERR,\eta} X_{TOI,\eta}$$
$$+ Y_{NPERR,\eta} Y_{TOI,\eta} + Z_{NPERR,\eta} Z_{TOI,\eta}) \quad (A.20)$$

One can reformulate Eq. (A.20) to attain the range history error due to nodal precession:

$$\Delta R_p(\eta) = 2 \cdot \frac{X_{NPERR,\eta} X_{TOI,\eta} + Y_{NPERR,\eta} Y_{TOI,\eta} + Z_{NPERR,\eta} Z_{TOI,\eta}}{2 R_S(\eta) + \Delta R_p(\eta)} \quad (A.21)$$

In accordance with Eq. (A.22), in conjunction with the Taylor series expansion, one can acquire:

$$\Delta R_p(\eta) \ll R_S(\eta) \quad (A.22)$$

$$\frac{1}{2 R_S(\eta) + \Delta R_p(\eta)} = \frac{1}{2 R_S(\eta)} - \frac{\Delta R_p(\eta)}{4 R_S^2(\eta)} + \cdots \approx \frac{1}{2 R_S(\eta)} \quad (A.23)$$

Building upon this foundation, Eq. (A.21) can be reformulated as:

$$\Delta R_p(\eta) \approx \frac{X_{NPERR,\eta} X_{TOI,\eta} + Y_{NPERR,\eta} Y_{TOI,\eta} + Z_{NPERR,\eta} Z_{TOI,\eta}}{R_S(\eta)}$$
$$= \frac{\mathbf{R}_{NPERR,\eta} \cdot \mathbf{R}_{TOI,\eta}}{\left\| \mathbf{R}_{SAR,\eta} - \mathbf{R}_{TOI,\eta} \right\|_2} \quad (A.24)$$

APPENDIX B: THE RELATIONS IN THE ORBITAL INCLINATION AND RAAN

In the context of nodal precession, the relationship between the orbital inclination and RAAN can be observed:

$$\cos \vartheta_{M,\eta} = \cos i_\eta \cos \vartheta_{E,\eta} + \sin i_\eta \sin \vartheta_{E,\eta} \cos \Omega_\eta \tag{A.25}$$

where $\vartheta_{M,\eta}$ represents the MBSAR's orbital inclination in relation to the Earth's ecliptic, $\vartheta_{E,\eta}$ denotes the Earth's obliquity, both vary with azimuth time η. Although Earth's obliquity fluctuates during the SAT, the effect of this change is negligibly small. Therefore, we substitute $\vartheta_{E,\eta}$ with ϑ_{E0}, which denotes the Earth's instantaneous obliquity at the zero-azimuth time.

In the event of any uncertainties in the orbital inclination and RAAN, Eq. (A.25) must be revised accordingly:

$$\cos \vartheta_{M,\eta} = \cos(i_\eta + \delta i_\eta)\cos \vartheta_{E0} + \sin(i_\eta + \delta i_\eta)\sin \vartheta_{E0} \cos(\Omega_\eta + \delta \Omega_\eta) \tag{A.26}$$

where $\delta \Omega_\eta$ and δi_η are the uncertainties in the RAAN and orbital inclination.

Through utilizing the Taylor series expansion, one can establish the following relationships:

$$\begin{cases} \cos(i_\eta + \delta i_\eta) \approx \cos i_\eta - \delta i_\eta \cdot \sin i_\eta \\ \sin(i_\eta + \delta i_\eta) \approx \sin i_\eta + \delta i_\eta \cdot \cos i_\eta \\ \cos(\Omega_\eta + \delta \Omega_\eta) \approx \cos \Omega_\eta - \delta \Omega_\eta \cdot \sin \Omega_\eta \end{cases} \tag{A.27}$$

Upon considering Eqs. (A.27), it is possible to further rephrase Eq. (A.26) as follows:

$$\begin{aligned} \cos \vartheta_{M,\eta} = &\cos i_\eta \cos \vartheta_{E0} + \sin i_\eta \cdot \cos \Omega_\eta \cdot \sin \vartheta_{E0} \\ &- \delta \Omega_\eta \cdot \sin \Omega_\eta \cdot \sin i_\eta \cdot \sin \vartheta_{E0} \\ &- \delta i_\eta \cdot \sin i_\eta \cdot \cos \vartheta_{E0} + \delta i_\eta \cdot \cos i_\eta \cdot \cos \Omega_\eta \cdot \sin \vartheta_{E0} \end{aligned} \tag{A.28}$$

Undertaking a comparison between Eqs. (A.26) and (A.30), it is apparent that:

$$\begin{aligned} &-\delta \Omega_\eta \cdot \sin \Omega_\eta \cdot \sin i_\eta \cdot \sin \vartheta_{E0} \\ &- \delta i_\eta \cdot \sin i_\eta \cdot \cos \vartheta_{E0} + \delta i_\eta \cdot \cos i_\eta \cdot \cos \Omega_\eta \cdot \sin \vartheta_{E0} = 0 \end{aligned} \tag{A.29}$$

It is possible to make further modifications to Eq. (A.29), then one can yield:

$$\delta \Omega_\eta = \delta i_\eta \cdot (\cot i_\eta \cdot \cot \Omega_\eta - \csc \Omega_\eta \cot \vartheta_{E0}) \tag{A.30}$$

When the azimuth time equals zero, Eq. (A.30) can be expressed as:

$$\delta \Omega_0 = S_1 \cdot \delta i_0 \tag{A.31}$$

with:

$$S_1 = \cot i_0 \cot \Omega_0 - \csc \Omega_0 \cot \vartheta_{E0} \tag{A.32}$$

We shall now inspect the correlation between the error rate in RAAN and that in orbital inclination. With this in mind, we can reformulate Eq. (A.30) as follows:

$$\delta \Omega_0 + \delta k_\Omega \cdot \eta = (\delta i_0 + \delta k_{\text{inc}} \cdot \eta) \cdot (\cot i_\eta \cot \Omega_\eta - \csc \Omega_\eta \cot \vartheta_{E0}) \tag{A.33}$$

By means of the Taylor series expansion, one can establish the following relationships:

$$\begin{cases} \cot i_\eta = \cot(i_0 + k_{\text{inc}} \cdot \eta) \approx \cot i_0 - k_{\text{inc}} \cdot \csc^2 i_0 \cdot \eta \\ \cot \Omega_\eta = \cot(\Omega_0 + k_\Omega \cdot \eta) \approx \cot \Omega_0 - k_\Omega \cdot \csc^2 \Omega_0 \cdot \eta \\ \csc \Omega_\eta = \csc(\Omega_0 + k_\Omega \cdot \eta) \approx \csc \Omega_0 - k_\Omega \cdot \cot \Omega_0 \csc \Omega_0 \cdot \eta \end{cases} \tag{A.34}$$

Upon substitution of Eq. (A.34) into Eq. (A.33), the resultant expression is obtained as:

$$\begin{aligned} \delta\Omega_0 + \delta k_\Omega \cdot \eta &\approx (\delta i_0 + \delta k_{\text{inc}} \cdot \eta) \cdot (\cot i_0 \cot \Omega_0 - \csc \Omega_0 \cot \vartheta_{E0} \\ &- k_{\text{inc}} \cdot \csc^2 i_0 \cot \Omega_0 \cdot \eta - k_\Omega \cdot \cot \Omega_0 \csc \Omega_0 \cot \vartheta_{E0} \cdot \eta \\ &+ k_\Omega \cdot \cot i_0 \cdot \csc^2 \Omega_0 \cdot \eta + k_{\text{inc}} \cdot k_\Omega \cdot \csc^2 i_0 \cdot \csc^2 \Omega_0 \cdot \eta^2) \end{aligned} \tag{A.35}$$

After grouping alike terms and removing any 2nd or higher-order expressions, Eq. (A.35) becomes:

$$\begin{aligned} \delta\Omega_0 + \delta k_\Omega \cdot \eta &\approx \delta i_0 \cdot (\cot i_0 \cot \Omega_0 - \csc \Omega_0 \cot \vartheta_{E0}) \\ &+ \delta k_{\text{inc}} \cdot (\cot i_0 \cot \Omega_0 - \csc \Omega_0 \cot \vartheta_{E0}) \cdot \eta \\ &- \delta i_0 \cdot (k_{\text{inc}} \cdot \csc^2 i_0 \cot \Omega_0 \\ &+ k_\Omega \cdot \cot i_0 \csc^2 \Omega_0 - k_\Omega \cdot \cot \Omega_0 \csc \Omega_0 \cot \vartheta_{E0}) \cdot \eta \end{aligned} \tag{A.36}$$

Subtracting Eq. (A.31) from Eq. (A.36) yields the following relation:

$$\begin{aligned} \delta k_\Omega &\approx \delta k_{\text{inc}} \cdot (\cot i_0 \cot \Omega_0 - \csc \Omega_0 \cot \vartheta_{E0}) \\ &- \delta i_0 \cdot (k_{\text{inc}} \cdot \csc^2 i_0 \cot \Omega_0 \\ &+ k_\Omega \cdot \cot i_0 \csc^2 \Omega_0 - k_\Omega \cdot \cot \Omega_0 \csc \Omega_0 \cot \vartheta_{E0}) \end{aligned} \tag{A.37}$$

One may express Eq. (A.37) as:

$$\delta k_\Omega = S_1 \cdot \delta k_{\text{inc}} + S_2 \cdot \delta i_0 \tag{A.38}$$

with:

$$S_2 = -k_{\text{inc}} \cdot \csc^2 i_0 \cot \Omega_0 + k_\Omega \cdot (\cot \Omega_0 \csc \Omega_0 \cot \vartheta_{E0} - \cot i_0 \csc^2 \Omega_0) \tag{A.39}$$

APPENDIX C: THE INDUCTION OF POSITION AND VELOCITY ERRORS UNDER UNCERTAINTIES CAUSED BY ORBITAL PRECESSIONS

To thoroughly comprehend how the MBSAR's position and velocity errors occur under orbital precessions, it is essential to analyze the nodal and apsidal precession effects separately. In this regard, we can isolate the uncertainties caused by a specific orbital precession and focus solely on its corresponding precession effects.

Regarding the nodal precession, there may be uncertainties in the orbital inclination and RAAN. In this scenario, the MBSAR's coordinates can be expressed as follows:

$$\begin{aligned} X_{\text{SAR},\eta} + \delta X_{\text{SAR},\eta} = R_{\text{SAR},\eta} \cdot \big[&\cos(\Omega_\eta + \delta\Omega_\eta) \cdot \cos u_\eta \\ &- \sin(\Omega_\eta + \delta\Omega_\eta) \cdot \cos(i_\eta + \delta i_\eta) \cdot \sin u_\eta \big] \end{aligned} \tag{A.40}$$

$$\begin{aligned} Y_{\text{SAR},\eta} + \delta Y_{\text{SAR},\eta} = R_{\text{SAR},\eta} \cdot \big[&\sin(\Omega_\eta + \delta\Omega_\eta) \cdot \cos u_\eta \\ &+ \cos(\Omega_\eta + \delta\Omega_\eta) \cdot \cos(i_\eta + \delta i_\eta) \cdot \sin u_\eta \big] \end{aligned} \tag{A.41}$$

$$Z_{\text{SAR},\eta} + \delta Z_{\text{SAR},\eta} = R_{\text{SAR},\eta} \cdot \sin(i_\eta + \delta i_\eta) \cdot \sin u_\eta \tag{A.42}$$

By employing the Taylor series expansion, we can obtain the following expression:

$$\begin{cases}
\cos(\Omega_\eta + \delta\Omega_\eta) \approx \cos\Omega_\eta - \delta\Omega_\eta \cdot \sin\Omega_\eta \\
\sin(\Omega_\eta + \delta\Omega_\eta) \approx \sin\Omega_\eta + \delta\Omega_\eta \cdot \cos\Omega_\eta \\
\cos(i_\eta + \delta i_\eta) \approx \cos i_\eta - \delta i_\eta \cdot \sin i_\eta \\
\sin(i_\eta + \delta i_\eta) \approx \sin i_\eta + \delta i_\eta \cdot \cos i_\eta
\end{cases} \tag{A.43}$$

Utilizing Eq. (A.43), we can further rephrase Eq. (A.40) as follows:

$$\begin{aligned}
X_{\text{SAR},\eta} + \delta X_{\text{SAR},\eta} &= R_{\text{SAR},\eta} \cdot (\cos\Omega_\eta \cos u_\eta - \sin\Omega_\eta \cos i_\eta \sin u_\eta) \\
&\quad + \delta i_\eta \cdot R_{\text{SAR},\eta} \cdot \sin i_\eta \sin\Omega_\eta \sin u_\eta \\
&\quad - \delta\Omega_\eta \cdot R_{\text{SAR},\eta} \cdot (\sin\Omega_\eta \cos u_\eta + \cos\Omega_\eta \cos i_\eta \sin u_\eta)
\end{aligned} \tag{A.44}$$

By subtracting Eq. (A.9) from Eq. (A.44), we are able to obtain:

$$\begin{aligned}
\delta X_{\text{SAR},\eta} &= \delta i_\eta \cdot R_{\text{SAR},\eta} \cdot \sin i_\eta \sin\Omega_\eta \sin u_\eta \\
&\quad - \delta\Omega_\eta \cdot R_{\text{SAR},\eta} \cdot (\sin\Omega_\eta \cos u_\eta + \cos\Omega_\eta \cos i_\eta \sin u_\eta)
\end{aligned} \tag{A.45}$$

Consequently, the MBSAR's location deviation along the x-axis can be expressed as:

$$\begin{aligned}
\delta X_{\text{SAR},\eta} &= \delta i_\eta \cdot \sin\Omega_\eta \cdot Z_{\text{SAR},\eta} - \delta\Omega_\eta \cdot Y_{\text{SAR},\eta} \\
&= \delta i_0 \cdot \sin\Omega_\eta \cdot Z_{\text{SAR},\eta} - \delta\Omega_0 \cdot Y_{\text{SAR},\eta} \\
&\quad + \delta k_{\text{inc}} \cdot \sin\Omega_\eta \cdot Z_{\text{SAR},\eta} \cdot \eta - \delta k_\Omega \cdot Y_{\text{SAR},\eta} \cdot \eta
\end{aligned} \tag{A.46}$$

The act of differentiating Eq. (A.46) over time leads to:

$$\begin{aligned}
\delta \dot{X}_{\text{SAR},\eta} &= \delta i_0 \cdot \sin\Omega_\eta \cdot \dot{Z}_{\text{SAR},\eta} - \delta\Omega_0 \cdot \dot{Y}_{\text{SAR},\eta} - \delta k_\Omega \cdot Y_{\text{SAR},\eta} \\
&\quad + \delta k_{\text{inc}} \cdot \sin\Omega_\eta \cdot Z_{\text{SAR},\eta} + \delta k_{\text{inc}} \cdot \sin\Omega_\eta \cdot \dot{Z}_{\text{SAR},\eta} \cdot \eta \\
&\quad + \cos\Omega_\eta \left[\delta i_0 \cdot \dot{\Omega}_\eta \cdot Z_{\text{SAR},\eta} + \delta k_{\text{inc}} \cdot \dot{\Omega}_\eta \cdot Z_{\text{SAR},\eta} \cdot \eta \right] - \delta k_\Omega \cdot \dot{Y}_{\text{SAR},\eta} \cdot \eta \\
&\approx \delta i_0 \cdot \sin\Omega_\eta \cdot \dot{Z}_{\text{SAR},\eta} - \delta\Omega_0 \cdot \dot{Y}_{\text{SAR},\eta} - \delta k_\Omega \cdot Y_{\text{SAR},\eta} \\
&\quad + \delta k_{\text{inc}} \cdot \sin\Omega_\eta \cdot Z_{\text{SAR},\eta} + \delta k_{\text{inc}} \cdot \sin\Omega_\eta \cdot \dot{Z}_{\text{SAR},\eta} \cdot \eta - \delta k_\Omega \cdot \dot{Y}_{\text{SAR},\eta} \cdot \eta
\end{aligned} \tag{A.47}$$

By utilizing Eqs. (A.46) and (A.47), combining with $\eta = 0$, one is ready to yield:

$$\delta X_{\text{SAR}} = -\delta\Omega_0 \cdot Y_{\text{SAR}} + \delta i_0 \cdot \sin\Omega_0 \cdot Z_{\text{SAR}} \tag{A.48}$$

$$\delta V_{\text{SX}} = \delta k_{\text{inc}} \cdot \sin\Omega_0 \cdot Z_{\text{SAR}} - \delta m \cdot Y_{\text{SAR}} + \delta i_0 \cdot \sin\Omega_0 \cdot V_{\text{SZ}} - \delta\Omega_0 \cdot V_{\text{SY}} \tag{A.49}$$

Likewise, the deviations in the MBSAR's position and velocity in the y- and z-axes are articulated as follows:

$$\begin{cases}
\delta Y_{\text{SAR}} = \delta\Omega_0 \cdot X_{\text{SAR}} - \delta i_0 \cdot \cos\Omega_0 \cdot Z_{\text{SAR}} \\
\delta Z_{\text{SAR}} = \delta i_0 \cdot \cot i_0 \cdot Z_{\text{SAR}}
\end{cases} \tag{A.50}$$

and:

$$\begin{cases}
\delta V_{\text{SY}} = \delta k_\Omega \cdot X_{\text{SAR}} - \delta k_{\text{inc}} \cdot \cos\Omega_0 \cdot Z_{\text{SAR}} + \delta\Omega_0 \cdot V_{\text{SX}} - \delta i_0 \cdot \cos\Omega_0 \cdot V_{\text{SZ}} \\
\delta V_{\text{SZ}} = \delta k_{\text{inc}} \cdot \cot i_0 \cdot Z_{\text{SAR}} + \delta i_0 \cdot \cot i_0 \cdot V_{\text{SZ}}
\end{cases} \tag{A.51}$$

When analyzing the apsidal precession, it is essential for the AOL to factor in the uncertainty of AOP, which is mathematically represented as:

$$u_{\eta(\text{uncertainty})} = v_\eta + \omega_\eta + \delta\omega_\eta = u_\eta + \delta\omega_\eta \tag{A.52}$$

Given this backdrop, the trajectory of the MBSAR can be elucidated by:

$$X_{\text{SAR},\eta} + \delta X_{\text{SAR},\eta} = R_{\text{SAR},\eta} \cdot \left[\cos\Omega_\eta \cos\left(u_\eta + \delta\omega_\eta\right) - \sin\Omega_\eta \cos i_\eta \sin\left(u_\eta + \delta\omega_\eta\right)\right] \tag{A.53}$$

$$Y_{\text{SAR},\eta} + \delta Y_{\text{SAR},\eta} = R_{\text{SAR},\eta} \cdot \left[\sin\Omega_\eta \cos\left(u_\eta + \delta\omega_\eta\right) + \cos\Omega_\eta \cos i_\eta \sin\left(u_\eta + \delta\omega_\eta\right)\right] \tag{A.54}$$

$$Z_{\text{SAR},\eta} + \delta Z_{\text{SAR},\eta} = R_{\text{SAR},\eta} \cdot \left[\sin i_\eta \sin\left(u_\eta + \delta\omega_\eta\right)\right] \tag{A.55}$$

In general, the level of uncertainty in AOP would be relatively minor. Therefore, we could utilize the subsequent relation:

$$\cos\left(u_\eta + \delta\omega_\eta\right) \approx \cos u_\eta - \delta\omega_\eta \cdot \sin u_\eta \tag{A.56}$$

$$\sin\left(u_\eta + \delta\omega_\eta\right) \approx \sin u_\eta + \delta\omega_\eta \cdot \cos u_\eta \tag{A.57}$$

By means of substituting Eqs. (A.56), (A.57) into (A.53), one can derive the ensuing result:

$$\begin{aligned} X_{\text{SAR},\eta} + \delta X_{\text{SAR},\eta} = R_{\text{SAR},\eta} \cdot \big[&\cos\Omega_\eta (\cos u_\eta - \delta\omega_\eta \cdot \sin u_\eta) \\ &- \sin\Omega_\eta \cos i_\eta (\sin u_\eta + \delta\omega_\eta \cdot \cos u_\eta)\big] \end{aligned} \tag{A.58}$$

Eq. (A.58) is subject to reconfiguration in the following manner:

$$\begin{aligned} X_{\text{SAR},\eta} + \delta X_{\text{SAR},\eta} = R_{\text{SAR},\eta} \cdot (\cos\Omega_\eta \cos u_\eta - \sin\Omega_\eta \cos i_\eta \sin u_\eta) \\ - \delta\omega_\eta \cdot R_{\text{SAR},\eta} \cdot (\cos\Omega_\eta \sin u_\eta + \sin\Omega_\eta \cos i_\eta \cos u_\eta) \end{aligned} \tag{A.59}$$

Subtracting Eq. (A.1) from Eq. (A.59) leads to the following result:

$$\delta X_{\text{SAR},\eta} = -\delta\omega_\eta \cdot R_{\text{SAR},\eta} \cdot (\cos\Omega_\eta \sin u_\eta + \sin\Omega_\eta \cos i_\eta \cos u_\eta) \tag{A.60}$$

Likewise, the MBSAR's position deviations along the y and z axes are articulated as follows:

$$\delta Y_{\text{SAR},\eta} = -\delta\omega_\eta \cdot R_{\text{SAR},\eta} \cdot (\sin\Omega_\eta \sin u_\eta - \cos\Omega_\eta \cos i_\eta \cos u_\eta) \tag{A.61}$$

$$\delta_{\text{SAR},\eta} = \delta\omega_\eta \cdot R_{\text{SAR},\eta} \cdot \sin i_\eta \cos u_\eta \tag{A.62}$$

To facilitate derivation, we shall introduce a vector denoted as $\mathbf{C}_\eta$. The components of this vector are delineated as follows:

$$\begin{cases} C_{X,\eta} = -R_{\text{SAR},\eta} \cdot \left(\cos\Omega_\eta \sin u_\eta + \sin\Omega_\eta \cos i_\eta \cos u_\eta\right) \\ C_{Y,\eta} = -R_{\text{SAR},\eta} \cdot \left(\sin\Omega_\eta \sin u_\eta - \cos\Omega_\eta \cos i_\eta \cos u_\eta\right) \\ C_{Z,\eta} = R_{\text{SAR},\eta} \cdot \sin i_\eta \cos u_\eta \end{cases} \tag{A.63}$$

The coordinate deviations resulting from the AOP error can be restated as follows:

$$\begin{cases} \delta X_{\text{SAR},\eta} = \delta\omega_\eta \cdot C_{X,\eta} \\ \delta Y_{\text{SAR},\eta} = \delta\omega_\eta \cdot C_{Y,\eta} \\ \delta Z_{\text{SAR},\eta} = \delta\omega_\eta \cdot C_{Z,\eta} \end{cases} \tag{A.64}$$

where the uncertainty of AOP is expressed by:

$$\delta\omega_\eta = \delta\omega_0 + \delta k_\omega \cdot \eta \tag{A.65}$$

By conducting temporal differential operations on Eq. (A.63), one can derive:

$$\begin{aligned} \dot{C}_{X,\eta} = &-\dot{R}_{\text{SAR},\eta} \cdot (\cos\Omega_\eta \sin u_\eta + \sin\Omega_\eta \cos i_\eta \cos u_\eta) \\ &+ R_{\text{SAR},\eta} \cdot (\dot{\Omega}_\eta \cdot \sin\Omega_\eta \sin u_\eta - \dot{u}_\eta \cdot \cos\Omega_\eta \cos u_\eta \\ &- \dot{\Omega}_\eta \cdot \cos\Omega_\eta \cos i_\eta \cos u_\eta + \dot{i}_\eta \cdot \sin\Omega_\eta \sin i_\eta \cos u_\eta \\ &+ \dot{u}_\eta \cdot \sin\Omega_\eta \cos i_\eta \sin u_\eta) \quad . \end{aligned} \tag{A.66}$$

Through reordering comparable elements, it is possible to articulate Eq. (A.66) in the following form:

$$\begin{aligned} \dot{C}_{X,\eta} = &-R_{\text{SAR},\eta} \cdot (\dot{u}_\eta \cdot \cos\Omega_\eta \cos u_\eta - \dot{u}_\eta \cdot \sin\Omega_\eta \cos i_\eta \sin u_\eta) \\ &+ R_{\text{SAR},\eta} \cdot (\dot{\Omega}_\eta \cdot \sin\Omega_\eta \sin u_\eta - \dot{\Omega}_\eta \cdot \cos\Omega_\eta \cos i_\eta \cos u_\eta) \\ &- \dot{R}_{\text{SAR},\eta} \cdot (\cos\Omega_\eta \sin u_\eta + \sin\Omega_\eta \cos i_\eta \cos u_\eta) \\ &+ \dot{i}_\eta \cdot R_{\text{SAR},\eta} \cdot \sin\Omega_\eta \sin i_\eta \cos u_\eta \quad . \end{aligned} \tag{A.67}$$

It is noteworthy that the first term dominates the magnitude of Eq. (A.67), while the succeeding terms have a relatively insignificant impact. Additionally, the temporal variation of AOL is more significantly influenced by the true anomaly compared to the AOP. Based on this premise, we can arrive at the following approximation.

$$\begin{aligned} \dot{C}_{X,\eta} &\approx -w_{\text{SAR},\eta} \cdot R_{\text{SAR},\eta} \cdot (\cos\Omega_\eta \cos u_\eta - \sin\Omega_\eta \cos i_\eta \sin u_\eta) \\ &= -w_{\text{SAR},\eta} \cdot X_{\text{SAR},\eta} \end{aligned} \tag{A.68}$$

Continuing in a similar token, one arrives at:

$$\dot{C}_{Y,\eta} \approx -w_{\text{SAR},\eta} \cdot Y_{\text{SAR},\eta} \tag{A.69}$$

$$\dot{C}_{Z,\eta} \approx -w_{\text{SAR},\eta} \cdot Z_{\text{SAR},\eta} \tag{A.70}$$

To elevate this analysis, we shall approximate the vector $\mathbf{C}_\eta$ as a linear function of azimuth time, formulated as:

$$\begin{aligned} C_{X,\eta} &\approx C_{X0} + V_{CX} \cdot \eta \\ \{C_{Y,\eta} &\approx C_{Y0} + V_{CY} \cdot \eta \\ C_{Z,\eta} &\approx C_{Z0} + V_{CZ} \cdot \eta \end{aligned} \tag{A.71}$$

where $\mathbf{C}_0 = [C_{X0}, C_{Y0}, C_{Z0}]^T$ is the vector $\mathbf{C}_\eta$ at zero-azimuth time. $\mathbf{V}_C = [V_{CX}, V_{CY}, V_{CZ}]^T$ is the change rate of $\mathbf{C}_\eta$ over time, which takes the following forms:

$$\begin{cases} V_{CX} \approx -w_{\text{SAR0}} \cdot X_{\text{SAR0}} \\ V_{CY} \approx -w_{\text{SAR0}} \cdot Y_{\text{SAR0}} \\ V_{CZ} \approx -w_{\text{SAR0}} \cdot Z_{\text{SAR0}} \end{cases} \tag{A.72}$$

Thus, in the MBSAR, the temporally varying coordinate drifts can be reformulated by:

$$\begin{cases} \delta X_{\text{SAR},\eta} = \delta\omega_0 \cdot C_{X0} + \delta k_\omega \cdot C_{X0} \cdot \eta + \delta\omega_0 \cdot V_{CX} \cdot \eta + \delta k_\omega \cdot V_{CX} \cdot \eta^2 \\ \delta Y_{\text{SAR},\eta} = \delta\omega_0 \cdot C_{Y0} + \delta k_\omega \cdot C_{Y0} \cdot \eta + \delta\omega_0 \cdot V_{CY} \cdot \eta + \delta k_\omega \cdot V_{CY} \cdot \eta^2 \\ \delta Z_{\text{SAR},\eta} = \delta\omega_0 \cdot C_{Z0} + \delta k_\omega \cdot C_{Z0} \cdot \eta + \delta\omega_0 \cdot V_{CZ} \cdot \eta + \delta k_\omega \cdot V_{CZ} \cdot \eta^2 \end{cases} \quad \text{(A.73)}$$

Upon setting the azimuth time equal to 0, the subsequent equation can be derived:

$$\begin{cases} \delta X_{\text{SAR}} = \delta\omega_0 \cdot C_{X0} \\ \delta Y_{\text{SAR}} = \delta\omega_0 \cdot C_{Y0} \\ \delta Z_{\text{SAR}} = \delta\omega_0 \cdot C_{Z0} \end{cases} \quad \text{(A.74)}$$

After conducting partial differential operations on Eq. (A.67) with respect to time and employing a zero-azimuth time, the error in MBSAR's velocity can be calculated by:

$$\begin{cases} \delta V_{SX} = \delta k_\omega \cdot C_{X0} + \delta\omega_0 \cdot V_{CX} \\ \delta V_{SY} = \delta k_\omega \cdot C_{Y0} + \delta\omega_0 \cdot V_{CY} \\ \delta V_{SZ} = \delta k_\omega \cdot C_{Z0} + \delta\omega_0 \cdot V_{CZ} \end{cases} \quad \text{(A.75)}$$

APPENDIX D: THE RANGE ERROR UNDER ORBITAL PRECESSION EFFECTS

In the presence of orbital precession effects, the MBSAR's range history can be expressed in the following manner:

$$R_{SP}(\eta) = \Big[\big(X_{\text{SAR},\eta} + \delta X_{\text{SAR},\eta} - X_{\text{TOI},\eta} \big)^2$$
$$+ \big(Y_{\text{SAR},\eta} + \delta Y_{\text{SAR},\eta} - Y_{\text{TOI},\eta} \big)^2 + \big(Z_{\text{SAR},\eta} + \delta Z_{\text{SAR},\eta} - Z_{\text{TOI},\eta} \big)^2 \Big]^{0.5} \quad \text{(A.76)}$$

Where $R_{SP}(\eta) = R_S(\eta) + \delta R_p(\eta)$.

The slant range given in Eq. (A.76) can be represented in a squared form as follows:

$$R_{SP}^2(\eta) = (X_{\text{SAR},\eta} + \delta X_{\text{SAR},\eta})^2 + (Y_{\text{SAR},\eta} + \delta Y_{\text{SAR},\eta})^2$$
$$+ (Z_{\text{SAR},\eta} + \delta Z_{\text{SAR},\eta})^2 + X_{\text{TOI},\eta}^2 + Y_{\text{TOI},\eta}^2 + Z_{\text{TOI},\eta}^2$$
$$- 2\Big[X_{\text{TOI},\eta} \cdot (X_{\text{SAR},\eta} + \delta X_{\text{SAR},\eta}) \quad \text{(A.77)}$$
$$+ Y_{\text{TOI},\eta} \cdot (Y_{\text{SAR},\eta} + \delta Y_{\text{SAR},\eta}) + Z_{\text{TOI},\eta} \cdot (Z_{\text{SAR},\eta} + \delta Z_{\text{SAR},\eta}) \Big]$$

Since the orbital precession exerts no effect on the distance from Earth to MBSAR. Therefore, Eq. (A.77) can be reformulated as follows:

$$R_{SP}^2(\eta) = R_{\text{SAR},\eta}^2 + R_{\text{TOI}}^2$$
$$- 2(X_{\text{SAR},\eta} \cdot X_{\text{TOI},\eta} + Y_{\text{SAR},\eta} \cdot Y_{\text{TOI},\eta} + Z_{\text{SAR},\eta} \cdot Z_{\text{TOI},\eta}) \quad \text{(A.78)}$$
$$- 2(\delta X_{\text{SAR},\eta} \cdot X_{\text{TOI},\eta} + \delta Y_{\text{SAR},\eta} \cdot Y_{\text{TOI},\eta} + \delta Z_{\text{SAR},\eta} \cdot Z_{\text{TOI},\eta})$$

The difference between Eqs. (A.78) and (A.18) can be expressed as follows:

$$\delta R_p(\eta) \cdot \Big[2R_S(\eta) + \delta R_p(\eta) \Big] = -2 \cdot (\delta X_{\text{SAR},\eta} X_{\text{TOI},\eta}$$
$$+ \delta Y_{\text{SAR},\eta} Y_{\text{TOI},\eta} + \delta Z_{\text{SAR},\eta} Z_{\text{TOI},\eta}) \quad \text{(A.79)}$$

Eq. (A.79) can be further expressed in the following manner:

$$\delta R_p\left(\eta\right) = -\frac{\delta X_{\text{SAR},\eta} X_{\text{TOI},\eta} + \delta Y_{\text{SAR},\eta} Y_{\text{TOI},\eta} + \delta Z_{\text{SAR},\eta} Z_{\text{TOI},\eta}}{R_S\left(\eta\right) + 0.5 \cdot \delta R_p\left(\eta\right)} \tag{A.80}$$

Given the fact that $\delta R_p(\eta) \ll R_S(\eta)$, together with Taylor series expansion, it is possible to obtain the subsequent approximation:

$$\frac{1}{R_S\left(\eta\right) + A_{\text{const}} \cdot \delta R_p\left(\eta\right)} \approx \frac{1}{R_S\left(\eta\right)} \tag{A.81}$$

By virtue of Eq. (A.81), Eq. (A.80) can be approximated in the following manner:

$$\delta R_p\left(\eta\right) = -\frac{\delta X_{\text{SAR},\eta} X_{\text{TOI},\eta} + \delta Y_{\text{SAR},\eta} Y_{\text{TOI},\eta} + \delta Z_{\text{SAR},\eta} Z_{\text{TOI},\eta}}{R_S\left(\eta\right)} \tag{A.82}$$

At the instant of zero-azimuth time, one can determine the central slant range error as follows:

$$r_{p0} = \delta R_p\left(\eta\right)|_{\eta=0} = -\frac{\delta X_{\text{SAR}} \cdot X_{\text{TOI}} + \delta Y_{\text{SAR}} \cdot Y_{\text{TOI}} + \delta Z_{\text{SAR}} \cdot Z_{\text{TOI}}}{R_c} \tag{A.83}$$

Upon conducting partial differential operations respecting time on Eq. (A.79), one can derive the subsequent expression:

$$\begin{aligned} R_{\text{temp}}\left(\eta\right) = &-2 \cdot \left(\delta X_{\text{SAR},\eta} \cdot \dot{X}_{\text{TOI},\eta} + \delta Y_{\text{SAR},\eta} \cdot \dot{Y}_{\text{TOI},\eta}\right) \\ &-2 \cdot \left(\delta \dot{X}_{\text{SAR},\eta} \cdot X_{\text{TOI},\eta} + \delta \dot{Y}_{\text{SAR},\eta} \cdot Y_{\text{TOI},\eta} + \delta \dot{Z}_{\text{SAR},\eta} \cdot Z_{\text{TOI},\eta}\right) \end{aligned} \tag{A.84}$$

with:

$$R_{\text{temp}}\left(\eta\right) = 2 \cdot \delta \dot{R}_p\left(\eta\right) \cdot \left[R_S\left(\eta\right) + \delta R_p\left(\eta\right)\right] + 2 \cdot \delta R_p\left(\eta\right) \cdot \dot{R}_S\left(\eta\right) \tag{A.85}$$

Henceforth, the first derivative of range error is given by:

$$\begin{aligned} \delta \dot{R}_p\left(\eta\right) = &-\frac{\delta \dot{X}_{\text{SAR},\eta} \cdot X_{\text{TOI},\eta} + \delta \dot{Y}_{\text{SAR},\eta} \cdot Y_{\text{TOI},\eta} + \delta \dot{Z}_{\text{SAR},\eta} \cdot Z_{\text{TOI},\eta}}{R_S\left(\eta\right) + \delta R_p\left(\eta\right)} \\ &-\frac{\delta X_{\text{SAR},\eta} \cdot \dot{X}_{\text{TOI},\eta} + \delta Y_{\text{SAR},\eta} \cdot \dot{Y}_{\text{TOI},\eta}}{R_S\left(\eta\right) + \delta R_p\left(\eta\right)} - \frac{\delta R_p\left(\eta\right) \cdot \dot{R}_S\left(\eta\right)}{R_S\left(\eta\right) + \delta R_p\left(\eta\right)} \end{aligned} \tag{A.86}$$

It is noteworthy that the approximation in Eq. (A.81) is also applicable to Eq. (A.86), leading to the representation of the range error's derivative:

$$\begin{aligned} \delta \dot{R}_p\left(\eta\right) = &-\frac{\delta \dot{X}_{\text{SAR},\eta} \cdot X_{\text{TOI},\eta} + \delta \dot{Y}_{\text{SAR},\eta} \cdot Y_{\text{TOI},\eta} + \delta \dot{Z}_{\text{SAR},\eta} \cdot Z_{\text{TOI},\eta}}{R_S\left(\eta\right)} \\ &-\frac{\delta X_{\text{SAR},\eta} \cdot \dot{X}_{\text{TOI},\eta} + \delta Y_{\text{SAR},\eta} \cdot \dot{Y}_{\text{TOI},\eta}}{R_S\left(\eta\right)} - \frac{\delta R_p\left(\eta\right) \cdot \dot{R}_S\left(\eta\right)}{R_S\left(\eta\right)} \end{aligned} \tag{A.87}$$

If we set the azimuth time to zero, one shall obtain the resulting outcome:

$$r_{p1} = -\frac{\delta V_{SX} \cdot X_{TOI} + \delta V_{SY} \cdot Y_{TOI} + \delta V_{SZ} \cdot Z_{TOI}}{R_c}$$
$$-\frac{\delta X_{SAR} \cdot V_{TX} + \delta Y_{SAR} \cdot V_{TY}}{R_c} - R_1 \frac{\delta X_{SAR} X_{TOI} + \delta Y_{SAR} Y_{TOI} + \delta Z_{SAR} Z_{TOI}}{R_c^2} \tag{A.88}$$

One may take the derivative of Eq. (A.85) to obtain:

$$\dot{R}_{temp}(\eta) = 2 \cdot \left\{ \delta \ddot{R}_p(\eta) \cdot \left[R_S(\eta) + \delta R_p(\eta) \right] \right.$$
$$\left. + \delta \dot{R}_p(\eta) \cdot \left[\dot{R}_S(\eta) + \delta \dot{R}_p(\eta) \right] + \delta \dot{R}_p(\eta) \cdot \dot{R}_S(\eta) + \delta R_p(\eta) \cdot \ddot{R}_S(\eta) \right\} \tag{A.89}$$

Evidently, Eq. (A.89) is tantamount to the following expression:

$$\dot{R}_{temp}(\eta) = -4\left(\delta \dot{X}_{SAR,\eta} \dot{X}_{TOI,\eta} + \delta \dot{Y}_{SAR,\eta} \dot{Y}_{TOI,\eta} \right)$$
$$-2\left(\delta X_{SAR,\eta} \ddot{X}_{TOI,\eta} + \delta Y_{SAR,\eta} \ddot{Y}_{TOI,\eta} \right) \tag{A.90}$$
$$-2\left(\delta \ddot{X}_{SAR,\eta} X_{TOI,\eta} + \delta \ddot{Y}_{SAR,\eta} Y_{TOI,\eta} + \delta \ddot{Z}_{SAR,\eta} Z_{TOI,\eta} \right)$$

Note the first term in both Eqs. (A.89) and (A.90) holds more sway in determining the rate of R_{temp}, while the remaining terms make minor contributions, they can be disregarded. Consequently, upon implementing Eq. (A.81), one can obtain:

$$\delta \ddot{R}_p(\eta) \approx -2 \frac{\delta \dot{X}_{SAR,\eta} \cdot \dot{X}_{TOI,\eta} + \delta \dot{Y}_{SAR,\eta} \cdot \dot{Y}_{TOI,\eta}}{R_S(\eta)} \tag{A.91}$$

Ultimately, we reach the following result:

$$r_{p2} = \frac{\delta \ddot{R}_p(\eta)|_{\eta=0}}{2} = -\frac{\delta V_{SX} \cdot V_{TX} + \delta V_{SY} \cdot V_{TY}}{R_c} \tag{A.92}$$

APPENDIX E: THE CRITERIA FOR THE ORBITAL DETERMINATION UNDER APSIDAL PRECESSION

Since the range shift is primarily contingent on the constant error of AOP, one can derive the ensuing relation:

$$\delta L_{rg} = |r_{ap0}| \leq T_{rg} \tag{A.93}$$

By employing Eqs. (A.74), (A.83), and (A.93), it is possible to obtain an upper bound for the constant error in AOP, which is expressed as follows:

$$|\delta \omega_0| \leq \frac{R_c \cdot T_{rg}}{C_{X0} \cdot X_{TOI} + C_{Y0} \cdot Y_{TOI} + C_{Z0} \cdot Z_{TOI}} \tag{A.94}$$

By contrast, the azimuth offset is intrinsically linked to the threshold value by:

$$\delta L_{az} = \Re_{az} \cdot |r_{p1}| \leq T_{az} \tag{A.95}$$

It is possible to express Eq. (A.95) in an alternative manner:

$$|R_c \cdot r_{p1}| \leq \Re_{az}^{-1} \cdot R_c \cdot T_{az} \tag{A.96}$$

On the other hand, the parameter $R_c \cdot r_{ap1}$ can be articulated as a function of $\delta\omega_0$ and δk as per Eqs. (A.74), (A.75), and (A.88), specifically:

$$
\begin{aligned}
R_c \cdot r_{ap1} = &-\delta\omega_0 \cdot \{ C_{X0} \cdot V_{TX} + C_{Y0} \cdot V_{TY} \\
&+ R_1 \cdot R_c^{-1} \cdot (C_{X0} \cdot X_{TOI} + C_{Y0} \cdot Y_{TOI} + C_{Z0} \cdot Z_{TOI}) \\
&- w_{SAR} \cdot (X_{SAR0} \cdot X_{TOI} + Y_{SAR} \cdot Y_{TOI} + Z_{SAR} \cdot Z_{TOI}) \} \\
&- \delta k_\omega \cdot \{ C_{X0} \cdot X_{TOI} + C_{Y0} \cdot Y_{TOI} + C_{Z0} \cdot Z_{TOI} \}
\end{aligned}
\tag{A.97}
$$

One could express Eq. (A.97) in a more succinct representation:

$$R_c \cdot r_{ap1} = -\delta\omega_0 \cdot \xi_{k\omega1} - \delta k_\omega \cdot \zeta_\omega \tag{A.98}$$

with:

$$
\begin{aligned}
\xi_{k\omega1} = &C_{X0} \cdot V_{TX} + C_{Y0} \cdot V_{TY} \\
&+ R_1 \cdot R_c^{-1} \cdot \left(C_{X0} \cdot X_{TOI} + C_{Y0} \cdot Y_{TOI} + C_{Z0} \cdot Z_{TOI} \right) \\
&- w_{SAR} \cdot \left(X_{SAR} \cdot X_{TOI} + Y_{SAR} \cdot Y_{TOI} + Z_{SAR} \cdot Z_{TOI} \right)
\end{aligned}
\tag{A.99}
$$

$$\zeta_\omega = C_{X0} \cdot X_{TOI} + C_{Y0} \cdot Y_{TOI} + C_{Z0} \cdot Z_{TOI} \tag{A.100}$$

Upon inspecting Eq. (A.98), it is evident that the magnitude of $\delta k \cdot \zeta_\omega$ always surpasses that of $\delta\omega_0 \cdot \xi_{k1}$, while the sign of ξ_{k1} is contrary to that of ζ_ω. As a result, one can conclude that:

$$|R_c \cdot r_{ap1}| = |\delta k_\omega| \cdot |\zeta_\omega| - |\delta\omega_0| \cdot |\xi_{k\omega1}| \tag{A.101}$$

In light of Eq. (A.101), it is feasible to express Eq. (A.96) in an alternative manner as:

$$|\delta k_\omega| \cdot |\zeta_\omega| - |\delta\omega_0| \cdot |\xi_{k\omega1}| \leq \Re_{az}^{-1} \cdot R_c \cdot T_{az} \tag{A.102}$$

Hence, the maximum limit for the error rate in AOP is:

$$|\delta k_\omega| \leq \frac{\Re_{az}^{-1} \cdot R_c \cdot T_{az} + |\delta\omega_0| \cdot |\xi_{k\omega1}|}{|\zeta_\omega|} \tag{A.103}$$

After identifying the threshold value for $|\delta\omega_0|$ (denoted by $T_{\omega0}$), it is conceivable to rephrase Eq. (A.103) as follows:

$$|\delta k_\omega| \leq \frac{\Re_{az}^{-1} \cdot R_c \cdot T_{az} + T_{\omega0} \cdot |\xi_{k\omega1}|}{|\zeta_\omega|} \tag{A.104}$$

To guarantee optimal azimuth image quality, the azimuth QPE ought to be smaller than $\pi/4$, specifically:

$$\text{QPE}_{az} = \frac{\pi}{\lambda} \cdot \left| -r_{p2} + \frac{3}{2} \cdot \frac{R_3}{R_2} \cdot r_{p1} \right| \cdot T_{\text{SAR}}^2 < \pi / 4 \tag{A.105}$$

Eq. (A.105) can be utilized to derive:

$$\left| \left(-r_{np2} + \frac{3}{2} \cdot \frac{R_3}{R_2} \cdot r_{np1} \right) \cdot R_c \right| < \frac{\lambda \cdot R_c}{4T_{\text{SAR}}^2} \tag{A.106}$$

Afterward, by utilizing Eqs. (A.74), (A.75), (A.88), and (A.92), one can deduce the ensuing correlation:

$$\left(-r_{np2} + \frac{3}{2} \cdot \frac{R_3}{R_2} \cdot r_{np1} \right) \cdot R_c = \delta k_\omega \cdot \zeta_{k\omega2} - \delta\omega_0 \cdot \xi_{k\omega2} \tag{A.107}$$

With:

$$\zeta_{k\omega2} = C_{X0} \cdot V_{\text{TX}} + C_{Y0} \cdot V_{\text{TY}} - 1.5 R_3 R_2^{-1} \cdot \left(C_{X0} \cdot X_{\text{TOI}} + C_{Y0} \cdot Y_{\text{TOI}} + C_{Z0} \cdot Z_{\text{TOI}} \right) \tag{A.108}$$

$$\begin{aligned} \xi_{k\omega2} = {} & w_{\text{SAR0}} \cdot \left(X_{\text{SAR}} \cdot V_{\text{TX}} + Y_{\text{SAR}} \cdot V_{\text{TY}} \right) \\ & + 1.5 R_3 R_2^{-1} \cdot \left[-w_{\text{SAR}} \cdot \left(X_{\text{SAR}} X_{\text{TOI}} + Y_{\text{SAR}} Y_{\text{TOI}} + Z_{\text{SAR}} Z_{\text{TOI}} \right) \right. \\ & \left. + C_{X0} V_{\text{TX}} + C_{Y0} V_{\text{TY}} + R_1 R_c^{-1} \left(C_{X0} X_{\text{TOI}} + C_{Y0} Y_{\text{TOI}} + C_{Y0} Z_{\text{TOI}} \right) \right] \end{aligned} \tag{A.109}$$

After conducting a thorough examination of Eq. (A.107), one can conclude that:

$$\left| \left(-r_{np2} + \frac{3}{2} \cdot \frac{R_3}{R_2} \cdot r_{np1} \right) \cdot R_c \right| = \left| \delta k_\omega \right| \cdot \left| \zeta_{k\omega2} \right| - \left| \delta\omega_0 \right| \cdot \left| \xi_{k\omega2} \right| \tag{A.110}$$

Then, Eq. (A.110) can be substituted into Eq. (A.106), and accounting for the threshold value of $\delta\omega_0$, the subsequent outcome can be formulated:

$$\left| \delta k_\omega \right| \cdot \left| \zeta_{k\omega2} \right| - T_{\omega0} \cdot \left| \xi_{k\omega2} \right| < \frac{\lambda \cdot R_c}{4T_{\text{SAR}}^2} \tag{A.111}$$

Ultimately, one can attain an additional threshold for the error rate of AOP that guarantees the azimuth focusing quality. The threshold can be mathematically expressed as:

$$\left| \delta k_\omega \right| < \frac{\lambda \cdot R_c}{4T_{\text{SAR}}^2 \cdot \left| \zeta_{k\omega2} \right|} + \frac{T_{\omega0} \cdot \left| \xi_{k\omega2} \right|}{\left| \zeta_{k\omega2} \right|} \tag{A.112}$$

REFERENCES

[1] J. Meeus, *Mathematical Astronomy Morsels*. Richmond, USA: Willmann-Bell, 1997.

[2] Z. Xu, K. S. Chen, Z. L. Li, and G. Y. Du, "Apsidal Precession Effects on the Lunar-Based Synthetic Aperture Radar Imaging Performance," *IEEE Geoscience and Remote Sensing Letters*, vol. 18, no. 6, pp. 1079–1083, Jun. 2021.

[3] O. Montenbruck, E. Gill, and F. Lutze, *Satellite Orbits: Models, Methods, and Applications*. Berlin: Springer, 2000.

[4] M. I. Skolnik, *Radar Handbook*, 3rd ed. New York: McGraw-Hill Education, 2008.

[5] M. Jiang, W. Hu, C. Ding, and G. Liu, "The Effects of Orbital Perturbation on Geosynchronous Synthetic Aperture Radar Imaging," *IEEE Geoscience and Remote Sensing Letters*, vol. 12, no. 5, pp. 1106–1110, Jan. 2015.

[6] I. G. Cumming and F. H. Wong, *Digital Signal Processing of Synthetic Aperture Radar Data: Algorithms and Implementation*. Boston: Artech House, 2005.

[7] D. Liang, et al., "Processing of Very High Resolution GF-3 SAR Spotlight Data with Non-Start–Stop Model and Correction of Curved Orbit," *IEEE Journal of Selected Topics in Applied Earth Observations and Remote Sensing*, vol. 13, pp. 2112–2122, May. 2020, doi: 10.1109/JSTARS.2020.2986862

[8] Z. Xu, K. S. Chen, and H. Guo, "On Azimuthal Resolution of the Lunar-Based SAR under the Orbital Perturbation Effects," *IEEE Transactions on Geoscience and Remote Sensing*, vol. 61, Apr. 2023, doi: 10.1109/TGRS.2023.3266548

[9] A. Moccia and A. Renga, "Synthetic Aperture Radar for Earth Observation from a Lunar Base: Performance and Potential Applications," *IEEE Transactions on Aerospace and Electronic Systems*, vol. 46, no. 3, pp. 1034–1051, Jul. 2010.

[10] W. M. Folkner, et al., "The Planetary and Lunar Ephemerides DE430 and DE431," *Interplanetary Network Progress Report*, vol. 196, no. 1, pp. 1–81, Feb. 2014.

[11] R. S. Park, W. M. Folkner, J. G. Williams, and D. H. Boggs, "The JPL Planetary and Lunar Ephemerides DE440 and DE441," *The Astronomical Journal*, vol. 161, no. 3, p. 105, Feb. 2021, doi: 10.3847/1538-3881/abd414

[12] J. D. Mulholland and P. J. Shelus, "Improvement of the Numerical Lunar Ephemeris with Laser Ranging Data," *The Moon*, vol. 8, no. 4, pp. 532–538, Oct. 1973, doi: 10.1007/BF00562077

[13] R. G. Cionco and D. A. Pavlov, "Solar Barycentric Dynamics from a New Solar-Planetary Ephemeris," *Astronomy & Astrophysics*, vol. 615, p. A153, Jul. 2018.

[14] Z. Xu, K. S. Chen, and G. Liu, "On Orbital Determination of the Lunar-Based SAR under Apsidal Precession," *IEEE Transactions on Geoscience and Remote Sensing*, vol. 60, May. 2022, doi: 10.1109/TGRS.2022.3176836

[15] Z. Xu, K. S. Chen, and G. Liu, "On Evaluating the Imaging Performance and Orbital Determination under Perturbations of Orbital Inclination and RAAN in the Lunar-Based SAR," *IEEE Transactions on Geoscience and Remote Sensing*, vol. 60, Jul. 2022, doi: 10.1109/TGRS.2022.3188294

[16] R. R. Bate, D. D. Mueller, J. E. White, and W. W. Saylor, *Fundamentals of Astrodynamics*. New York: Courier Dover Publications, 2020.

[17] M. C. Gutzwiller, "Moon-Earth-Sun: The Oldest Three-Body Problem," *Review of Modern Physics*, vol. 70, no. 2, pp. 589–639, Apr. 1998.

[18] S. E. Urban and P. K. Seidelmann, *Explanatory Supplement to the Astronomical Almanac*, 3rd ed. University Science Books, 2013.

[19] H. M. Urbassek, "Precession of the Earth–Moon System," *European Journal of Physics*, vol. 30, no. 6, p. 1427, Nov. 2009, doi: 10.1088/0143-0807/30/6/020

[20] M. Chapront-Touzé and J. Chapront, *Lunar Tables and Programs from 4000 BC to AD 8000*. Richmond: Willmann-Bell, Inc., 1991.

[21] Z. Xu, K. S. Chen, and G. Q. Zhou, "Effects of the Earth's Irregular Rotation on the Moon-Based Synthetic Aperture Radar Imaging," *IEEE ACCESS*, vol. 7, pp. 155014–155027, Oct. 2019.

[22] B. Boashash, *Time-Frequency Signal Analysis and Processing: A Comprehensive Reference*. Oxford: Academic Press, 2015.

[23] K. S. Chen, *Principles of Synthetic Aperture Radar: A System Simulation Approach*. Boca Raton: CRC Press, 2015.

[24] J. C. Curlander and R. N. McDonough, *Synthetic Aperture Radar: Systems and Signal Processing*. New York: Wiley, 1991.

[25] M. A. Richards, *Fundamentals of Radar Signal Processing*. New York: McGraw-Hill Education, 2014.

[26] I. D. Haigh, M. Eliot, and C. Pattiaratchi, "Global Influences of the 18.61 Year Nodal Cycle and 8.85 Year Cycle of Lunar Perigee on High Tidal Levels," *Journal of Geophysical Research: Oceans*, vol. 116, no. C6, Jun. 2011.

[27] Z. Xu, K. S. Chen, and H. Guo, "Doppler Estimation with "Non-Stop-and-Go" Assumption in Moon-Based SAR Imaging," *Proc. 2018 IEEE International Geoscience and Remote Sensing Symposium*, Valencia, Spain, pp. 7809–7812, Jul. 2018.

[28] Z. Xu and K. S. Chen, "Effects of the Stop-and-Go Approximation on the Lunar-Based SAR Imaging," *IEEE Geoscience and Remote Sensing Letters*, Apr. 2021, doi: 10.1109/LGRS.2021.3070323

[29] S. V. Tsynkov, "On the Use of Start–Stop Approximation for Spaceborne SAR Imaging," *SIAM Journal on Imaging Sciences*, vol. 2, no. 2, pp. 646–669, 2009.

8 Ionospheric Effects on the L-Band MBSAR Imaging

8.1 INTRODUCTION

The Moon-Based SAR (MBSAR) provides an extensive coverage swath with a short revisit time, rendering it the ideal tool for monitoring Earth [1]. Relevant studies suggest that the L-band aligns with the necessities of optimal coverage and system performance in the MBSAR [2] and provides a greater penetration depth that is beneficial for characterizing vegetation profiles and soil moisture [3]. However, the propagation of the L-band radio signal encounters substantial disruption owing to the ionospheric effects [4]. Furthermore, the spatial inhomogeneity of the ionosphere across swath width exacerbates the difficulties in MBSAR imaging [5].

The investigation concerning ionospheric effects has been focused mainly on the LEO SAR, whose SAT is short, permitting line path assumption for SAR trajectory and time-freezing assumption for ionosphere [6]. In other words, the conventional analysis of the ionospheric effect is predicated on the linear path trajectory and ionospheric freezing model. Regarding the case of MBSAR, its imaging performance is more susceptible to ionospheric effects due to its specialized scenario: 1) the SAT spans hundreds of seconds, requiring the ionosphere to be treated as temporally varying; 2) the extensive MBSAR's coverage exacerbates the degree of ionospheric spatial inhomogeneity; 3) the extended SAT and non-stop-and-go echo demands considering the curved trajectory when examining ionospheric effects. Therefore, the conventional ionospheric analysis may be unsuitable in the MBSAR.

When propagating through the ionosphere, the radio signal encounters two groups of effects [7]: one is the effect of the background ionosphere (the large-scale and non-randomly varying ionosphere), and another is the effect stemming from the ionospheric irregularities (the small-scale turbulence in the ionosphere). This chapter takes the background ionosphere as an example to examine its effects on MBSAR imaging. It is worth noting that the lunar ionosphere also exists, but its electron content is significantly less, approximately two orders of magnitude smaller than that of Earth [8]. Furthermore, the lunar ionospheric thickness is considerably minuscule, making its impact on the MBSAR signal remarkably less pronounced [5]. Therefore, we shall confine this analysis to the Earth's background ionosphere and its spatiotemporal variation characteristics on the MBSAR's imaging performance.

In light of the spatiotemporal variation characteristics of the background ionosphere and the curved trajectory of MBSAR, a methodology for analyzing the background ionospheric effects is derived. From this basis, we probe into the corresponding imaging degradation in terms of the image offset and defocusing. Simulations of target responses under diverse conditions are performed to assess the MBSAR's imaging deterioration. Finally, we explore diverse characteristics of the background ionosphere across the swath width and how those diversities affect the image performance of the MBSAR.

8.2 CHARACTERISTICS OF THE BACKGROUND IONOSPHERE IN THE MBSAR

As the radio signal propagates through the ionosphere, it encounters phase dispersion due to background ionospheric effects, which imparts a phase error to the signal [9]. This, in turn, leads to the image offset, degradation in resolution, and defocusing of images in the SAR system [10]. Such effects are highly correlated to the Total Electron Content (TEC) along the signal path, generally

DOI: 10.1201/9781003308430-8

referred to as slant TEC (STEC), which is formally defined as the cumulative quantity of free electrons [11]:

$$STEC = \int_{path} \rho_e(l)\,dl \qquad (8.1)$$

where ρ_e is the electron density along the ray path.

Another relevant term is the vertical TEC (VTEC), the TEC along the vertical path above a particular position, and it is defined by [11]:

$$VTEC = \int \rho_e(h)\,dh \qquad (8.2)$$

According to the geometry shown in Figure 8.1, the Single-Layer Model (SLM), which assumes the ionosphere as a thin spherical shell at a fixed height, relates the STEC to VTEC as [12]:

$$STEC \approx \frac{VTEC}{\cos\theta_z} = VTEC \cdot \left\{ 1 - \left[\frac{R_E}{R_E + H_{iono}} \sin\theta_i \right]^2 \right\}^{-0.5} \qquad (8.3)$$

where:

θ_z is the zenith angle at the Ionospheric Pierce Point (IPP),

θ_i is the incident angle, and

H_{iono} is the altitude of a single layer above Earth, usually set between 350 and 500 km.

Drawing upon an approximation of the JPL's extended slab model mapping function, a Modified Single-Layer Model (MSLM) was proposed [13, 14]:

$$STEC \approx VTEC \cdot \left\{ 1 - \left[\frac{R_E}{R_E + 506.7\,(km)} \sin(\alpha_{iono}\theta_i) \right]^2 \right\}^{-0.5} \qquad (8.4)$$

where $\alpha_{iono} = 0.9782$.

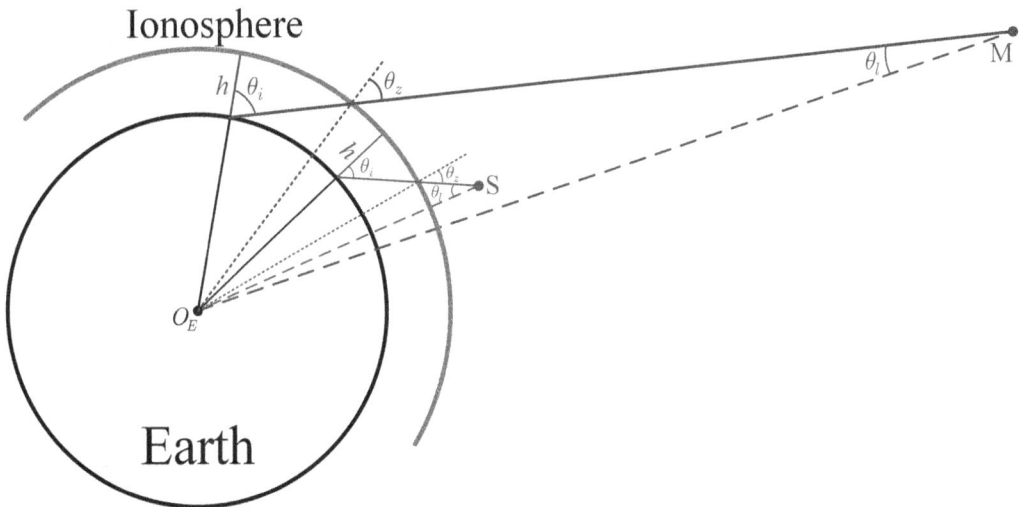

FIGURE 8.1 The diagram of the ionospheric single-layer model in the scenarios of the spaceborne SAR and MBSAR, where "S" stands for spaceborne SAR, while "M" represents the MBSAR.

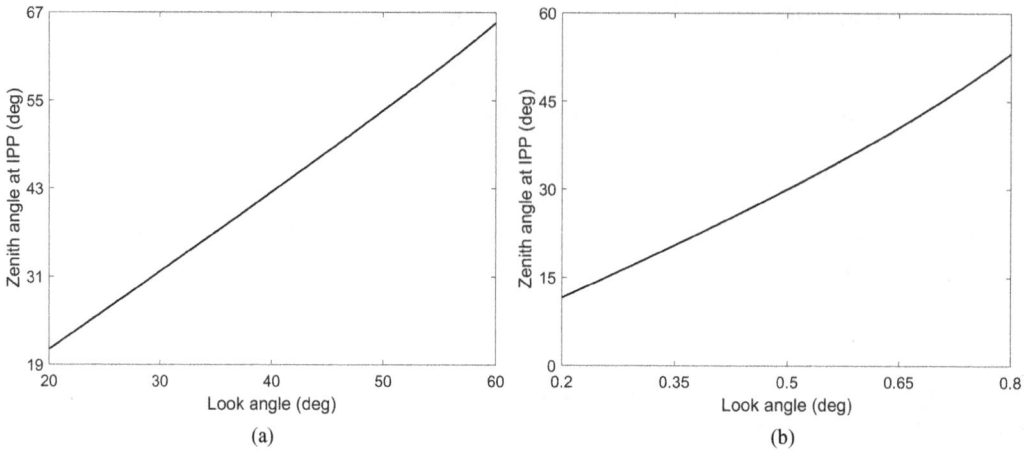

FIGURE 8.2 The comparison of the zenith angle at IPP versus the look angle in scenarios of (a) spaceborne SAR, (b) MBSAR. In this case, we respectively set the platform's altitude to 6.95×10^2 km and 3.75×10^5 km in the spaceborne SAR and MBSAR, while $h_{\text{iono}} = 350$ km for both scenarios.

In the spaceborne SAR, the look angle approaches the corresponding zenith angle at IPP, as shown in Figure 8.2. As a result, the STEC can be further approximated as a function of the VTEC and look angle, as [15]:

$$STEC \approx VTEC \cdot \sec \theta_l \qquad (8.5)$$

In the MBSAR, however, the look angle is far smaller than the zenith angle at IPP, the difference between the look and zenith angles increases with the increasing look angle. Hence, it is unattainable to approximate the STEC using Eq. (8.5). Conversely, we can reasonably employ either of Eqs. (8.3) or (8.4) to obtain the STEC from VTEC, in conjunction with the incident angle.

Referring to the electron density reported by the International Reference Ionosphere 2012 (IRI 2012) at 12: 00:00 on Mar. 20, 2010 UTC, Table 8.1 compares the numerical STEC and that approximated by SLM and MSLM under various look angles. The results reveal that the relative discrepancy between the numerical STEC and these approximated STECs ranges from 4.98 % to 7.29 %. While it is prudent to minimize this disparity when compensating for ionospheric effects, such a deviation is unlikely to exert a substantial impact on evaluating the background ionospheric effects. Also, we observe that the SLM demonstrates slightly better performance in characterizing STEC than the MSLM in the MBSAR. Consequently, the SLM is adopted in the following analysis.

The ionospheric freezing model, typically applicable to LEOSAR, apparently loses its effect due to the long SAT [5]; to be more precise, the background ionosphere is both spatial and

TABLE 8.1

The Comparison of STECs Acquired by Different Methods

Look Angle	0.2°	0.3°	0.4°	0.5°	0.6°
Numerical STEC	23.24 TECU	16.43 TECU	13.73 TECU	12.65 TECU	12.03 TECU
STEC (SLM)	21.73 TECU	15.30 TECU	12.99 TECU	12.10 TECU	11.40 TECU
STEC (MSLM)	21.69 TECU	15.23 TECU	12.89 TECU	11.94 TECU	11.15 TECU

1) TECU refers to the TEC Unit, 1 TECU = 10^{16} electron/m²;
2) The altitude for SLM is set to 350 km in this case.

temporal-varying in the MBSAR [16–18]. To understand such characteristics, we analyze its spatial and temporal variations referring to the ionospheric VTEC.

We showed the MBSAR's extensive spatial coverage in Chapter 4. In this context, the VTEC within the MBSAR's coverage exhibits spatial heterogeneity at any given instant. We depict the global distribution of VTEC at 22:00:00 UTC on Apr.23 and 24, 2021, in Figure 8.3. Both instances correspond to the normal ionosphere at low solar activity. An ionospheric storm, in response to a geomagnetic storm, occurred between April 23 and 24, 2023, whereby the ionospheric VTEC exceeded 150 TECU at certain locations. To conduct a comparative inspection, Figure 8.3 also portrays global VTEC maps at 22:00:00 UTC on April 23 and 24, 2023. In both cases, the ionospheric data are adapted from [19].

Figure 8.3 shows that the magnitude of ionospheric VTEC and its spatial variability are dependent on solar activity. At low solar activity, corresponding to a normal ionospheric condition, the VTEC has a mean value that fluctuates around 10 TECU, while its maximum value can be up to several dozen TECU. When solar activity is high, the average and maximum values of VTEC exhibit an upward surge: the average VTEC ranges within tens of TECU, whereas the maximum value can surpass one hundred TECU.

When contemplating the background ionospheric effects, it is noteworthy that phase errors affecting the MBSAR imaging performance predominantly emerge from the constant part, 1st-order and 2nd-order derivatives of the VTEC. Conversely, the phase errors of higher orders engendered by background ionosphere are negligible, obviating the need to consider them [5, 18]. From this perspective, our analysis shall only consider the 1st-order and 2nd-order derivatives of the ionospheric VTEC.

The spatial-varying characteristics of VTEC could potentially affect the SAR's Doppler properties under the normal ionosphere, while it induces more significant Doppler errors at high solar

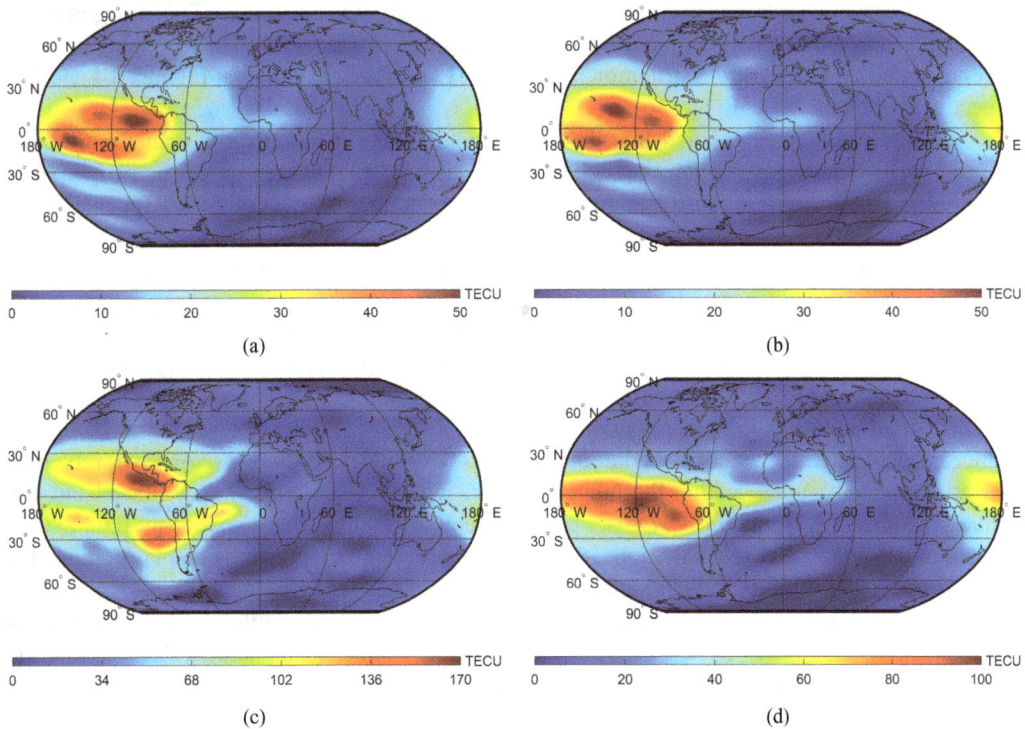

FIGURE 8.3 The global distribution of the ionospheric VTEC at 22:00:00 UTC on (a) April 23, 2021; (b) April 24, 2021; (c) Apr. 23, 2023; (d) Apr. 24, 2023.

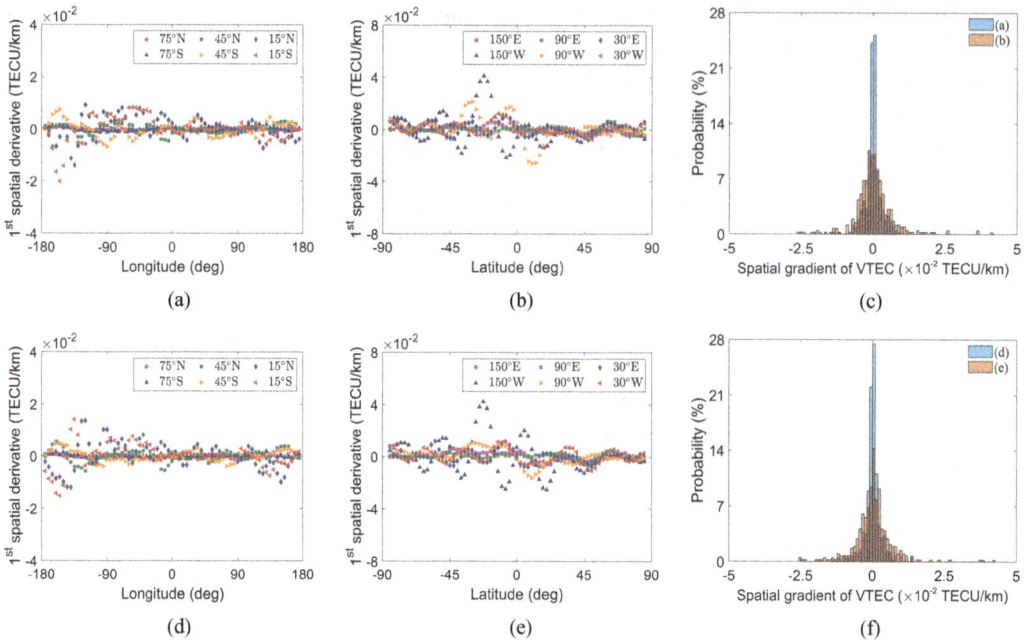

FIGURE 8.4   At 22:00:00 UTC on Apr. 23, 2021, the spatial gradient of VTEC along (a) longitudinal direction, (b) latitudinal direction, (c) the PDHs of spatial gradient; at 22:00:00 UTC on Apr. 24, 2021, the spatial gradient of VTEC along (d) longitudinal direction, (e) latitudinal direction, (f) the PDHs of spatial gradients.

activity [20, 21]. For further inspection, we inspect the spatial gradients of the VTEC along both the latitudinal and longitudinal directions, using the data taken from Figure 8.3. The resultant outcomes and corresponding probability distribution histograms (PDHs) are presented in Figures 8.4 and 8.5, respectively.

As depicted in Figures 8.4 and 8.5, the spatial gradients of VTEC exhibit spatial variability between positive and negative values. Such variability is highly contingent upon the geographical coordinates, and further, it is also modulated by the level of solar activity. A more specific discussion is given below.

Under normal ionospheric conditions, the VTEC's spatial gradients manifest a greater magnitude in the longitudinal direction than in the latitudinal direction. Specifically, the spatial gradient stands between -5×10^{-3} to 5×10^{-3} TECU/km in the longitudinal direction. Regarding the latitudinal VTEC, most spatial gradients fall between -1×10^{-2} and 1×10^{-2} TECU/km along this direction, with the maximum magnitude exceeding 0.03 TECU/km.

When solar activity is high, the spatial gradient of VTEC is amplified, resulting in more pronounced spatial fluctuations. Still, the latitudinal VTEC exhibits a stronger variation, with the bulk of spatial gradients falling within the range of -0.06 to 0.06 TECU/km. On the other hand, the spatial gradients of longitudinal VTEC are primarily confined within the scope from -0.03 to 0.03 TECU/km. Notably, the spatial variations in the VTEC concerning solar activity display comparable tendencies on other days. To avoid redundancy, we have omitted repetitive analysis.

Next, we examine the 2nd-order spatial derivatives of VTEC along the latitudinal and longitudinal directions, Figures 8.6 and 8.7 showcase the corresponding results and associated PDHs at low and high solar activities, respectively. In both cases, the background ionospheric data are taken from Figure 8.3.

The results in Figures 8.6 and 8.7 show that the 2nd-order spatial derivatives of VTEC exhibit discernible variations depending on the geographical location. Particularly, these spatial derivatives

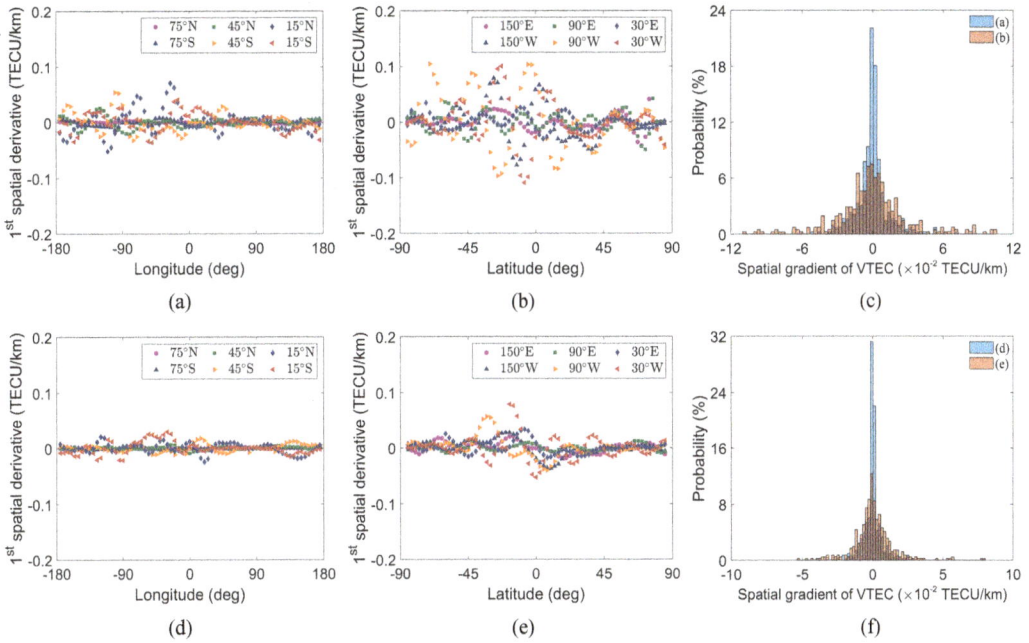

FIGURE 8.5 At 22:00:00 UTC on Apr. 23, 2023, the spatial gradient of VTEC along (a) longitudinal direction, (b) latitudinal direction, (c) the PDHs of spatial gradient; at 22:00:00 UTC on Apr. 24, 2023, the spatial gradient of VTEC along (d) longitudinal direction, (e) latitudinal direction, (f) the PDHs of spatial gradients.

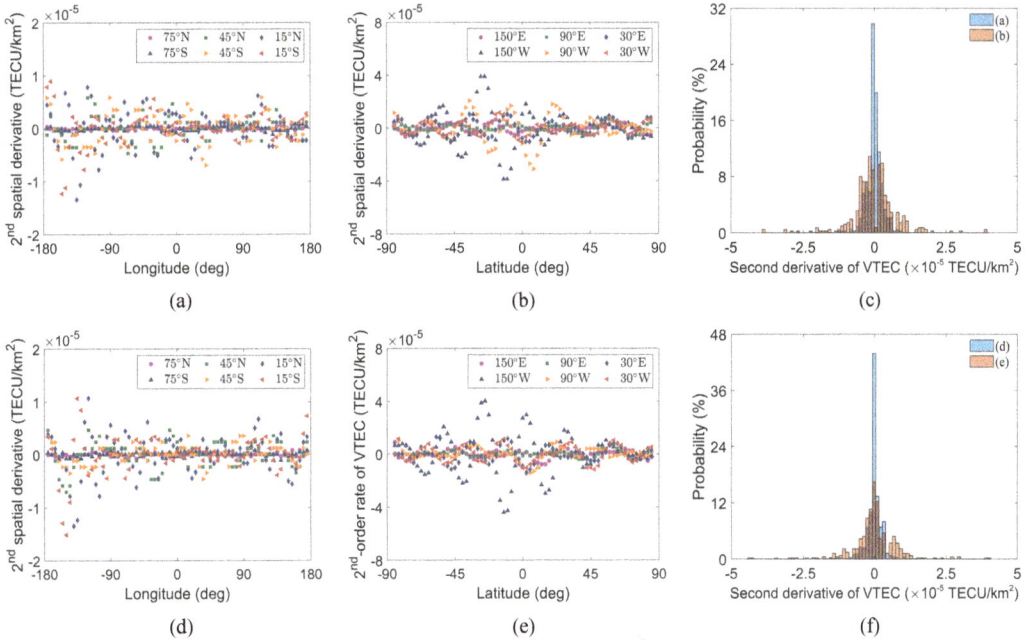

FIGURE 8.6 At 22:00:00 UTC on Apr. 23, 2021, the 2nd-order spatial derivates of VTEC along (a) longitudinal direction, (b) latitudinal direction, (c) the PDHs of 2nd-order spatial derivates; at 22:00:00 UTC on Apr. 24, 2021, the 2nd-order spatial derivates over distance along (d) longitudinal direction, (e) latitudinal direction, (f) the PDHs of 2nd-order spatial derivates.

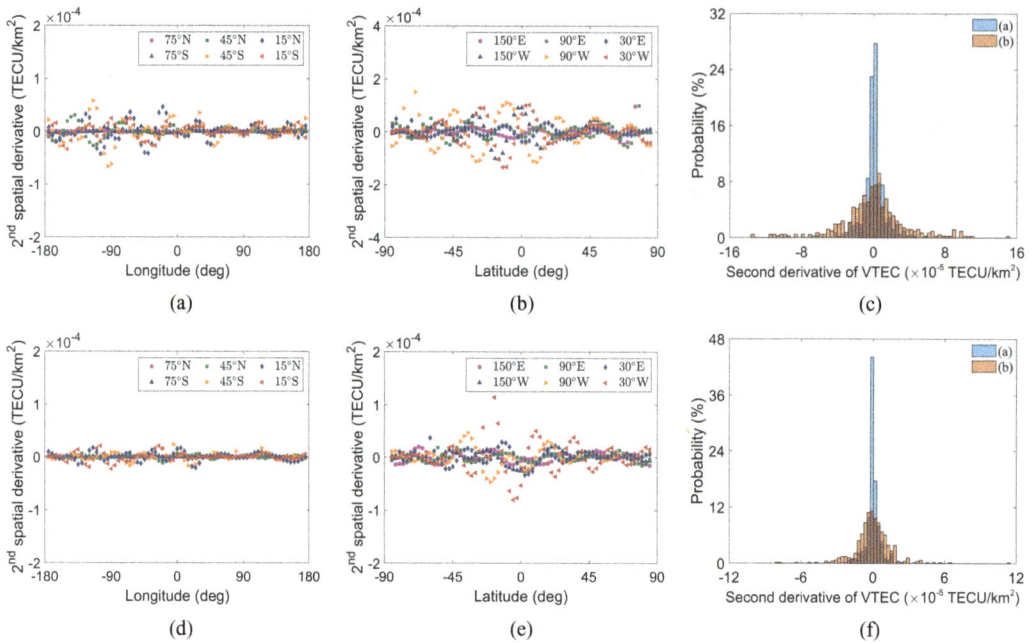

FIGURE 8.7   At 22:00:00 UTC on Apr. 23, 2023, the 2nd-order spatial derivates of VTEC along (a) longitudinal direction, (b) latitudinal direction, (c) the PDHs of 2nd-order spatial derivates; at 22:00:00 UTC on Apr. 24, 2023, the 2nd-order spatial derivates over distance along (d) longitudinal direction, (e) latitudinal direction, (f) the PDHs of 2nd-order spatial derivates.

have smaller magnitudes along the longitudinal direction in contrast to the latitudinal direction. In both directions, the change in solar activity could lead to variations in the 2nd-order spatial derivatives of VTEC.

At the low solar activity, most of the 2nd-order longitudinal derivatives lie within the range of -1×10^{-5} TECU/km^2 and -1×10^{-5} TECU/km^2. In comparison, the 2nd-order derivatives of VTEC along the latitudinal direction exhibit a more extensive spreading, with the majority falling between -1.5×10^{-5} TECU/km^2 and -1.5×10^{-5} TECU/km^2.

Concerning the high solar activity, it is observed the largest proportion of the 2nd-order longitudinal derivatives ranges from -5×10^{-5} TECU/km^2 to -5×10^{-5} TECU/km^2. In the context of latitudinal direction, the 2nd-order derivatives exhibit a range spanning between -1×10^{-4} TECU/km^2 and 1×10^{-4} TECU/km^2, with the peak magnitude exceeding 1.5×10^{-4} TECU/km^2.

The temporal variations of ionospheric VTEC can also influence the MBSAR's imaging performance. In this case, the real-time ionospheric data, as reported by [22], are employed to depict such variations; Figures 8.8 and 8.9 illustrate the temporal variations of VTEC on Aug. 23, 2021 (low solar activity) and Apr. 23, 2023 (high solar activity), respectively. In each case, the global VTEC at 10:00:00 UTC serves as a benchmark, the temporal variations reflect the difference of VTEC at the given instant relative to the reference VTEC, with an epoch of one hour and a five-minute interval.

Figures 8.8 and 8.9 show that there are discernible temporal variations in the ionospheric VTEC when the time span is sufficiently long. Besides, the temporal variations exhibit spatial heterogeneities, which are highly contingent upon solar activity. Therefore, the magnitude of VTEC and its temporal variation are more prominent at the high solar activity compared to the low solar activity. Specifically, the maximal variation of VTEC is several TECU per hour under the low solar activity, while it can increase to dozens of TECU within one hour at the high solar activity.

Now we investigate the temporal gradients of VTEC in both latitudinal and longitudinal directions, utilizing the data taken from Figures 8.8 and 8.9. The corresponding results and related PDHs

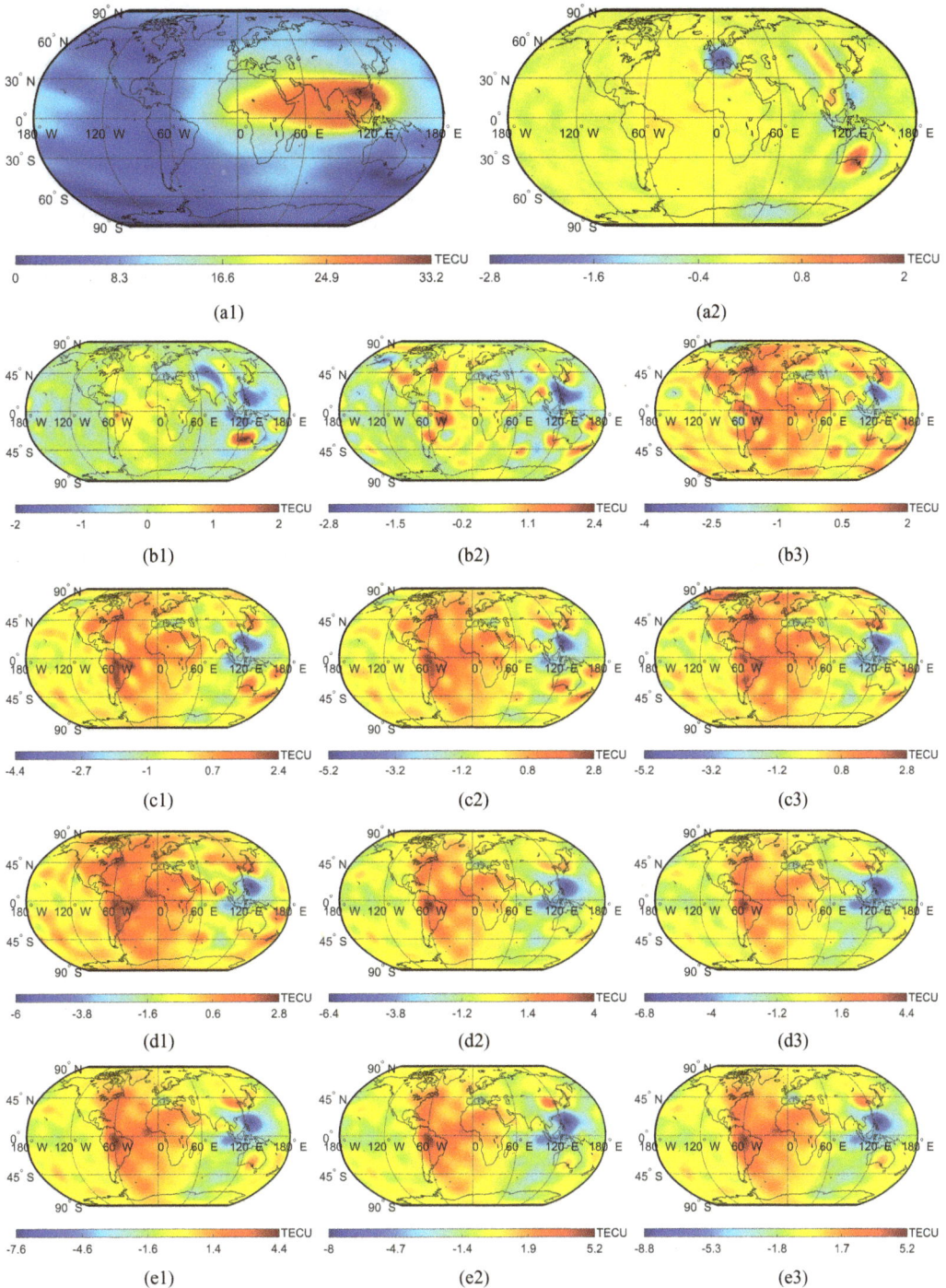

FIGURE 8.8 On Aug. 23, 2021, referring to the global VTEC map at 10:00:00 UTC, the temporal variation in the global VTEC from 10:05:00 UTC to 11:05:00 UTC with an interval of 5 minutes.

at low and high solar activities are depicted in Figures 8.10 and 8.11, respectively. Both cases illustrate the temporal gradients of VTEC at two instants: 10:10:00 UTC and 10:50:00 UTC.

The examination of Figures 8.10 and 8.11 shows that the temporal gradient of VTEC exhibits no discernible patterns between the longitudinal and latitudinal directions. By contrast, the magnitudes

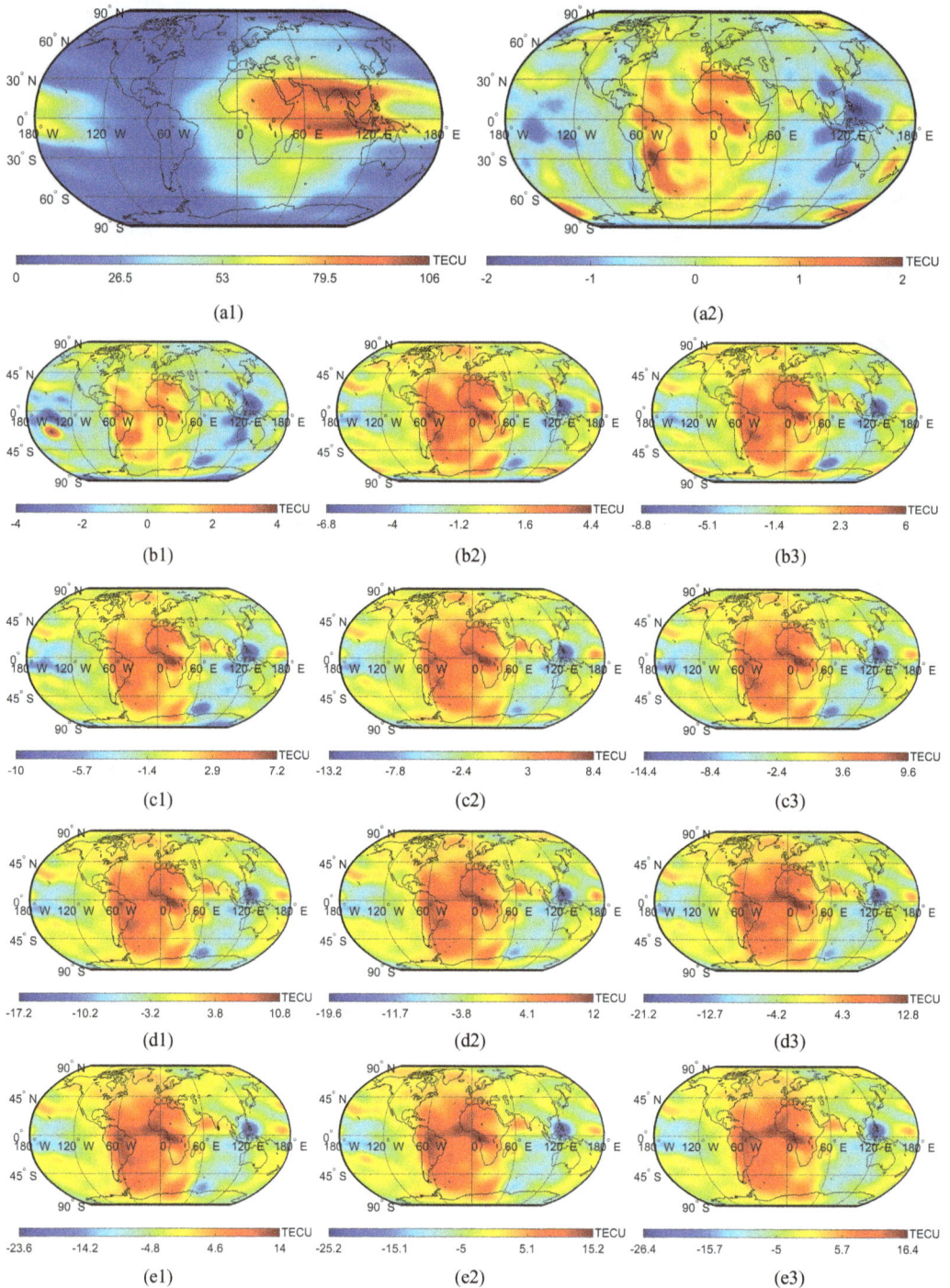

FIGURE 8.9 On Apr. 23, 2023, referring to the global VTEC map at 10:00:00 UTC, the temporal variation in the global background ionosphere from 10:05:00 UTC to 11:00:00 UTC with an interval of 5 minutes.

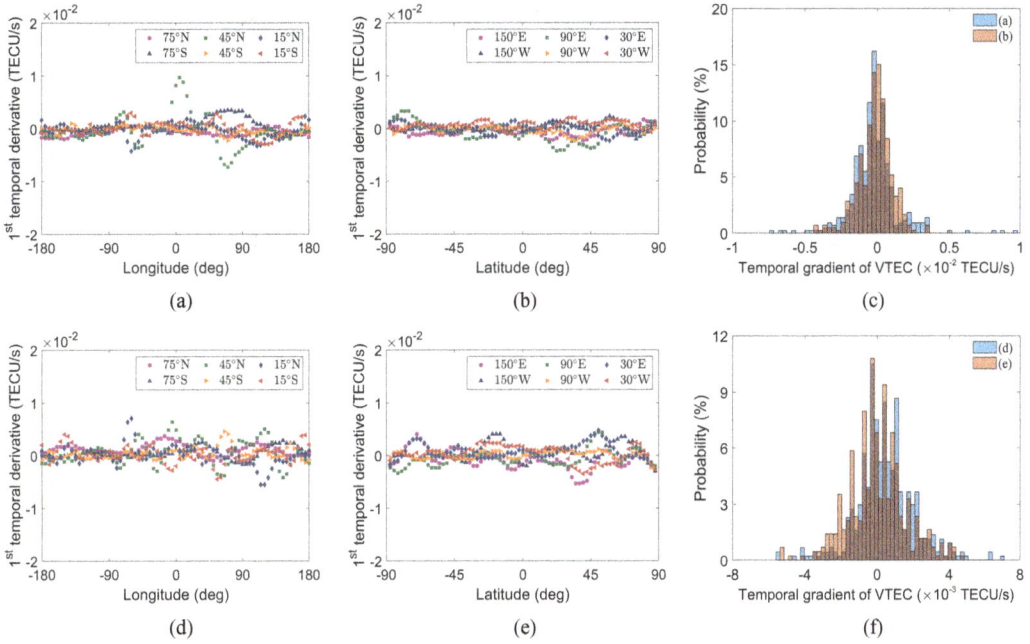

FIGURE 8.10 On Aug. 23, 2021, the temporal gradient of VTEC along (a) longitudinal direction, (b) latitudinal direction, (c) the PDHs of temporal gradients, at 10:10:00 UTC; the temporal gradient of VTEC along (d) longitudinal direction, (e) latitudinal direction, (f) the PDHs of temporal gradients, at 10:50:00 UTC.

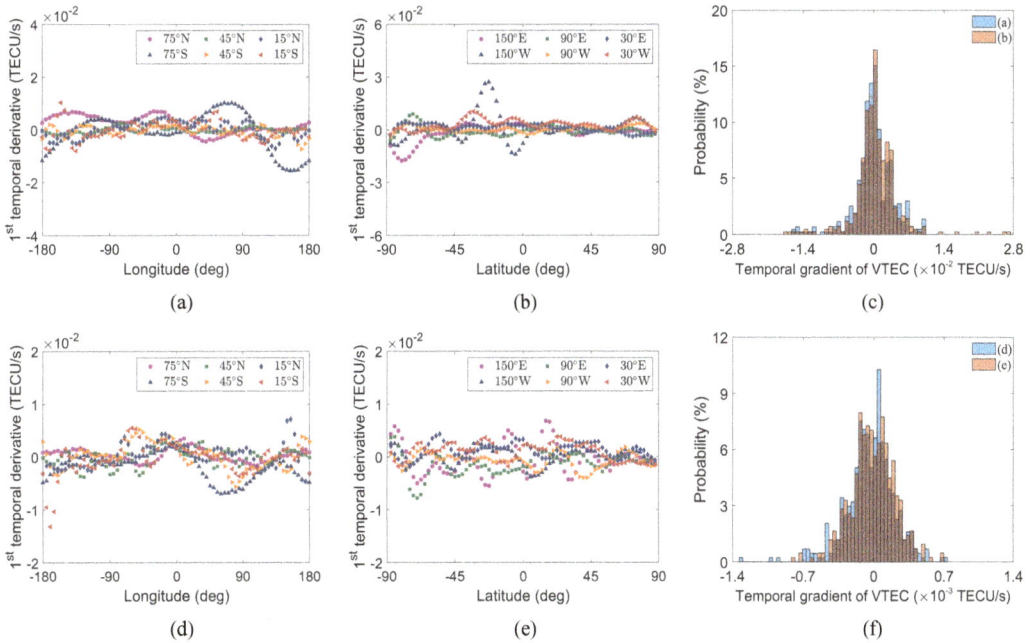

FIGURE 8.11 On Apr. 23, 2023, the temporal gradient of VTEC along (a) longitudinal direction, (b) latitudinal direction, (c) the PDHs of temporal gradients, at 10:10:00 UTC; the temporal gradient of VTEC along (d) longitudinal direction, (e) latitudinal direction, (f) the PDHs of temporal gradients. at 10:50:00 UTC.

and variations of temporal gradients are contingent upon the level of solar activity. In particular, the temporal gradient of VTEC predominantly varies from −0.003 to 0.003 TECU/s at low solar activity. During the high solar activity, the temporal variation experiences a substantial increase: most temporal gradients of VTEC fall within the range of −0.008 to 0.008 TECU/s, with the maximum one possibly exceeding 0.02 TECU/s. Moreover, the results indicate that the VTEC's temporal gradient exhibits both temporal and spatial variations, thereby aggravating the intricacies of background ionospheric effects in the MBSAR.

The 2nd-order temporal derivatives of VTEC in latitudinal and longitudinal directions are gleaned from Figures 8.8 and 8.9. The corresponding outcomes and associated PDHs under low and high solar activities are depicted in Figures 8.12 and 8.13, respectively. Each scenario presents the 2nd-order temporal derivatives of VTEC at 10:10:00 UTC and 10:50:00 UTC.

As delineated in Figures 8.12 and 8.13, there are discernible variabilities in the 2nd-order temporal derivatives of VTEC that can be attributed to the level of solar activity. At low solar activity, most of the 2nd-order temporal derivatives fall between $−8 \times 10^{-6}$ and 8×10^{-6} TECU/s^2. By comparison, at the high solar activity, the 2nd-order temporal derivative extends to a broader scope, ranging from $−2 \times 10^{-5}$ to 2×10^{-5} TECU/s^2, with the maximum magnitude exceeding 8×10^{-5} TECU/s^2. We see that the 2nd-order temporal derivatives of VTEC also show spatial and temporal variations. Further, comparable trends can be observed in the spatiotemporal variations of background ionosphere across other periods. We shall refrain from further repetition analysis to maintain brevity and avoid redundancy.

The spatial and temporal variations are strongly coupled, rendering the spatiotemporal variations in the background ionosphere more complex. For example, the VTEC's spatial and temporal gradients may exhibit temporal variability over extended periods, and both gradients show spatial variation when the coverage is sufficiently extensive. For the sake of simplicity, when modeling the ionospheric VTEC in the MBSAR, we focus on a single TOI within the image scene. Also, we

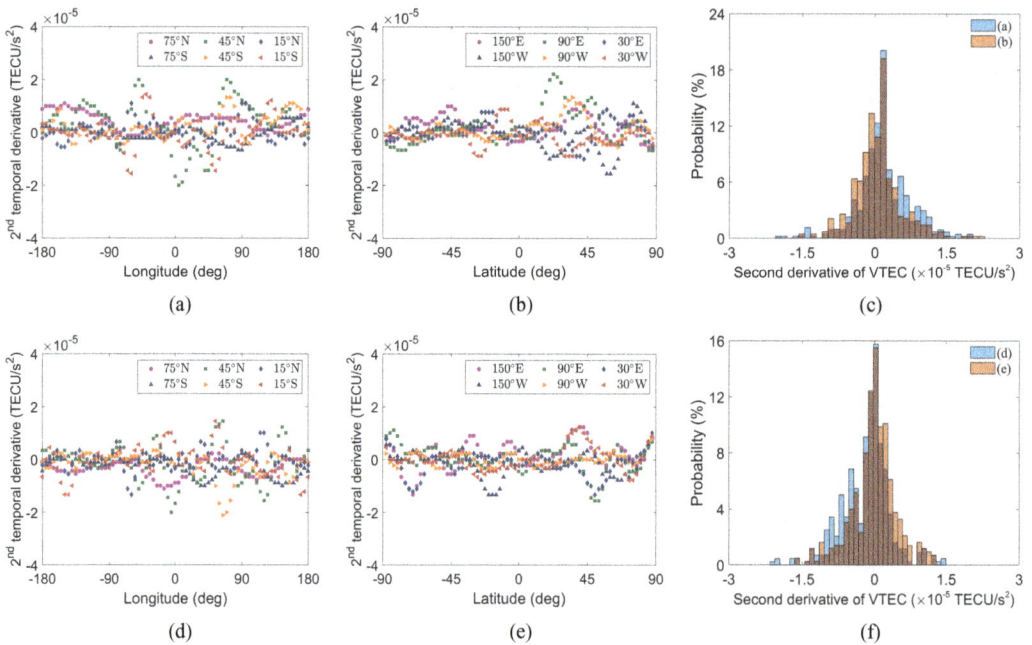

FIGURE 8.12 On Aug. 23, 2021, the 2nd-order temporal derivatives of VTEC along (a) longitudinal direction, (b) latitudinal direction, (c) the PDHs of 2nd-order temporal derivatives, at 10:10:00 UTC; the 2nd-order temporal derivatives of VTEC along (c) longitudinal direction, (d) latitudinal direction, (c) the PDHs of 2nd-order temporal derivatives, at 10:50:00 UTC.

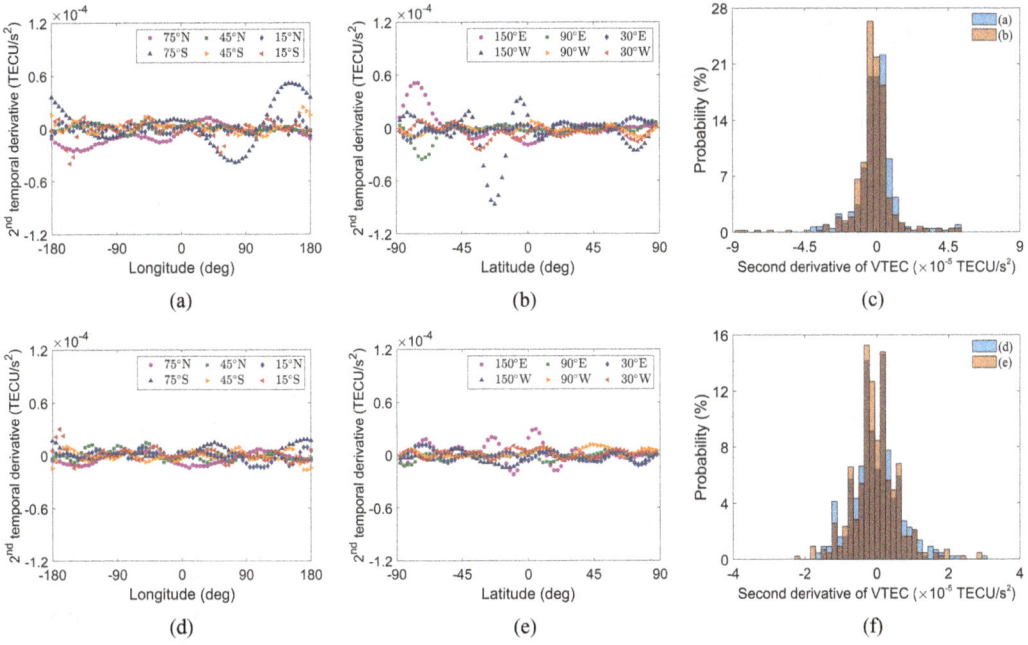

FIGURE 8.13 On Apr. 23, 2023, the 2nd-order temporal derivatives of VTEC along (a) longitudinal direction, (b) latitudinal direction, (c) the 2nd-order temporal PDHs of derivatives, at 10:10:00 UTC; the 2nd-order temporal derivatives of VTEC along (c) longitudinal direction, (d) latitudinal direction, (c) the 2nd-order temporal PDHs of derivatives, at 10:50:00 UTC.

assume that the temporally and spatially varying components of TEC remain independent during the SAT. The reason for adopting this assumption is that while the SAT of MBSAR can surpass that of the LEOSAR, it is still shorter than the period of temporal variation in the large-scale background ionosphere. The spatiotemporally varying VTEC within the SAT is given by:

$$\text{VTEC}(\eta) = \text{VTEC}_0 + \underbrace{t_{\text{vt1}} \cdot \eta + t_{\text{vt2}} \cdot \eta^2 + \cdots}_{\text{Temporally-varying part}} + \underbrace{s_{\text{vt1}} \cdot r_a(\eta) + s_{\text{vt2}} \cdot r_a^2(\eta) + \cdots}_{\text{Spatially-varying part}} \tag{8.6}$$

where:

VTEC$_0$ is the VTEC at zero-azimuth time;

t_{vt1} and t_{vt2} are 1st and 2nd-order polynomial coefficients for the temporal-varying part of VTEC, respectively;

s_{vt1} and s_{vt2} are the 1st and 2nd-order polynomial coefficients for the spatial-varying part of VTEC, respectively; and

r_a is the azimuth coordinate that relates to the azimuth time by:

$$r_a = R_E^{-1}\left(R_E + H_{\text{iono}}\right) V_g \eta \tag{8.7}$$

By referring to the VTEC at the MBSAR's nadir point (reported by the IRI 2012), Table 8.2 delineates the parameters for the spatiotemporally varying VTEC within an SAT of 1200 s. In this case, the zero-azimuth time is set to 01:00:00 on Mar. 20, 2010, UTC. Upon this basis, the temporally varying parameters are extrapolated from the reported VTEC data at the nadir point within an SAT of 1200 s, while the spatially varying parameters are acquired by the reported VTEC along the MBSAR's trajectory at the zero-azimuth time. The corresponding comparisons are graphically

TABLE 8.2

Constant and Spatiotemporally Varying Parameters of the VTEC

VTEC Parameters	Symbol	Quantity	Unit
Constant TEC	VTEC_0	41.209	TECU
Temporally varying parts of VTEC	t_{vt1}	5.567×10^{-3}	TECU/s
	t_{vt2}	-5.314×10^{-7}	TECU/s^2
Spatially varying parts of VTEC	s_{vt1}	-8.438×10^{-4}	TECU/km
	s_{vt2}	-9.687×10^{-8}	TECU/km^2

represented in Figure 8.14, parts (a) and (b). For further comparison, the spatiotemporally varying VTEC reported by the IRI 2012, alongside those simulated based on Eq. (8.3) and parameters in Table 8.2, are depicted in Figure 8.14(c). Note that the SAR cannot perform the nadir imaging; the comparisons are merely utilized to confirm the validity of Eq. (8.6).

As illustrated in Figure 8.14, there is a small disparity between the VTEC encountered by MBSAR's signal and the linear superposition of spatially and temporally varying components of VTEC. Notwithstanding, the results suggest the degree of divergence is insignificant over a short interval, such as an SAT below 600 s. Under this premise, the spatial and temporal components of the VTEC may be conveniently considered as mutually independent within the SAT.

To further expound the background ionospheric effects on the signal of the MBSAR, we expand the ionospheric TEC in terms of VTEC and incident angle, such that:

$$\text{TEC}(\eta) \approx \text{TEC}_0 + k_{t1}\eta + k_{t2}\eta^2 + \cdots \tag{8.8}$$

with:

$$\begin{cases} \text{TEC}_0 \approx \text{VTEC}_0 \cdot \sec\theta_z \\ k_{t1} \approx \sec\theta_z \left(t_{vt1} + s_{vt1}V_{\text{iono}}\right) \\ k_{t2} \approx \sec\theta_z \left(t_{vt2} + s_{vt2}V_{\text{iono}}^2\right) \end{cases} \tag{8.9}$$

As the radio signal travels through the spatiotemporally varying background ionosphere, it may incur a phase error due to dispersive effects that arise [23]. The reciprocity principle posits that the dispersive effects on the round-trip propagation are twice those experienced in unidirectional transmission [24]. As a result, the phase error can be characterized by:

$$\Delta\varphi_{\text{iono}} = \frac{-4\pi A_{\text{iono}}}{c(f_\tau + f_c)}\text{TEC}(\eta), \quad -\frac{B_r}{2} \le f_\tau \le \frac{B_r}{2} \tag{8.10}$$

where A_{iono}, a constant, equals 40.28 m^3/s^2.

At present, a thorough analysis has been conducted on the background ionosphere and its spatiotemporal variation characteristics. It has been observed that the phase error is, in fact, temporally and spatially varying under such effects. Leveraging insights gleaned from the pertinent research outlined in [5, 18, 25], coupled with the discussions on orbital perturbation effects in Chapter 7, we are equipped to devise a methodology for assessing the influence of the background ionosphere on MBSAR imaging. The specifics are detailed in the subsequent section.

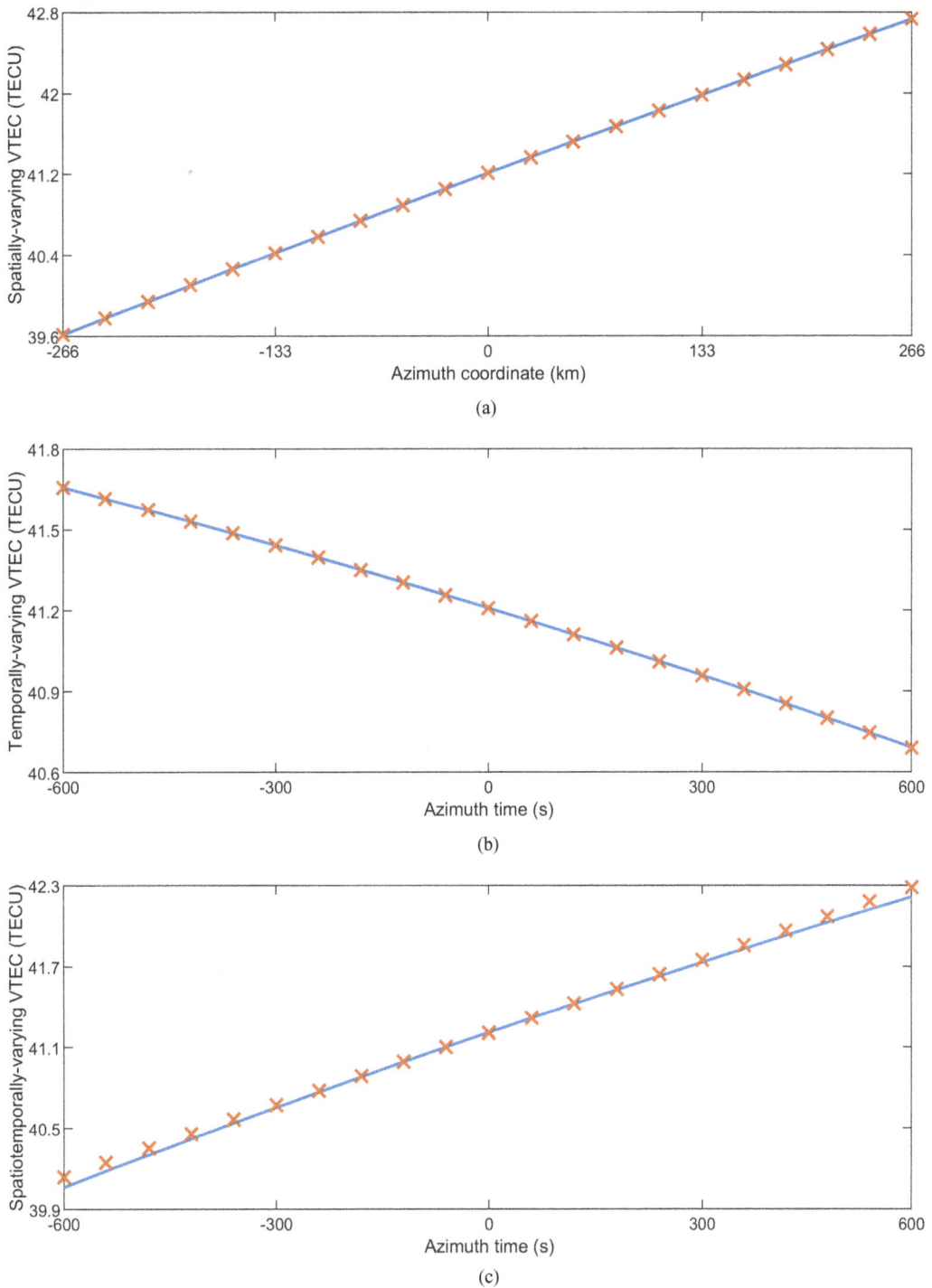

FIGURE 8.14 The comparison of the VTEC reported by IRI 2012 (red cross symbol) to the VTEC simulated by Table 8.2 (blue solid line), (a) the spatially varying VTEC versus azimuth coordinate, (b) the temporally varying VTEC versus azimuth time, (c) the spatiotemporally varying VTEC versus azimuth time.

8.3 THE METHODOLOGY FOR EVALUATING BACKGROUND IONOSPHERIC EFFECTS

To assess the background ionospheric effects, we formulate a framework in line with the MBSAR's signal model. Recall the MBSAR's signal in the range frequency domain, as elucidated by Eq. (6.32) in Chapter 6, which is repeated here for ease of reference:

$$S_r\left(f_\tau,\eta\right) = A_0 W_r\left(f_\tau\right) w_a\left(\eta\right) \exp\left(-j\pi \frac{f_\tau^2}{K_r}\right) \exp\left[-j4\pi \frac{\left(f_c + f_\tau\right) R_S\left(\eta\right)}{c}\right] \tag{6.32}$$

It is important to note that the background ionospheric effects differ substantially from the orbital perturbation effects: under the orbital perturbation effects, the phase error is non-dispersive. By contrast, the phase error resulting from the ionosphere is dispersive in nature [26]. Thus, under the background ionospheric effects, the corresponding signal can be acquired by incorporating the phase error from Eq. (8.10) into the signal model in Eq. (6.32):

$$S_{r\_\text{iono}}\left(f_\tau,\eta\right) = A_0 W_r\left(f_\tau\right) w_a\left(\eta\right) \exp\left(-j\pi \frac{f_\tau^2}{K_r}\right)$$
$$\times \exp\left[-j4\pi \frac{\left(f_c + f_\tau\right) R_S\left(\eta\right)}{c} - j4\pi \frac{A_\text{iono} \text{TEC}\left(\eta\right)}{c\left(f_\tau + f_c\right)}\right] \tag{8.11}$$

The above equation can be re-expressed as:

$$S_{r\_\text{iono}}\left(f_\tau,\eta\right) = A_0 W_r\left(f_\tau\right) w_a\left(\eta\right) \exp\left(-j\pi \frac{f_\tau^2}{K_r}\right)$$
$$\times \exp\left\{-j4\pi \frac{\left(f_c + f_\tau\right)}{c}\left[R_S\left(\eta\right) + \Delta R_\text{iono}\left(f_\tau,\eta\right)\right]\right\} \tag{8.12}$$

with:

$$\Delta R_\text{iono}\left(f_\tau,\eta\right) = \frac{A_\text{iono}}{c} \cdot \frac{1}{\left(f_\tau + f_c\right)^2} \cdot \left(\text{TEC}_0 + k_{t1}\eta + k_{t2}\eta^2 + \cdots\right) \tag{8.13}$$

By Taylor series expansion, the range error is given by:

$$\Delta R_\text{iono}\left(f_\tau,\eta\right) = \frac{A_\text{iono}}{c} \frac{1}{\left(f_\tau + f_c\right)^2} \cdot \text{TEC}_0 + \frac{A_\text{iono}}{c} \cdot \frac{1}{f_c^2} \cdot \left(k_{t1}\eta + k_{t2}\eta^2\right)$$
$$+ \frac{A_\text{iono}}{c} \cdot \frac{1}{f_c^2} \cdot \left\{\frac{f_\tau}{f_c}\left[-2\left(k_{t1}\eta + k_{t2}\eta^2\right) + 3\left(k_{t1}\eta + k_{t2}\eta^2\right) \cdot \frac{f_\tau}{f_c}\right]\right\} + \cdots \tag{8.14}$$

As $f_\tau \ll f_c$, the term enclosed in the brace is negligible compared to the first two terms, so Eq. (8.14) can be simplified to:

$$\Delta R_\text{iono}\left(f_\tau,\eta\right) \approx \frac{A_\text{iono}}{c} \frac{1}{\left(f_\tau + f_c\right)^2} \cdot \text{TEC}_0 + \frac{A_\text{iono}}{c} \cdot \frac{1}{f_c^2}\left(k_{t1}\eta + k_{t2}\eta^2\right) \tag{8.15}$$

By means of the POSP and MSR [27–29], we take the Fourier transform of Eq. (8.12) in the azimuth direction to obtain a two-dimensional (2D) spectrum considering background ionospheric effects:

$$S_{r\_iono}\left(f_\tau, f_\eta\right) = A_0 \cdot W_r\left(f_\tau\right) \cdot W_a\left(f_\eta\right) \cdot \exp\left\{j\Phi_{iono}\left(f_\tau, f_\eta\right)\right\} \tag{8.16}$$

with:

$$\begin{aligned}
\Phi_{iono}\left(f_\tau, f_\eta\right) = -\pi \frac{f_\tau^2}{K_r} - \frac{4\pi R_c\left(f_c + f_\tau\right)}{c} - \frac{2\pi}{c}\frac{2A_{iono}TEC_0}{\left(f_c + f_\tau\right)} \\
+ \frac{2\pi c}{2\left(f_c + f_\tau\right)}\frac{1}{8\left(A_{iono}k_{t2}f_c^{-2} + R_2\right)}\left[f_\eta + 2\left(\frac{A_{iono}k_{t1}}{f_c^2} + R_1\right)\frac{\left(f_c + f_\tau\right)}{c}\right]^2 \\
+ \frac{2\pi c^2}{4\left(f_c + f_\tau\right)^2}\frac{R_3}{32\left(A_{iono}k_{t2}f_c^{-2} + R_2\right)^3}\left[f_\eta + 2\left(\frac{A_{iono}k_{t1}}{f_c^2} + R_1\right)\frac{\left(f_c + f_\tau\right)}{c}\right]^3 + \cdots
\end{aligned} \tag{8.17}$$

A detailed derivation of the 2D signal spectrum follows those presented in Chapter 6.

To discern the imaging performance of MBSAR under the background ionospheric effects, we expand Eq. (8.17) as follows:

$$\begin{aligned}
\Phi_{iono}\left(f_\tau, f_\eta\right) = \Phi_{rc\_iono}\left(f_\tau\right) + \Phi_{ac\_iono}\left(f_\eta\right) \\
+ \Phi_{rcm\_iono}\left(f_\tau, f_\eta\right) + \Phi_{src\_iono}\left(f_\tau, f_\eta\right) + \Phi_{res\_iono}
\end{aligned} \tag{8.18}$$

Like the orbital perturbation effects, as discussed in Chapter 7, the range cell migration, secondary range compression, and residual phase terms impact little on the MBSAR imaging under background ionospheric effects [16]. Hence, these terms are excluded from the subsequent analysis. In contrast, the range and azimuth compression terms primarily influence the imaging performance of MBSAR. Correspondingly, the expressions for both terms are provided below.

$$\begin{aligned}
\Phi_{rc\_iono}(f_\tau) = 2\pi\left[\left(-2R_c + \frac{2A_{iono}TEC_0}{f_c^2} + \frac{P_1^2}{2P_2}\right)\frac{f_\tau}{c}\right. \\
\left. - \frac{1}{2K_r}f_\tau^2 - \frac{2A_{iono}TEC_0}{cf_c^3}f_\tau^2 + \frac{2A_{iono}TEC_0}{cf_c^4}f_\tau^3\right]
\end{aligned} \tag{8.19}$$

$$\begin{aligned}
\Phi_{ac\_iono}(f_\eta) = 2\pi\left[\left(\frac{P_1}{2P_2} + \frac{3P_1^2 R_3}{8P_2^3}\right)f_\eta\right. \\
\left. + \frac{c}{f_c}\left(\frac{1}{8P_2} + \frac{3P_1 R_3}{16P_2^3}\right)f_\eta^2 + \frac{c^2}{f_c^2}\left(\frac{R_3}{32R_2^3} + R_1\frac{9R_3^2 - 4R_2 R_4}{64R_2^5}\right)f_\eta^3\right]
\end{aligned} \tag{8.20}$$

where $P_i = A_{iono}f_c^{-2}k_{ti} + R_i$, $i = 1, 2, \cdots$.

From Eqs. (8.19) and (8.20), it is recognized that the background ionosphere induces geometric deviations and defocusing phenomena, to be detailed below.

8.3.1 Range Imaging Distortion due to Background Ionospheric Effects

Following the procedure in Section 7.2, Chapter 7, we now treat range compression under background ionospheric effects. In doing so, we establish the signal model considering the phase error in fast time:

$$
\begin{aligned}
s_{\text{SAR\_iono}}(\tau,\eta) &= \text{IFFT}_{\text{range}}\left\{\text{FFT}_{\text{range}}\left\{s_{\text{SAR}}(\tau,\eta)\right\}\exp\left[-j\left(\Phi_{\text{rc}}(f_\tau)-\Phi_{\text{rc\_iono}}(f_\tau)\right)\right]\right\} \\
&= \text{IFFT}_{\text{range}}\left\{\text{FFT}_{\text{range}}\left\{s_{\text{SAR}}(\tau,\eta)\right\}\exp\left[-j2\pi\frac{1}{c}\left(\frac{R_1^2}{2R_2}-\frac{2A_{\text{iono}}\text{TEC}_0}{f_c^2}-\frac{P_1^2}{2P_2}\right)\right. \right. \\
&\qquad\left.\left. f_\tau+\frac{2A_{\text{iono}}\text{TEC}_0}{cf_c^3}f_\tau^2-\frac{2A_{\text{iono}}\text{TEC}_0}{cf_c^4}f_\tau^3\right]\right\}
\end{aligned}
\tag{8.21}
$$

As shown in Eq. (8.21), the background ionospheric effects might give rise to both the image offset and defocusing along the range direction, wherein the range offset comes from the linear phase error and is given by:

$$
L_{rg}=\frac{c}{2}\cdot\frac{1}{c}\left(\frac{R_1^2}{2R_2}-\frac{2A_{\text{iono}}\text{TEC}_0}{f_c^2}-\frac{P_1^2}{2P_2}\right)
\tag{8.22}
$$

which can be further expressed as:

$$
L_{rg}=L_{rg0}+L_{rg1}
\tag{8.23}
$$

with:

$$
L_{rg0}=-\frac{A_{\text{iono}}\text{TEC}_0}{f_c^2},\quad L_{rg1}=\frac{R_1^2}{4R_2}-\frac{\left(R_1+A_{\text{iono}}f_c^{-2}k_{t1}\right)^2}{4\left(R_2+A_{\text{iono}}f_c^{-2}k_{t2}\right)}
\tag{8.24}
$$

Eqs. (8.23) and (8.24) represent two components of range offset: one is the offset instigated by the constant TEC, which is the same as that under the ionospheric freezing model. Another offset is caused by the spatiotemporally varying components of the background ionosphere, which is connected to the Doppler errors related to TEC's change rates (1st and 2nd orders) and Doppler parameters of the MBSAR.

The DFMR error caused by the background ionosphere is far smaller than the MBSAR's DFMR, allowing the following approximation:

$$
\frac{1}{R_2+A_{\text{iono}}f_c^{-2}k_{t2}}\approx\frac{1}{R_2}
\tag{8.25}
$$

As a result, the range offset induced by the Doppler error can be estimated by:

$$
L_{rg1}\approx-\frac{R_1}{2R_2}\frac{A_{\text{iono}}}{f_c^2}k_{t1}-\frac{k_{t1}^2}{4R_2}\frac{A_{\text{iono}}^2}{f_c^2}
\tag{8.26}
$$

In the context of range focusing, the influences from the background ionosphere are distinct from those instigated by orbital perturbations, which have little impact on the range imaging. This difference stems from the fact that the SAR transmits signals with a specific bandwidth; all frequency components are subject to the same range history error if we consider only the orbital perturbation. By contrast, when the radio signal passes through the ionosphere, the duration required for different frequency components is non-uniform due to the dispersion, leading to a stretch or compression of the radio signal [30], which, in turn, impacts the range focusing of the MBSAR.

The background ionospheric effects on the range focusing can be inspected through the range Quadratic Phase Error (QPE) and Cubic Phase Error (CPE) in Eq. (8.21). To a certain extent, the QPE can broaden the main lobe and elevate the side lobe. By comparison, the CPE can potentially bring about asymmetric distortion to the signal [31, 32]. Both errors, in effect, lead to the degradation of range imaging. Note that only the range frequency encompassed within the range of $[-0.5B_r, 0.5B_r]$ could influence the range focusing. Hence, the maximal magnitudes of both QPE and CPE appear at $f_r = 0.5B_r$. In this regard, to quantify the extent of range defocusing, we can define the magnitudes of range QPE and CPE as follows:

$$QPE_{rg} = 2\pi \frac{2A_{iono}TEC_0}{cf_c^3}\left(\frac{B_r}{2}\right)^2 = \frac{\pi A_{iono}B_r^2}{cf_c^3}TEC_0 \tag{8.27}$$

$$CPE_{rg} = 2\pi \frac{2A_{iono}TEC_0}{cf_c^4}\left(\frac{B_r}{2}\right)^3 = \frac{\pi A_{iono}B_r^3}{2cf_c^4}TEC_0 \tag{8.28}$$

Eqs. (8.27) and (8.28) show that both the range QPE and CPE are associated with the constant part of TEC but show no dependency on the spatiotemporally varying component of TEC. Moreover, the signal bandwidth and carrier frequency substantially influence the range QPE and CPE. As a rule of thumb, the range QPE and CPE threshold values stand at $\pi/4$ and $\pi/8$, respectively. As long as either phase error surpasses its threshold, it can result in the filter mismatch, thereby leading to the range defocusing in the MBSAR.

8.3.2 AZIMUTH IMAGING DISTORTION DUE TO BACKGROUND IONOSPHERIC EFFECTS

The constant TEC has no impact on azimuth imaging. By contrast, the 1st and 2nd order TEC derivatives, associated with errors in Doppler frequency and DFMR induced by the background ionospheric effects, give rise to both the image offset and defocusing in the azimuth imaging. We explore these by recognizing the following signal model:

$$
\begin{aligned}
s_{SAR\_iono}(\tau,\eta) &= IFFT_{azimuth}\left\{FFT_{azimuth}\left\{s_{SAR}(\tau,\eta)\right\}\exp\left[-j\left(\Phi_{ac}(f_\tau)-\Phi_{ac\_iono}(f_\tau)\right)\right]\right\} \\
&= IFFT_{azimuth}\left\{FFT_{azimuth}\left\{s_{SAR}(\tau,\eta)\right\}\exp\left[-j2\pi\left[\left(\frac{R_1}{2R_2}+\frac{3R_1^2R_3}{8R_2^3}-\frac{P_1}{2P_2}-\frac{3P_1^2P_3}{8P_2^3}\right)\right.\right.\right. \\
&\qquad f_\eta + \frac{c}{f_c}\left(\frac{1}{8R_2}+\frac{3R_1R_3}{16R_2^3}-\frac{1}{8P_2}-\frac{3P_1P_3}{16P_2^3}\right)f_\eta^2 \\
&\qquad \left.\left.\left.\frac{c^2}{f_c^2}\left(\frac{R_3}{32R_2^3}+R_1\frac{9R_3^2-4R_2R_4}{64R_2^5}-\frac{R_3}{32P_2^3}-P_1\frac{9R_3^2-4P_2R_4}{64P_2^5}\right)f_\eta^3\right]\right]\right\}
\end{aligned}
\tag{8.29}
$$

In Eq. (8.29), the linear phase term represents an image offset caused by the spatiotemporally varying component of the background ionosphere. The azimuth offset is given by:

$$
\begin{aligned}
L_{az} &= V_g\left(\frac{R_1}{2R_2}+\frac{3R_1^2R_3}{8R_2^3}-\frac{P_1}{2P_2}-\frac{3P_1^2P_3}{8P_2^3}\right) \\
&= -V_g\left[\left(\frac{R_1+A_{iono}f_c^{-2}k_{t1}}{2\left(R_2+A_{iono}f_c^{-2}k_{t2}\right)}-\frac{R_1}{2R_2}\right)+\left(\frac{3R_3\left(R_1+A_{iono}f_c^{-2}k_{t1}\right)^2}{8\left(R_2+A_{iono}f_c^{-2}k_{t2}\right)^3}-\frac{3R_1^2R_3}{8R_2^3}\right)\right]
\end{aligned}
\tag{8.30}
$$

In the zero-Doppler steering mode, Eq. (8.30) can be further approximated to:

$$L_{az} \approx -V_g \frac{\left(2R_2^2 + 3R_1R_3\right)}{4R_2^3} \frac{A_{\text{iono}}}{f_c^2} k_{t1} \tag{8.31}$$

Eq. (8.31) reveals a correlation between the azimuth offset and TEC gradient. Further, the beam-crossing velocity, Doppler parameters, and carrier frequency of the MBSAR could contribute to the azimuth offset.

To determine the individual contributions of both the spatially varying and temporally varying components of ionospheric TEC to the azimuth offset, we may separate the azimuth offset in Eq. (8.31) into two components:

$$L_{az} = L_{azs} + L_{azt} \tag{8.32}$$

with:

$$L_{azs} \approx V_g \frac{\left(2R_2^2 + 3R_1R_3\right)}{4R_2^3} \frac{A_{\text{iono}}}{f_c^2} \frac{s_{\text{vt1}} \cdot V_{\text{iono}}}{\cos\theta_z} \tag{8.33}$$

$$L_{azt} \approx V_g \frac{\left(2R_2^2 + 3R_1R_3\right)}{4R_2^3} \frac{A_{\text{iono}}}{f_c^2} \frac{t_{\text{vt1}}}{\cos\theta_z} \tag{8.34}$$

Concerning the MBSAR azimuth focusing quality, we know that the azimuth QPE and azimuth CPE broaden the main lobe symmetrically and asymmetrically, respectively. Note that only the azimuth frequency within the Doppler bandwidth in Eq. (8.35) can influence the azimuth focusing.

$$-0.5B_D \leq f_\eta \leq 0.5B_D, \quad B_D = f_{\text{dr}}T_{\text{SAR}} \tag{8.35}$$

Hence, we can determine the azimuth QPE from Eqs. (8.29) and (8.35):

$$\begin{aligned}
\text{QPE}_{az} &= 2\pi \cdot \left| \frac{c}{f_c}\left(\frac{1}{8P_2} + \frac{3P_1R_3}{16P_2^3}\right) - \frac{c}{f_c}\left(\frac{1}{8R_2} + \frac{3R_1R_3}{16R_2^3}\right)\right|\left(\frac{B_D}{2}\right)^2 \\
&= \frac{\pi}{32}\frac{c}{f_c}f_{\text{dr}}^2 T_{\text{SAR}}^2 \left| \frac{2P_2^2 + 3P_1R_3}{P_2^3} - \frac{2R_2^2 + 3R_1R_3}{R_2^3}\right|
\end{aligned} \tag{8.36}$$

For a more rigorous analysis, it is necessary to include the azimuth CPE resulting from the background ionosphere:

$$\begin{aligned}
\text{CPE}_{az} &= 2\pi \frac{c^2}{f_c^2}\left| \frac{R_3}{32P_2^3} + P_1\frac{9R_3^2 - 4P_2R_4}{64P_2^5} - \frac{R_3}{32R_2^3} - R_1\frac{9R_3^2 - 4R_2R_4}{64R_2^5}\right|\left(\frac{B_D}{2}\right)^3 \\
&= \frac{\pi}{256}\frac{c^2}{f_c^2}f_{\text{dr}}^3 T_{\text{SAR}}^3 \left| \frac{2P_2^2R_3 + 9P_1R_3^2 - 4P_1P_2R_4}{P_2^5} - \frac{2R_2^2R_3 + 9R_1R_3^2 - R_14R_2R_4}{R_2^5}\right|
\end{aligned} \tag{8.37}$$

The thresholds pertaining to the azimuth QPE and CPE are ascertained at $\pi/4$ and $\pi/8$, respectively. As shown in Eqs. (8.36) and (8.37), the azimuth focusing is subject to various factors, including the spatiotemporally varying part of the background ionosphere, the Doppler parameters, carrier frequency, and SAT.

Thus far, we have looked into the background ionospheric effects on MBSAR imaging. The image distortions in the MBSAR that arise from the background ionospheric effects are in order. Recall that the background ionospheric effects are subject to spatial variation within the swath width. Such spatial-variation characteristics stem from two aspects: the spatially varying factors embedded in the MBSAR's geometry (including the beam-crossing velocity, Doppler parameters, and look angle), and diversities of both ionospheric TEC and its derivates at disparate locations within the image scene. We first examine the background ionospheric effects under the specific VTEC parameters.

8.4 BACKGROUND IONOSPHERIC EFFECTS UNDER SPECIFIC TEC PARAMETERS

From Figure 8.3 through Figure 8.13, together with data reported in [5, 16–18], Table 8.3 lists VTEC parameters at varying solar activities. By employing the provided ionospheric parameters, the background ionospheric effects on the MBSAR imaging can be evaluated in terms of the range and azimuth distortions. The orbital cycle T_{lc1}, whose corresponding epoch in TDB is specified in Table 2.4, is selected to facilitate the analysis.

8.4.1 BACKGROUND IONOSPHERIC EFFECTS ON RANGE IMAGING

To begin with, we shall delve into the background ionospheric effects on range imaging. The range offset L_{rg} and its components, i.e., L_{rg0} and L_{rg1}, are inspected under two scenarios: the range offset versus the look angle at three MBSAR locations, and that versus the AOL under three look angles. The results at low and high solar activities are presented in Figures 8.15 and 8.16, respectively.

Figures 8.15 and 8.16 reveal that the magnitude of range offset given rise by the constant VTEC increases with the increasing look angle, and the upward tendency is affected by the position of MBSAR. Specifically, the magnitude of L_{rg0} is observed to ascend first with respect to the AOL, peaking at the AOL of 165°, followed by a subsequent descent. Further, the magnitude of L_{rg0} is markedly substantial, with a minimum magnitude of several meters at low solar activity, while its maximum magnitude can be up to dozens of meters.

We find that the range offset caused by VTEC gradients exhibits different tendencies regarding the AOL when compared to the range offsets L_{rg0} under specified look angles. By contrast, an upward trend can be observed in the magnitude of range offset L_{rg1} concerning the look angle. Despite the spatial variations, the range offset L_{rg1} consistently remains below 3×10^{-3} m, even during periods of high solar activity. We may infer that the constant component of the background ionosphere is the principal factor impacting the range positioning accuracy of TOI, whereas the influence of spatio-temporally varying components can be reasonably ignored in the range offset.

TABLE 8.3
Parameters for the Spatiotemporally Varying Ionospheric VTEC

Ionospheric VTEC Components		Low Solar Activity	High Solar Activity
Constant VTEC	$VTEC_0$	25 TECU	100 TECU
Temporally varying part of VTEC	t_{vt1}	1.0×10^{-3} TECU/s	4.0×10^{-3} TECU/s
	t_{vt2}	8.0×10^{-7} TECU/s^2	3.0×10^{-6} TECU/s^2
Spatially varying part of VTEC	s_{vt1}	2.5×10^{-3} TECU/km	1.5×10^{-2} TECU/km
	s_{vt2}	3.0×10^{-6} TECU/km^2	2.0×10^{-5} TECU/km^2

Note: t_{vt2} and s_{vt2} are half the magnitudes of the 2nd-order temporal and spatial derivates of the VTEC.

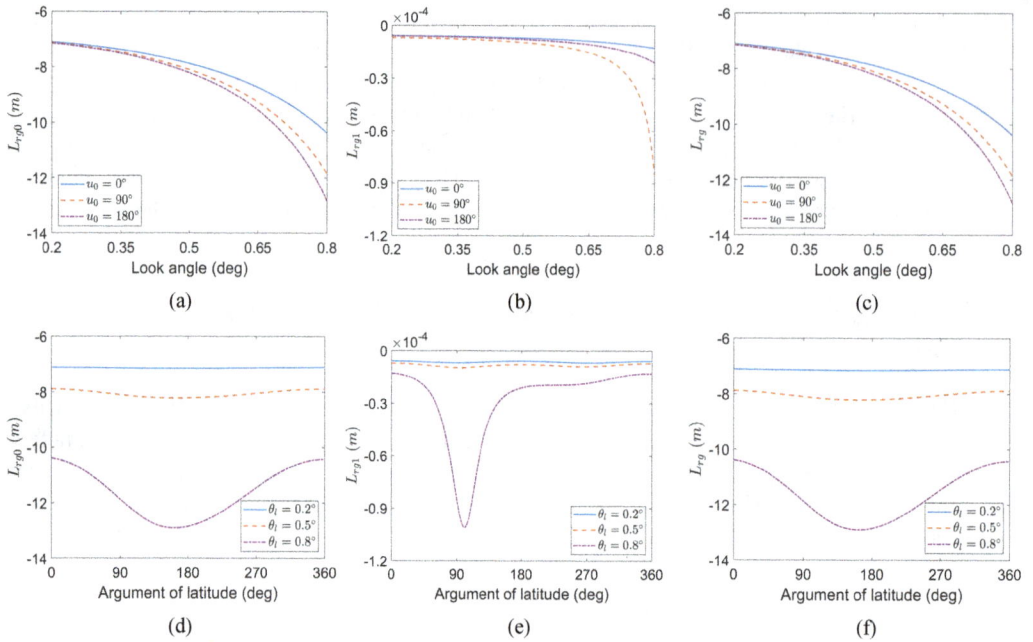

FIGURE 8.15 The range offset and its components at low solar activity, where the 1st (a, b, c) and 2nd (d, e, f) rows represent the range offset versus look angle and that against the AOL, while the 1st (a, d), 2nd (b, e), and 3rd (c, f) columns stand for the range offset components L_{rg0}, L_{rg1}, and entire range offset L_{rg}.

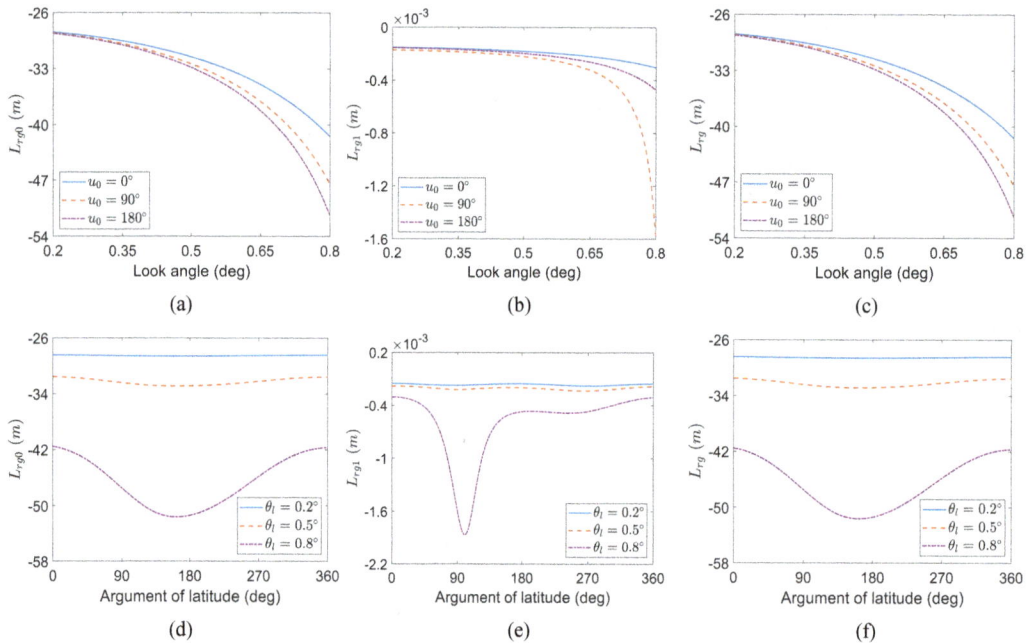

FIGURE 8.16 The range offset and its components at high solar activity, where the 1st (a, b, c) and 2nd (d, e, f) rows represent the range offset versus look angle and that against the AOL, while the 1st (a, d), 2nd (b, e), and 3rd (c, f) columns stand for the range offset components L_{rg0}, L_{rg1}, and entire range offset L_{rg}.

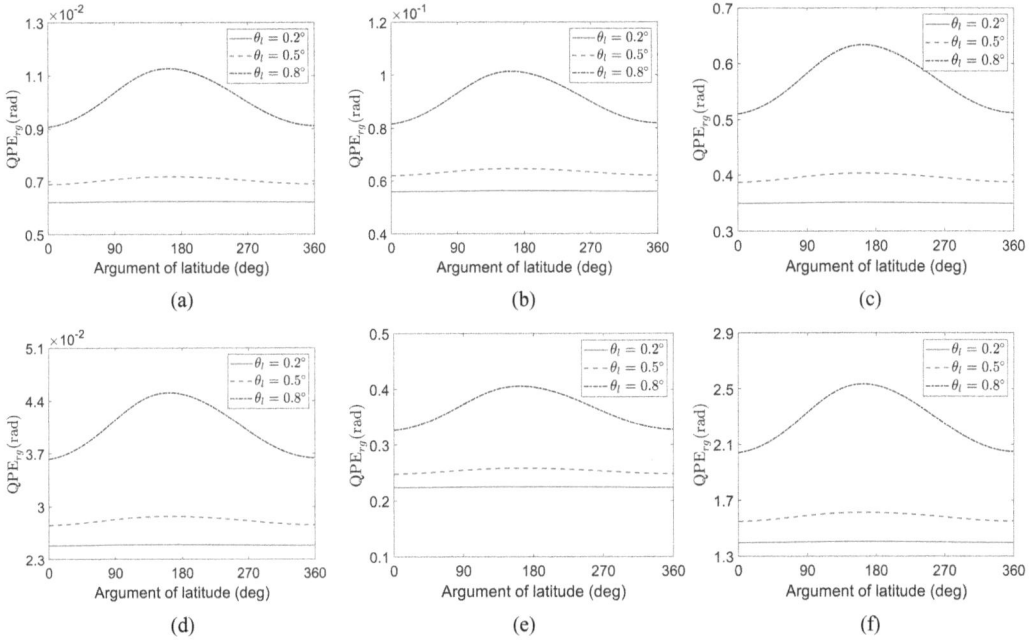

FIGURE 8.17   The range QPEs with various look angles versus the AOL under background ionospheric effects, where the 1st (a–c) and 2nd (d–f) rows represent the low and high solar activity, while the 1st (a, d), 2nd (b, e), and 3rd (c, f) columns stand for the signal bandwidth of 10 MHz, 30 MHz, and 75 MHz, respectively.

To evaluate the range focusing, we look into the range QPEs at both low and high levels of solar activity. Figure 8.17 illustrates the range QPEs as a function of AOL under various signal bandwidths for the look angles of $0.2°$, $0.5°$, and $0.8°$.

When the signal bandwidth is at or below 30 MHz, the range QPE never surpasses the threshold of $\pi/4$, irrespective of the solar activity. Hence, we can infer that background ionospheric effects have little impact on the range focusing quality for a resolution coarser than 5 m. When employing a higher signal bandwidth, for instance, 75 MHz, the range imaging is less influenced only if the solar activity is low. In other scenarios, the range QPE may exceed the threshold of $\pi/4$, thereby deteriorating the range imaging in the MBSAR.

Using the same set of parameters as in Figure 8.17, Figure 8.18 shows the range CPE as a function of AOL at low and high solar activity periods. We see that the range CPE has a far smaller magnitude than the range QPE, regardless of the look angle and position of the MBSAR. The range CPE remains consistently below its threshold of $\pi/8$, indicating that it exerts a negligible impact on the range focusing of MBSAR. Accordingly, we may simply focus on the QPE when considering the background ionospheric effects on the range focusing.

8.4.2   BACKGROUND IONOSPHERIC EFFECTS ON AZIMUTH IMAGING

We now analyze the geometric accuracy and focusing quality in the azimuth direction. As a measure of the TOI's geometric deviation along the azimuth direction, the image offset can be further inspected by considering each contribution from the spatially and temporally varying components of the VTEC. The azimuth offset is examined in two scenarios: one versus the look angle at three locations and the other versus AOL under three look angles at the low and high solar activities, as shown in Figures 8.19 and 8.20.

The magnitude and variation regularity of L_{az} results from those of two components: L_{azs} and L_{azt}, the image offsets caused by the spatially and temporally varying components of VTEC. Both

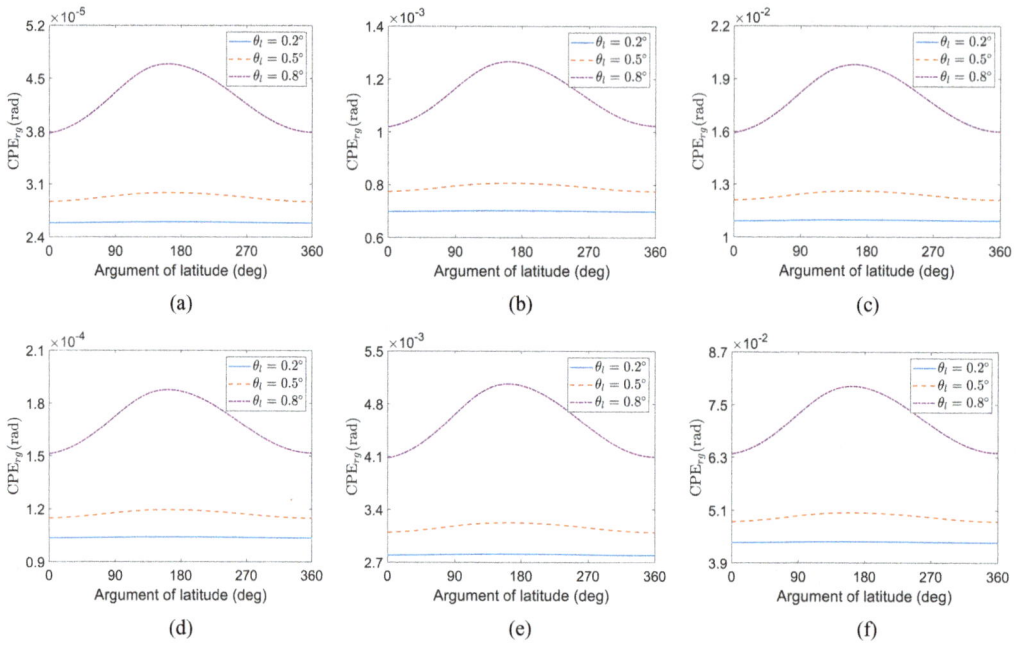

FIGURE 8.18 The range CPEs with various look angles versus the AOL under background ionospheric effects, where the 1st (a–c) and 2nd (d–f) rows represent the low and high solar activity, while the 1st (a, d), 2nd (b, e), and 3rd (c, f) columns stand for the signal bandwidth of 10 MHz, 30 MHz, and 75 MHz, respectively.

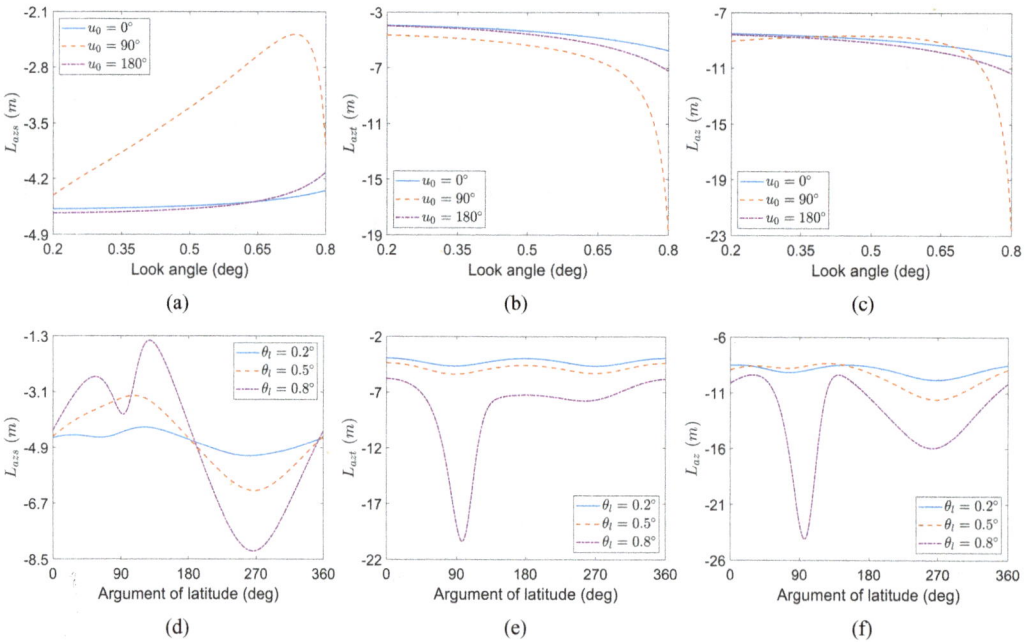

FIGURE 8.19 At the low solar activity, the azimuth offset and its components under different conditions, where the upper (a, b, c) and low (d, e, f) rows represent the azimuth offset versus look angle and that against the AOL, while the 1st (a, d), 2nd (b, e), and 3rd (c, f) columns stand for the azimuth offsets caused by spatially varying, temporally varying, and spatiotemporally varying parts of VTEC.

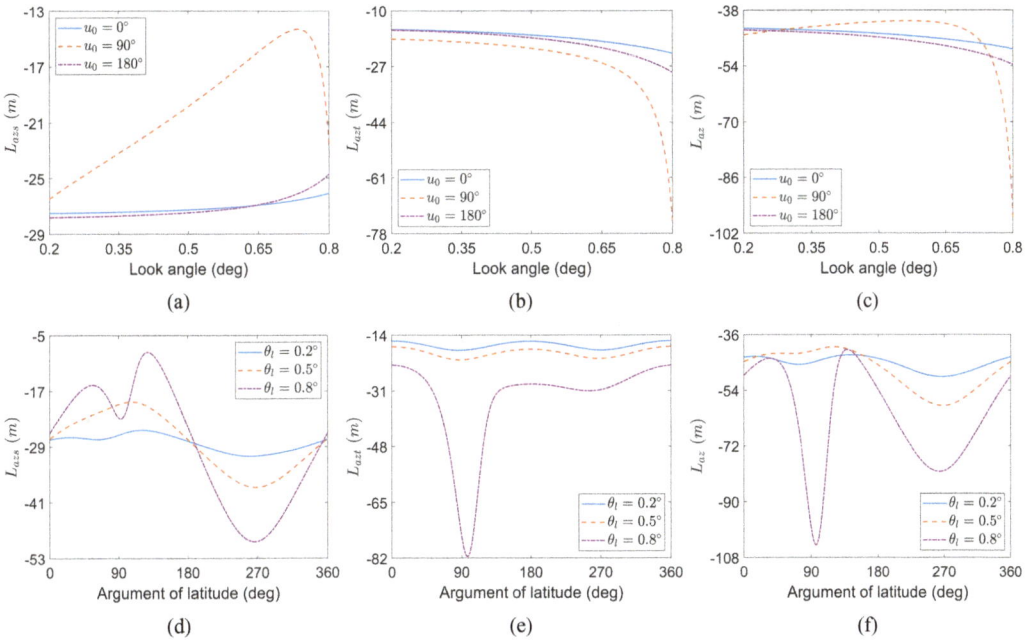

FIGURE 8.20 At the high solar activity, the azimuth offset and its components under different conditions, where the upper (a, b, c) and low (d, e, f) rows represent the azimuth offset versus look angle and that against the AOL, while the 1st (a, d), 2nd (b, e), and 3rd (c, f) columns stand for the azimuth offsets caused by spatially varying, temporally varying, and spatiotemporally varying parts of VTEC.

components vary depending on solar activity, look angle, and MBSAR's location. The image offset L_{azs} varies non-monotonically versus the look angle and the corresponding variation pattern is dependent on the MBSAR's location. The magnitude of this image offset varies from several to tens of meters during low solar activity. At the high solar activity, there is an increase in the image offset L_{azs}, with the majority of magnitudes on the order of several dozen meters.

As for the azimuth offset component stemming from the temporal variability of VTEC, its magnitude monotonically increases with the increasing look angle regardless of MBSAR's location. Moreover, the image offset L_{azs} fluctuates between several and tens of meters within the swath width at low solar activity, whereas it can extend to several dozen meters in the presence of high solar activity.

Recall that the total azimuth offset L_{az} consists of L_{azs} and L_{azt}, where the variation regularity of image offset L_{azs} versus the AOL differs from that of L_{azt} under a specified look angle. Further, in both components of azimuth offset, the tendencies of variation, with respect to the AOL, exhibit different patterns across look angles. Consequently, diversity variation patterns are observed in the azimuth offset, necessitating the focus on the spatiotemporal variations of the background ionosphere in the case of MBSAR.

The azimuth focusing is dependent on the azimuth QPEs. Figure 8.21 displays the azimuth QPEs as a function of AOL under different SATs, both at low and high solar activities. The azimuth QPE shows a clear dependency on the look angle, SAT, and solar activity. Specifically, when the MBSAR's SAT falls at or below 100 s, the azimuth QPE never exceeds the threshold of $\pi/4$ even at high solar activity. Once the SAT increases to 200 s, the azimuth QPE remains below $\pi/4$ during low solar activity but exceeds this threshold for most MBSAR's locations at high solar activity. In contrast, for a longer SAT (e.g., 400 s), the azimuth QPE is observed to exceed $\pi/4$ at some MBSAR locations during low solar activity. It is far larger than the threshold $\pi/4$ at high solar activity, resulting in degrading azimuth imaging.

We investigate the azimuth CPEs induced by the background ionosphere, using the scenarios depicted in Figure 8.21, as shown in Figure 8.22. The azimuth CPE exhibits variability in the look

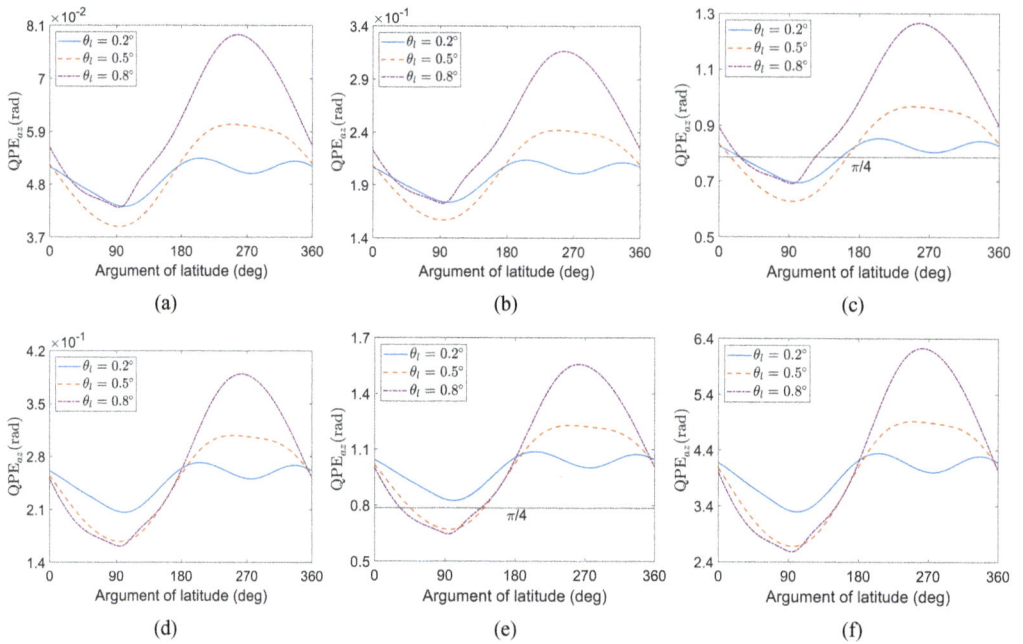

FIGURE 8.21 The azimuth QPEs with various look angles versus the AOL under background ionospheric effects, where the 1st (a–c) and 2nd (d–f) rows represent the low and high solar activity, while the 1st (a, d), 2nd (b, e), and 3rd (c, f) columns stand for the SATs of (b) 100 s, (c) 200 s, (d) 400 s.

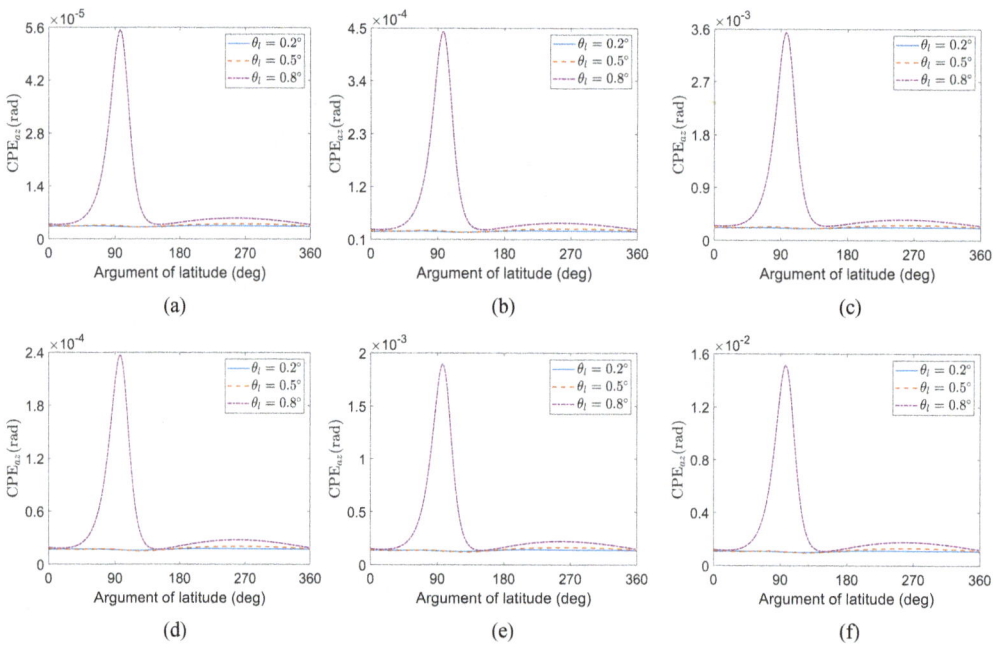

FIGURE 8.22 The azimuth CPEs with various look angles versus the AOL under background ionospheric effects, where the 1st (a–c) and 2nd (d–f) rows represent the low and high solar activity, while the 1st (a, d), 2nd (b, e), and 3rd (c, f) columns stand for the SATs of (b) 100 s, (c) 200 s, (d) 400 s.

angle and location of the MBSAR. Moreover, it is noteworthy that high solar activity yields greater magnitude and more pronounced variations in the azimuth CPE. Notwithstanding, the azimuth CPE is negligibly small, consistently remaining below the threshold of $\pi/8$. These findings suggest that the challenges related to azimuth focusing posed by the azimuth CPE are negligible in the MBSAR.

We have analyzed the imaging distortions, i.e., geometric deviation and defocusing, under the background ionospheric effects. It is apparent that both the range and azimuth offsets can affect the geometric fidelity to some extent. Regarding the focusing quality, it is found that neither the range CPE nor the azimuth CPE exerts certain impacts on focusing quality, but both the range and azimuth QPEs need to be adequately tackled.

8.4.3 SIMULATION OF TARGET RESPONSE UNDER BACKGROUND IONOSPHERIC EFFECTS

Next, we shall simulate the target responses to illustrate the background ionospheric effects on the MBSAR imaging performance with various signal bandwidths (10 MHz and 75 MHz) and different SATs (100 s and 400 s). We set the zero-azimuth time to 00:00:00 on March 20, 2024, TDB. Further, for ease of identifying image offset, the central position of the image scene, disregarding the background ionospheric effects, is normalized to the coordinate (0, 0) in each instance. Building upon the above foundations, we simulate and present the point target responses at look angles of 0.2°, 0.5°, and 0.8° in Figures 8.23–8.25.

Inspecting Figures 8.23–8.25, we note a correlation between solar activity and image distortions manifested as image offset and defocusing, depending on the MBSAR's spatial resolution. When employing a signal bandwidth of 10 MHz and an SAT of 100 s, corresponding to spatial resolutions between 10 and 20 m, the PSLR and ISLR are acceptable along the azimuth and range directions, regardless of the solar activity. This implies that we can yield well-focused images in the MBSAR at such a spatial resolution scale. If a larger signal bandwidth and a longer SAT, e.g., 75 MHz and 400 s, are applied, the range PSLR and ISLR suffer deteriorations, which are further amplified at a larger look angle. For azimuth focusing, it is challenging, if not impossible, to obtain a focused image. Hence, the finer spatial resolution poses more challenges to yield satisfactory focusing quality under the background ionospheric effects.

The image offsets remain unacceptable for most Earth observation applications even at low solar activity, whereas their magnitude becomes more pronounced during high solar activity. Interestingly, the range offset becomes slightly more severe under a wider signal bandwidth. This phenomenon can be attributed to the influence of range CPE, which brings about asymmetric distortion to the target response. Generally, the spatiotemporal variation of the background ionosphere seldom exhibits the same degree of variation as that experienced during high solar activity. Thus, employing coarse spatial resolutions (short SAT and narrow signal bandwidth) could circumvent image distortions resulting from the background ionospheric effects for most conditions.

The range and azimuth offsets prevail under all spatial resolutions regardless of the level of solar activity, thus impairing the MBSAR imaging performance. Further, the spatial inhomogeneity of the background ionosphere across the swath width could give rise to additional image distortions, as will be discussed in the subsequent section.

8.5 IMAGE DISTORTION DUE TO THE INHOMOGENEOUS BACKGROUND IONOSPHERE

In fact, the background ionosphere exhibits spatial inhomogeneity, specifically, there are spatiotemporal variations in TEC that are dependent on locations across the imaging swath. In what follows, we look into such an effect. To exemplify the inhomogeneous TEC across the imaging swath, we select a beamwidth of 0.5° and a near-look angle of 0.2° with the antenna azimuth angle regulated through the zero-Doppler steering. We designate four distinct epochs, each encompassing an SAT

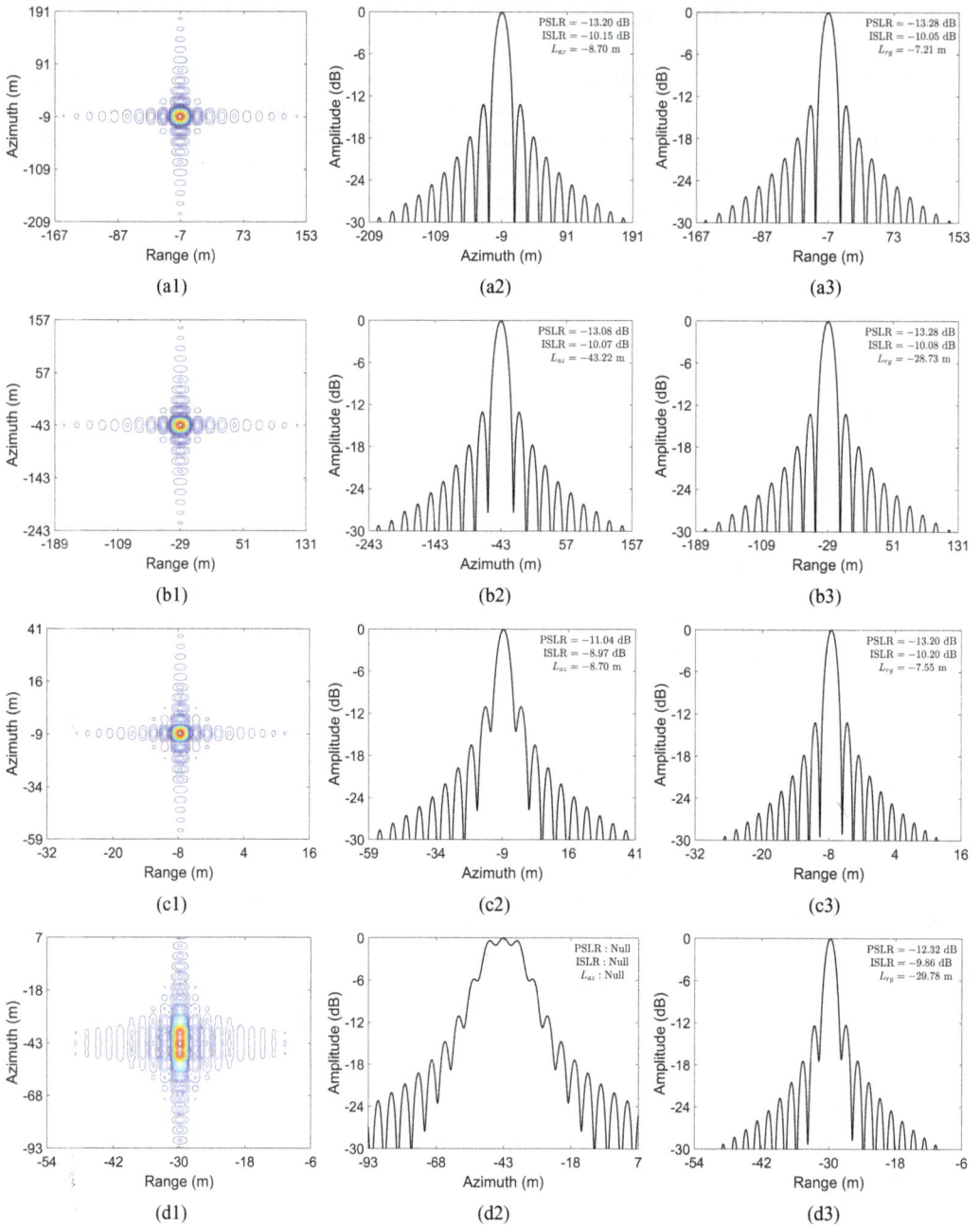

FIGURE 8.23 At a look angle of 0.2°, the point target response under background ionospheric effects, where the 1st (a1–c1), 2nd (a2–c2), and 3rd (a3–c3) columns respectively represent the response contour, azimuth, and range profiles, while 1st (a1–a3) and 2nd (b1–b3) rows indicate the responses under $B_r = 10$ MHz, $T_{SAR} = 100$ s at low and high solar activities, and 3rd (c1–c3) and 4th (d1–d3) rows stands for the responses under $B_r = 75$ MHz, $T_{SAR} = 400$ s at low and high solar activities.

of 600 s, with the zero-azimuth time alignment coinciding with the following specific instants: 08:00:00, 12:00:00, 16:00:00, and 20:00:00 on March 20, 2004, UTC. In this context, the IRI 2012 model is adopted to provide ionospheric data, such data includes two components: the constant part of TEC as a function of look angle, and the spatiotemporally varying portion of TEC versus look angle and SAT, as displayed in Figure 8.26.

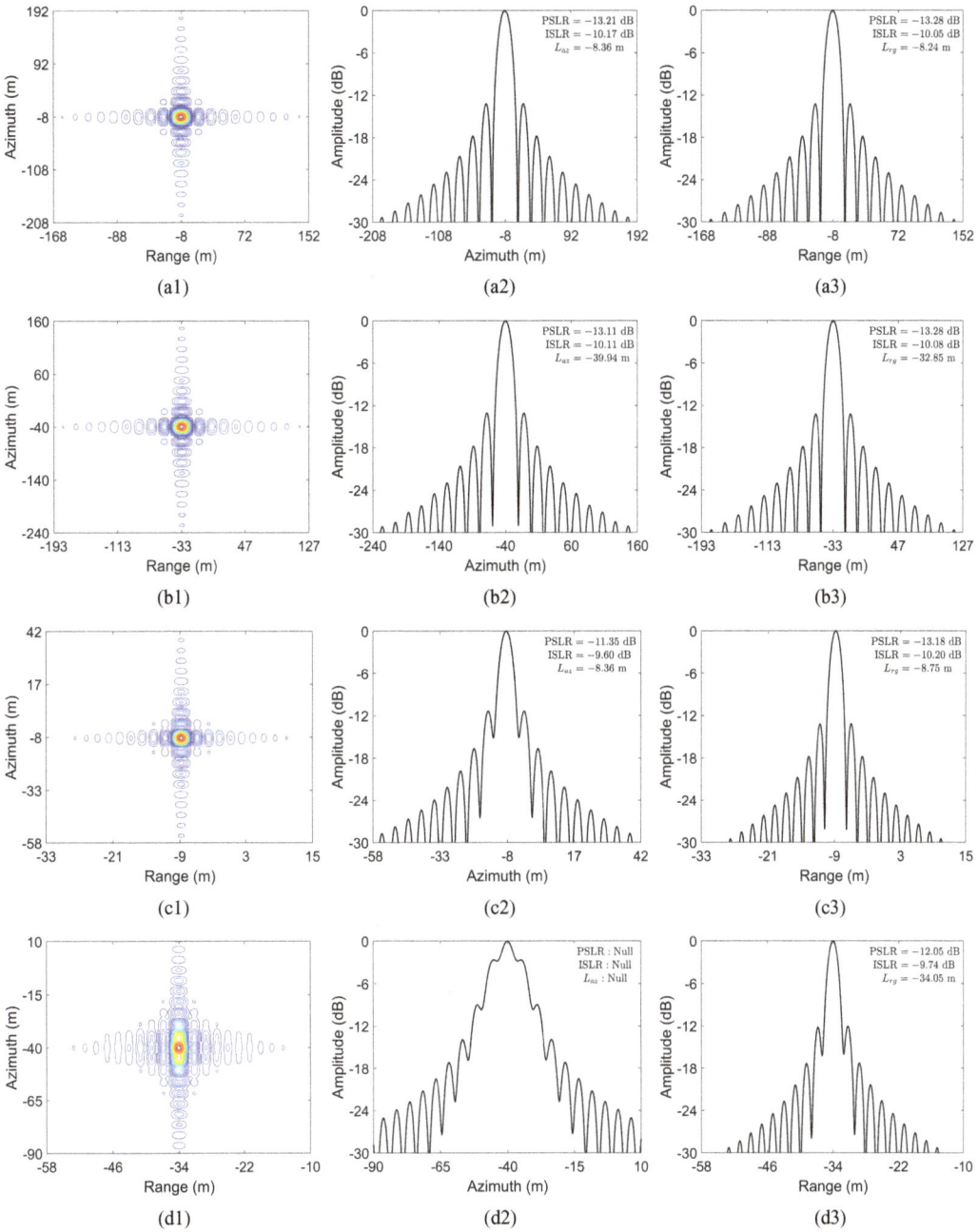

FIGURE 8.24 At a look angle of 0.5°, the point target response under background ionospheric effects, where the 1st (a1–c1), 2nd (a2–c2), and 3rd (a3–c3) columns respectively represent the response contour, azimuth, and range profiles, while 1st (a1–a3) and 2nd (b1–b3) rows indicate the responses under $B_r = 10$ MHz, $T_{SAR} = 100$ s at low and high solar activities, and 3rd (c1–c3) and 4th (d1–d3) rows stands for the responses under $B_r = 75$ MHz, $T_{SAR} = 400$ s at low and high solar activities.

Figure 8.26 demonstrates that disparities occur in the background ionosphere across different look angles at different instants. Specifically, for an SAT of 600 s, the maximum disparity of the TEC within the swath width is about 0.6 TECU when the zero-azimuth time corresponds to 08:00:00 UTC, while this disparity can be up to approximately 1.1 TECU when the zero-azimuth time aligns with 16:00:00 UTC. Notably, the influence of spatiotemporally varying components in TEC on the MBSAR

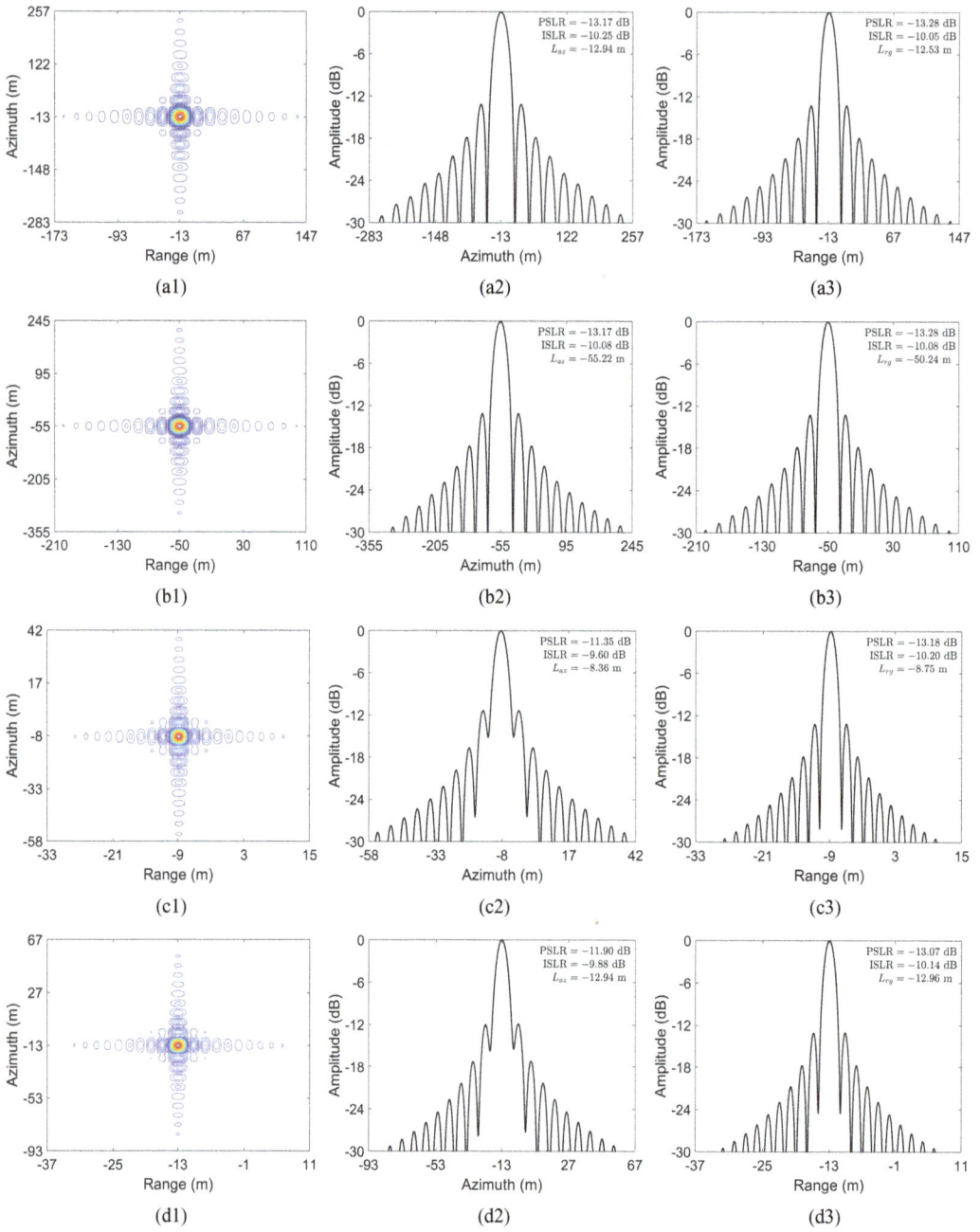

FIGURE 8.25 At a look angle of 0.8°, the point target response under background ionospheric effects, where the 1st (a1–c1), 2nd (a2–c2), and 3rd (a3–c3) columns respectively represent the response contour, azimuth, and range profiles, while 1st (a1–a3) and 2nd (b1–b3) rows indicate the responses under $B_r = 10$ MHz, $T_{SAR} = 100$ s at low and high solar activities, and 3rd (c1–c3) and 4th (d1–d3) rows stands for the responses under $B_r = 75$ MHz, $T_{SAR} = 400$ s at low and high solar activities.

imaging can be attributed to k_{t1} and k_{t2}, the factors associated with the 1st-order and 2nd-order rates of TEC. Therefore, it is necessary to examine both k_{t1} and k_{t2} at the given epochs, utilizing the ionospheric data obtained from Figure 8.26, as illustrated in Figure 8.27 as a function of the look angle.

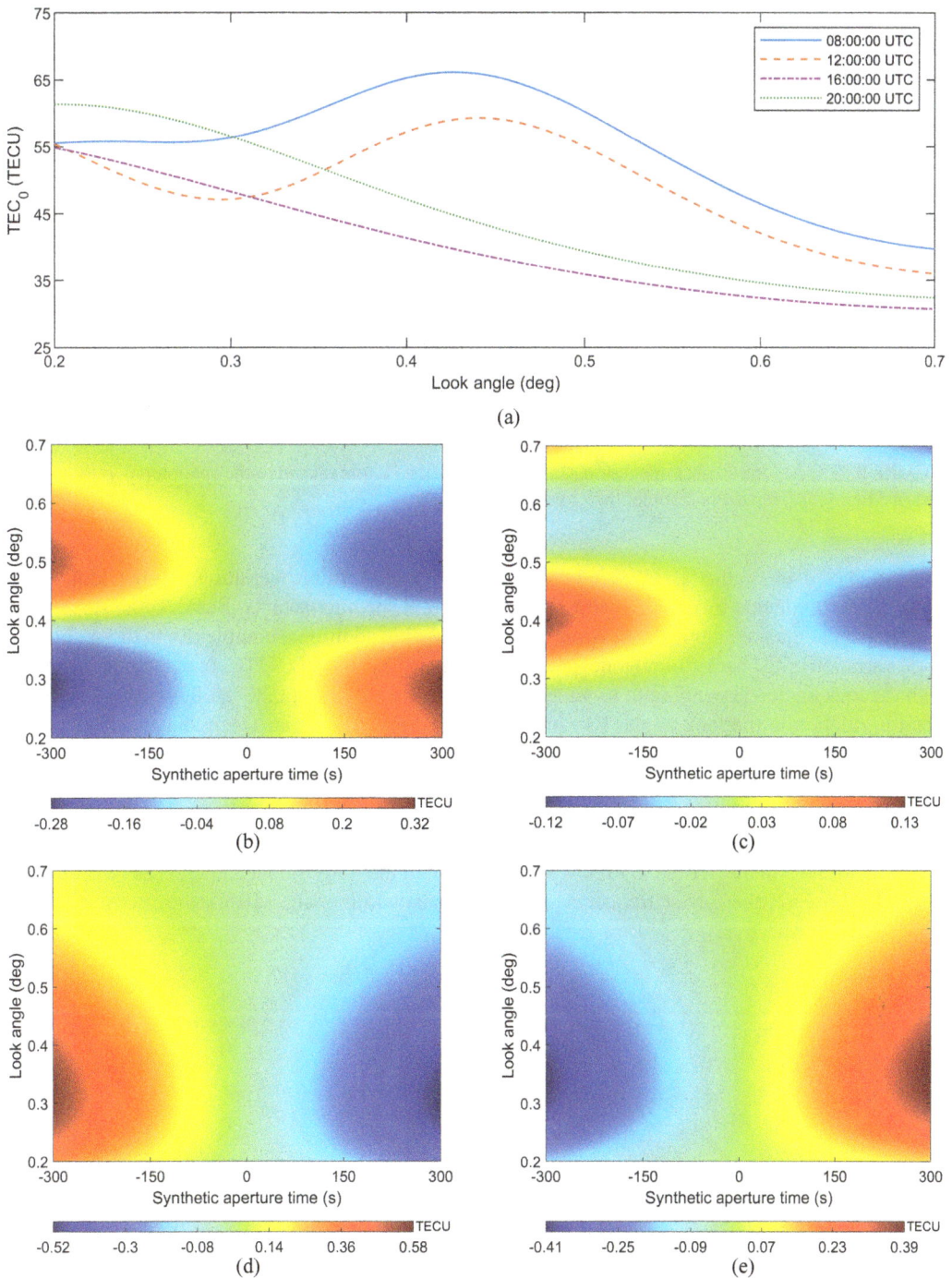

FIGURE 8.26 The inhomogeneous spatiotemporally varying TEC within the swath width of the MBSAR at four distinct epochs on March 20, 2004, (a) the constant part of TEC versus look angle; the spatiotemporally varying portion of TEC versus look angle and SAT with the zero-azimuth time corresponding to: (b) 08:00:00 UTC, (c) 12:00:00 UTC, (d) 16:00:00 UTC, (e) 20:00:0 UTC.

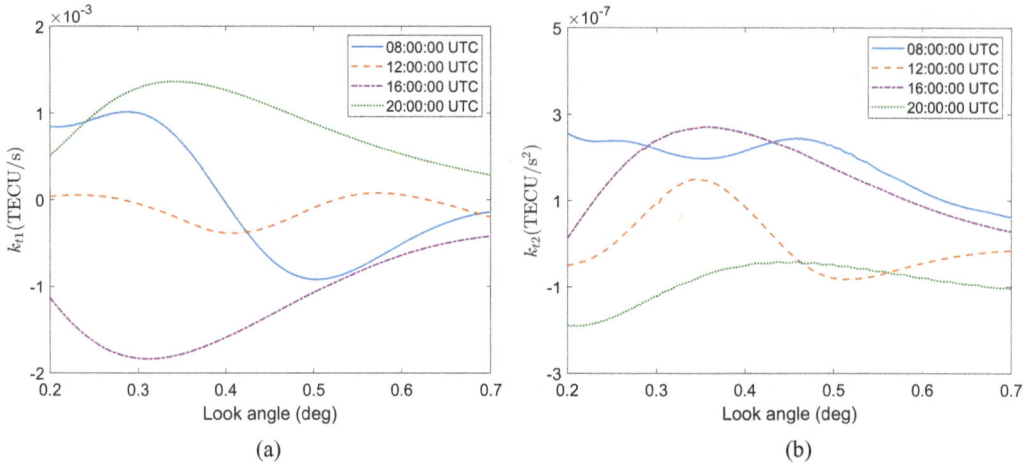

FIGURE 8.27 At four distinct epochs on March 20, 2004, the factors related to the spatiotemporally varying part of TEC, (a) k_{t1} versus look angle, (b) k_{t2} versus look angle.

As depicted in Figure 8.27, both k_{t1} and k_{t2} exhibit discernible variations as the look angle changes, accompanied by distinct patterns of variation at specific instants. Besides, the slope of k_{t1} with respect to the look angle differs from that of k_{t2}, adding additional complexity to the spatiotemporal variation of the background ionosphere. For instance, at 16:00:00 UTC, the factor k_{t1} decreases with the look angle, descending to its valley point at the look angle of 0.30°, followed by an upward trend in relation to the look angle. By contrast, the factor k_{t2} shows an ascending trend within the look angle range of 0.2° to 0.35°, followed by a monotonous decline as the look angle increases. Consequently, the diversities in the constant and spatiotemporally varying components of TEC are far from ignorable when considering the swath width.

The effects of an inhomogeneous background ionosphere could give rise to different levels of image distortions in the MBSAR. Figure 8.28 shows both the range offset and QPE against the look angle. In this case, the carrier frequency and signal bandwidth of the MBSAR are set to 1.2 GHz and 30 MHz, respectively.

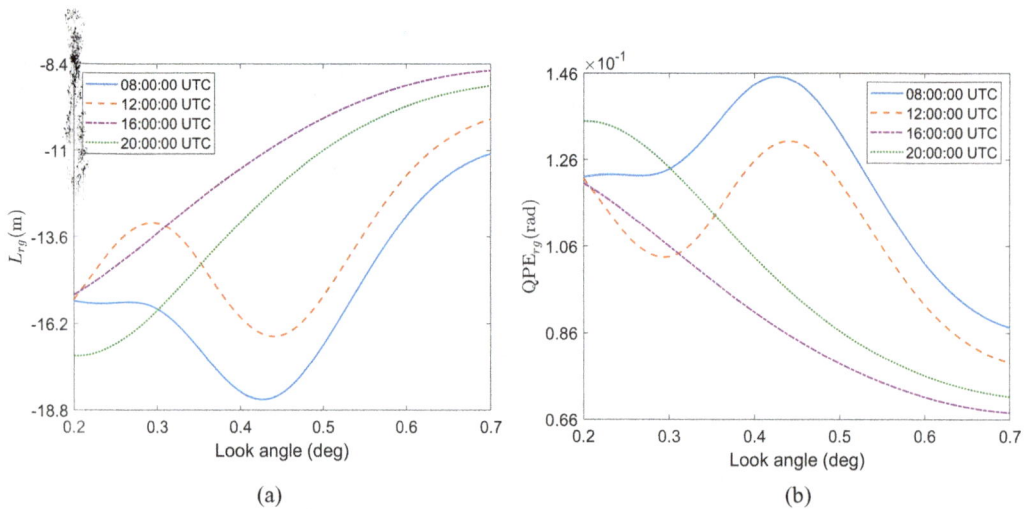

FIGURE 8.28 Under diversities of background ionosphere, (a) the range offset versus look angle, (b) the range QPE versus look angle.

Figure 8.28 indicates that both the range offset and QPE are spatial disparities within the swath width of the MBSAR. For a range resolution of 5 m, the range QPE is persistently below the threshold of $\pi/4$, albeit with distinct discrepancies at different epochs. As a result, the issue concerning the focusing quality poses little impact on the range imaging of MBSAR under the provided configurations. In contrast, the range offset remains evident under the configured signal bandwidth. Additionally, the inhomogeneity of the TEC causes spatial disparities of range offsets across the swath width; the maximum difference in the range offset approaches two range resolution cells in the given epochs. Furthermore, the image distortion resulting from the spatial discrepancies of range offsets will be amplified under high solar activity conditions. Henceforth, the inhomogeneous TEC within the swath width must be considered in MBSAR.

The spatiotemporally varying component of TEC, along with its inherent diversities within the swath width, can impact the azimuth imaging of the MBSAR to a certain extent. To clarify these impacts on azimuth imaging in terms of geometric fidelity and focusing quality, Figure 8.29 illustrates the azimuth offset and QPE using the ionospheric data provided by Figure 8.27, both as a function of look angle. The SAT is set to 200 s in this example.

Figure 8.29 shows the azimuth offset and QPE exhibiting spatial disparities across the imaging swath. Moreover, the variations in both factors with respect to the look angle display distinct patterns at different epochs. Interestingly, the azimuth QPE, similar to the range QPE, falls below the $\pi/4$, indicating that the azimuth focusing is little impacted when employing a coaster azimuthal resolution. On the other hand, the azimuth offsets reveal spatial disparities at different look angles. Such spatial variations in azimuth offset can result in additional image distortions in the MBSAR. Hence, we have to consider the spatial variations in the TEC gradients within the swath width.

The spatial resolution on a decameter level is suggested for the MBSAR [33–35]. In this case, the focusing quality may not be problematic, but the range and azimuth offsets are not ignorable. Figures 8.28 and 8.29 indicate that the disparities in the constant and spatiotemporally varying components of TEC can result in discrepancies in range and azimuth offsets. To further investigate the effects of inhomogeneous background ionosphere on the geometric fidelity, we examine the 2-D image offsets in terms of range and azimuth offsets within the coverage of MBSAR over one Earth day. In this case, the TEC parameters are obtained by employing real-time ionospheric data reported from [22] in combination with the spatial interpolation method [36, 37]. The 2-D image offsets on Apr. 24, 2022, and 2023 are presented in Figures 8.30 and 8.31, respectively.

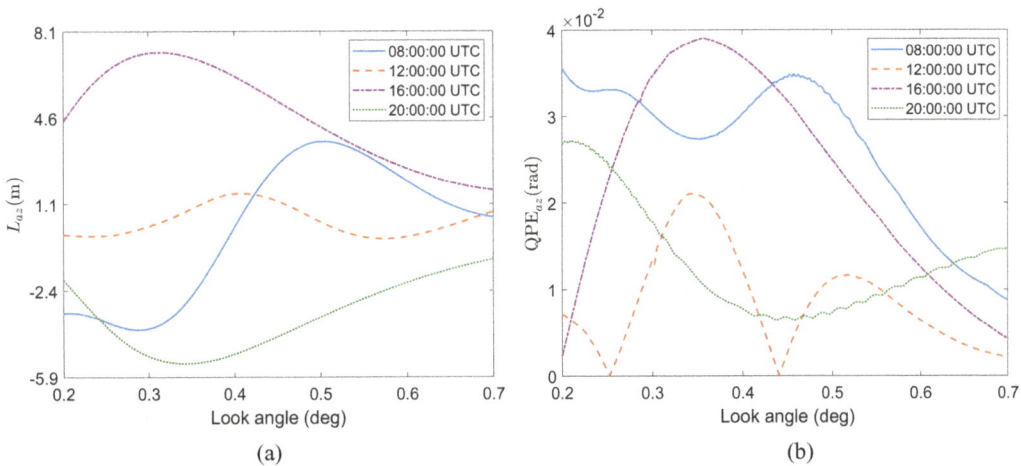

FIGURE 8.29 Under the effects of diversities in background ionosphere, (a) the azimuth offset versus look angle, (b) the azimuth QPE versus look angle.

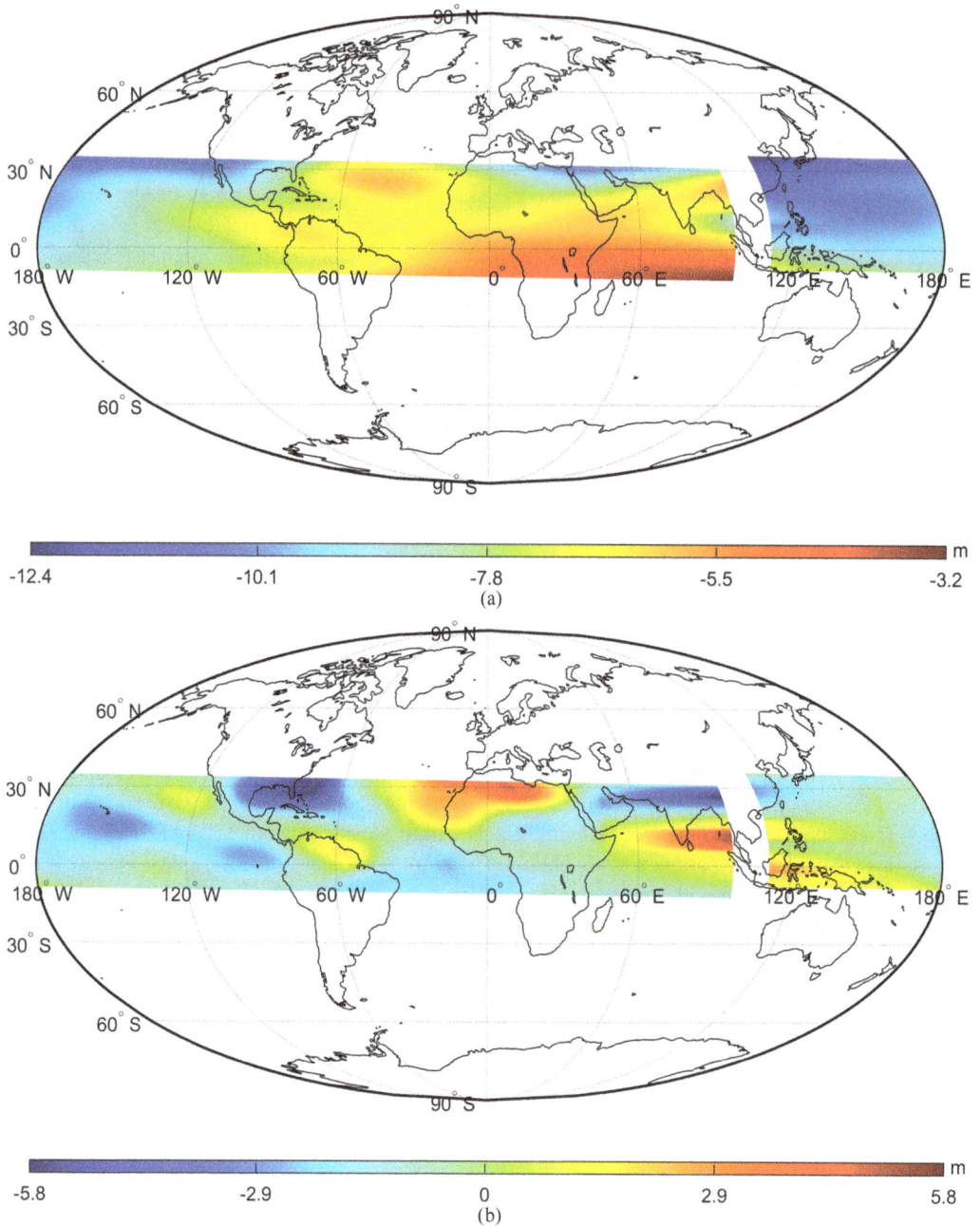

FIGURE 8.30 On Apr. 24, 2022, the 2-D image offsets under the effects of diversities in background ionosphere, (a) the range offset within the coverage of MBSAR, (b) the azimuth offset within the coverage of MBSAR.

We see that both the range and azimuth offsets vary as geographical location changes. Besides, within the coverage region, the variation pattern of azimuth offset exhibits little synchronicity compared to that of the range offset. The shift direction of range offset remains consistent, while that of azimuth offset varies depending on the geographical location. Thus, the variation of the azimuth offset within the swath is more pronounced and complicated than the range offset. The rationale behind

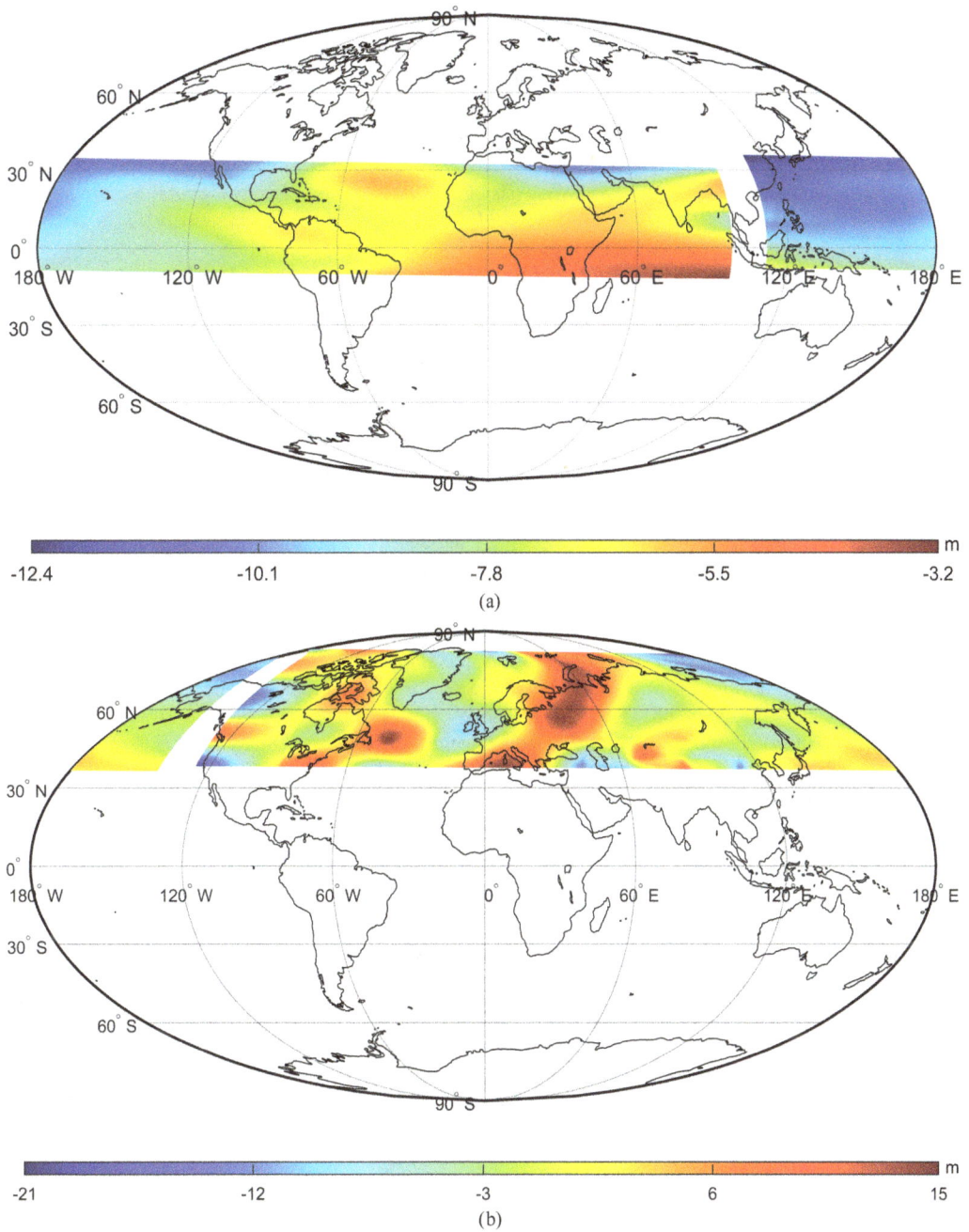

FIGURE 8.31 On Apr. 24, 2023, the 2-D image offsets under the effects of diversities in background ionosphere, (a) the range offset within the coverage of MBSAR, (b) the azimuth offset within the coverage of MBSAR.

this phenomenon is that the range offset depends on the constant component of TEC and carrier frequency. By contrast, the azimuth offset depends on the carrier frequency, beam-crossing velocity, and Doppler parameters of the MBSAR and TEC gradients, which can be negative or positive.

Until now, we have analyzed the spatiotemporally varying background ionospheric effects on the imaging performance of MBSAR. The inhomogeneous characteristics of the background ionosphere substantially worsen the image distortion in the MBSAR. Hence, it would be challenging to detect the TOI's precise location in the presence of the background ionosphere if no phase compensation is made, which should be a pivotal topic to pursue in the future study of the MBSAR system.

REFERENCES

[1] A. Renga and A. Moccia, "Preliminary Analysis of a Moon-based Interferometric SAR System for Very High Resolution Earth Remote Sensing," *Proc. 9th ILEWG International Conference on Exploration and Utilisation of the Moon*, Sorrento, Italy, pp. 22–26, Oct. 2007.

[2] H. Guo, et al., "Conceptual Study of Lunar-Based SAR for Global Change Monitoring," *Science China Earth Sciences*, vol. 57, no. 8, pp. 1771–1779, Aug. 2014.

[3] S. Huber, et al., "Tandem-L: A Technical Perspective on Future Spaceborne SAR Sensors for Earth Observation," *IEEE Transactions on Geoscience and Remote Sensing*, vol. 56, no. 8, pp. 4792–4807, Jun. 2018.

[4] A. Ishimaru, et al., "Ionospheric Effects on Synthetic Aperture Radar at 100 MHz to 2 GHz," *Radio Science*, vol. 34, no. 1, pp. 257–268, Jan.-Feb. 1999.

[5] Z. Xu and K. S. Chen, "Temporal-Spatial Varying Background Ionospheric Effects on the Moon-Based Synthetic Aperture Radar Imaging: A Theoretical Analysis," *IEEE ACCESS*, vol. 6, pp. 66767–66786, Jul. 2018.

[6] W. Liu, et al., "A Modified CSA Based on Joint Time-Doppler Resampling for MEO SAR Stripmap Mode," *IEEE Transactions on Geoscience and Remote Sensing*, vol. 56, no. 6, pp. 3573–3586, Jun. 2018, doi: 10.1109/TGRS.2018.2802545

[7] Z. W. Xu, J. Wu, and Z. S. Wu, "A Survey of Ionospheric Effects on Space-Based Radar," *Waves in Random Media*, vol. 14, no. 2, pp. S189–S273, Apr. 2004.

[8] S. A. Stern, "The Lunar Atmosphere: History, Status, Current Problems, and Context," *Reviews of Geophysics*, vol. 37, no. 4, pp. 453–491, Nov. 1999.

[9] L. Jun, et al., "Ionospheric Effects on SAR Imaging: A Numerical Study," *IEEE Transactions on Geoscience and Remote Sensing*, vol. 41, no. 5, pp. 939–947, May. 2003, doi: 10.1109/TGRS.2003.811813

[10] Z. W. Xu, J. Wu, and Z. S. Wu, "Potential Effects of the Ionosphere on Space-Based SAR Imaging," *IEEE Transactions on Antennas and Propagation*, vol. 56, no. 7, pp. 1968–1975, Jul. 2008, doi: 10.1109/TAP.2008.924695

[11] B. Hofmann-Wellenhof, H. Lichtenegger, and J. Collins, *Global Positioning System: Theory and Practice*. New York: Springer, 2012.

[12] G. Xu and Y. Xu, *GPS: Theory, Algorithms and Applications*. Berlin: Springer, 2007.

[13] J. Böhm and H. Schuh, *Atmospheric Effects in Space Geodesy*. Berlin: Springer, 2013.

[14] J. Chen, X. Ren, S. Xiong, and X. Zhang, "Modeling and Analysis of an Ionospheric Mapping Function Considering Azimuth Angle: A Preliminary Result," *Advances in Space Research*, vol. 70, no. 10, pp. 2867–2877, Nov. 2022.

[15] M. Jehle, et al., "Measurement of Ionospheric Faraday Rotation in Simulated and Real Spaceborne SAR Data," *IEEE Transactions on Geoscience and Remote Sensing*, vol. 47, no. 5, pp. 1512–1523, 2009, doi: 10.1109/TGRS.2008.2004710

[16] K. S. Chen and Z. Xu, "Ionospheric Effects on Satellite and Moon-Based SAR: Current Situation and Prospects," *Journal of Nanjing University of Information Science and Technology*, vol. 12, no. 2, pp. 135–149, Apr. 2020, doi: 10.13878/j.cnki.jnuist.2020.02.001

[17] Z. Xu and K. S. Chen, "Numerical Study of the Spatiotemporally-Varying Background Ionospheric Effects on P-Band Satellite SAR Imaging," *IEEE ACCESS*, vol. 8, pp. 123182–123199, Jul. 2020.

[18] Z. Xu, K. S. Chen, P. Xu, and H. D. Guo, "Ionospheric Effects on the Lunar-Based Radar Imaging," *Proc. IEEE International Geoscience and Remote Sensing Symposium*, Fort Worth, USA, pp. 5390–5393, Jul. 2017.

[19] IGS Ionosphere Associate Analysis Center of Wuhan University. *Global Ionospheric VTEC Maps*. Accessed: Jun. 01, 2023. [Online]. Available: ftp://igs.gnsswhu.cn/pub/whu/MGEX/ionosphere

[20] H. B. Vo and J. C. Foster, "A Quantitative Study of Ionospheric Density Gradients at Midlatitudes," *Journal of Geophysical Research: Space Physics*, vol. 106, no. A10, pp. 21555–21563, Oct. 2001.

[21] J. C. Foster and W. Rideout, "Midlatitude TEC Enhancements during the October 2003 Superstorm," *Geophysical Research Letters*, vol. 32, no. 12, Jun. 2005.

[22] IGS Ionosphere Associate Analysis Center of Wuhan University. *Real-Time Global Ionospheric VTEC Maps*. Accessed: Jun. 01, 2023. [Online]. Available: ftp://igs.gnsswhu.cn/pub/whu/MGEX/realtime-ionex

[23] S. Quegan and J. Lamont, "Ionospheric and Tropospheric Effects on Synthetic Aperture Radar Performance," *International Journal of Remote Sensing*, vol. 7, no. 4, pp. 525–539, May. 1986.

[24] M. Jehle, O. Frey, D. Small, and E. Meier, "Measurement of Ionospheric TEC in Spaceborne SAR Data," *IEEE Transactions on Geoscience and Remote Sensing*, vol. 48, no. 6, pp. 2460–2468, Jun. 2010, doi: 10.1109/TGRS.2010.2040621

[25] Z. Xu and K. S. Chen, "On Signal Modeling of Moon-Based Synthetic Aperture Radar (SAR) Imaging of Earth," *Remote Sensing*, vol. 10, no. 3, p. 486, Mar. 2018.

[26] A. Ishimaru, *Wave Propagation and Scattering in Random Media*. New York: Academic Press, 1978.

[27] E. Key, E. Fowle, and R. Haggarty, "A Method of Designing Signals of Large Time-Bandwidth Product," *IRE Int. Conv. Rec*, vol. 4, pp. 146–155, Mar. 1961.

[28] Y. L. Neo, F. Wong, and I. G. Cumming, "A Two-Dimensional Spectrum for Bistatic SAR Processing Using Series Reversion," *IEEE Geoscience and Remote Sensing Letters*, vol. 4, no. 1, pp. 93–96, Jan. 2007.

[29] Y. L. Neo, F. H. Wong, and I. G. Cumming, "Processing of Azimuth-Invariant Bistatic SAR Data Using the Range Doppler Algorithm," *IEEE Transactions on Geoscience and Remote Sensing*, vol. 46, no. 1, pp. 14–21, Jan. 2008, doi: 10.1109/TGRS.2007.909090

[30] A. Ishimaru, *Electromagnetic Wave Propagation, Radiation, and Scattering: From Fundamentals to Applications*. Hoboken: Wiley, 2017.

[31] C. Wang, et al., "Cubic Phase Distortion and Irregular Degradation on SAR Imaging Due to the Ionosphere," *IEEE Transactions on Geoscience and Remote Sensing*, vol. 53, no. 6, pp. 3442–3451, 2015.

[32] C. Wang, et al., "Effects of Anisotropic Ionospheric Irregularities on Space-Borne SAR Imaging," *IEEE Transactions on Antennas and Propagation*, vol. 62, no. 9, pp. 4664–4673, Sep. 2014, doi: 10.1109/TAP.2014.2333055

[33] Z. Xu, K. S. Chen, G. Liu, and H. Guo, "Spatiotemporal Coverage of a Moon-Based Synthetic Aperture Radar: Theoretical Analyses and Numerical Simulations," *IEEE Transactions on Geoscience and Remote Sensing*, vol. 58, no. 12, pp. 8735–8750, 2020.

[34] Z. Xu, K. S. Chen, and G. Liu, "On Orbital Determination of the Lunar-Based SAR under Apsidal Precession," *IEEE Transactions on Geoscience and Remote Sensing*, vol. 60, May. 2022, doi: 10.1109/TGRS.2022.3176836

[35] Z. Xu, K. S. Chen, and G. Liu, "On Evaluating the Imaging Performance and Orbital Determination under Perturbations of Orbital Inclination and RAAN in the Lunar-Based SAR," *IEEE Transactions on Geoscience and Remote Sensing*, vol. 60, Jul. 2022, doi: 10.1109/TGRS.2022.3188294

[36] N. S.-N. Lam, "Spatial Interpolation Methods: A Review," *The American Cartographer*, vol. 10, no. 2, pp. 129–150, Jan. 1983.

[37] J. Li and A. D. Heap, "Spatial Interpolation Methods Applied in the Environmental Sciences: A Review," *Environmental Modelling and Software*, vol. 53, pp. 173–189, Mar. 2014.

Index

Pages in *italics* refer to figures and pages in **bold** refer to tables.

For Product Safety Concerns and Information please contact our EU
representative GPSR@taylorandfrancis.com
Taylor & Francis Verlag GmbH, Kaufingerstraße 24, 80331 München, Germany

www.ingramcontent.com/pod-product-compliance
Lightning Source LLC
Chambersburg PA
CBHW080910220326
41598CB00034B/5530

* 9 7 8 1 0 3 2 3 1 1 7 1 5 *